Essentials of Microbiology

Essentials of Microbiology

Amita Jain, MD, PhD, FAMS, FRCPath

Professor and Head, Department of Microbiology
King George's Medical University
Lucknow, Uttar Pradesh
India

Parul Jain, MD, PhD

Assistant Professor, Department of Microbiology
Super Speciality Cancer Institute & Hospital
Lucknow, Uttar Pradesh
India

ELSEVIER

ELSEVIER

RELX India Pvt. Ltd.
Registered Office: 818, 8th Floor, Indraprakash Building, 21, Barakhamba Road, New Delhi 110001
Corporate Office: 14th Floor, Building No. 10B, DLF Cyber City, Phase II, Gurgaon-122002, Haryana, India

Content Strategist - Education Solutions: Arvind Koul
Content Project Manager: Goldy Bhatnagar
Sr Production Executive: Ravinder Sharma
Sr Graphic Designer: Milind Majgaonkar

Typeset by Thomson Digital
Printed and bound in India at EIH Limited - Unit Printing Press, IMT Manesar, Gurgaon (Haryana)

Dedicated to

My Father
Late Mr. P.C. Jain, who left us for his final abode recently
He made me what I am today

My Mother
Mrs. Urmila Jain, who is struggling with her chronic illness
for the past 20 years, still is the source of my strength

Amita Jain

My Parents and First Teachers
Mr. Y.K. Jain and Mrs. Sugandha Jain

My Strength
Husband Atin Singhai and children Nandini and Siddhant

Parul Jain

PREFACE

We take immense pleasure in introducing a much-required book *Essentials of Microbiology*. This book is intended to be the primary textbook for undergraduate students of medical, dental, paramedical and nursing courses in India and rest of Asia. The content has specially been designed for the Indian students. Therefore, aetiological agents important in India have been dealt with in greater details. We had made an effort to map the content of the book as per new competency-based curriculum for medical students.

This book provides meaningful introduction of infectious agents and diseases in a systematic manner. The text is written in easily understandable language while maintaining its relevance. The content that directly relates with the clinical practice has been included.

The text is organised into three sections. Section I deals with general microbiology and immunology; Section II deals with aetiological agents including bacteriology, virology, parasitology and mycology; and Section III deals with laboratory approach to major clinical syndromes, a unique feature of this book. Chapter outline given in the beginning of each chapter provides the clear layout of the text. An attractive feature of the book is MCQs covering all the chapters as a source material for rapid revision and for self-assessment of the students. Further reading is given at the end of the book so that any other latest and relevant detail of a topic can be read in case of need.

We are sure that this book will provide simple and complete content for undergraduate students who find microbiology to be an uphill task.

In addition, complimentary access to online videos along with full e-book is also provided.

We will appreciate the suggestions or comments of the readers (at indiacontact@elsevier.com).

It would not have been possible to write this book without the help and support of the people around us. We thank Dr. Suruchi Shukla (MD, PhD, Microbiology, King George's Medical University) and Dr. Atin Singhai (MD, Associate Professor, Pathology, King George's Medical University) for their valuable contribution to the book.

We are greatly thankful to our families, who have understood our commitment for the project and supported us to their best.

We gratefully acknowledge the help and cooperation received from the staff of RELX India Pvt. Ltd., especially Arvind Koul (Content Strategist), and Goldy Bhatnagar (Content Project Manager), who were instrumental in pushing this book and provided the much needed guidance and inspiration.

Last but least we thank almighty to show us the light, and giving the strength to complete the task.

Enjoy reading!

Amita Jain
Parul Jain

CONTENTS

General Microbiology

Introduction and History of Microbiology

Amita Jain

LEARNING OBJECTIVES

- Definitions in relation to microbiology and microorganisms
- History and branches of microbiology
- Classification and nomenclature of microbes
- Koch's postulates

INTRODUCTION

Microbiology is a diverse science encompassing almost every sphere of life such as environment, agriculture, industry, food, water, health and so on. In this chapter, we will briefly discuss the history of microbiology with emphasis on medical microbiology.

MICROBIOLOGY

The prefix 'micro' generally refers to an object sufficiently small to be visualised only under microscope. Microbiology is the branch of science that deals with the organisms too small to be visualised with the naked eye. Viruses, bacteria, fungi and protozoa are mainly included in this category. Box 1.1 lists the branches of microbiology.

BOX 1.1 Branches of Microbiology

Important disciplines of microbiology:
1. **General microbiology**: Study of broad range of microbes.
2. **Medical microbiology**: Study of microbes that cause human disease.
3. **Immunology:** Study of the host immune system.
4. **Agricultural microbiology**: Study of microbes that can have an impact on agriculture.
5. **Microbial ecology**: Study of relationships between microbes and their habitats.
6. **Environmental microbiology**: Study of microorganisms that influence environment.
7. **Food microbiology**: Study of preventive methods of foodborne disease.
8. **Industrial microbiology**: Study of microbes to produce commercial products.
9. **Biotechnology**: Study of methods of manipulation of organisms to form useful products such as drugs and vaccines.

MICROORGANISMS

Microorganisms are microscopic organisms that exist as unicellular/multicellular organisms, and are found in nature, almost everywhere. Some of them are beneficial, while some can cause injury to the host.

Microorganisms are mainly divided into bacteria, fungi, protozoa and viruses.

CLASSIFICATION AND NOMENCLATURE OF MICROBES

Taxonomy is the science of classifying organisms. Hence, the hierarchical groups in classification are called taxa. The morphological characteristics, differential staining, biochemical testing, DNA base composition etc. are usually the basis of classifying organisms.

Microorganisms are named using a binomial nomenclature (using two words):
1. **Genus** (Latin noun): The first letter of the genus name is always capitalised and whole name is *italicised*, for example *Staphylococcus aureus*.
2. **Species**: They are written in small italicised letters.

It is customary to classify microorganisms, as shown in Flowchart 1.1.

Domain

Domain is the first and the most inclusive group under which all organisms are classified. It has three subgroups: such as bacteria, archaea and eukarya. The domain is further divided on the basis of similarities in ribosomal RNA sequences of microorganisms.

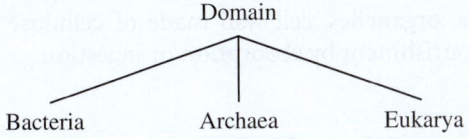

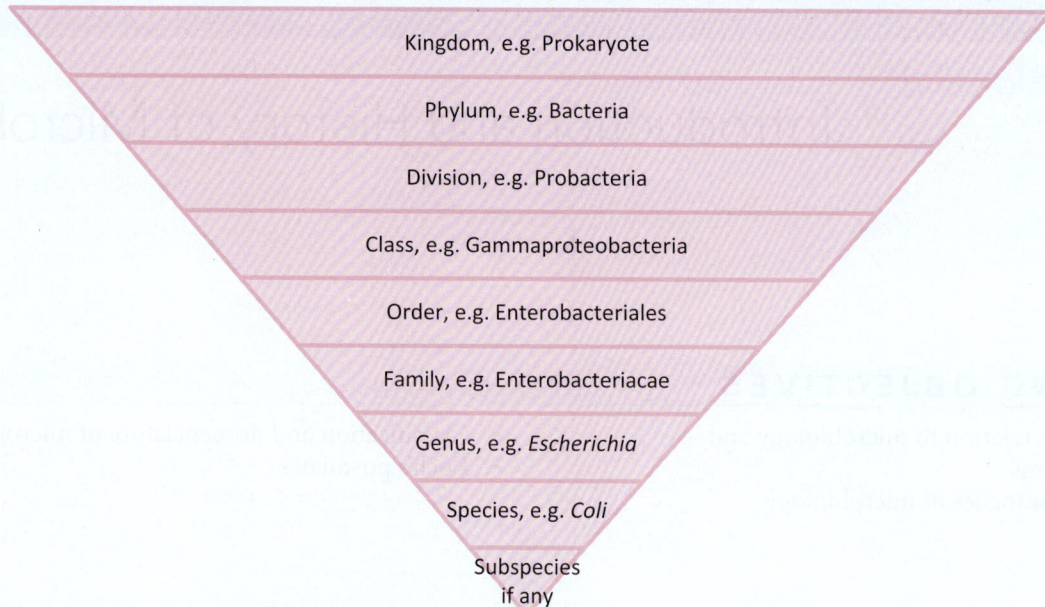

Flowchart 1.1 Taxonomic Classification of *Escherichia coli*.

Kingdom

Kingdom is the second largest group. Five major kingdoms include prokaryote (e.g. archaea and bacteria), protoctista (e.g. protozoa and algae), fungi, plantae and animalia.

A kingdom is further divided into phylum, division, class, order, family, genus and species in the same sequence (Flowchart 1.1).

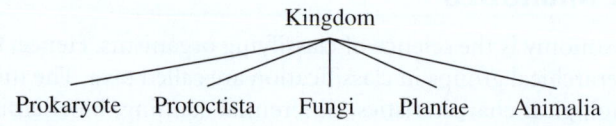

The kingdoms of medical microbiological importance are prokaryote (archaea and bacteria), protoctista (protozoa and algae), fungi, and animalia (multicellular parasites).

Archaea

Archaea lack peptidoglycans in their cell wall (point of difference from bacteria). They are prokaryotic cells and can survive in extreme environmental conditions. They exist in nature, but their pathogenic potential is not known.

Bacteria

Bacteria are prokaryotic unicellular organisms. They lack a nucleus. The presence of peptidoglycan cell wall is a unique character of bacteria. They divide by binary fission.

Protozoa

Protozoa are unicellular aerobic eukaryotic cells. They have a nucleus, organelles, cell wall made of cellulose and they obtain nourishment by absorption or ingestion.

Algae

Algae are unicellular/multicellular eukaryotic organisms that derive food and energy by photosynthesis. They live in water, soil, vegetation, rocks etc., produce oxygen and carbohydrates and do not cause human infections.

Multicellular Parasites

Multicellular parasites are a group of eukaryotes commonly referred to as worms. They usually live their life cycle in both macroscopic and microscopic form. Some of them are pathogenic to humans.

Fungi

Fungi are eukaryotes. They contain a true nucleus. Most fungi are multicellular and their cell wall is composed of chitin. They absorb organic material from their environment, and may have a symbiotic or parasitic relationship with their host. They usually exist in three forms such as: mushroom, moulds and yeasts.

Viruses

Viruses are acellular microbes. They do not have a well-formed nucleus. They consist of a nucleic acid (either DNA or RNA) surrounded by a protein coat. They lack many essential molecules, hence cannot reproduce and metabolise on their own, extracellularly. Viruses were not included in any kingdom because they are acellular.

HISTORY OF MICROBIOLOGY

History of Microbial Description

It is not well known who made the first observations of microorganisms, but during the mid-1600s an English scientist named **Robert Hooke** observed strands of fungi. After 1670s, a Dutch merchant named **Antonie van Leeuwenhoek** made an observation of small microscopic organisms and called them **animalcules**. He accurately described protozoa, fungi and bacteria. Table 1.1 shows the important milestones, which are of importance in naming microbes.

TABLE 1.1 Historical Milestones Describing Nomenclature of Microbes

Scientist	Year	Discovery
A. Leeuwenhoek	1676	Described the microorganisms for the first time
C. Chamberland	1884	Made a filter which was used to separate bacteria from viruses
Loeffler and Frosch	1898	Discovered that foot and mouth disease in animals is caused by an agent which cannot be filtered
M. Beijerinck	1898–1900	Discovered tobacco mosaic virus
S. Prusiner	1982	Discovered that prions are infectious communicable protein molecules that lack nucleic acids

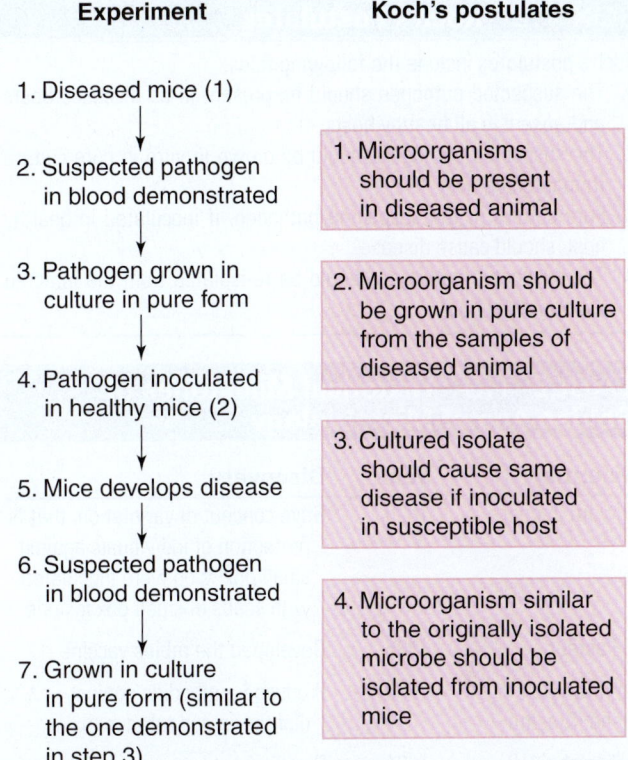

Flowchart with two columns: "Experiment" and "Koch's postulates"

Experiment:
1. Diseased mice (1)
2. Suspected pathogen in blood demonstrated
3. Pathogen grown in culture in pure form
4. Pathogen inoculated in healthy mice (2)
5. Mice develops disease
6. Suspected pathogen in blood demonstrated
7. Grown in culture in pure form (similar to the one demonstrated in step 3)

Koch's postulates:
1. Microorganisms should be present in diseased animal
2. Microorganism should be grown in pure culture from the samples of diseased animal
3. Cultured isolate should cause same disease if inoculated in susceptible host
4. Microorganism similar to the originally isolated microbe should be isolated from inoculated mice

Flowchart 1.2 Experiment Developed by Robert Koch, Which Made Basis for Koch's Postulates.

History of Spontaneous Generation of Microbes

In earlier years, scientists believed that microorganisms arise from lifeless matter such as beef broth. **Francesco Redi** later demonstrated that fly maggots do not arise from decaying meat if the meat is covered to prevent the entry of flies. **Lazzaro Spallanzani** also demonstrated that boiled broth would not give rise to microscopic forms of life (Table 1.2).

Louis Pasteur discovered that bacteria are essential to brew wine and give sour taste to dairy products. Based on his observations in winery he proposed the **germ theory of disease**, which means microorganisms make people sick. His theory was later proved by **Robert Koch**, a German scientist. He cultivated anthrax bacteria, injected pure cultures of the bacilli into mice and showed that mice developed anthrax. This was developed as **Koch's postulates** (Flowchart 1.2,

Box 1.2). Koch's postulates later became the standard to establish causative association of microbes with disease.

Golden Age of Microbiology

The period between 1800 and 1970 was called the golden era of microbiology as many agents of different infectious diseases were identified during this period (Table 1.3).

TABLE 1.2 Historical Milestones Describing Spontaneous Generation of Microbes and Microorganism–Disease Association

Scientist	Year	Discovery
F. Redi	1688	First one to reject theory of spontaneous regeneration of microorganisms. He demonstrated that maggots do not form in rotten meat if it is kept away from flies
R. Needham	1748	Supported theory of spontaneous generation of microbes. He showed that even after boiling mutton broth, growth of microbes occurred in it
L. Spallanzani	1776	Rejected theory of spontaneous generation. He showed that sealed and boiled containers do not produce microbes
L. Pasteur	1861	Disapproved theory of spontaneous generation. He demonstrated that air contained microbial organisms. To demonstrate that he used specially designed flasks with curved neck which allowed only sterile air to pass through the neck. He used this flask to boil broth and later showed that till broth was held within the flask no growth occurred
Lucretius (BC) and Fracastoro	1546	Proposed the idea that invisible organisms caused disease
A. Bassi	1835	Demonstrated that a fungus is responsible for disease in silkworm
J. Lister	1867	Demonstrated that good antiseptic procedures reduced the frequency of surgical site infections
R. Koch	1876–1884	His experiments proved that *Bacillus anthracis* caused anthrax (Fig. 1.2, Box 1.2). He also proved *Mycobacterium tuberculosis* caused tuberculosis. Koch's postulates were thereafter named after him

BOX 1.2   Koch's Postulates

Koch's postulates include the following rules:
1. The suspected pathogen should be present in *all* diseased hosts and *absent* in all healthy hosts.
2. The suspected pathogen should be grown in vitro in pure culture from infected host.
3. Pure culture of the suspected pathogen, if inoculated in healthy host, should cause disease.
4. The suspected pathogen should be re-isolated from the infected experimental host.

TABLE 1.3   Historical Milestones of Golden Age of Microbiology

Scientist	Year	Discovery
Edward Jenner	1820	Gave concept of variolation, that is protection of individuals against small pox, who were inoculated with scabs of small pox lesions
L. Pasteur	1885	Developed the rabies vaccine
Emil von Behring and Kitasato	1890	Produced antibodies against diphtheria and tetanus toxins
E. Metchnikoff	1884	Described phagocytosis of bacteria by macrophages

TABLE 1.4   Historical Milestones of Modern Microbiology and Biotechnology

Scientist	Year	Discovery
L. Pasteur	1856	Invented that wine gets its taste because of lactic acid fermentation
S. Winogradsky and M. Beijerinck	1887–1900	Studied that soil microbes are involved in biochemical cycles of sulphur, carbon, nitrogen and so on
Beadle and Tatum	1941	Proposed that one gene codes for the synthesis of one enzyme
Luria and Delbruck	1943	Proposed that mutations in microbes occur spontaneously in nature
Avery, MacLeod and McCarty	1944	Proposed that DNA is the genetic material which codes for proteins

Modern Microbiology and Biotechnology

Microorganisms are exploited in biotechnical processes and industries to produce insulin, interferon, numerous blood clotting factors, clot-dissolving enzymes, number of vaccines, vitamins, amino acids, enzymes, growth supplements, fermented dairy products (sour cream, yogurt and buttermilk), pickles, breads, alcoholic beverages etc. (Table 1.4).

Many microbiologists and scientists have been awarded Nobel Prize in science. Box 1.3 lists a few of them.

The antibiotics and vaccines were introduced which decreased the incidence of many diseases such as pneumonia, tuberculosis, meningitis, syphilis etc. In 1940s, the electron microscope and cultivation methods for viruses were developed.

BOX 1.3   List of Famous Nobel Prize Winner Microbiologists and Immunologists

1901   Emil Adolf Von Behring: Developed serum treatment, especially in diphtheria.
1902   Sir Ronald Ross: Discovered the life cycle of the malaria parasites in humans and mosquitoes.
1905   Robert Koch: For his discoveries in relation to tuberculosis.
1907   Charles Louis Alphonse Laveran: Showed that the mosquito is the agent of transmission for malaria and the identification of the malaria parasite.
1908   Ilya Ilyich Mechnikov (Elie Metchnikoff: Father of Natural Immunity), Paul Ehrlich: Studied immune reactions and phagocytic cells.
1919   Jules Bordet: Made fundamental discoveries in immunity.
1928   Charles Nicolle: Discovered that epidemic typhus is transmitted by lice.
1930   Karl Landsteiner: Discovered the ABO human blood groups.
1939   Gerhard Domagk: Discovered sulpha drugs.
1945   Sir Alexander Fleming, Sir E. B. Chain and Lord H. W. Florey: Discovered and developed penicillin.
1951   Max Theiler: Developed vaccine for yellow fever.
1952   Selman A. Waksman: For his discovery of streptomycin.
1954   John Franklin Enders, Thomas Huckle Weller and Frederick Chapman Robbins: For their discovery of the ability of poliomyelitis viruses to grow in cultures of various types of tissue and for making polio vaccine possible.
1962   F. H. C. Crick, J. D. Watson and H. F. Wilkins: Elucidated the molecular structure of DNA.
1976   B. S. Blumberg and D. C. Gajdusek: Discovered antigen important in diagnosing serum hepatitis.
1983   Kary Mullis: Developed the polymerase chain reaction.
1997   Stanley B. Prusiner: Discovered and characterised prions.
2005   Barry Marshall and Robin Warren: For the identification of *Helicobacter pylori* and its role in gastritis and peptic ulcer disease.
2008   Harald zur Hausen, for his discovery that human papillomaviruses can cause cervical cancer, and Françoise Barré-Sinoussi and Luc Montagnier, for their discovery of HIV.
2011   Bruce A. Beutler, Jules A. Hoffmann and Ralph M. Steinman: For their discoveries on the activation of innate immunity and also for discovering the role of dendritic cell in adaptive immunity.
2015   One half to Youyou Tu for discovering Artemisinin, a novel therapy against malaria and the other half jointly to William C. Campbell and Satoshi Omura for discovering Avermectin, a new drug against roundworms causing river blindness and lymphatic filariasis.

Microbial Genetics

Amita Jain

LEARNING OBJECTIVES

- Structure and organisation of nucleic acid
- Genetic elements in a bacterial cell
- DNA replication

- Transfer of genetic material in bacteria
- Genetic mutations
- Gene expression

INTRODUCTION

The phenotypic characters of any living being are coded by the genetic material found in cells in the form of nucleic acid. Nucleic acids are of two types: deoxyribonucleic acid (DNA) and ribonucleic acid (RNA).

STRUCTURE OF NUCLEIC ACIDS (Fig. 2.1)

Nucleic acids are large molecules made of nucleotides. DNA is composed of two complementary strands placed parallel but opposite to each other. The phosphodiester backbones of both the strands lie on the exterior of the molecule. Watson and Crick proposed the double helix model and base pairing for DNA structure.

DNA/RNA molecules are made up of the following:

- **Nitrogenous bases:** (1) pyrimidines (cytosine and thymine), and (2) purines (adenine and guanine). In a RNA molecule, thymine is replaced with uracil. The nitrogenous bases of each strand face each other and complementary bases of hydrogen bond to each other, stabilising the double helix. Complementary base pairing occurs as: (1) adenine pairs with thymine (with a double bond) and (2) cytosine pairs with guanine (with a triple bond).
- **Nucleotide** is made up of (1) a pentose sugar, (2) a phosphate group and (3) a nitrogenous base. The sugar present in a RNA molecule is ribose, while the sugar present in a DNA molecule is deoxyribose.
- 3'-hydroxyl group of first nucleotides gets linked to 5'-phosphate group of next phosphodiester bonds.

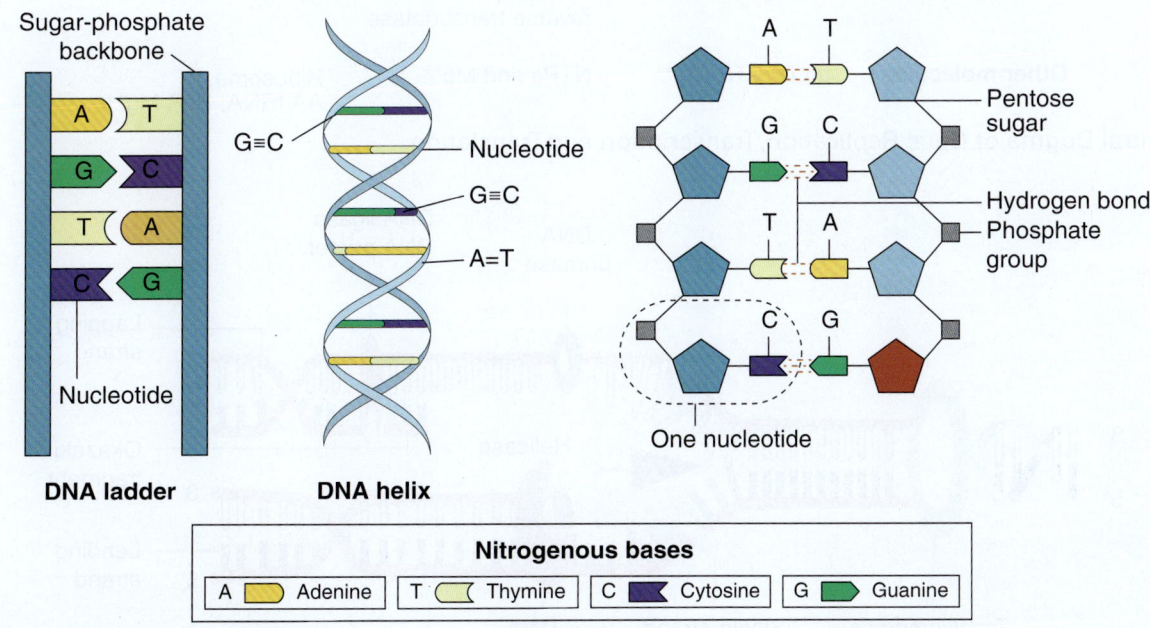

Fig. 2.1 Structure of DNA Molecule.

Hence, the chain has free 3′-hydroxyl group at one end and 5′-phosphate group at other end.

GENETIC ELEMENTS PRESENT IN BACTERIAL CELLS

Chromosome: A circular piece of DNA, single in number.

Plasmid: It is a small, circular piece of DNA present in bacterial cytoplasm in high copy number. These are double-stranded DNA molecules which replicate independently of the bacterial chromosome. They can also provide a mechanism for genetic exchange.

Conjugative plasmids (F factor): F factor carries genes encoding the sex pilus for physical transfer of genetic material. F factor (fertility factor) from *Escherichia coli* is the most studied.

R-factors: Resistance or R-plasmids confer resistance to antibiotics, hence are named so. R-plasmid usually has two types of genes:

1. RTF (resistance transfer factor): Genes for both plasmid replication and conjugation are present.
2. R-determinant: These are resistance genes for resistance, but not for conjugation.

Hfr: Conjugative plasmid integrated into the chromosome is called Hfr and has the characteristic of high frequency of recombination. It passes this trait to any receptive bacteria with which the host cell conjugates.

Transposons: They are simple, insertional sequences, named as jumping gene, and can move from one location on the chromosome to another. They can also move from plasmid to chromosome or vice versa or from one plasmid to another (jumping gene). These elements are flanked by palindromic or reverted repeat sequences that make it easy to insert foreign DNA into the bacterial genome.

DNA REPLICATION

The central dogma of genetics is shown in Fig. 2.2. DNA either replicates to make its own copies as a part of cell division or else is transcribed to RNA which is translated into protein. DNA possesses all the information required for survival as well as replication of the cell.

DNA Replication in a Prokaryote Cell (Fig. 2.3)

DNA replication is part of cell division in all organisms, both prokaryote and eukaryote. It is a complicated but rapid process, for example, about 4.6 million base pairs present in *E. coli* chromosome get replicated in approximately 42 min. It means approximately 1000 nucleotides are added per second.

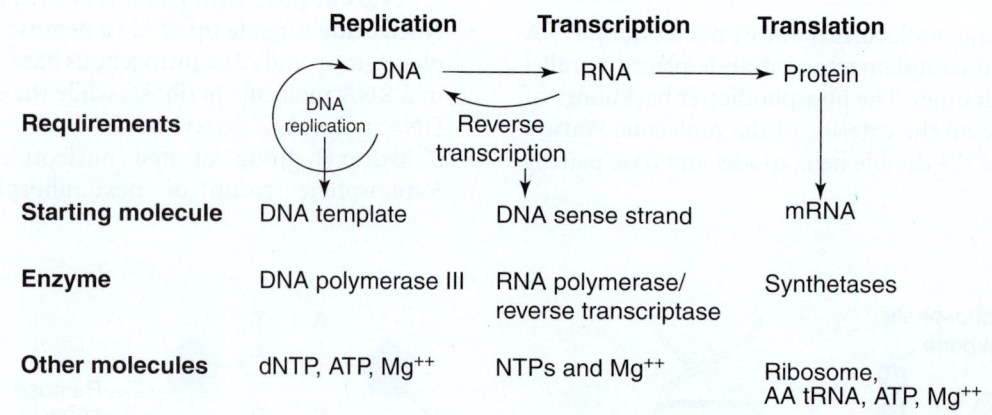

Fig. 2.2 Central Dogma of Gene Replication, Transcription and Translation.

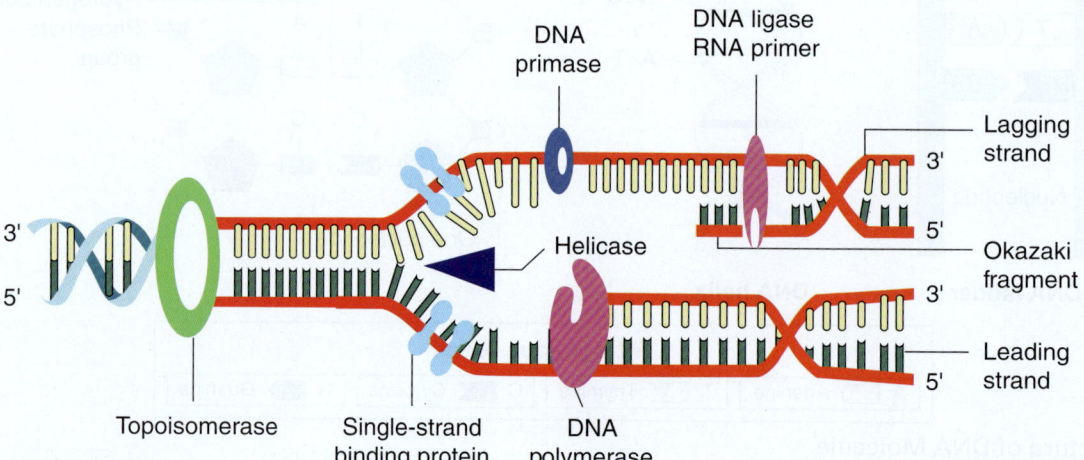

Fig. 2.3 Replication of DNA.

TABLE 2.1 Enzymes Needed for DNA Replication and Their Function

Enzyme/ Protein	Functions
Primase	Synthesises RNA primers. First step to start replication primers actually. Primes the DNA synthesis, hence, called the primer. It is about 5–10 nucleotides long and complementary to the DNA
DNA polymerase I	Exonuclease activity, that is, removing RNA primer and replacing with newly synthesised DNA
DNA polymerase II	Repairs the newly formed DNA strand
DNA polymerase III	Adds nucleotides in the 5′ to 3′ direction
Helicase	Opens the DNA helix by breaking hydrogen bonds between the nitrogenous bases. A DNA ladder forms
Ligase	Seals the gaps between the Okazaki fragments. Hence continuous DNA strand forms
Sliding clamp	Holds the DNA polymerase at the site where nucleotides are being added
Topoisomerase	Causing nicks in DNA molecule to relieve the stress when unwinding. Later reseals the DNA nicks
Single-strand binding proteins (SSB)	Binds to single-stranded DNA to avoid DNA rewinding back

DNA replication needs a large number of enzymes (Table 2.1). Many steps in DNA replication like addition of nucleotides require energy. This energy is released when the bond between the phosphates is broken.

Box 2.1 lists the steps of DNA replication in sequential order.

TRANSFER OF GENETIC MATERIAL

There are three types of genetic transfer methods found in bacteria: transformation, conjugation and transduction.

Transformation

This is the process where bacteria uptake a piece of external naked DNA. Usually, this process is used in the laboratory to introduce a plasmid into a bacterial cell (Fig. 2.4).

Natural Transformation

Frederick Griffith reported in 1928 that bacteria are capable of transferring genetic information through transformation. *Streptococcus pneumoniae* expresses DNA-binding proteins on the cell surface in stationary phase growth conditions. This natural *competent cell* allows uptake of 'naked DNA'. Frederick Griffith's experiments (Fig. 2.5) were conducted with *S. pneumoniae*. Griffith injected smooth (virulent)

BOX 2.1 Steps of DNA Deplication

- **DNA unwinding**: It starts at the origin of replication.
- **Origin of replication**: Specific nucleotide sequences on a DNA molecule where replication begins. It is rich in AT sequences and is approximately 245 base pairs long. Most prokaryotes have a single origin of replication on its one chromosome.
- **Forming replication forks**: Helicase opens up the DNA molecule in two strands by breaking the hydrogen bonds between the nitrogenous base pairs which extend bidirectionally and form a fork-like structure. Later, to prevent rewinding of the DNA a protein coat binds on single strand of DNA around the replication fork.
- **Prevent supercoiling**: Topoisomerases prevent supercoiling by binding at the region ahead of the replication fork.
- **Primase synthesis**: RNA primers complementary to the DNA strand are synthesised by primase.
- **Adding nucleotides**: DNA polymerase starts to add the nucleotides at 3′-OH end of the primer; hence, elongation of both the lagging and the leading strand goes on.
  a. **Synthesis of leading strand**: The synthesis of this strand requires a free 3′-OH group to which it can add nucleotides. A phosphodiester bond forms between the 3′-OH end and the 5′-phosphate of the next nucleotide. It cannot add nucleotides if a free 3′-OH group is not available. Only one strand can be extended in this manner and is known as leading strand.
  b. **Synthesis of lagging strand:** The synthesis of this strand occurs in small pieces known as Okazaki fragments. RNA primase synthesises a RNA primer which gets attached to the other strand at various places. DNA polymerase can now extend this RNA primer, adding nucleotides one by one that are complementary to the template strand (Fig. 2.3). After degrading RNA primers, ligase enzyme seals the gaps between Okazaki fragments.
- **Removing RNA primers**: It occurs by exonuclease activity of DNA pol I.
- **Filling the gaps**: DNA pol II fills the gap by adding dNTPs.
- **Sealing gap**: It seals the gap between the two DNA fragments – It occurs by DNA ligase.

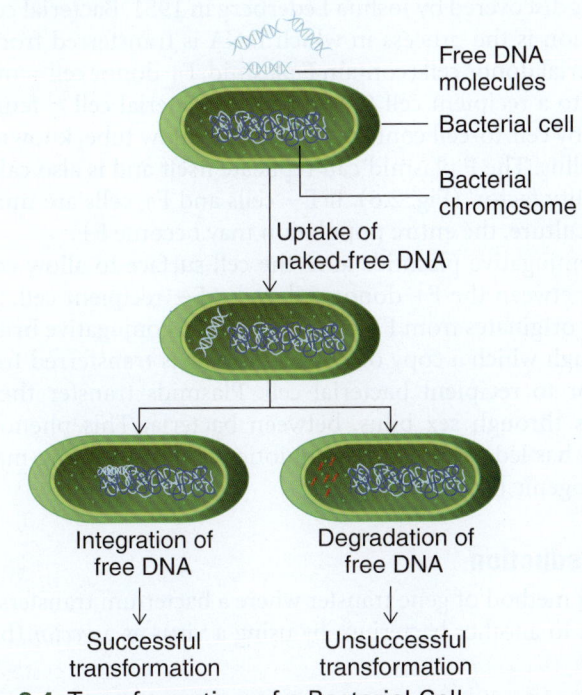

Fig. 2.4 Transformation of a Bacterial Cell.

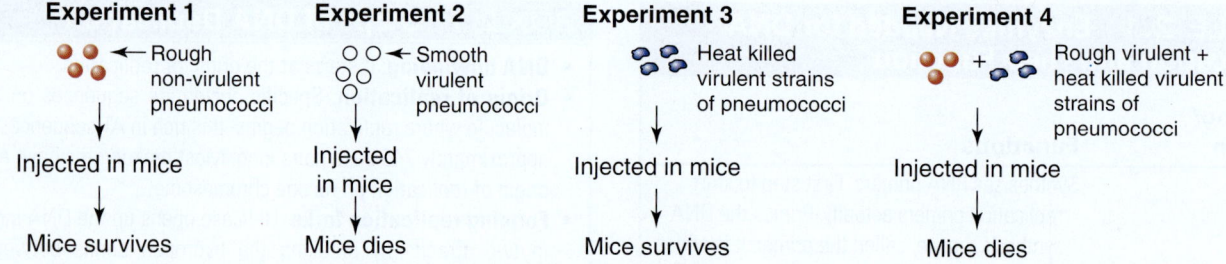

Fig. 2.5 Griffith Experiment Demonstrating Natural Transformation.

strains of pneumococci to one mice and rough (non-virulent) strain to another mice. The one which was infected with the smooth strain developed pneumonia and died, while the one infected with the rough strain stayed alive. After that, Griffith heat killed the smooth strain of bacteria and injected into mice. Mice did not develop pneumonia and survived. Then, he mixed the heat killed smooth and live rough strains together and injected this mixture into mice. The mice died. A smooth strain of bacteria was later isolated from dead mice. It proved that DNA from killed smooth bacteria (virulent) was taken up by rough strains, which converted into the smooth variety.

In Vitro Transformation

It can be artificially induced in a laboratory after treating cell with chemicals like Ca^{2+} or with physical methods like heat shock, treatment with polyethylene glycol (PEG) and electroporation competence due to increased membrane permeability. Such cells are called competent cells and are ready for transformation.

Conjugation

It was discovered by Joshua Lederberg in 1951. Bacterial conjugation is the process in which DNA is transferred from a bacterial donor cell (contain F plasmid, F+ donor cell = male cell) to a recipient cell (F− recipient bacterial cell = female cell) by cell-to-cell contact, through a hollow tube, known as sex pilus. The F plasmid can replicate itself and is also called 'fertility factor' (Fig. 2.6). If F− cells and F+ cells are mixed in a culture, the entire population may become F+.

Conjugative plasmids alter the cell surface to allow contact between the F+ donor cell and a F− recipient cell. Sex pilus originates from F+ cell and makes a conjugative bridge through which a copy of DNA molecule is transferred from donor to recipient bacterial cell. Plasmids transfer themselves through sex pilus, between bacteria. This phenomenon has led to spread of antibiotic resistance among many pathogenic bacteria.

Transduction

It is a method of gene transfer where a bacterium transfers its DNA to another bacterium by using a virus as a *vector* (bac-

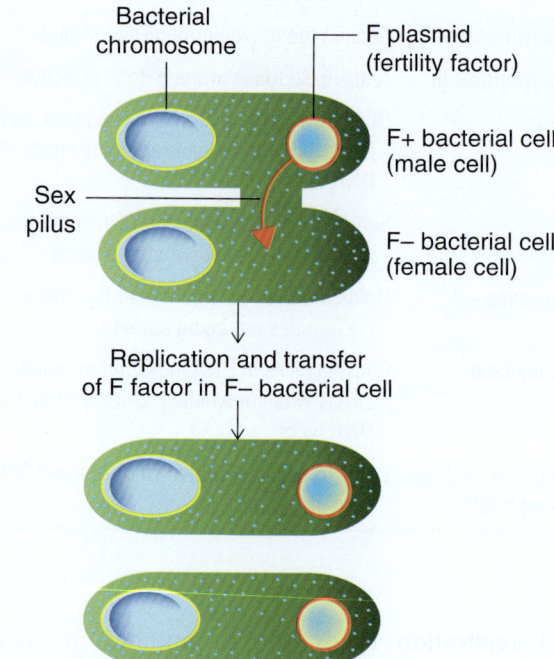

Fig. 2.6 Bacterial Conjugation and Transfer of F Plasmid.

teriophage). Bacteriophages are often used to transfer the DNA molecules in bacteria, within laboratory. Transduction is of two kinds: generalised and specialised transduction (Figs. 2.7 and 2.8).

Generalised Transduction

It occurs in lytic virus cycle. This is a method of transduction that allows genes to be transferred from one bacterium to the other. Normally, when a bacteriophage infects a new bacterial cell, the bacterial DNA disintegrates, viral nucleic acid multiplies and new viral particles are made. Accidentally, during packing of genetic material in virus particles, bacterial DNA is incorporated in the viral capsid. These virus particles are released and can infect new host cell and transfer bacterial DNA packed in them to new bacterial cell (Fig. 2.8). Many different viral and bacterial genes can be transmitted in this way to other bacterial population. This is a rare natural event.

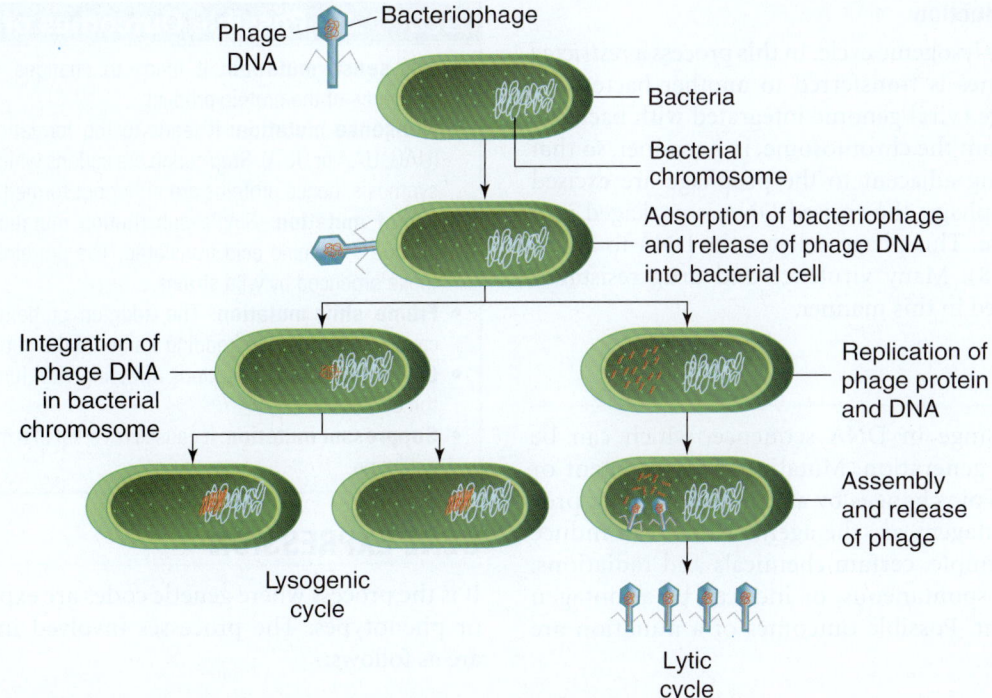

Fig. 2.7 Lytic and Lysogenic Cycles of Viral Replication in a Bacterial Cell.

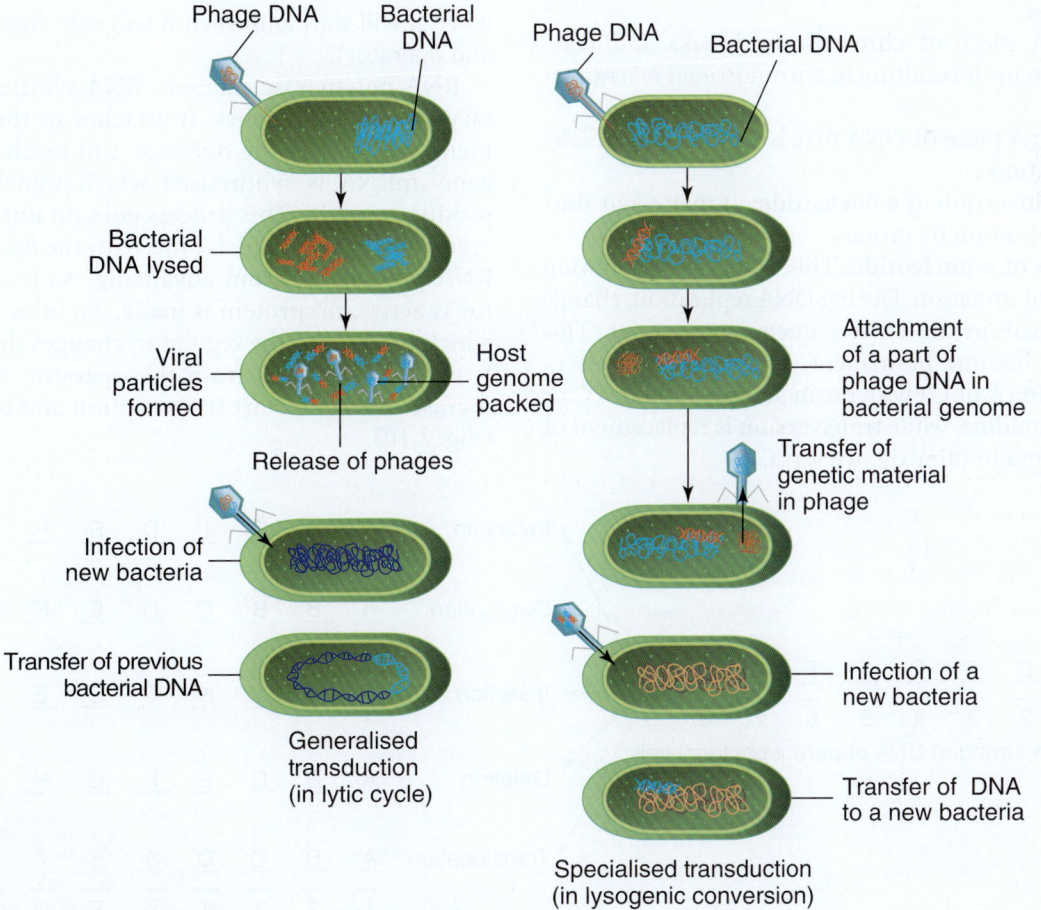

Fig. 2.8 Bacterial Gene Transfer by Generalised and Specialised Transduction.

Specialised Transduction

It occurs during the lysogenic cycle. In this process a *restricted* set of bacterial genes is transferred to another bacterium. When the prophage (viral genome integrated with bacterial genome) excises from the chromosome, in a manner, so that bacterial genes lying adjacent to the prophage are excised along with the prophage, the excised DNA is packaged into a new virus particle. This phage delivers the DNA to a new bacterium (Fig. 2.8). Many virulence and drug resistance genes are transferred in this manner.

MUTATIONS

A mutation is change in DNA sequence, which can be passed on to next generation. Mutations may be silent or may show phenotypic changes by altering the type of protein produced. Mutagens are the agents which can induce mutations, for example, certain chemicals and radiations. Mutations can be spontaneous, or induced by a mutagen in the environment. Possible outcomes of a mutation are listed in Box 2.2.

Types of Mutations (Fig. 2.9)

- **Translocation:** When the nucleotide sequence on leading strand of DNA is replaced by complementary nucleotide sequence on lagging strand, during replication, it is called translocation.
- **Inversion:** A piece of chromosome breaks and rearranges within itself resulting in chromosomal rearrangement.
- **Duplication:** A piece of DNA may be copied abnormally one or more times.
- **Deletion or insertion of a nucleotide:** It may occur during DNA replication by mistake.
- **Substitution of a nucleotide:** This is the most common mechanism of mutation. During DNA replication, change of a single base in the DNA sequence may occur. This change may become permanent. **Transition** is replacement of purin (A or G) with purin and pyramidine (C or T) with pyramidine, while **transversion** is replacement of purine with pyramidine or vice versa.

BOX 2.2   Possible Outcomes of a Mutation

- **Mis-sense mutation:** It leads to changes in the amino acid sequence of the protein product.
- **Nonsense mutation:** It leads to the formation of a stop codon (UAG, UAA or UGA). Stop codon are codons which terminate protein synthesis, hence proteins are either not formed or are incomplete.
- **Silent mutation:** Single substitution mutation which does not change the amino acid translated. The proteins produced are like those produced by wild strains.
- **Frame shift mutation:** The addition or deletion of base pairs causing a shift in the 'reading frame' of the gene.
- **Lethal mutation:** Mutations affecting vital function, hence killing the cell.
- **Suppressor mutation:** It causes reversal of a mutant phenotype to wild type.

GENE EXPRESSION

It is the process where genetic codes are expressed as proteins or phenotypes. The processes involved in gene expression are as follows:

Transcription

Synthesis of RNA from a DNA template. Transcription is regulated by operons. **An operon** is a gene which does not code for protein, but regulates the protein synthesis as to when it will start and when it will stop (regulator, promoter and operator).

RNA polymerase catalyses RNA synthesis complementary to structural genes. It attaches to the promoter segment, passes over the operator and reaches the structural gene. mRNA is synthesised which signals ribosomes to produce protein. This process goes on until cell stops it by *regulator proteins*, which bind with the operator, and stops RNA polymerase from advancing. As long as the regulator is active, no protein is made. An inducer is a molecule which binds with the regulator, changes the shape of regulator and releases it from the operator. Now RNA polymerase can again start transcription and protein synthesis (Fig. 2.10).

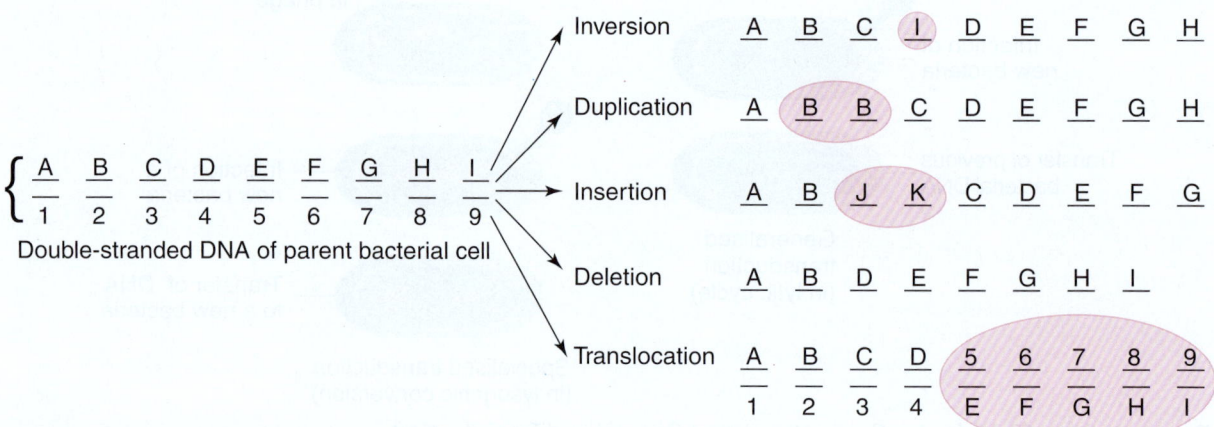

Fig. 2.9 Types of Chromosomal Mutations.

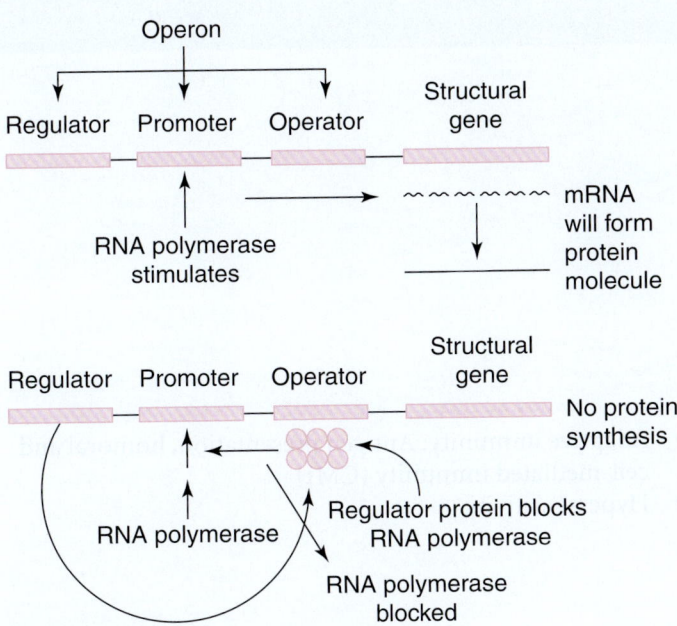

Fig. 2.10 Transcription Regulation.

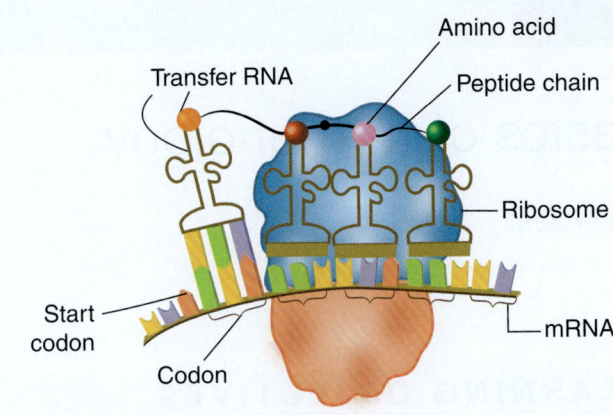

Fig. 2.11 Translation Process.

Translation

Initiation of the translation process occurs when ribosomes bind to mRNA and move along it in a 5′ to 3′ direction until it reaches AGU, that is the start codon. The anticodons present on transfer RNA (tRNA) molecules bind to complementary codons. The amino acid carried by the tRNA binds to the polypeptide chain by a peptide bond. This process continues until the ribosome reaches a stop codon (Fig. 2.11).

Basics of Immunology

Amita Jain

INTRODUCTION

Immunology is a science that deals with the study of host defence mechanisms. The immune system deals with cellular (e.g. lymphocytes, macrophages etc.) and molecular components (e.g. antibodies, MHC etc.) present in the host body. It contributes to host defence against foreign substances (non-self) and pathogens. Basic immunology is not covered in this chapter. However, before we begin, we will revise certain important terms, which will be used in this chapter repeatedly.

IMPORTANT TERMINOLOGY

Antibody (Ab)/Immunoglobulin (Ig)

It is a protein molecule produced by B cells as a result of the response to an antigen. It is usually specific to the antigen that stimulated its production and can bind this antigen molecule specifically. There are five classes (IgG, IgM, IgA, IgD and IgE) and many subclasses of the antibody. These classes and subclasses vary in their structure and function (Table 3.1). A diagrammatic representation of these molecules is shown in Fig. 3.1. An antibody molecule consists of two identical heavy chains and two identical light chains. The light chain in all the classes is same which can be either κ or λ. Heavy chain varies among classes and is the basis of their classification. The heavy chains are linked together by two disulphide bonds.

Antigen (Ag)

Antigen is a substance that can react with an antibody. Some antigens can induce antibody production and are called immunogens. These can be proteins, sugars or fat. Proteins make most potent antigens.

Hapten

Haptens are the small molecules on immunogen, which are immunogenic once combined with a larger molecule.

TABLE 3.1	Characteristics of Different Classes of Immunoglobulins				
Characteristics	**IgG**	**IgM**	**IgA**	**IgD**	**IgE**
Serum concentration (%)	85	5	7–15	0.3	0.02
Structure	Monomer	Pentamer	Dimer/monomer	Monomer	Monomer
Valency	2	10	4/2	2	2
Heavy chain	γ (Gamma)	μ (Mu)	α (Alpha)	δ (Delta)	ε (Epsilon)
Distribution in body	Intravascular and extravascular	Mostly intravascular	Intravascular and body secretions, such as tears	Bound on B-lymphocyte membrane	Basophils and mast cells
Biological properties	Crosses placenta and blood–brain barrier; opsonisation	Primary antibody to be produced in response to infection; binds strongly to antigens	First line of defence on body surfaces	Unknown	Immunity to parasites; hypersensitivity

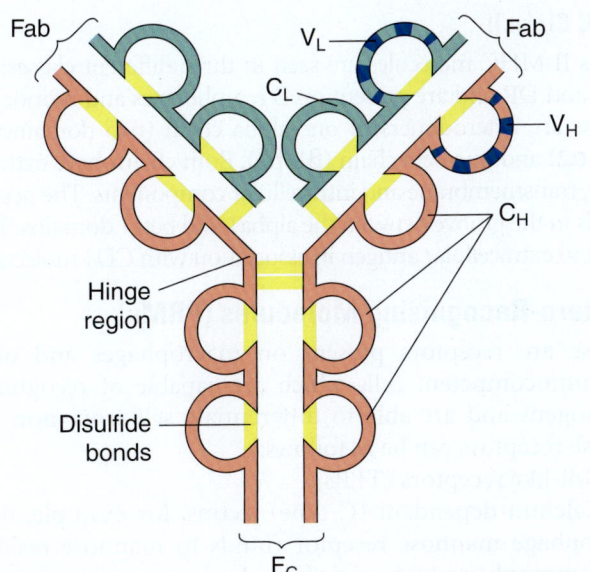

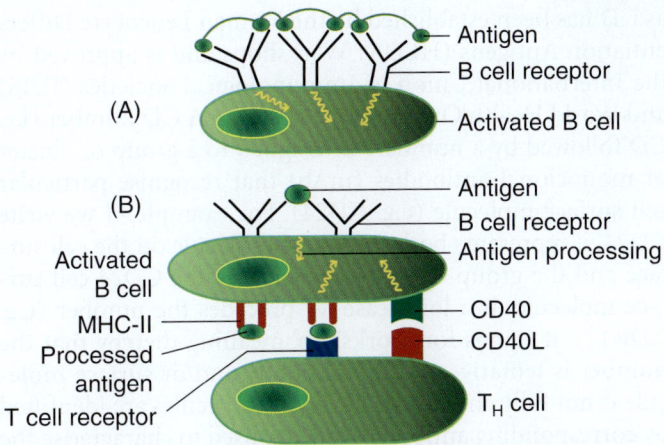

Fig. 3.2 Activation of B Cells. (A) T cell-independent and (B) T cell-dependent antigens.

Fig. 3.1 Shapes and Structure of IgG Class of Antibody Molecule.
Fab = fragment, antigen-binding; F$_C$ = fragment, crystallisable; C$_L$ = constant domain, light chain; C$_H$ = constant domain, heavy chain; V$_L$ = variable domain, light chain; V$_H$ = variable domain, heavy chain.

They are too small to induce antibody production on their own.

Epitope (Antigenic Determinant)
Epitope is the site on an antigen molecule recognised by an antibody.

Adjuvants
Adjuvants are the additional carrier molecules which are injected with haptens or antigens to increase their potency as antigens. Antibodies formed do not react with adjuvants.

Salient Properties of Antigen
Salient properties of antigen are: (1) foreignness: non-self, (2) molecular size: usually large proteins, (3) chemical and structural complexity: chemically complex and (4) antigenic determinants: at least one or more.

Types of Antigens
Antigens capable of activating B cells may be categorised into two groups:
  Thymus-independent antigens
  Thymus-dependent (TD) antigens
Thymus-independent antigens. (Fig. 3.2A): These are of two types – thymus-independent type 1 and thymus-independent type 2. They do **not** stimulate B cells through interaction with the immunoglobulin receptor and give rise to relatively low-affinity antibody responses. Type 1 antigens are molecules of microbial origin (e.g. bacterial polysaccharide and lipopolysaccharides [LPSs]) that bind to B cells and stimulate their proliferation and differentiation. These substances induce numerous B-cell divisions and are

therefore called mitogens. They also stimulate various clones of B cells and therefore result in production of multiclonal antibodies. IgM is the predominant antibody produced with little IgG and IgA response. Immunologic memory is often weak or short-lived. In contrast, TD antigens tend to lead to the generation of higher-affinity antibodies and lasting immunologic memory. Type 2 thymus – independent antigens mainly elicit antigen-specific antibody responses, although these are similar to type 1 antigen in structure. They are capable of cross-linking specific immunoglobulin receptors on B cells. Examples include epitopes of encapsulated bacteria, such as *Streptococcus pneumoniae* capsular polysaccharide.

Thymus-dependent antigens. (Fig. 3.2B): Most of the antigens are TD antigens. In order to present antigen to B cells, T helper cells recognize the antigen molecule bound with histocompatibility molecules present on surface of an antigen presenting cell (APC). T cell gets activated and supports a B cell for antibody production (T helper or Th 'effector' cell). Later B cells internalise antigen through immunoglobulin receptor, process it and present it in association with MHC class II molecules. This action makes the response of Th cell specific, so that only appropriate B cells get stimulated.

Mitogens
Mitogens are usually polymers containing a repeating structural unit. They can be polysaccharides, lectins or proteins that agglutinate cells or precipitate certain macromolecules. At low concentrations, they stimulate specific antibody responses and at higher concentrations, they induce non-specific B-cell proliferation. Few known potent mitogens are *Staphylococcus aureus* Cowan (SAC) strain (bacterial protein A that aggregates surface immunoglobulin receptors) and a plant lectin, known as pokeweed mitogen.

Cluster of Differentiation
CD molecules are antigens present on cell surface of immunocompetent cells. Nomenclature of these antigen molecules

as CD has been established by the Human Leucocyte Differentiation Antigens (HLDA) Workshops and is approved by the International Union of Immunological Societies (IUIS) and World Health Organization (WHO). A CD number (i.e. CD followed by a number) is assigned to a group or cluster of monoclonal antibodies (mAb) that recognise particular cell surface molecule (e.g. CD24). For example, if we write CD24, it represents both the CD24 molecule on the cell surface and the group of mAbs recognising the CD24 cell surface molecule. If a lowercase 'w' precedes the number (e.g. CDw12), it stands for 'workshop' meaning thereby that the number is tentative and the antibody and/or surface molecule is not fully characterised. These molecules are identified by corresponding antibodies and are used to characterise the cell types.

Major Histocompatibility Complex

MHC is a cluster of genes located in close proximity, on small arm of human chromosome 6. These genes encode the **histocompatibility antigens** (MHC molecules). Histocompatibility molecules play a very important role in immune response and transplant rejections (hence also known as human leucocyte antigen [HLA] complex). The MHC molecules bind to antigens (derived from pathogens) and make them recognisable by specific T cells by displaying them on the cell surface. These molecules are of two types: **MHC class I and MHC class II**. Their structure is diagrammatically represented in Fig. 3.3.

MHC Class I

Class I MHC molecules are present on all nucleated cells and platelets and are found in three different classes: A, B and C. These are heterodimers of an alpha (or heavy) chain and a smaller beta-2-microglobulin chain. Alpha chain has extracellular components (three domains: $\alpha 1$, $\alpha 2$ and $\alpha 3$), and a transmembrane part, which travels intracellularly. The beta-2-microglobulin molecule is totally extracellular. A groove formed by the alpha1 and alpha2 domains is the site for protein binding. They bind to intracellular proteins generated within the cytoplasm, with CD8 molecule.

MHC Class II

Class II MHC molecules are seen in three different classes: DP, DQ and DR and are present on B lymphocytes and monocytes. These are heterodimers of one alpha chain (two domains: $\alpha 1$ and $\alpha 2$) and one beta chain ($\beta 1$, $\beta 2$). Both chains have extracellular, transmembrane and intracellular components. The peptide binds in the groove between the alpha1 and beta1 domains. They bind to extracellular antigen in association with CD4 molecule.

Pattern-Recognising Molecules (PRMs)

These are receptors present on macrophages and other immunocompetent cells which are capable of recognising pathogens and are able to differentiate self from non-self. These receptors can be as follows:
1. Toll-like receptors (TLRs)
2. Calcium-dependent (C type) lectins, for example, macrophage mannose receptor (binds to mannose residues commonly present on surface of microorganisms), found on phagocytes
3. f-Met-Leu-Phe receptor (binds to *N*-formyl peptides and when present, attracts neutrophils)
4. Complement receptors: Designated CRs (binds to complement components, such as C3b and C4b which opsonise microorganims as a consequence of the activation of the complement cascade)
5. Scavenger receptors: At least six different scavenger receptors with different specificities have been described (recognise certain anionic polymers and acetylated low-density lipoproteins)
6. CD14 (receptor on the surface of phagocytes which allows for the recognition of LPS)

Cells of Immune System (Flowchart 3.1)
Antigen-Presenting Cells

These cells display antigen complexed with MHCs on their surfaces. Almost all cell types can serve as APCs. There are two types of APCs: **professional APCs** (macrophages, B cells, dendritic cells), which present foreign antigens to helper T cells and **non-professional cells**, which can only present antigens originating inside the cell, for example virus-infected host cells or cancer cells can present antigens to cytotoxic T cells (CTLs). **Professional APCs** express MHC class II molecules and non-professional APCs express MHC class I molecules.

B Cell (B Lymphocyte)

It is a type of small lymphocyte (bursa derived) which can be stimulated to produce antibody. They play a very important role in adaptive humoral immune response and are responsible for production of antibodies. Early B-cell development occurs in the foetal liver prenatally, and in the bone marrow later in life. Several distinct B-cell subsets are known to exist with distinct functions in both adaptive and innate humoral immune responses.

T Cell (T Lymphocyte) (Thymus-Derived)

It plays a very important role in immune response and participates in a variety of cell-mediated immune reactions.

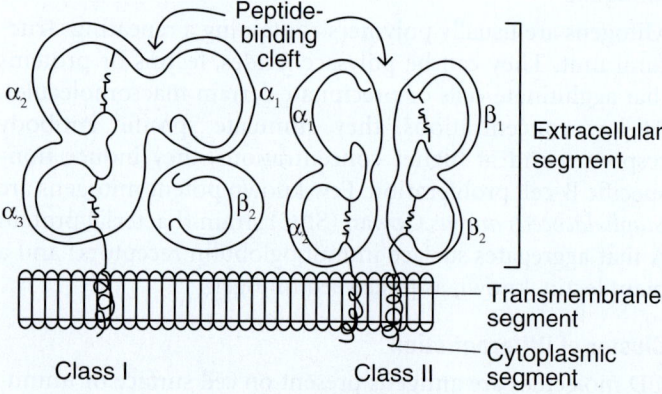

Fig. 3.3 Class I and Class II Molecules.

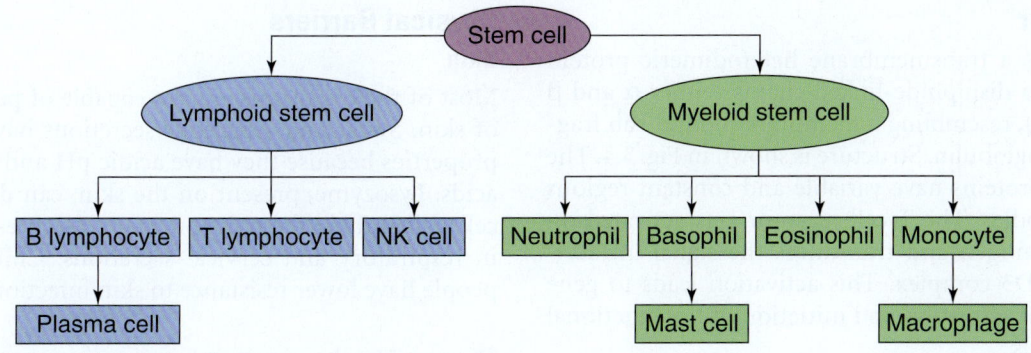

Flowchart 3.1 Cells of Immune System.

They are of several subtypes and each subtype plays a different role. T cells can be distinguished from other lymphocytes, by the presence of a T-cell receptor on cell surface. Antigen-naïve T cells differentiate into memory and effector T cells. Effector T cells are the superset of T cells that actively respond immediately to a stimulus. They may be helper, killer or regulatory. Memory cells are longer-lived to target future infections.

1. **T helper cells** (T_H cells, also known as **CD4$^+$ T cells**) help in immunologic processes, including antibody production. These cells express the CD4 antigen on their surfaces.
2. **Cytotoxic T cells** (also known as T_C cells, **CD8$^+$ T cells**, T-killer cells, killer T cells) destroy virus-infected cells and tumour cells. They express the CD8 glycoprotein at their surfaces.
3. **Memory T cells** are long-lived and can quickly expand to large numbers of effector T cells upon re-exposure to their cognate antigen, hence, providing 'memory' against previously encountered pathogens. Memory T cells may be either CD4$^+$ or CD8$^+$ and usually express CD45RO.
4. **Regulatory T cells** (suppressor T cells) play a very important role in maintenance of immunological tolerance. They suppress T cell-mediated immunity and autoreactive T cells in the end of an immune reaction.
5. **Natural killer T cells** (NKT cells) can recognise glycolipid antigen presented by a molecule called CD1d. They can function as both T_H cells and T_C cells. They are also able to recognise and destroy tumour cells and cells infected with herpes viruses. They are different from natural killer (NK) cells which play role in innate immunity.

Natural Killer Cells

NK cells are lymphocytes classified as group I innate lymphocytes (ILCs). They efficiently kill virally infected cells and control early cancer. NK cells do not need any priming or activation to kill tumour cells (CTLs need priming by APCs). They secrete cytokines, such as interferon gamma (IFN-γ) and tumour necrosis factor-α (TNF-α), which help in potentiating immune response.

Plasma Cell (Also Known as Plasma B Cells, Plasmocytes, Plasmacytes, Effector B Cells)

Plasma cell is a terminally differentiated B cell that secretes antibody. They originate in the bone marrow; however, their CD markers are different from that of mature B cell. Plasma cells do not express common pan-B-cell markers, such as CD19 and CD20. Instead, they express CD27 and CD138 at high levels.

B-Cell Receptor

B-cell receptors are antibody molecules, which can be a form of IgM or IgD. These membrane-bound Ig molecules bind with the same antigen-binding specificity as a membrane-bound receptor (IgM or IgD) interacts with other cell surface molecules causing their aggregation, and transduce signals intracellularly by interacting with tyrosine kinase molecules and the other components of the signal transduction machinery. These signals result in cell activation. Diagrammatic representation of B- and T-cell receptors is shown in Fig. 3.4.

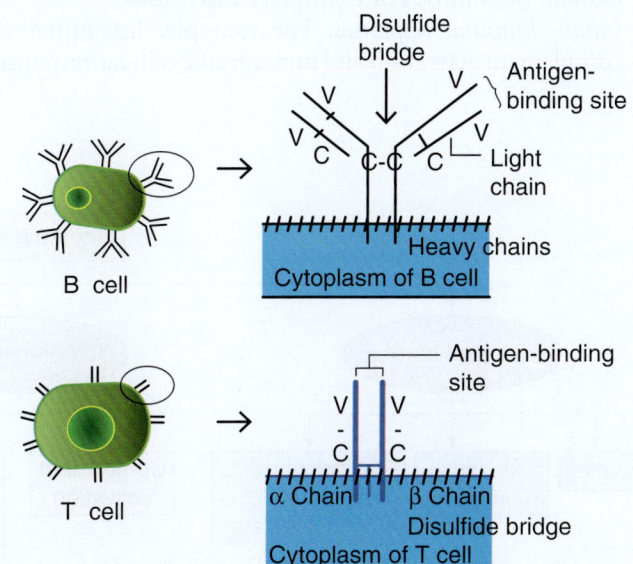

Fig. 3.4 B- and T-Cell Receptors. V = variable regions; C = constant regions; TMR = transmembrane regions.

T-Cell Receptor

T-cell receptor is a transmembrane heterodimeric protein composed of two disulphide-linked chains (either α and β or rarely γ and δ), resembling a membrane-bound Fab fragment of immunoglobulin. Structure is shown in Fig. 3.4. The T-cell receptor proteins have variable and constant regions similar to antibodies. The T-cell receptor gets activated on recognition of antigen and transduces the signal intracellularly by the CD3 complex. This activation leads to gene transcription, cell activation and initiation of the functional activities of T cells.

The CD4 and CD8 molecules that differentiate the two major functional classes of T cell function as co-receptor molecules. During recognition of antigen, the CD4 and CD8 molecules interact with the T-cell receptor complex and with MHC molecules. CD4 binds to MHC class II molecules and CD8 binds to MHC class I molecules.

INNATE IMMUNITY

The innate immunity is provided by defence systems present in host body against infection, since birth. These systems are present inherently, are non-specific and can be activated immediately after contact with non-self. Characteristics of innate immune response include the following:
1. Broad spectrum (non-specific) responses
2. No memory or lasting protective immunity
3. Limited repertoire of recognition molecules
4. Responses are phylogenetically ancient

The innate immune system is made up of the following (Flowchart 3.2):
1. *Physical barriers and defence mechanisms*: For example, skin, mucosa of gastrointestinal tract, respiratory tract and nasopharynx, eyelashes, cilia and body secretions, such as mucus, bile, gastric acid, saliva, tears and sweat
2. *Cellular immunity*: For example, *phagocytosis*
3. *Innate immune responses*: For example, inflammation, complement activation and non-specific cellular responses

Physical Barriers

Skin

Most of the organisms are not capable of penetrating layers of skin. Sweat and sebaceous secretions have antimicrobial properties because they have acidic pH and are rich in fatty acids. Lysozyme, present on the skin, can dissolve bacterial cell wall and kill them. Lysozyme is also present in tears and in respiratory and cervical secretions. Children and older people have lower resistance to skin infections.

Mucous Membranes

Provide great resistance to infections due to one or more of the following properties:
1. Microbes getting trapped in mucus present on surface
2. Ciliary movement of ciliary surface cells expel the trapped pathogens as seen on respiratory mucosa
3. Presence of lysozyme in secretions
4. Presence of specific IgA antibody on surfaces
5. Presence of numerous hydrolytic enzymes as seen in saliva
6. Acidity of the stomach kills many ingested bacteria
7. Phagocytosis by phagocytes present on surface and transport to regional lymph nodes
8. Presence of constant normal microbial flora that itself opposes establishment of pathogenic microorganisms

Innate Immune Responses

Innate immune response is the first line of defence against non-self, occurs rapidly and generally holds the spread of pathogen until a specific adaptive response is initiated. The responses which help in the process are as follows:
1. Phagocytosis
2. Activation of complement by the alternative pathway
3. Release of cytokines from macrophages and of other mediators that trigger the **inflammatory response**
4. Release of interferon
5. NK cells
6. Apoptosis

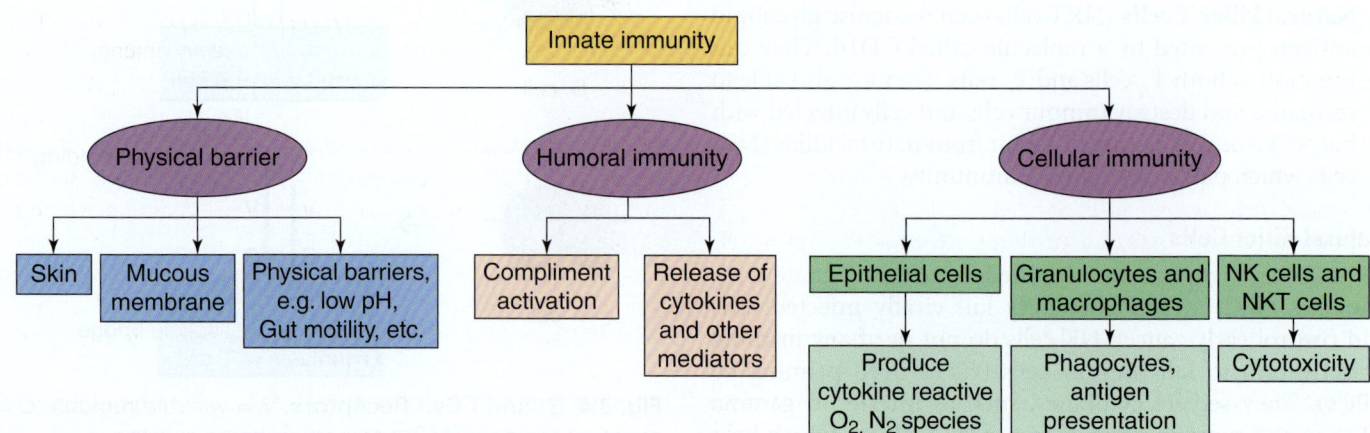

Flowchart 3.2 Components of Innate Immunity.

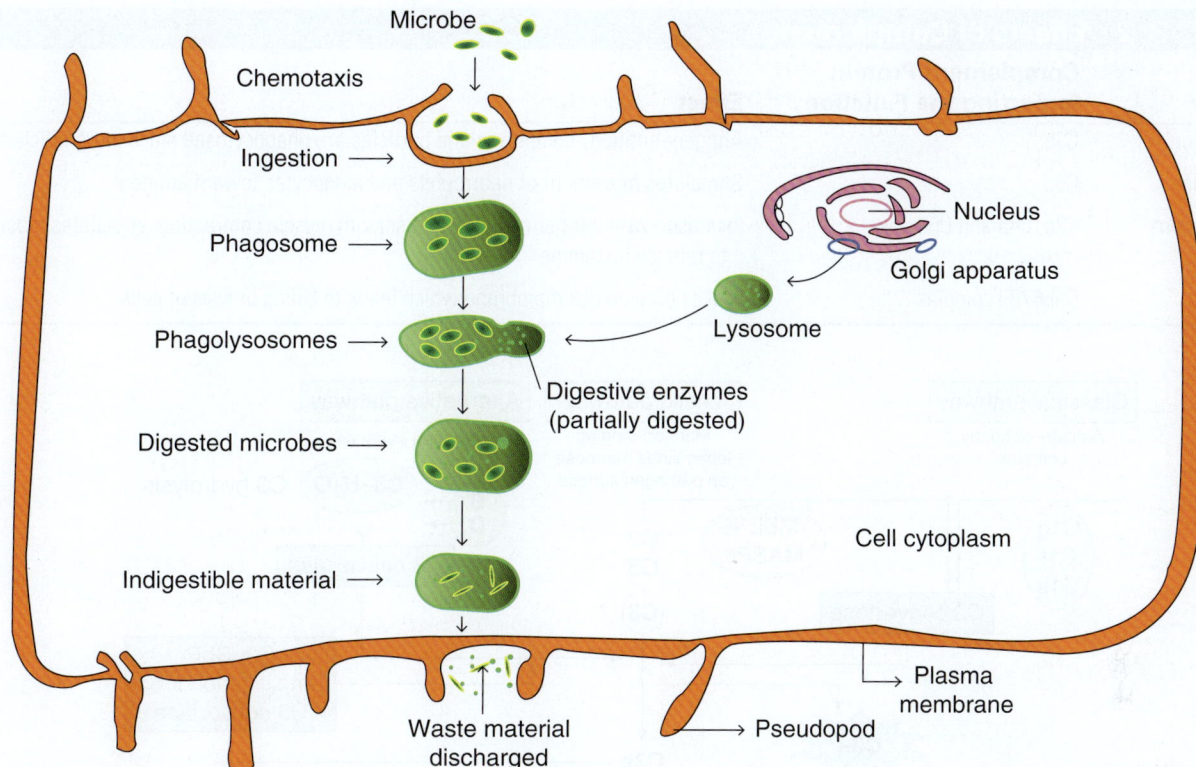

Fig. 3.5 Process of Phagocytosis.

Phagocytosis (Fig. 3.5)

Phagocytic cells are mononuclear cells present in blood, lymphoid tissue, liver, spleen, lung and other tissues; cells lining blood and lymph sinuses (e.g. Kupffer cells in liver) and macrophages. Functions of phagocytic cells are as follows:

Migration and chemotaxis. Whenever a foreign particle or a pathogen enters a host body, phagocytes are attracted towards it. During bacterial infection, the number of circulating phagocytic cells increases. Microorganisms may also release or express chemotactic factors that attract phagocytic cells. When macrophages recognise microbes, through PRMs, they get stimulated and release cytokines, which act as chemotactic factors. Defects in chemotaxis may be acquired or inherited and cause hypersusceptibility to bacterial infections.

Attachment and ingestion. The phagocytic cells attach to the microbe and engulf it making a vesicle called phagosome. The wall of the vesicle is derived from cell membrane. Phagocytes contain intracellular granules known as lysosomes, which contain lysozyme, other hydrolytic enzymes, several cationic proteins, defensins, lactoferrin and toxic nitrogen oxides. Phagosome fuses with lysosome to form phagolysosome.

Phagocytosis is enhanced by antibodies (opsonins) that coat the surface of bacteria and facilitate their ingestion by phagocytes. Macrophages have receptors on their membranes for the Fc portion of antibody and for the C3 component of complement. Opsonisation can occur by three mechanisms:

(1) antibody alone; (2) antibody plus antigen can activate complement via the classic pathway and (3) a heat-labile system in which immunoglobulin or other factors activate C3 via the alternative pathway.

Microbial killing. Intracellular pathogen killing can occur by: (1) non-oxidative mechanisms (e.g. increased glycolysis via hexose monophosphate shunt; and discharge and activation of hydrolytic enzymes of lysozome in contact with microorganisms, action of antimicrobial peptides) and (2) oxidative mechanisms (increased generation of superoxide anion (O_2^-) and increased release of H_2O_2).

Complement System

The complement system is a group of serum and membrane-bound proteins that function in a series of proteolytic cascades. These proteins complement (augment) the immune system (e.g. antibody), hence the name 'complement'. The alternative complement pathway can be activated by microbial surfaces and proceeds in the absence of antibody. The functions of complement system are listed in Table 3.2.

Complement activation. (**Flowchart 3.3**): The components of the complement proteins are numbered from C1 to C9. The complement activation sequence is C1-C4-C2-C3-C5-C6-C7-C8-C9. Several complement components are proenzymes for the next step. C2–C5 proteins are cleaved liberating smaller fragments which are denoted by the letter a (e.g. C4a) and the larger fragments which are denoted

TABLE 3.2   Effects of Complement

Function	Complement Protein Imparting the Function	Effect
Opsonisation	C3b	Antigen–antibody complexes and particles are phagocytosed much more efficiently
Chemotaxis	C5a	Stimulates movement of neutrophils and monocytes toward antigen
Anaphylatoxins	C3a, C4a and C5a	Increased vascular permeability and smooth muscle contraction; stimulates mast cells to release histamine
Cytolysis	C5b6789 complex	Makes pore on cell membrane which leads to killing or lysis of cells

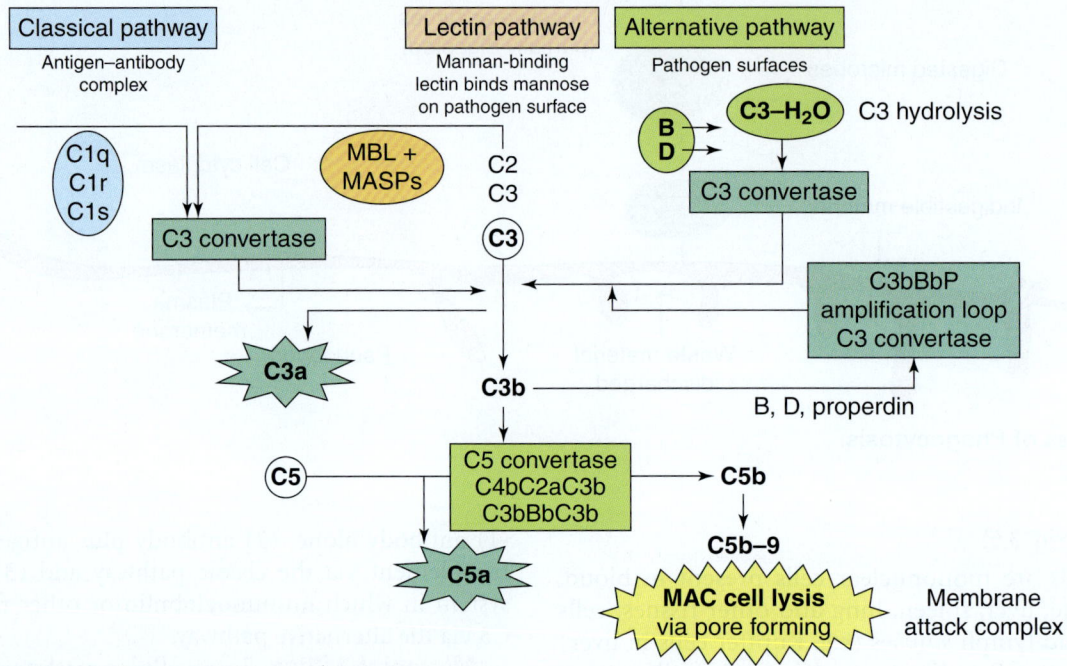

Flowchart 3.3 Complement Activation Pathways.

by letter b (e.g. C4b). Complement activation is regulated by several factors shown in Box 3.1. There are three pathways of complement activation:

The classic pathway. Only antibodies (IgM and IgG subclasses 1, 2 and 3) can activate complement via the classic pathway. C1 is composed of three proteins: C1q, C1r and C1s. C1q is an aggregate of polypeptides that binds to the Fc portion of IgG/IgM. This complex activates C1s, which cleaves C4 and C2 to form C4b2b. C4b2b is C3 convertase, which cleaves C3 molecules into C3a and C3b. C3a is an anaphylatoxin. C3b forms a complex with C4b2b to form

C4b2b3b, which acts as C5 convertase and cleaves C5 to form C5a and C5b. C5a is an anaphylatoxin and a chemotactic factor. C5b binds to C6 and C7 to form a complex which binds C8, followed by the polymerisation of up to 16 C9 molecules. This forms a membrane attack complex (MAC) that generates a pore in the membrane and causes cytolysis by allowing passage of water across the cell membrane.

The alternative pathway. Some substances, for example endotoxin converts C3 via the action of factors B, D and properdin. C3 convertase (C3bBb) generates more C3b and forms C3bBbC3b, which is an alternative C5 convertase that generates C5b, leading to production of the MAC as described in classical pathway.

Mannan-binding lectin (MBL) pathway. A plasma protein termed MBL binds to sugar residues, such as mannose, found in microbial surface polysaccharides, such as LPS and can activate C4 and C2. The rest of this pathway is the same as the classic pathway of complement activation.

Release of Cytokines and Other Mediators of Inflammation

When macrophages come in contact with antigens, they release cytokines interleukin-1 (IL-1) and TNF-α along with

TABLE 3.3 Mediators of Inflammation and Their Actions

Inflammation Indicator	Mediators
Vasodilation, increased vascular permeability	Histamine, serotonin, bradykinin, complement proteins (C3a, C5a), leukotrienes (LTC$_4$, LTD$_4$), prostaglandins (PGI$_2$, PGE$_2$, PGD$_2$, PGF$_2$), activated Hageman factor, kininogen fragments, fibrinopeptides
Vasoconstriction	TXA$_2$, LTB$_4$, LTC$_4$, LTD$_4$, C5a
Smooth muscle contraction	C3a, C5a, histamine, LTB$_4$, LTC$_4$, LTD$_4$, TXA$_2$, serotonin, PAF, bradykinin
Mast cell degranulation	C5a, C3a
Stem cell proliferation	IL-3, G-CSF, GM-CSF, M-CSF
Chemotaxis	C5a, LTB$_4$, IL-8, PAF, 5-HETE, histamine, others
Lysosomal granule release	C5a, IL-8, PAF
Phagocytosis	C3b, iC3b
Platelet aggregation	TXA$_2$, PAF
Endothelial cell stickiness	IL-1, TNF-α, LTB$_4$
Granuloma formation	IL-1, TNF-α
Pain	PGE$_2$, bradykinin, histamine, serotonin
Fever	IL-1, IL-6, TNF-α, PGE$_2$

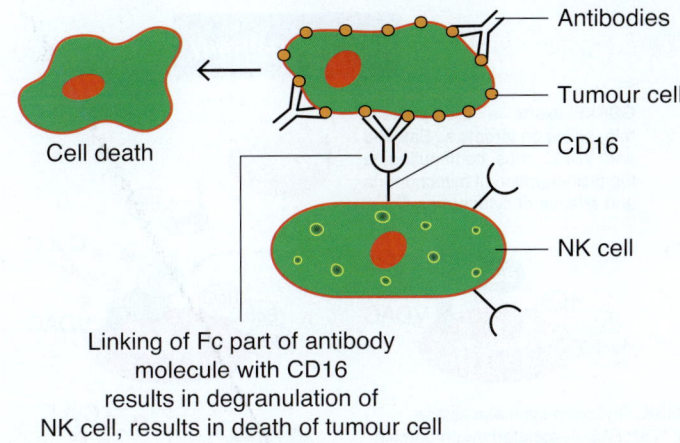

Linking of Fc part of antibody molecule with CD16 results in degranulation of NK cell, results in death of tumour cell

Fig. 3.6 Antibody-Dependent Cellular Cytotoxicity.

prostaglandins and leukotrienes. These substances elicit several inflammatory responses, of which most important are: (1) dilatation of local arterioles and capillaries causing escape of plasma in the area of injury and (2) induction of changes in expression of various adhesion molecules (selectins and integrins) on endothelial cells and leucocytes. Adhesion molecules increase attachment of leucocytes to the endothelial cells of the blood vessels. Polymorphonuclear leucocytes' concentration in vessels increases and they migrate out of the capillaries at the site of injury. Chemokines like IL-8 play an important role in recruiting monocytes and neutrophils from the blood into sites of infection, which phagocytose and kill the microorganisms. Soon the pH of the inflamed area becomes acidic causing lysis of the leucocytes. Macrophages scavenge cellular debris and inflammation starts resolving. Various mediators of inflammation and their role are listed in Table 3.3.

Interferons

Interferons are antiviral proteins released as a response of viral infections. Alpha and beta interferons control viral replication by inhibiting protein synthesis.

Natural Killer Cell

NK cell is a type of large lymphocytes, related to T cells, which play a role in antibody-dependent cellular cytotoxicity

(ADCC). They provide a strong defence against intracellular pathogens. Their mode of action is non-specific immunity as they do not possess specific antigen receptors. They can recognise carbohydrate ligands in association with MHC class I molecules. They can lyse tumour and virus-infected cells. High levels of alpha and beta interferons potentiate their action.

ADCC (Antibody-Dependent Cellular Cytotoxicity)

It is a cytotoxic response. Specific antibodies produced by B cells bind to antigens and Ag–Ab complex (Fc) is recognised by NK cells. This interaction causes release of cytotoxic substances and cause destruction of the pathogenic organism (Fig. 3.6). ADCC allows NK cells to recognise pathogens, which do not express pathogen-associated molecular patterns (PAMPs). It plays an important role in the immune response to parasites. The specificity of binding antibodies makes the reaction specific.

Cells, which can cause killing of antibody-coated cells and parasites are listed in Table 3.4.

TABLE 3.4 ADCC Effector Cells

Cell Type	Functions	Mechanism
Macrophages	Killing of pathogens	Release of enzymes and toxic substances
Natural killer cells	Detection of abnormal cells without MHC expression, e.g. virus-infected and tumour cells	Release of enzymes and perforin
Neutrophils	Antibacterial	Release of enzymes and generation of reactive oxygen species
Eosinophils	Antiparasitic	Release of major basic protein and other toxic substances

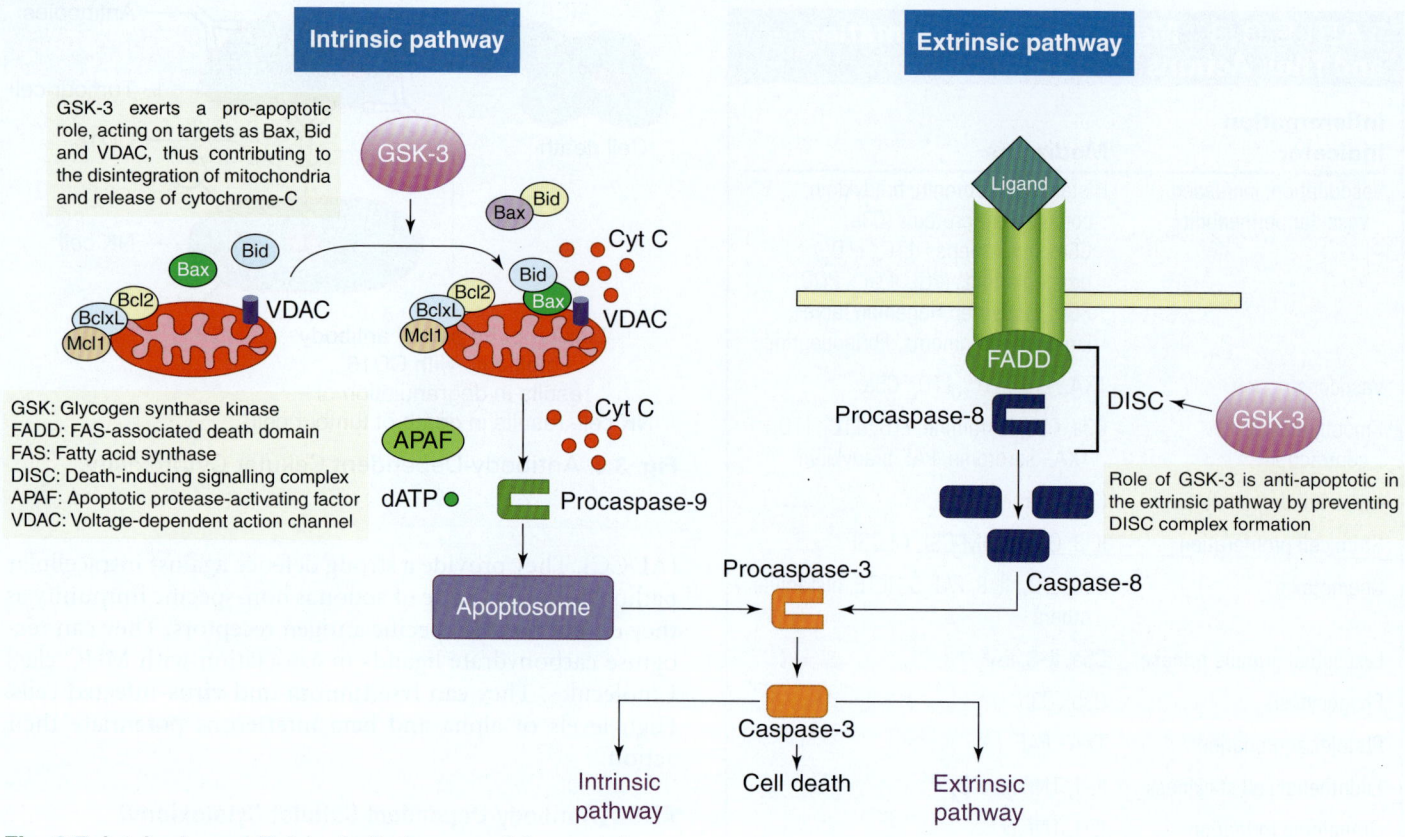

Fig. 3.7 Intrinsic and Extrinsic Pathways of Apoptosis.

Apoptosis

Apoptosis is referred as 'programmed' cell death. In contrast to necrosis, which occurs due to external injury, apoptosis happens due to programming in the cell's DNA. Apoptosis can kill pre-cancerous cells. Mutations that prevent apoptosis may cause cancer. There are two major types of apoptosis pathways: 'intrinsic pathways' and 'extrinsic pathways' (Fig. 3.7).

1. **Extrinsic pathway** (Flowchart 3.4): Signal is received from outside to kill the cell if the cell is no longer needed or if it is diseased. FAS and TRAIL, two common types of chemical messengers, excreted by neighbouring cells trigger the extrinsic pathway.
2. **Intrinsic pathway**: Intrinsic pathway is given in Flowchart 3.5.

Role of Antibodies in Innate Immunity

Opsonisation

Opsonins are host proteins like antibodies; antigen-binding sites (Fab) of which bind to pathogen (opsonise bacteria), enhancing their phagocytic uptake by phagocytes.

Natural Antibodies

These are antibodies that are present since neonatal life and form without any antigen exposure. Natural antibodies are

---

Starts with a signal molecule binding (FAS and TRAIL) to a receptor ('FASR' for 'FAS receptor') on the outside of the cell membrane

↓

Changes in the intracellular domain of TRAILR or FASR changes FADD (FAS-associated death domain" protein), an intracellular protein

↓

Activated FADD interacts with pro-caspase-8 and pro-caspase-10, which start the process of cell death

↓

Parts of the pro-caspase-8 and pro-caspase-10 are 'cleaved' to become caspase-8 and caspase-10 'the beginning of the end'

↓

Caspases-8 and 10 trigger changes to several other molecules to start the breakdown of DNA

↓

After BID is transformed into tBID, tBID moves to the mitochondria and activates BAX and BAK. (Here onwards steps are shared by both the extrinsic and intrinsic pathways)

Flowchart 3.4 Extrinsic Pathway of Apoptosis.

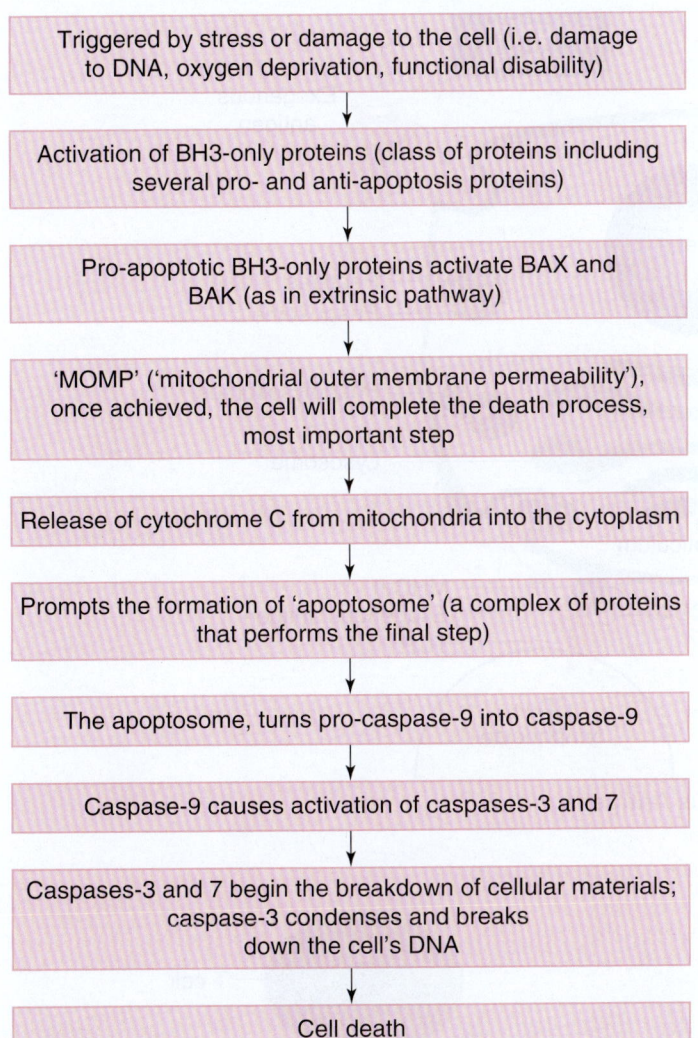

Flowchart 3.5 Intrinsic Pathway of Apoptosis.

Properties	Innate Immunity	Adaptive Immunity
Time taken to respond	Fast: minutes or hours	Few days to weeks
Specificity	Differentiate self from non-self	Highly specific
Cell types	Macrophages, neutrophils, natural killer cells, dendritic cells, basophils, eosinophils	T cells, B cells and antigen-presenting cells
Immunological memory	None	Leads to faster response to recurrent or subsequent infections with same pathogen

TABLE 3.5 Differences in Innate and Adaptive Immunity

often poly-specific and of IgM isotype. They are produced by a distinct subset of B cells (B-1 cells), which produce antibody without prior antigenic stimulation. Natural antibodies often recognise epitopes, which are generated during the oxidative processes involved in metabolism, ageing and inflammation. Oxidation-specific epitopes are ubiquitous on both microbes and on ageing and apoptotic host cells and represent a class of PAMP or damage-associated molecular pattern (DAMP) recognised by receptors of the innate immune system. Thus, natural antibodies appear to function as a type of pattern recognition receptor (PRR) for these molecular structures.

For infants, these antibodies, along with maternal IgG are important for protection. Antigen-stimulated antibody production is rapidly superimposed on natural antibodies because every newborn starts receiving new antigenic exposures.

ADAPTIVE IMMUNE RESPONSE

The adaptive immune response can be humoral (antibody-mediated) and cell-mediated (cellular). Mostly both types of responses are seen against an antigen. The main components of adaptive immune response are as follows:

1. Antigen presentation to immunocompetent cells
2. Humoral response, that is antibody production
3. Cell-mediated immune response (CMI), that is cellular and cytokine response
4. Regulation of immune response

Differences in innate and adaptive immunity are shown in Table 3.5.

Antigen Presentation

1. **Exogenous antigens** (inhaled, ingested or injected) are internalised by APCs, that is phagocytic cells, such as dendritic cells, macrophages, and B lymphocytes. APCs engulf the antigen by endocytosis, endosome fuses with a lysosome and antigen is degraded into fragments (e.g. short peptides, approximately 10–30 amino acid residues). These peptides may be recognised by CD4+ T cells. Class II MHC molecules are synthesised in the rough endoplasmic reticulum and then move on to cell membrane through the Golgi apparatus. The **invariant chain (Ii)**, which is a small polypeptide chain, protects the binding site of the class II αβ dimer by blocking the antigen-binding site, till the acidic pH created after fusion with an endosomal vesicle causes a dissociation of the Ii chain. The MHC class II–peptide antigen complex is then transported to the cell surface for display and recognition by a T-cell receptor of a CD4+ T cell (Fig. 3.8).

2. **Endogenous antigens** are generated within a cell (e.g. viral proteins in any infected cell) and are degraded into fragments (e.g. peptides, approximately 8–11 residues) by a peptidase complex known as the **proteasome** within the cell. Peptides gain access to MHC class I molecules in the rough endoplasmic reticulum via peptide transporter systems (transporters associated with antigen processing; TAPs). Within the lumen of the endoplasmic

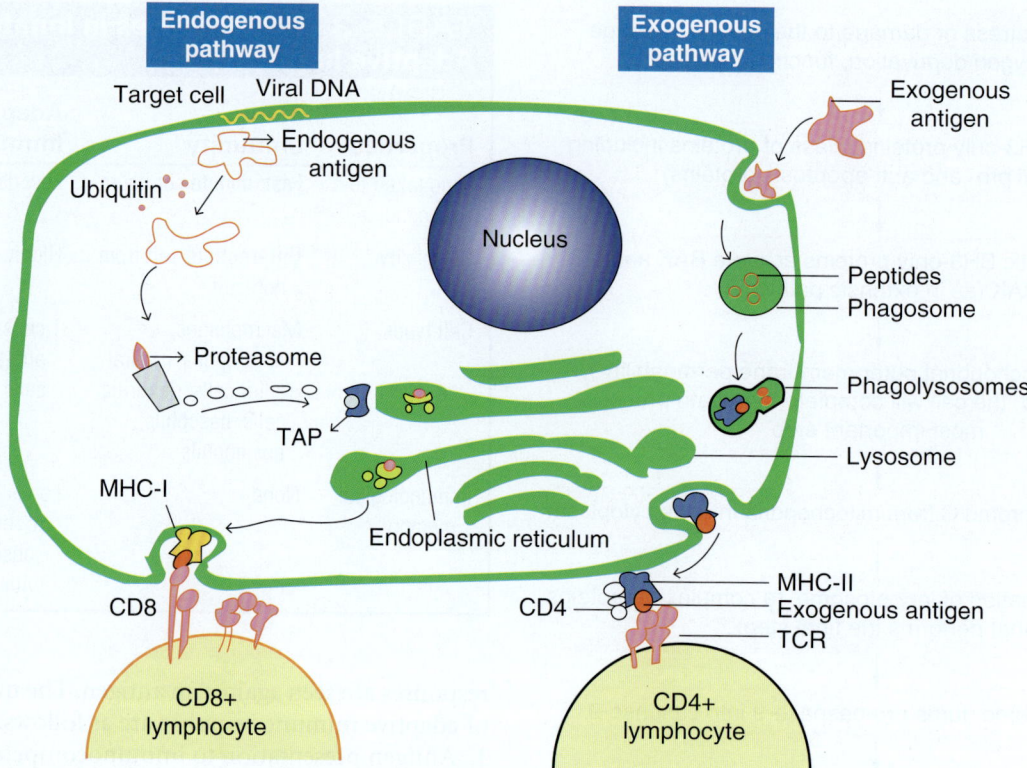

Fig. 3.8 Antigen Presentation of Endogenous and Exogenous Antigens.

reticulum, peptide antigens complex with nascent MHC class I proteins and cooperate with β_2-microglobulin to create a stable, fully folded MHC class I–peptide antigen complex that is then transported to the cell surface for display and recognition by CD8 CTLs. Most CD8$^+$ T cells are cytotoxic. The binding groove of the class I molecule is smaller than that of the class II molecule, hence, shorter peptides are found in class I than in class II MHC molecules.

3. Some **superantigens** are able to bind to MHC molecules outside the peptide-binding cleft, hence, can activate T cells non-specifically (Fig. 3.9). These antigens bind to the 'outside' of the MHC protein (not the cleft where other antigens bind) and to the T-cell receptor. They are called superantigens as very low concentration can cause release of cytokines, including IL-1 and TNF. Release of large amounts of cytokines decides pathogenesis of diseases by organisms expressing superantigens. Certain bacterial toxins, including the staphylococcal enterotoxins, toxic shock syndrome toxin and group A streptococcal pyrogenic exotoxin A are examples of superantigen and can activate about 10% of T cells non-specifically.

Antibody Production (Humoral Immune Response)

The humoral immunity is mediated by antibodies. Humoral immunity includes the primary and secondary immune responses to antigen. Following antigen exposure, there is a lag period before antibody is detectable, correspond-

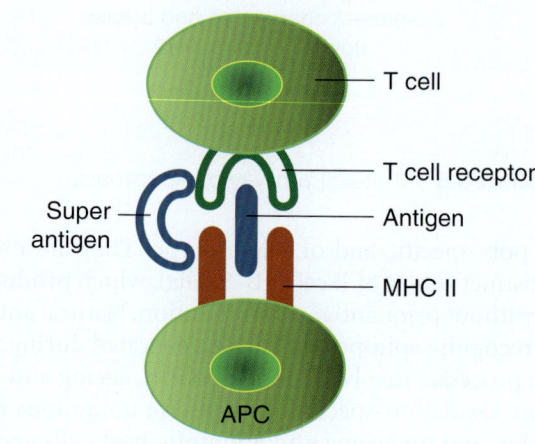

Fig. 3.9 Non-Specific Antigen Presentation by Superantigen.

ing to the time taken in the activation, proliferation and differentiation of B cells to plasma cells. The duration of the lag period varies, depending on the dose and route of administration of antigen, and on **antigen priming of host**. In a naïve host, primary response to a TD antigen shows logarithmic increase in serum IgM antibodies. This increase reflects clonal expansion and antibody production. Naïve B cells express IgM and IgD on their surface but are negative for CD27. CD27 is a receptor that is important in T-cell co-stimulation.

Primary and Secondary Immune Response

Primary immune response. It occurs when an antigen is encountered by the host for the first time. Naïve B cells need more time for activation and proliferation than primed B cells. Hence, primary antibody response may take longer than secondary immune response and is slow to protect against pathogens. Poly-specific 'natural antibodies' (low affinity) and the innate immune system provide first line of defence. IgM antibodies are the first to appear, which are later replaced by IgG antibodies (isotype switching). Both IgG and IgM have low affinity and quantity (Fig. 3.10).

Secondary immune response. It occurs on subsequent encounters with the antigen and involves activation of already primed memory B cell. It is faster and more effective than the primary response. IgG is the principal antibody type produced and have high affinity and quantity (Fig. 3.10).

Passive and Active Humoral Immunity

Passive humoral immunity. It is acquired by transfer of preformed antibodies from an external source to a new host, for example injection of intramuscular or intravenous human immunoglobulin. Passive immunity is naturally acquired during gestation by the transplacental transfer of maternal immunoglobulin G (IgG) via the neonatal Fc receptor (FcRn). Immunoglobulin A (IgA) and immunoglobulin M (IgM) are not transported across the placenta. After birth, the breast-fed neonate continues to receive IgA passively via colostrums and breast milk, which affords local protection in the gut. Breast milk immunoglobulins are not absorbed by the infant's gastrointestinal tract.

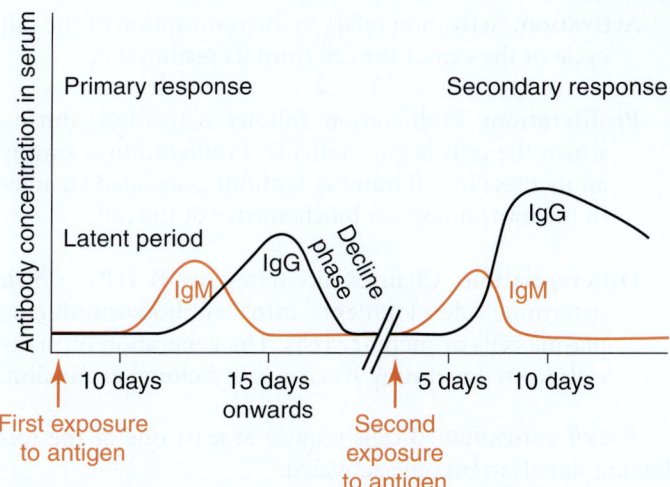

Fig. 3.10 Primary and Secondary Humoral Immune Response.

Active (or adaptive) immunity. It is the response generated during the encounter of the immune system with antigen. This may occur during the course of a natural infection or after intentional antigen administration (vaccination).

B-Cell Development for Antibody Production (Antigen-Dependent)

B-cell development for antibody production occurs in three stages: activation, proliferation, and differentiation (Fig. 3.11).

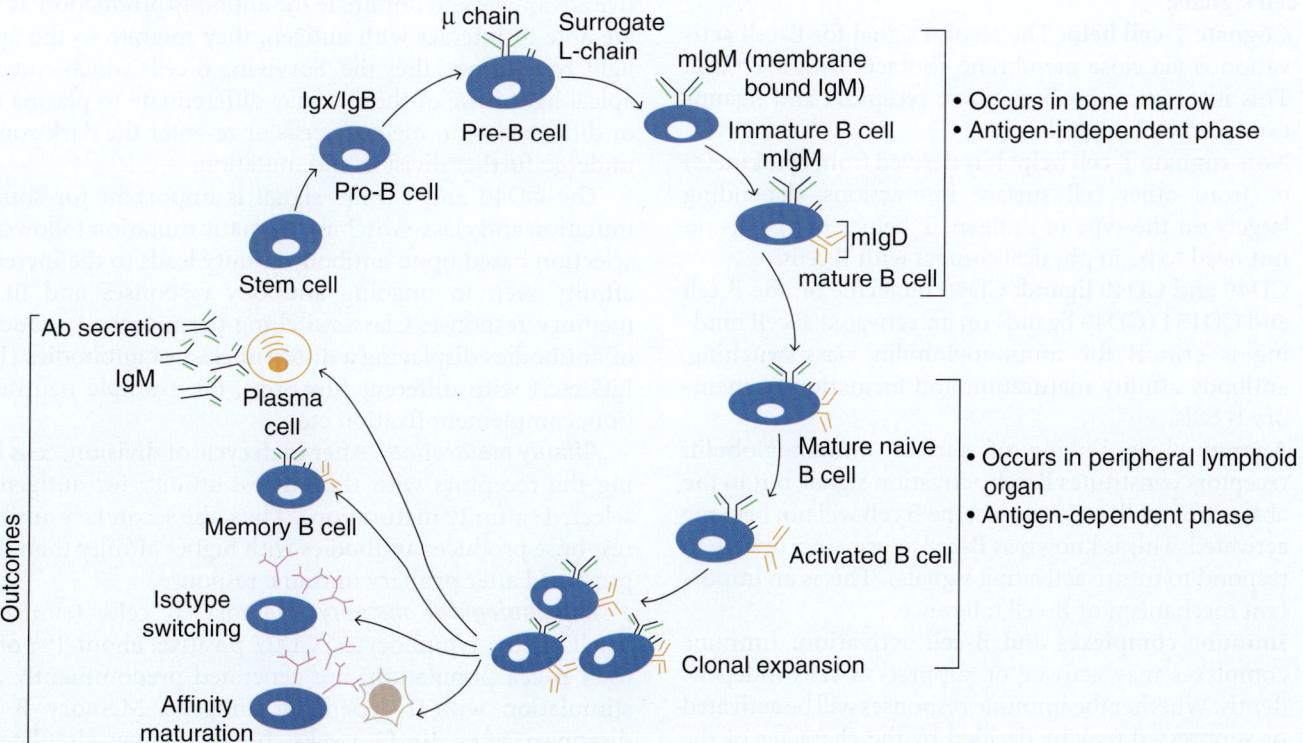

Fig. 3.11 B-Cell Activation and Antibody Production.

Activation: Activation refers to the resumption of the cell cycle or the exit of the cell from its resting state.

↓

Proliferation: Proliferation follows activation, during which the cells begin to divide. Proliferation is simply an increase in cell number without associated changes in the morphology or biochemistry of the cell.

↓

Differentiation: Changes occurring in B cells, which determine development into antibody-producing plasma cells or memory cells. The generation of many cells from one resting B cell is called **clonal expansion**.

B-cell activation. B cells require at least one of the following signals to become activated.

1. **Cross-linking of immunoglobulin receptor**: It is important for thymus-independent antigens, which may not be required for TD antigens. Antigens which are monovalent (e.g. soluble protein and glycoprotein) cannot cross-link immunoglobulin receptors. Immune complexes can cross-link immunoglobulin receptors and stimulate strong immune response.

   Cross-linking of B-cells' surface receptors causing B-cell activation is also possible by:
   a. Antibody directed against antibody, binding to the immunoglobulin receptor: Anti-immunoglobulin antibodies activate B cells regardless of their antigen specificity.
   b. Antibody directed against one of several B-cell surface molecules: For example class II histocompatibility molecules.
   c. Binding of toll-like receptors on B cells by their ligands.

2. **T-cell signals:**
   a. **Cognate T-cell help:** The second signal for B-cell activation is via close membrane contacts with a Th cell. This interaction involves many receptors and ligands expressed by both cells.
   b. **Non-cognate T-cell help:** It is derived from cytokine(s) or from other cell surface interactions, depending largely on the type of antigen. T helper (Th) cells do **not** need to be in physical contact with B cells.
   c. **CD40 and CD40 ligand:** CD40 molecule on the B cell and CD154 (CD40 ligand) on an activated T-cell binding is critical for immunoglobulin class-switching, antibody affinity maturation and formation of memory B cells.
   d. **Anergy:** Cross-linking of surface immunoglobulin receptors constitutes B-cell activation signal, but in the absence of additional signals, the B cell will not become activated. This is known as B-cell anergy (i.e. unable to respond to future activating signals). This is an important mechanism of B-cell tolerance.
   e. **Immune complexes and B-cell activation:** Immune complexes may activate or suppress B cells independently. Whether the immune responses will be activated or suppressed may be decided by the character of the immune complexes, the timing of administration and the isotype of the antibodies in the complex. Antigen-specific antibodies administered passively with antigen may suppress immune responses by masking the antigen and interfering with its recognition (binding) by B cells or by interfering with T–B-cell cooperation. At times as seen in dengue virus infection, pre-existing antibodies can enhance antigen-specific responses.

B-cell proliferation and development. Within lymph nodes and spleen, B-cell maturation and cellular events for immunologic memory take place in germinal centre (GC). Antigen naïve B cells form spherical collections of cells called primary lymphoid follicles, present in peripheral lymph node. Primary follicles contain follicular dendritic cells (FDCs), interspersed with small resting B cells. The FDCs are APCs; they present antigen–antibody complexes on their surface and stimulate B-cell growth and differentiation. Once activated, numerous intracellular signalling pathways in lymphocytes (B and T) are monitored via tyrosine, serine or threonine phosphorylation of components of pathway. The primary lymphoid follicle becomes populated with few activated B cells (may be 2–3, coming from peripheral circulation) that will proliferate and give rise to progeny cells. These B cells develop into plasma cells within GC. Plasma cell produces high-affinity antibodies or memory B cells. The pre-existing resting B cells are pushed to the periphery of GC to form the follicular mantle.

B-cell differentiation. B-cell differentiation within the GC is highly regulated. B cells migrate to one end of the lymphoid follicle (centroblasts) to form the dark zone of the GC. These centroblasts divide and give rise to centrocytes that migrate to the basal light zone. Somatic mutation of immunoglobulin V genes then takes place during cell division. The cells with receptors having the highest affinity for antigen have a selective advantage and dominate the antibody production. If they are able to interact with antigen, they migrate to the apical light zone; if not, they die. Surviving B cells which enter the apical light zone of the GC may differentiate to plasma cells or differentiate to memory cells or re-enter the dark zone to undergo further division and mutation.

The CD40 and CD40L signal is important for somatic mutation and class-switching. Somatic mutation followed by selection based upon antibody affinity leads to the increased affinity seen in ongoing antibody responses and in the memory response. Class-switching permits the production of antibodies displaying a different class of antibodies (IgM, IgG etc.) with different functions, for example neutralisation, complement fixation etc.

Affinity maturation. After each cycle of division, cells having the receptors with the highest affinity for antigen are selected (affinity maturation). Thus, the secondary antibody response produces antibodies with higher affinity than those produced after primary immune response.

Immunological memory. Memory B cells (similar to small resting lymphocytes, CD27 positive, about 1% of the total B-cell population) are generated predominantly after stimulation with T-dependent antigens. Memory B cells disappear after 10–12 weeks; however, may circulate for years or decades, in case antigenic stimulation persists, and

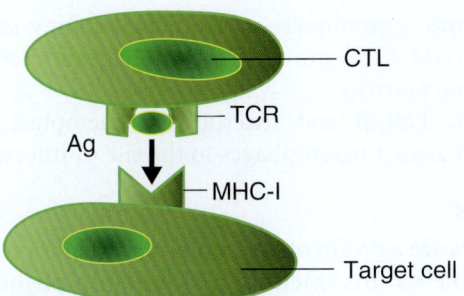

Fig. 3.12 Recognition of Foreign Cells by CTL. CTL = cytotoxic T cell; MHC = major histocompatibility; TCR = T cell receptor.

preferentially localise on mucosal surfaces. By virtue of their immunoglobulin receptors' high affinity for antigen, memory cells are stimulated by much lower antigen concentrations than virgin B cells. Memory cells are more efficient at antigen presentation and are more receptive to T-cell help. They can be stimulated without T-cell help.

Cell-Mediated Immunity

CMI is especially important for destroying intracellular bacteria, eliminating viral infections and destroying tumour cells. CMI is responsible for transplant rejection, delayed hypersensitivity and tumour surveillance. Effectors of this response are various immune-competent cells. CMI includes following reactions, which may overlap with each other:
- Phagocytosis and killing of intracellular pathogens (see Section: Innate Immunity)
- Cell killing by CTLs
- Cell killing by NK and K cells
- Macrophage activation by Th1 cells
- Cytokine release by immune-competent cells

Cytotoxic T Cells (CTL)

CTLs are antigen- and MHC-restricted cells, which require the following for stimulation:
1. Recognition of a specific antigenic determinant
2. Recognition of 'self' MHC I

CTLs recognise antigen via T-cell receptor (Fig. 3.12); this receptor interacts with antigenic determinant and class I MHC molecule on the surface of target cell. CD8 molecule expressed on CTLs may assist antigen recognition. CTLs principally act to eliminate endogenous antigens; hence, CTL 'programs' the target cell for self-destruction once antigen recognition is complete. This process occurs in one of following ways (Fig. 3.13):
1. CTLs may release perforin in the space between the CTL and target cell. Perforin is a monomer, which polymerises, forming channels in the target's cell membrane, in the presence of calcium ions. These holes in the cell membrane of target cell lyse the cell.
2. CTLs release various enzymes that pass through the polyperforin channels, causing target cell damage.

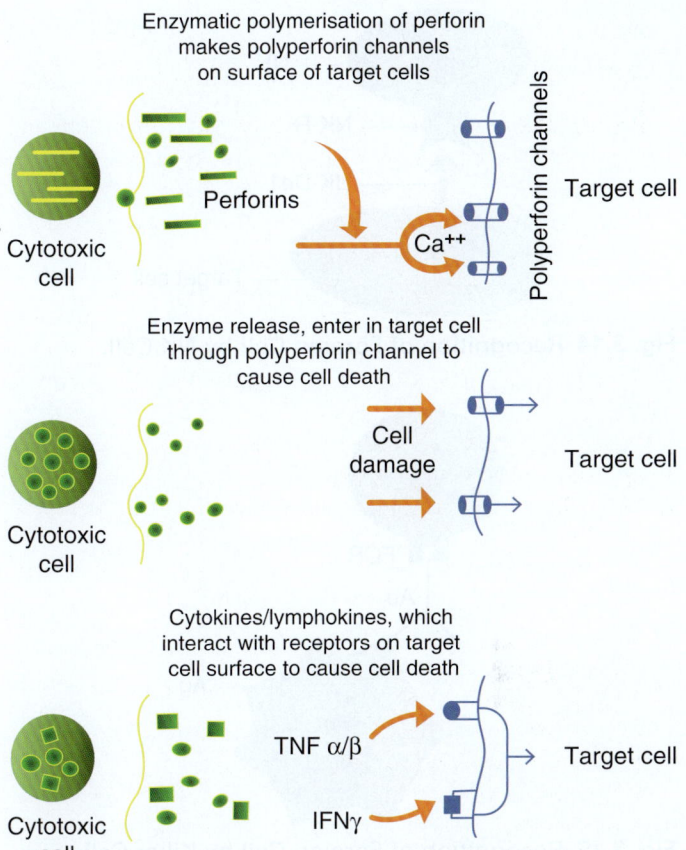

Fig. 3.13 Mechanism of Cell Destruction Mediated by CTLs.

3. CTLs release cytokines that interact with specific receptors on the target cell surface, causing destruction of the target cell.

Natural Killer Cells (NK)

NK cells are a type of large granular lymphocytes. They are generally non-specific, MHC-unrestricted cells, involved in elimination of virus-infected and tumour cells. Precise mechanism of their action is not clear. Probably, some NK-determinant antigen (NK-De), expressed by target cells, is recognised by NK receptor on NK cell surface. Once the target cell is recognised, killing occurs (similar to that caused by the CTL) (Fig. 3.14).

Killer Cell

Killer cell is not a separate cell type, but a separate function of the NK group of cells. It contains immunoglobulin Fc receptors on their surface and is involved in **antibody-dependent cell-mediated cytotoxicity** (**ADCC**). Antibody-bound target cell are recognised by K cell, via specific antigenic determinants expressed by target cell. Once bound, Fc portion of immunoglobulin can be recognised by K cell (Fig. 3.15). Killing occurs in a manner similar to that of CTLs. This type of cell-mediated immune response also results in Type II hypersensitivity.

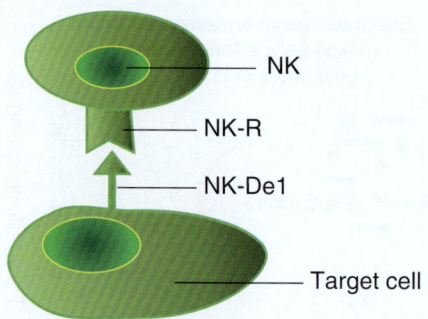

Fig. 3.14 Recognition of Foreign Cell by NK Cell.

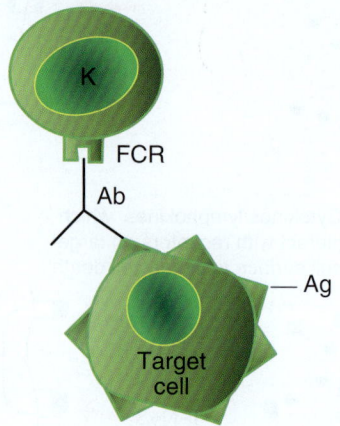

Fig. 3.15 Recognition of Foreign Cell by Killer Cell.

Macrophage Activation by Th1 (Inflammatory T) Cells

Macrophages phagocytose and kill pathogens using lysosomal enzymes. However, some pathogens evade killing by using one of the following mechanisms:

- Preventing fusion of lysosomes with phagocytic vesicle
- Preventing activation of the lysosomal enzymes by acidification of the phagolysosome
- Escaping the phagolysosome but live in the macrophage cytoplasm

Such pathogens are protected from complement and antibodies. Their peptides are also not presented to CD8 T cells. Macrophage activation by Th1 cells is required to eliminate these pathogens. Macrophage activation requires T-cell binding to antigen peptide on macrophage class II MHC, co-stimulation of the macrophage via CD40 and activation of the macrophage by IFN-α. Macrophage synthesise and secrete IFN-α. Macrophage activation is slower than CTL-mediated cytotoxicity because synthesis of IFN-α takes several hours.

When pathogens are large and can not be phagocytosed (large parasites), activated macrophages release oxygen radicals, nitric oxide and proteases into the extracellular fluid to kill pathogen. Excretion of these compounds also damages surrounding host tissues. The local inflammatory response resulting from activated Th1 cells and macrophages is called delayed-type or Type 4 hypersensitivity (DTH). Activated Th1 cells also secrete a variety of cytokines that regulate cellular immunity, for example:

- IL-2 stimulates clonal proliferation of T cells

- IL-3 and granulocyte-macrophage colony-stimulating factor (GM-CSF) stimulate macrophage differentiation in the bone marrow
- TNF-α, TNF-β and macrophage chemotactic protein (MCP) attract macrophages to the site of infection

Cytokines

Cytokines are a group of low molecular weight proteins that regulate the nature, intensity and duration of the immune response. They function as mediators of inflammation and immune responses. Originally they were called lymphokines because initially it was thought that they were produced only by lymphocytes. Then they were referred to as monokines because they were secreted by monocytes and macrophages. Later the name interleukins was given to them because they are produced by leucocytes and affect leucocytes. Finally term cytokine was used to cover lymphokines, monokines and interleukins. Term cytokine was proposed by Cohen (1974) for family of polypeptides that engage in immunologically mediated inflammatory reaction (Box 3.2).

BOX 3.2   Properties of Cytokines

1. Most cytokines are low molecular weight polypeptides or glycoprotein (8–40 kDa in size); most of them are monomer.
2. Single type of cytokine can be produced by different cells and single type of cell can secrete different cytokines.
3. They are highly potent, that is small quantity of cytokines are needed to elicit biological effects.
4. Cytokines can act in three different manners:
   a. Autocrine: Cytokine binds to receptor on the cell that secreted it (Fig. 3.16).
   b. Paracrine: Cytokine binds to receptors on nearby cells (Fig. 3.16).
   c. Endocrine: Cytokine binds cells in distant parts of the body (Fig. 3.16).
5. Cytokines initiate their actions by binding to specific membrane receptors on target cells (Fig. 3.17).

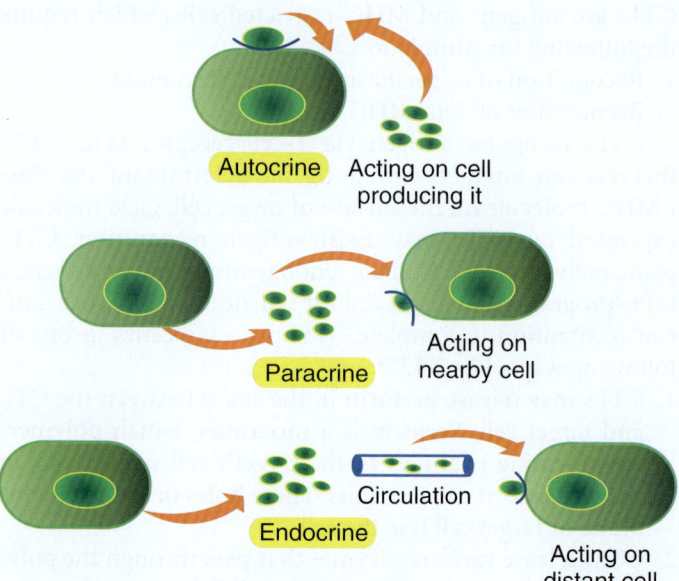

Fig. 3.16 Action Sites of Cytokines.

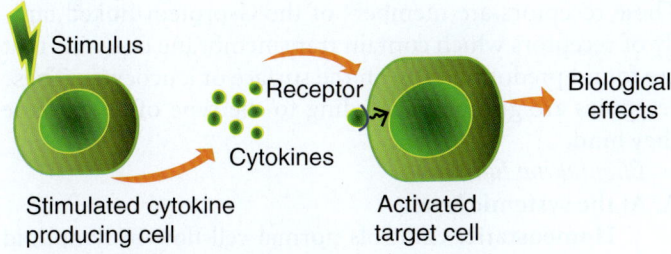

Fig. 3.17 Mechanism of Action of Cytokines.

Effects of Cytokine. **Pleiotropism:** Refers to the ability of one cytokine having multiple effects on diverse cell types (Fig. 3.18).

Redundancy: Refers to the property of multiple cytokines having the same or overlapping functional effects (Fig. 3.19).

Synergy: Refers to the property of two or more cytokines having greater than additive effects (Fig. 3.20).

Antagonism: Refers to the ability of one cytokine inhibiting the action of another (Fig. 3.21).

Cascade effect: One cytokine stimulates the production of other cytokines (Fig. 3.22).

Cytokine Families. Cytokines are classified into four groups as follows:

A. **Cytokines:**
1. Colony-stimulating factor (CSF) family
2. Interferon family
3. TNF family

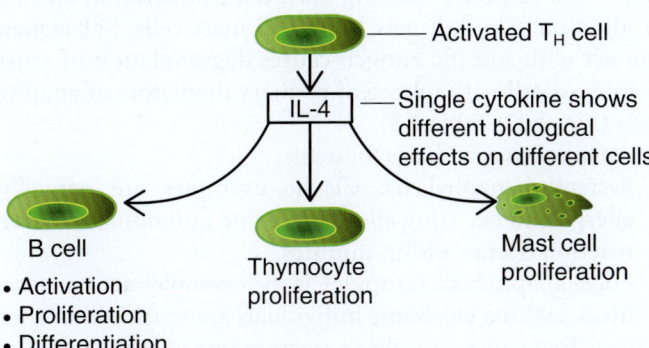

Fig. 3.18 Pleiotropism as Seen With Cytokines.

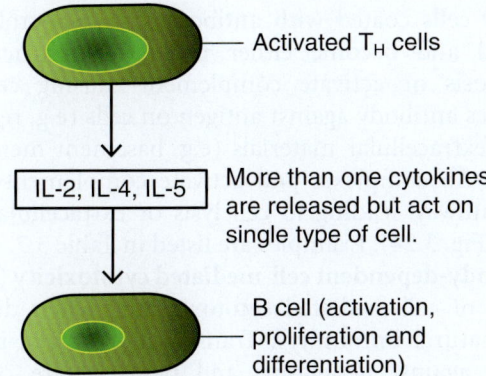

Fig. 3.19 Redundancy as Seen With Cytokines.

B. **Chemokines:**
4. Chemokine family

Colony-stimulating factor. Cytokines that stimulate proliferation or differentiation of pluripotent haematopoietic stem cell and different progenitors. CSFs are as follows:
- IL-2, IL-3, IL-4, IL-5, IL-6, IL-7, IL-9, IL-11, IL-12, IL-15
- GM-CSF
- Macrophage-CSF (M-CSF)
- Granulocyte-CSF (G-CSF)
- Stem cell factor (SCF)
- Erythropoietin (EPO)
- Thrombopoietin

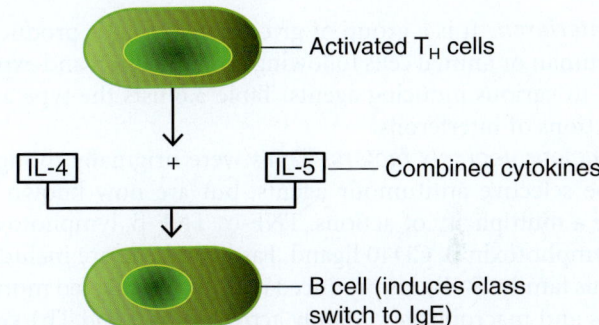

Fig. 3.20 Synergy as Seen With Cytokines.

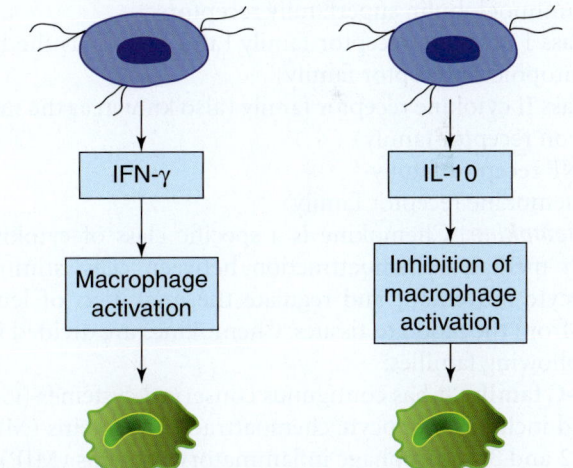

Fig. 3.21 Antagonism as Seen With Cytokines.

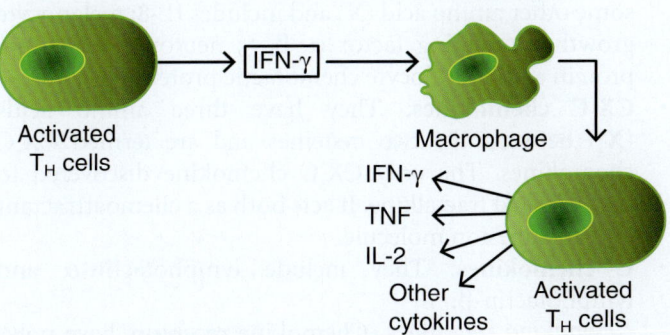

Fig. 3.22 Cascade Effect of Cytokines.

TABLE 3.6 Types and Functions of Interferon

Types	Produced by Cells	Main Functions
IFN-α	Leucocyte	Antiviral, immune regulation
IFN-β	Fibroblast	antiviral, immune regulation
IFN-γ	Th1,NK	Antiviral effect, immune regulation effect, antitumour effect

Interferon. It is a group of glycoprotein that is produced by human or animal cells following viral infection and exposure to various inducing agents. Table 3.6 lists the type and functions of interferons.

Tumour necrosis factors. TNFs were originally thought to be selective antitumour agents, but are now known to have a multiplicity of actions. TNF-α, TNF-β, lymphotoxin α, lymphotoxin β, CD40 ligand, Fas ligand etc. are included in this family. TNF-α is produced mainly by activated monocytes and macrophages. Mainly activated Th0 and Th1 cells produce TNF-β.

Cytokine receptors. Cytokine receptors fall within five families:

1. Immunoglobulin superfamily receptors
2. Class I cytokine receptor family (also known as the haematopoietin receptor family)
3. Class II cytokine receptor family (also known as the interferon receptor family)
4. TNF receptor family
5. Chemokine receptor family

Chemokine. Chemokine is a specific class of cytokines, which mediates chemoattraction between cells, stimulate leucocyte movement and regulate the migration of leucocytes from the blood to tissues. Chemokines are divided into the following families:

- **C–C family:** It has contiguous conserved cysteines (C–C) and includes monocyte chemoattractant proteins (MCP) 1, 2 and 3; macrophage inflammatory proteins (MIP) 1α and 1β; and RANTES (regulated upon activation, normal T cell expressed and secreted).
- **CXC family:** It has conserved cysteines (C), separated by some other amino acid (X) and includes IL-8; melanocyte growth-stimulating factor α, β, γ; neutophil-activating protein and granulocyte chemotactic protein-2.
- **CX_3C chemokines:** They have three amino acids (X_3) between the two cysteines and are termed CX_3C chemokines. The only CX_3C chemokine discovered to date is called fractalkine. It acts both as a chemoattractant and an adhesion molecule.
- **C chemokines:** They include lymphotactin-α and lymphotactin-β.

Chemokine receptors. Chemokine receptors have polypeptide chain, which traverses the membrane seven times.

These receptors are members of the G-protein-linked family of receptors which contain transmembrane domains that are found predominantly on the surface of leucocytes. These receptors are grouped according to the type of chemokine they bind.

Chemokine functions
A. **At the systemic level:**
   Homeostatic: Controls normal cell flow in lymphoid organs
   Inflammatory: Recruits cells to inflammatory sites
B. **At the cellular level:**
   Leucocyte chemoattraction
   Cytokine antagonist: Cytokine activity can be blocked by molecules which bind cytokines or their receptors

HYPERSENSITIVITY

It is a tissue damaging reaction produced by the normal immune system and includes allergic and autoimmune response. They are over-reactions of the immune system and cause tissue damage causing considerable morbidity and occasional mortality. Pre-sensitised immune status is a prerequisite for hypersensitivity.

There are four types of hypersensitivity reactions as shown in Table 3.7.

Type I Hypersensitivity (Anaphylaxis)

Pre-requisite of anaphylaxis is mast cells coated with IgE. Prior sensitisation of host with the antigen results in Th2 helper cell response, causing mast cell proliferation and IgE production. The IgE gets bound to mast cells. Subsequent contact with specific antigen causes degranulation of sensitised mast cell with release of primary mediators of anaphylaxis (Fig. 3.23; Table 3.8).

Anaphylaxis can be as follows:
1. Systemic anaphylaxis: Classic examples are penicillin allergy and bee sting allergy. In some individuals, a severe reaction occurs within minutes.
2. Local anaphylaxis (atopy): Classic examples are hay fever, hives, asthma etc. Some individuals are sensitised to allergens that cause a localised reaction on contact.

Type II Hypersensitivity (Cytotoxic Antibody/ Complement Dependent)

Normally cells coated with antibodies (e.g. microbes) get opsonised and become either prone to destruction by phagocytosis or activate complement causing cell lysis. Sometimes antibody against antigen on cells (e.g. red blood cells) or extracellular materials (e.g. basement membrane) make Ag–Ab complexes and activate complement via the classic pathway, leading to cell lysis or extracellular tissue damage (Fig. 3.24). Examples are listed in Table 3.7.

Antibody-dependent cell-mediated cytotoxicity (ADCC) is a type of cell-mediated cytotoxicity which is discussed earlier (natural immunity). Transplant rejection, immune reactions against neoplasms and parasites are common examples. Myasthenia gravis (acetylcholine receptor

TABLE 3.7 Types and Important Characteristics of Hypersensitivity Reactions

Types	Alternative Names	Common Clinical Examples	Mediators	Mechanism
I	Allergy (immediate)	• Atopy • Anaphylaxis • Asthma	IgE	IgE on mast cells and basophils' cell surface attaches with free antigen followed by release of vasoactive biomolecules; fast response (occurs in minutes)
II	Cytotoxic, antibody-dependent	• Autoimmune haemolytic anaemia • Rheumatic heart disease • Thrombocytopenia • Erythroblastosis fetalis • Goodpasture's syndrome • Graves' disease • Myasthenia gravis	Antibodies (IgM/IgG); complement; MAC	Circulating antibody binds to antigen on a target cell, leading to cellular destruction via the complement activation
III	Immune complex-mediated	• Serum sickness • Rheumatoid arthritis • Arthus reaction • Post-streptococcal glomerulonephritis • Membranous nephropathy • Reactive arthritis • Lupus nephritis • Systemic lupus erythematosus	Circulating IgG, soluble antigen, complement, neutrophils	Circulating IgG binds to soluble antigen forming a circulating immune complex, which may get deposited in the vessel walls of the joints and kidney and initiates a local inflammatory reaction
IV	Delayed-type hypersensitivity, cell-mediated hypersensitivity	• Contact dermatitis • Mantoux test • Chronic transplant rejection • Multiple sclerosis • Coeliac disease • Hashimoto's thyroiditis	Primed T helper cells	Primed helper T cells coming in contact with same antigen again activate macrophages and cause an inflammatory response causing tissue damage

TABLE 3.8 Primary Mediators of Anaphylaxis

Important Reactions	Mediators
Vasodilation and increased vascular permeability	• Histamine • Leukotriene C4, D4 and E4 • Prostaglandin D2 • Platelet-activating factor • Neutral proteases
Smooth muscle spasm	• Histamine • Platelet-activating factor • Leukotriene C4, D4 and E4 • Prostaglandin
Leucocyte extravasation	• Cytokines • Leukotriene B4 • Chemotactic factors for neutrophils and eosinophils

(*Source:* Wikipedia, the free encyclopedia. Type I hypersensitivity: List of a few mediators, released by mast cells in type 1 hypersensitivity and their actions.)

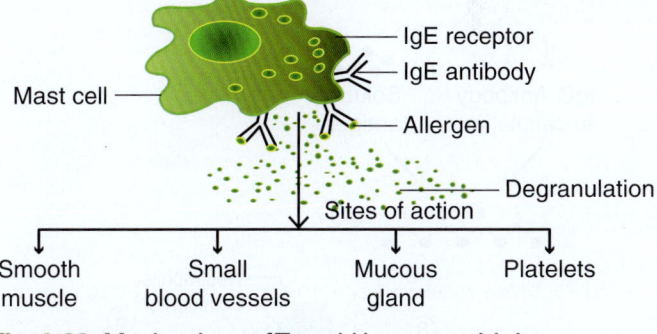

Fig. 3.23 Mechanism of Type I Hypersensitivity.

Type III Hypersensitivity

It is mediated by immune (Ag–Ab) complexes, which promote tissue damage primarily through complement activation (alternate pathway). Circulating antigen–antibody complexes get trapped in tissues, for example beneath the basement membrane of a small blood vessel, glomeruli, skin, joints, pleura and pericardium; activate complement and generate components that attract polymorphs to generate an ongoing inflammatory response. C3b, an opsonin and C5a, a chemoattractant attracts neutrophils, which release lysosomal enzymes (Fig. 3.25). Serum complement is used up in this process; hence serum complement level is reduced.

antibody), Graves's disease (thyrotoxicosis due to anti-TSH receptor antibody) and pernicious anaemia (anti-parietal cell antibody) are some of the diseases caused by antireceptor antibodies.

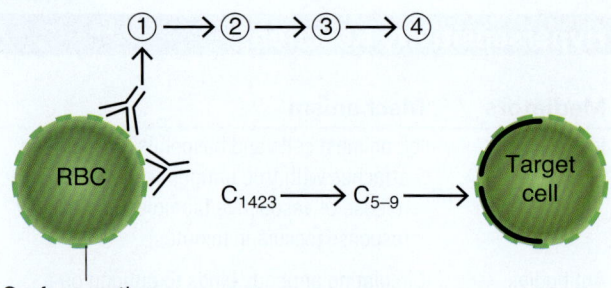

(A) Complement-dependent reactions:
① Antibody attaches to surface antigen;
② Complement cascade activated by Ag–Ab complex (via classical pathway); ③ Formation of membrane attack complex (MAC); ④ Cell lysis.

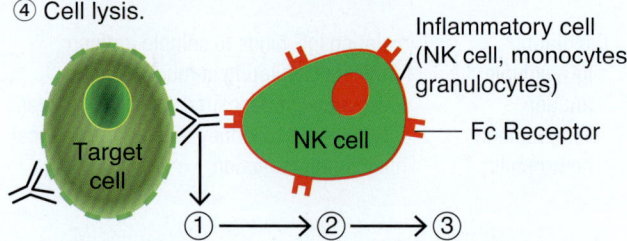

(B) Antibody-dependent cell-mediated cytotoxicity (ADCC): ① IgG/IgE Ab coat target cells; ② Fc receptor of inflammatory cells recognise Fc portion of antibody attached to a target cell; ③ Lysis of target cells by secretion of proteases.

Fig. 3.24 Mechanism of Type II Hypersensitivity.

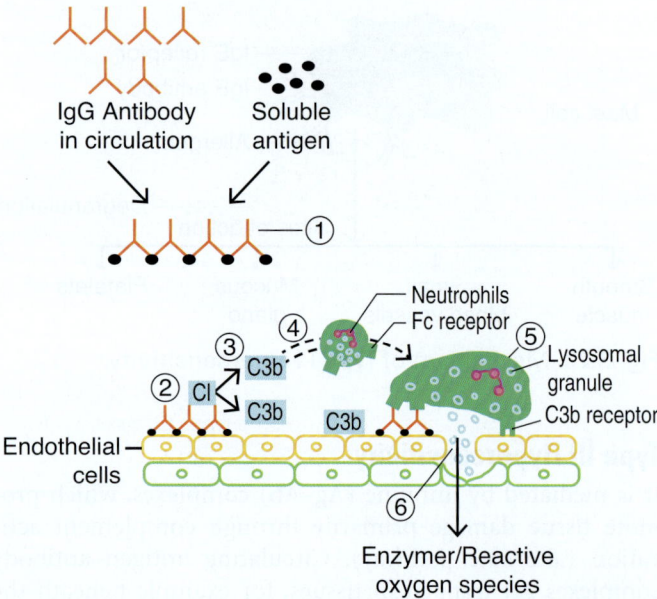

Fig. 3.25 Immune Complex-Mediated Hypersensitivity.
① Formation of intermediate size immune complexes; ② Ag–Ab complex get deposited in tissue; ③ Complement get activated (via alternate pathway); ④ C5a and C3a act as chemotactic factor for neutrophils; C3b acts as opsonin; ⑤ Neutrophil adherence and degranulation/release of lytic enzymer; ⑥ Cell destruction.

Immune complex-mediated diseases may be systemic or local.

Systemic immune complex-mediated diseases, (e.g. glomerulonephritis, serum sickness, vasculitis): Ag–Ab complexes form in the circulatory system and are deposited in tissues. Larger immune complexes are quickly phagocytised by macrophages and removed, but small-to-intermediate complexes formed with antigen excess may escape removal and cause inflammatory damage to involved organs.

Local immune complex disease (Arthus reaction): It occurs with local injection of the antigen and leads to focal vasculitis. Hypersensitivity pneumonitis (farmer's lung) is also a local disease caused due to immune complexes.

Type IV Cell-Mediated Hypersensitivity (Fig. 3.26)

It is initiated by T lymphocytes sensitised to antigen. Typical prototype is immunologic response to intracellular pathogens, for example, *Mycobacterium tuberculosis*, viruses, protozoa and fungi. After contact with antigen, the reaction is visible after 24–48 h, hence is called **delayed hypersensitivity**. Reaction is mediated by CTLs that directly cause cell injury. In direct cytotoxicity, recognition of a cell-surface antigen by the corresponding receptor on a T cell triggers the

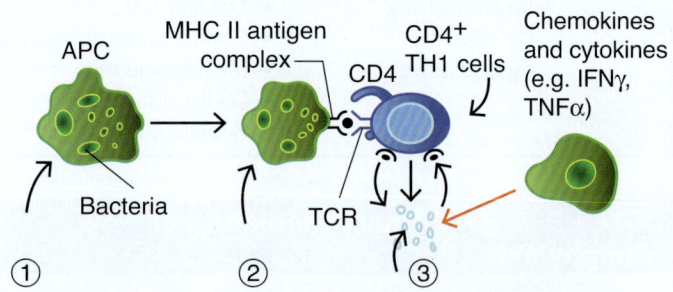

(A) Sensitisation phase: ① APC phagocytose bacteria; ② APC presents antigen to CD4 cells that release chemokines and cytokines; ③ These cytokines further induce release of other cytokines

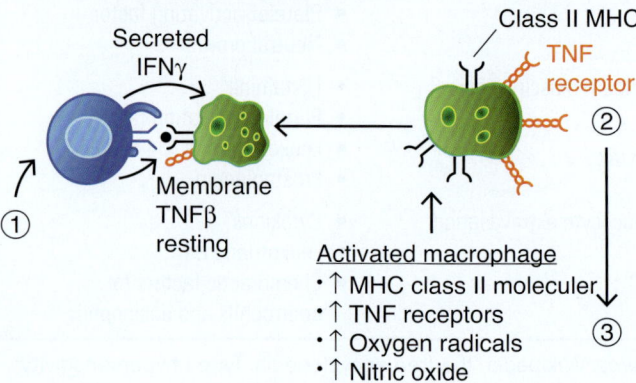

(B) Effector phase: ① Sensitised CD4+ TH1 cells release, cytokines (IFN, TNF, IL-2, IL-3, GM-CSF), chemokines (IL-8, MCAF, MIF); ② Macrophage recruitment and activation; ③ Localised reactivety manifested by erythema and induration.

Fig. 3.26 Delayed Hypersensitivity.

TABLE 3.9 Some Common Immunodeficiency Diseases

Disorders	Examples	Functional Defects	Common Infections
Severe combined immunodeficiency (SCID)	ADA	Both cellular responses and antibody production	Bacteria, viruses, fungi
Primary T-cell deficiency	Di George syndrome	Cellular immunity	Bacteria, viruses, fungi
Predominantly antibody deficiency	X-linked agammaglobulinaemia	Humoral immunity	Bacteria, viruses, fungi, *Giardia lamblia*
Others	Ataxia telangiectasia	Humoral immunity (overproduction of IgM and decreased production of other antibody classes)	Bacteria
	Hyper-IgE syndrome	Overproduction of IgE	Bacteria

ADA = adenosine deaminase deficiency; SCID = severe combined immunodeficiency.

TABLE 3.10 Some Common Autoimmune Diseases

Diseases	Organ Affected	Symptoms
Rheumatoid arthritis	Joint cartilage leading to joint destruction	Off and on episodes of joint pain, swelling, stiffness and joint function
Systemic lupus erythematosis (SLE)	Tissues throughout body	Joints, lungs, blood cells, kidneys and nerves may be affected
Inflammatory bowel disease (ulcerative colitis and Crohn's disease)	Intestinal lining	Episodes of diarrhoea, rectal bleeding, abdominal pain, fever and weight loss
Multiple sclerosis	Nervous tissue	Pain, blindness, weakness and poor coordination
Hashimoto's thyroiditis	Thyroid gland	Hypothyroidism

SLE = systemic lupus erythematosus.

T cell to release perforin and granzymes which cause apoptosis of the target cell.

This type of hypersensitivity has two effector mechanisms:
1. Delayed-type: CD4$^+$ cells mediate delayed-type hypersensitivity reactions, for example, granuloma formation in fungal and mycobacterial reactions.
2. Direct cytotoxicity: Mediated by CD8$^+$ cells, for example, contact hypersensitivity and poison ivy.

Both types of hypersensitivity reactions are necessarily immunologic reactions; and, at times, can cause extensive tissue damage.

IMMUNE SYSTEM DISORDERS

Immune system disorders include either low activity (primary immunodeficiency diseases) or over-activity (autoimmune diseases).

Primary Immunodeficiency Diseases

The immune system reacts inappropriately to the pathogens thereby making a person prone to infections. It results from defects in genes coding for the immune system. These diseases include defects in antibody production, defects in T-cell function, severe combined immunodeficiency and others. Some common immunodeficiency diseases are mentioned in Table 3.9.

Autoimmune Diseases

The immune system reacts abnormally in response to an unknown trigger (exogenous or endogenous) producing antibodies and destroying the body's cells instead of fighting the invading organisms. Some examples of such diseases are listed in Table 3.10. These diseases can be:
1. **Organ-specific diseases:** Type 1 diabetes mellitus, pemphigus vulgaris, vitiligo, autoimmune thrombocytopenic purpura, pernicious anaemia, myasthenia gravis, Guillain–Barré syndrome, acute rheumatic fever etc.
2. **Systemic diseases** (organ non-specific): Sjogren's syndrome, Wegener's granulomatosis, anti-phospholipid syndrome etc.

4

Host-Pathogen Interaction

Amita Jain

LEARNING OBJECTIVES

- Definitions of some general terms
- Classification of infectious diseases
- Classification of pathogens
- Reservoirs of infection

- Modes of disease transmission
- Phases of disease development
- Virulence factors and escape from host defence

INTRODUCTION

Outcome of interaction between microbe and host is determined by several factors such as host immunity and susceptibility, microbial pathogenicity and virulence, and role of environment. Host-pathogen interaction may or may not result in disease. This chapter is an effort to understand the types of host-pathogen relationship. The terms which are often used to understand the host parasite relationship are listed in Box 4.1.

CLASSIFICATION OF DISEASES

1. **Based on communicability, infectious diseases are classified as follows:**
   a. **Communicable infectious diseases** are contagious and easily spread from one host to another either directly or through vectors, for example influenza, tuberculosis, AIDS/HIV etc. Their spread depends on several host-related, microbe-related and environmental factors.
   b. **Non-communicable infectious diseases:** An infectious disease that does not spread from one host to another, but each individual contacts the disease independently, for example tetanus, food poisoning, subacute bacterial endocarditis and so on.

2. **Based on their disease burden, diseases in a population can be described as follows:**
   a. **Sporadic disease:** The scattered distribution of the disease in low frequency of occurrence is called sporadic. The cases occur occasionally and infrequently, for example tetanus cases in India.
   b. **Outbreak:** The occurrence of infection with a particular pathogen in a small, localised group (e.g. population of a village) in excess of the baseline incidence is called an outbreak, for example outbreak of measles in a school.
   c. **Endemic disease:** The presence of a disease within a given geographic area at a constant rate/stable

> **BOX 4.1 Definitions of Some General Terms Often Used to Describe Host–Pathogen Relationship**
>
> - **Disease: (Dis + ease)** abnormal state of health, state of discomfort.
> - **Infection:** Pathogens increase in number either within or on the body surface of the host.
> - **Contamination:** The presence of microorganisms in a sample/product.
> - **Pathogenicity:** The ability of an organism to cause a disease.
> - **Pathogens:** Organisms with high probability of causing an infection.
> - **Virulence:** It is the degree of pathogenicity. Some strains always cause infection, if they get access in host body. They are referred to as highly virulent.
> - **Parasitism:** It is a relationship in which pathogen derives benefits at the cost of host. Host is always harmed.
> - **Commensalism:** It is a relationship in which the pathogen gets benefits and the host is unaffected i.e. neither benefited nor harmed.
> - **Mutualism/symbiosis:** Refers to a close and prolonged interaction between organisms of different species. The term is restricted to a mutualistic relationship wherein both organisms benefit from the interaction, for example *Escherichia coli* in the colon; they produce vitamin K and some B vitamins. They also break down waste products.

incidence is called endemic, for example malaria is endemic in India.
   d. **Hyperendemic disease:** If the disease is present at a high incidence or prevalence rate in a population round the year, the disease is said to be hyperendemic in that population, for example tuberculosis is hyperendemic in India.
   e. **Epidemic disease:** It is an occurrence of a disease, in a community or region, clearly in excess of the normal expectancy. There is sudden increase in the number of cases of a particular disease in a particular area or place, for example epidemic of hepatitis C in Punjab state of India.

f. **Pandemic disease:** The global distribution of a disease affecting all population is called a pandemic. It is an epidemic that affects the entire world, for example influenza virus subtype H1N1 caused a pandemic in 2009–2010.

3. **Based on duration of illness, diseases are classified as follows:**
   a. **Acute disease:** It develops rapidly, usually within 1–2 weeks and may last for short duration, for example influenza, dengue etc.
   b. **Chronic disease:** It develops slowly, usually takes more than 2 weeks, and lasts for longer duration, for example tuberculosis, hepatitis B etc.
   c. **Latent disease:** Causative agent remains inactive in body of the host for a long time and at times can get reactivated, especially when the immunity gets lower, for example tuberculosis, toxoplasmosis, herpes zoster etc.

4. **Based on extent of illness, diseases are classified as follows:**
   a. **Local infection:** Infectious agent remains localised in one area of the body, for example abscesses in skin, fungal infection of skin, mycetoma and so on.
   b. **Systemic infection:** Infectious agent spreads throughout the body through blood or lymphatic vessels, for example septicaemia, typhoid and so on.

Diseases are also classified as follows:
1. **Primary infection:** First encounter of host with a pathogen may cause primary infection, usually manifestations are acute.
2. **Secondary infection:** Caused by a pathogen in a host who already has a primary infection, for example mycobacterial infections in a HIV/AIDS patient.

Reservoir of infections can be humans/animals/environment (Box 4.2).

CLASSIFICATION OF PATHOGENS

1. **Based on virulence, microbes are classified as follows:**
   a. **Pathogens:** Pathogens are microbes that can produce disease. Typically, the term is used to describe an infectious agent such as a virus, bacterium, protozoa, fungus or any other disease causing microorganism.
   b. **Opportunistic pathogens:** These are organisms that normally do not cause disease but can cause it if given an opportunity i.e. in an immuno-compromised host. They can also cause secondary infection in a host with low resistance, for example *Pneumocystis jiroveci* causes pneumonia in immunocompromised hosts.
   c. **Commensals:** The word originated from Latin [*com*-(together) *mensa* (table), meaning eating together]. Organisms that benefit from commensalism and does not bring any considerable injury/harm (or benefit) to the host are called commensals.

2. **Based on biosafety level** required for handling the pathogens, due to their transmissibility and disease severity, pathogens are divided into biosafety levels (BSL) (Table 4.1).

MODES OF DISEASE TRANSMISSION (Fig. 4.1)

1. **Transmission through contact**
   a. Direct contact: Disease spreads by direct contact with the body of infected host through touch, kiss, sexual contact for example hepatitis B.

BOX 4.2 Reservoirs of Infection
- **Human source:**
  a. **Case:** A case of active disease can be a reservoir of infection, e.g. influenza, chicken pox, cholera etc.
  b. **Carrier:** An individual, who has infection but has not developed signs and symptoms of the disease, can be a reservoir, e.g. tuberculosis, syphilis, HIV infection, hepatitis B infection etc.
- **Animal source:** Usually wild animal (sylvatic) **but at times** domestic animals can be reservoir of infection, e.g. pig is a reservoir of Japanese encephalitis virus.
- **Soil:** Harbours organisms that can cause infections, e.g. tetanus, botulism, anthrax, fungal infections etc.
- **Water:** If contaminated may serve as a vehicle of infection, e.g. giardiasis, typhoid fever, amoebiasis etc.

TABLE 4.1 Classification of Pathogens Based on Biosafety Levels

Biosafety Levels	Details	Examples	Pathogenic Potential
BSL-4	Posing a high risk of aerosol transmitted infection by dangerous pathogens, which produce fatal infections, and do not have a treatment/vaccine	Ebola virus, CCHF	*Highest risk*
BSL-3	Easily transmissible exotic pathogen posing a high risk of severe disease through respiratory route	*M. tuberculosis*	
BSL-2	Pathogens posing moderate risk of disease of varying severity to workers and environment	*S. aureus*	
BSL-1	Pathogens posing minimal risk to healthy host and environment	Nonpathogenic, *E. coli*	*Lowest risk*

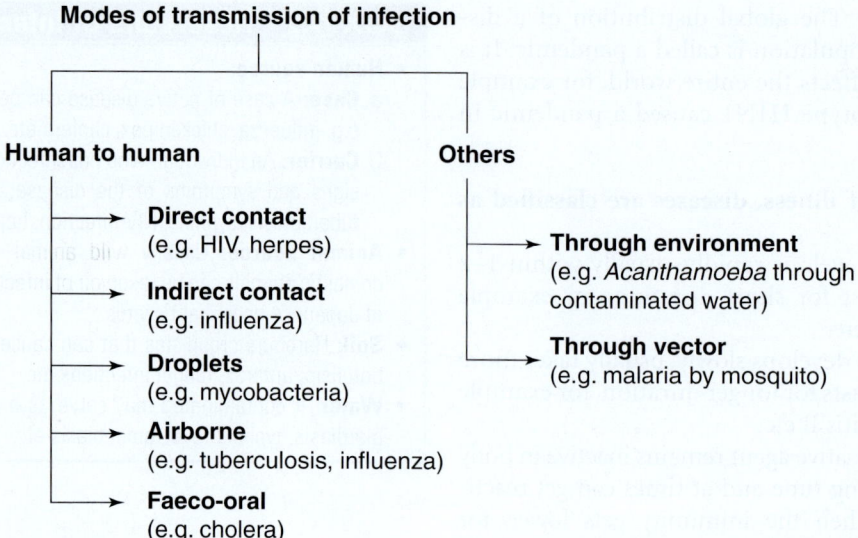

Fig. 4.1 Modes of Disease Transmission.

b. Indirect contact: Disease spreads by contact through fomites on inanimate objects, for example doorknobs, towels, phones, medical equipment etc., for example influenza.

c. Droplet transmission: Disease spreads by contact with droplet nuclei, which are medium sized, that is 3–5 μm in size. They can travel short distances and infect susceptible individuals. They are mainly important in causing respiratory infections, for example measles.

2. **Transmission through vehicle**
   a. Waterborne: Oral–fecal transmission through water contaminated with feces, for example cholera.
   b. Foodborne: Oral transmission through food which is raw or undercooked, or cooked but contaminated, for example salmonella infections.
   c. Airborne: It occurs through droplet nuclei in dust. Medium-sized nuclei are able to travel more than 1 m and cause disease, for example tuberculosis.

3. **Transmission through vector (usually an insect) and infected animals**
   a. Biological transmission: Microbe spend a part of its life cycle in host for some metabolic or other essential process, for example malaria parasite and mosquito; *Borrelia* and tick; *Trypanosoma* and tsetse fly.
   b. Mechanical transmission: The animal host picks up the microbes from the environment and spreads it to newer areas, for example flies landing on feces or dead animals can later sit on food and contaminate it.

4. **Zoonosis:** It is a disease of animal which can be transmitted to humans, for example rabies, plague, leptospirosis, lyme disease, toxoplasmosis and so on.

PHASES OF DISEASE DEVELOPMENT (Fig. 4.2)

1. **Incubation period**: Time duration between contact of host with infectious agent and the appearance of first signs or symptoms (S/S).

2. **Prodromal period**: Period of initial mild sign or symptoms, usually of short duration.
3. **Period of illness**: Period of presentation of disease manifestations.
4. **Period of decline (effervescence):** Symptoms start declining and patient start recovering.
5. **Period of convalescence**: Patient recovers and regains strength.

DEVELOPMENT OF DISEASE

1. **Portal of entry can be**
   a. **through mucous membranes** of mouth, nose, eyes, respiratory tract, gastrointestinal and genitourinary tract;
   b. through breech in skin and/or
   c. through parenteral route i.e. injection in body.
2. **Adherence at the site of infection**: Adhesins help in adherence of microbes at the site of infection
   a. Capsule present on bacterial surface helps in adherence and also impedes phagocytosis, for example *Streptococcus pneumoniae* and *Klebsiella pneumoniae* have capsule which helps in adherence.
   b. Fimbriae (pili) are often site specific, for example pili of *Escherichia coli* strains causing urinary tract infection, i.e. uropathogenic *E. coli* differ from those of enteropathogenic strains of *E. coli*. *Bordetella pertussis* attaches to respiratory epithelium cilia. Pili and outer membrane proteins of *Neisseria gonorrhoeae* allow adherence to urethral and vaginal epithelium, fallopian tube, sperms and neutrophils. Protein F of *Streptococcus pyogenes* attaches to pharyngeal epithelium. Protein M of *S. pyogenes* attaches to keratinocytes in the skin only. Both the proteins F and M are part of the fimbriae.
3. **Virulence factors and escape from host defence:** Pathogens are armed with mainly invasive factors and toxins, which are listed in Table 4.2.

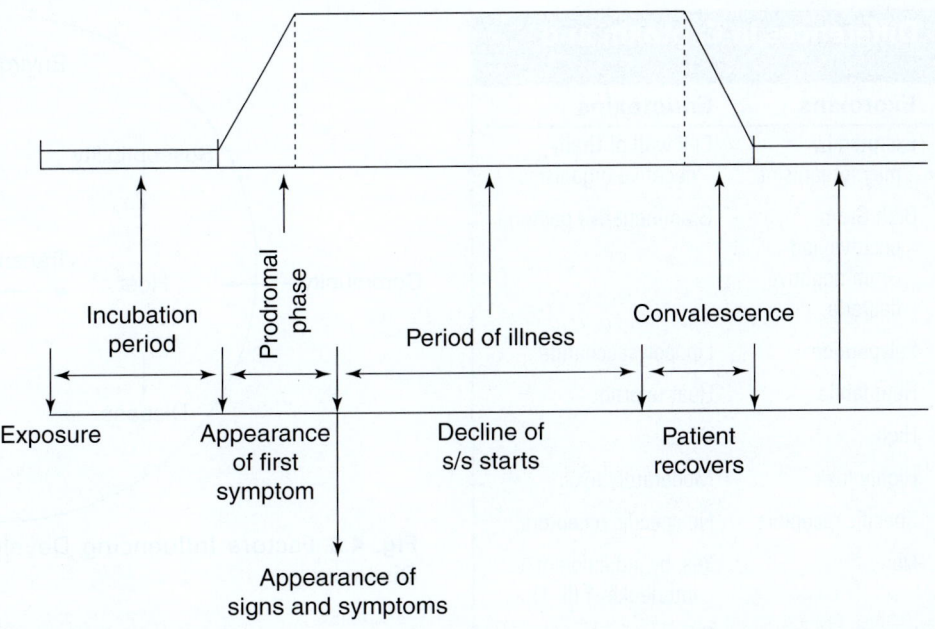

Fig. 4.2 Phases of Disease Development.

TABLE 4.2 Virulence Factors		
Virulence Factors	**Examples**	**Actions**
Bacterial cell wall	• Protein G of *Streptococcus pyogenes* • Protein A of *Staphylococcus aureus*	Binds to the antibodies, thus preventing phagocytosis
	Wax D and sulpholipids of *Mycobacterium tuberculosis*	Inhibit the killing mechanisms of phagocytes
Outer membrane of Gram-negative organisms (known as endotoxin)	Lipopolysaccharide containing lipid A	Released only when organism dies. These are toxins which initiate release of interleukin-1 (IL-1) and tumour necrosis factor (TNF)-alpha, resulting in fever, shock, haemorrhage, disseminated intravascular coagulation (DIC). Fever and toxic shock are hallmark
Bacterial enzymes	Leucocidins of *S. aureus* and *S. pyogenes*	Destroy phagocytes. Release of their digestive enzymes causes cell damage
	Hemolysins of *S. aureus*, *S. pyogenes* and *C. perfringens*	Destroy host RBCs
	Coagulase of *S. aureus*	Surround the organism in a fibrin clot
	Bacterial kinases, e.g. streptokinase and staphylokinase	Digest fibrin clots. Therapeutic use is to dissolve blood clots
	Hyaluronidase	Disintegrate hyaluronic acid
	Collagenase (gelatinase) of *C. perfringens*	Disintegrate tissue collagen
	Siderophore	Scavenge iron
Bacterial toxin (exotoxin): It can be classified as cytotoxin/neurotoxin/enterotoxin	Diphtheria toxin	Inhibits protein synthesis and is produced by bacteria which have lysogenic conversion with TOX gene
	Erythrogenic toxin	Acts as superantigen. Fever, rash and damage to capillaries caused
	Botulinum toxin	Causes flaccid paralysis by inhibiting acetylcholine release at neuromuscular junction
	Tetanus toxin (tetanospasmin)	Inhibitory motor neurons damage causing spastic paralysis
	Cholera toxin *E. coli* enterotoxin	Diarrhoea due to adenyl cyclase stimulation causing water and electrolytes disbalance
	S. aureus enterotoxin	Is a superantigen. Activates T-cells; which release interleukin-2 (IL-2) and tumour necrosis factor (TNF)

TABLE 4.3 Differences in Exotoxin and Endotoxin		
Properties	**Exotoxins**	**Endotoxins**
Production	Excreted by microorganisms	Cell wall of Gram-negative organisms
Found in	Both Gram-positive and Gram-negative bacteria	Gram-negative bacteria
Chemical nature	Polypeptide	Lipopolysaccharide
Heat stability	Heat labile	Heat tolerant
Antigenicity	High	Weak
Toxicity	Highly toxic	Moderately toxic
Binds to	Specific receptors	No specific receptors
Causes fever	No	Yes, by induction of interleukin 1 (IL-1)
Gene location	Usually plasmids	Chromosomal genes
Filterable	Yes	No
Enzymatic activity	No	Yes
Molecular weight	~10 kDa	50–1000 kDa
Denaturation at 100°C	Yes	No
Tests for detection	Neutralisation, precipitation etc.	Detected by Limulus lysate assay
Diseases produced	Tetanus, diphtheria, botulism	Meningococcemia, sepsis by Gram-negative rods

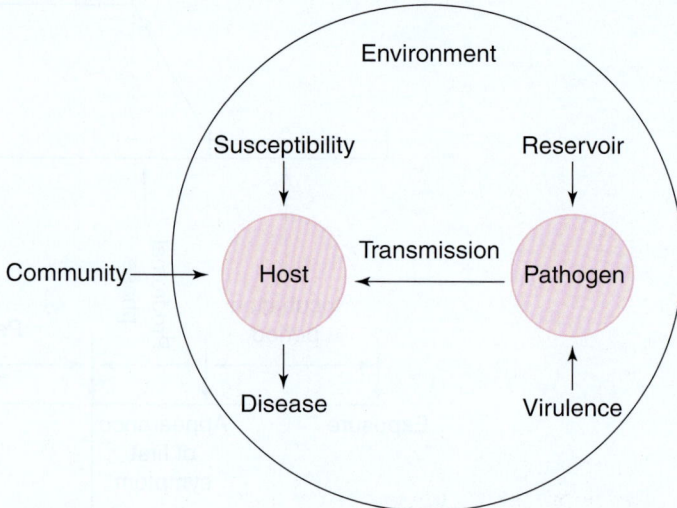

Fig. 4.3 Factors Influencing Development of Infectious Diseases.

4. Some bacteria have enzymes and properties that allow them to survive inside phagocytic cells, for example *Mycobacterium tuberculosis*, *Salmonella typhi* and *Neisseria gonorrhoeae*.

Endotoxins and Exotoxins are important bacterial virulence factors, as shown in Table 4.2. The differences in endotoxin and exotoxin are listed in Table 4.3.

As shown in Fig. 4.3, development of infectious disease is influenced by both host- and pathogen-related factors as well as environment. Pathogenic potential and transmission of pathogen to a susceptible host is essential for infection.

5

Normal Flora of Human and Microbiome

Amita Jain

LEARNING OBJECTIVES

- Normal human microbial flora and its significance
- Number of bacteria on different sites in human body
- Development of human normal flora

- Anatomical site-wise normal flora
- Common infections caused by normal flora

INTRODUCTION

Throughout life, body surfaces of every human being are colonised with a diverse microbial flora. This microbial population constitutes **the normal microbial flora.**

MICROBIAL FLORA

The number of bacterial cells (10^{14}) on human body outnumbers the cells in human body (about 10^{13}cells). The majority of the microbial flora in humans live in gut, especially in large intestine. Many scientists do not consider viruses and parasites as normal microbial flora because they are usually not commensals.

Microbiome

It is the genetic material of all the microbes that live on and inside the human body, including bacteria, fungi, protozoa and viruses. The term 'microbiome' was not used until the late 1990s. The microbiome may weigh as much as 5 pounds, and the number of microbial genes in one human body is at least ten times the number of genes in the human genome.

Human microbiome project (HMP), sponsored by the National Human Genome Research Institute (NHGRI), is mapping the human microbiome, giving insight into uncharted species and genes. The project started in 2008 as an extension of the Human Genome Project.

Significance of Normal Flora

Normal flora on a human body protects human from pathogens and helps the host by competing for microenvironments more effectively than pathogens. It may exist as **commensal** living on host body without causing any harm. At times it may harm the host by causing certain ailments, for example dental caries, abscesses or other infectious diseases (**pathogenic potential**). Generally, these organisms cause disease in immunocompromised hosts (**opportunistic pathogen**) without affecting healthy individuals.

The bacteria in the microbiome help protect against the disease, digest food, regulate immune system and produce vitamins, for example vitamins B12, thiamine, riboflavin and vitamin K. The microbiome configuration of an individual is associated with the development, immunity and nutrition. Dysfunction of microbiome is now supposedly associated with several diseases, for example diabetes, rheumatoid arthritis, muscular dystrophy, multiple sclerosis, obesity fibromyalgia and so on.

Development of Microbial Flora

The normal flora in human starts developing soon after birth. Stable normal adult flora is established by the adolescence. The first exposure of an infant with microorganisms occurs during birth as it moves through the birth canal. In breast-fed babies, Gram-positive bacteria, for example bifidobacteria and lactobacilli, predominate in the gastrointestinal tract, which is displaced by Enterobacteriaceae once the baby starts feeding on other food. Type of gastrointestinal flora is controlled by the type of food given to the baby.

Factors Determining the Composition of Normal Flora in Human Body

- Physical factors at the colonisation site: e.g. pH, temperature and redox potential
- Chemical factors at the colonisation site: e.g. oxygen, water and nutrient levels
- Other site-specific factors: for example, peristalsis, saliva secretion, lysozyme secretion and secretion of immunoglobulins

Number of Bacteria on Different Sites in Human Body

It varies from site to site. Some of the sites like oral cavity and large intestine are heavily colonised (Table 5.1).

TABLE 5.1 Number of Bacteria on Different Sites in Human Body	
Anatomical Site	Approximate Number of Organisms per square inch of Surface or per gram of Tissue in Human Body
Skin	10^{5-6}
Conjunctiva	10^1
Oral cavity	10^9
Respiratory mucosa	10^2
Stomach	10^3
Small intestine	10^5
Large intestine	10^{11}
Genitourinary mucosa	10^6

Normal Flora of Skin

It varies from site to site. Skin sites with folds like axilla, perineum and toe webs are heavily colonised in comparison to rest of the areas. Common microbes present on skin are as follows:

1. *Staphylococcus epidermidis* is a major inhabitant of the skin. It makes around 90% of aerobic flora.
2. *Staphylococcus aureus* colonises nose and perineum commonly. Level of colonisation varies with age, being greater in the newborn than in adults.
3. **Micrococci** are less common than staphylococci and diphtheroids. *Micrococcus luteus* is the commonest species.
4. **Diphtheroids (coryneforms)** bacteria belonging to the genus *Corynebacterium* are referred to as diphtheroids. *Corynebacterium acnes*, commonest anaerobic diphtheroids, is now classified as *Propionibacterium acne*.
5. **Streptococci** are rarely seen on normal skin. α-Hemolytic streptococci exist in the mouth, from where they may spread to the skin.
6. **Gram-negative bacilli:** *Enterobacter, Klebsiella, E. coli, Acinetobacter* spp. and *Proteus* spp. are present on skin in small numbers. They are predominant in moist skin areas, for example toe webs.

Nail Flora

It is similar to that of skin. In addition, fungi like *Aspergillus, Penicillium, Cladosporium* and *Mucor* may be found in dirt under the nails.

Oral and Upper Respiratory Tract Flora

The oral cavity, pharynx and trachea are colonised by α- and β-hemolytic streptococci, anaerobes, staphylococci, neisseriae, diphtheroids and many other bacteria. Some

of them can be potentially pathogenic like *Haemophilus*, mycoplasmas and pneumococci. The upper respiratory tract is often colonised by pathogens such as *Neisseria meningitides, Corynebacterium diphtheriae, Bordetella pertussis* and so on. Lower respiratory tract is usually sterile. The oral flora is involved in dental caries and periodontal disease. Anaerobes in the oral flora may cause infections in brain, face and lung.

Gastrointestinal Tract Flora

The stomach has a very acidic environment keeping bacterial number and types to its minimal (approximately 10^3–10^6 organisms/g of contents). *Helicobacter* species is a common stomach coloniser and plays an important role in causation of peptic ulcer. Aspirates of duodenal or jejunal fluid also contain approximately 10^3 organisms/mL, mainly streptococci, lactobacilli, bifidobacteria and *Bacteroides*. People with achlorhydria and slow peristalsis have high bacterial count in small intestine. Large intestine and faeces have highest number of organisms (10^9–10^{11} bacteria/g of contents). More than 400 species have been identified from colon of which 95%–99% belong to anaerobic genera such as *Bacteroides, Bifidobacterium, Eubacterium, Peptostreptococcus* and *Clostridium. E. coli* is commonest aerobic coloniser of colon.

Urogenital Flora

Vaginal pH in female infants during the first month of life is around 5, hence *Lactobacillus* spp. predominate. After 1 month of age, vaginal pH changes to around 7 and diphtheroids, *S. epidermidis*, streptococci and *E. coli* predominate. At puberty again the vaginal pH changes to acidic and girls acquire microbial flora as seen in an adult women. Predominantly, *Lactobacillus acidophilus*, corynebacteria, peptostreptococci, staphylococci, streptococci and *Bacteroides* make adult microbial vaginal flora. After menopause once again pH rises, and yeasts predominate (*Torulopsis* and *Candida*).

S. epidermidis, enterococci and diphtheroids are found frequently and *E. coli, Proteus* and *Neisseria* (nonpathogenic species) are reported occasionally in the anterior urethra of humans.

Conjunctival Flora

Approximately half of conjunctival cultures are sterile. Occasionally, staphylococci, streptococci, corynebacteria, neisseriae, *Haemophilus parainfluenzae* and Moraxellae may colonise.

Common Host Infections Caused by Normal Flora

Normal microbial flora provides protection from many infections. However, occasionally they may play a role in causing some infections (Table 5.2). Any disruption in normal flora may cause some infections. The normal flora may

TABLE 5.2 Common Host Infections Caused by Normal Flora

Site	Normal Flora	Pathogenic Potential
Intestinal tract	Anaerobes	Intra-abdominal abscesses and peritonitis
Colon after treatment with antibiotics	Anaerobes especially *Clostridium difficile*	Produce pseudomembranous colitis
Upper small intestine	Anaerobic bacteria	Deconjugate bile acids and bind available vitamin B12 causing malabsorption of vitamin and fats
Oral cavity	Anaerobes and other bacteria	Caries, periodontal disease, abscesses, foul-smelling mouth and endocarditis, brain infections
Vagina	Candida and other yeasts	Vaginitis
Any site	Microbial flora	Opportunistic infections in immune-compromised individuals

TABLE 5.3 Normal Flora of Community and Hospital Residents

Site	Normal Pedominant Flora in Normal Healthy Adults in Community	Predominant Flora in Hospitalised Adults
Upper respiratory tract	• *Staphylococcus* spp. • *Streptococcus* spp.: *pneumococci, viridans streptococci* • *Haemophilus* • *Anaerobes*	• *Staphylococcus* spp. • *Anaerobes* • *Enterobacteriacae: E. coli, Klebsiella* • *Pseudomonas* • *Candida*
Skin	• *Staphylococcus* spp. • *Diphtheroids* • *Propionibacterium*	• *Staphylococcus* spp. • *Enterobacteriacae*
Gastrointestinal tract	• *Anaerobes* • *Enterobacteriacae* • *Enterococci* • *Streptococci* • *Lactobacilli* • *Candida*	• *Anaerobes* • *Enterobacteriacae* • *Enterococci* • *Streptococci* • *Lactobacilli* • *Candida* • *Pseudomonas*
Genital tract	• *Lactobacilli* • *Streptococcus agalactiae*	*Candida*

get altered under several circumstances, for example after admission to a hospital or a long-term health care facility (Table 5.3). Antibiotic administration usually depletes the human gut normal microflora, which might cause overgrowth of pathogens and development of life-threatening infections.

6

Sterilisation and Disinfection

Parul Jain and Atin Singhai

LEARNING OBJECTIVES

- Sterilisation and disinfection
  - Definition and differences
  - Common methods
- Principle and use of autoclave
- Plasma sterilisation, principle and use

INTRODUCTION

Disinfection and sterilisation are essential procedures for ensuring that medical and surgical procedures and instruments do not transmit infectious pathogens (especially hepatitis B and C viruses, human immunodeficiency virus [HIV]) to patients/caregivers. Sterilisation of all patient care items may not be necessary. Definitions of cleaning, disinfection, sterilisation and antiseptics are given in Box 6.1, and important differences in disinfection and sterilisation are shown in Table 6.1.

BOX 6.1 Definitions

- **Sterilisation** is a process of making an article, surface or medium free of all microorganisms both in vegetative and spore form.
- **Disinfection** means destruction of all kinds of pathogenic organisms and inactivation of viruses. Spores may not be destroyed.
- **Antiseptic** is a substance that may not kill the microorganisms but does not allow their growth and multiplication.
- **Cleaning** is removal of visible soil/dirt from objects and surfaces.

IMPORTANCE OF CLEANING

1. The number of microorganisms is decreased
2. Organic matter like blood, tissue and other debris is removed, thereby making the sterilisation and disinfection procedures effective

SPAULDING'S CLASSIFICATION FOR INSTRUMENTS

The medical, surgical and dental instruments have been classified into three categories depending on the body site they come in contact with and the risk of infection that can be transmitted by them (Table 6.2).

TABLE 6.1 Differences Between Sterilisation and Disinfection

	Disinfection	Sterilisation
Definition	To eliminate pathogenic microorganisms with or without spores from surfaces or objects	To destroy all microbes and their spores present on a surface or object
Types/Methods	Phenolic disinfectants, chlorine compounds, iodine preparation, bleach, alcohols, hydrogen peroxide, detergents, quaternary ammonium compound, peracetic acid, aldehydes, heating, boiling and pasteurisation	Dry heat, wet heat (steam), irradiation (gamma), high pressure steam, and microfiltration, chemical sterilisation
Application	Decontaminate surfaces like table and chair, decontaminate instruments and objects used for patient care, used linen, appliances etc.	Injectable medicines, intravenous fluids, surgical instruments, used instruments, culture media etc.

TABLE 6.2	Spaulding's Classification for Instruments		
Category	**Definition**	**Examples**	**Sterilisation/Disinfection**
Critical items	Instruments that penetrate the mucosa or skin	Surgical instruments, periodontal knives, scaling instruments, burs and so on	Sterilised/discarded after single use
Semicritical items	Instruments that come in contact with mucosa or non-intact skin	• Dental hand-pieces, amalgam condensors, mouth mirrors • Plastic impression trays, plastic instruments, amalgam carriers (heat sensitive instruments)	Sterilised/high-level disinfection
Non-critical items	Instruments that come in contact with intact skin	Blood pressure cuff, pulse oximeter, facebow, radiograph head/cone	Cleaning/low-level disinfection/barrier protection

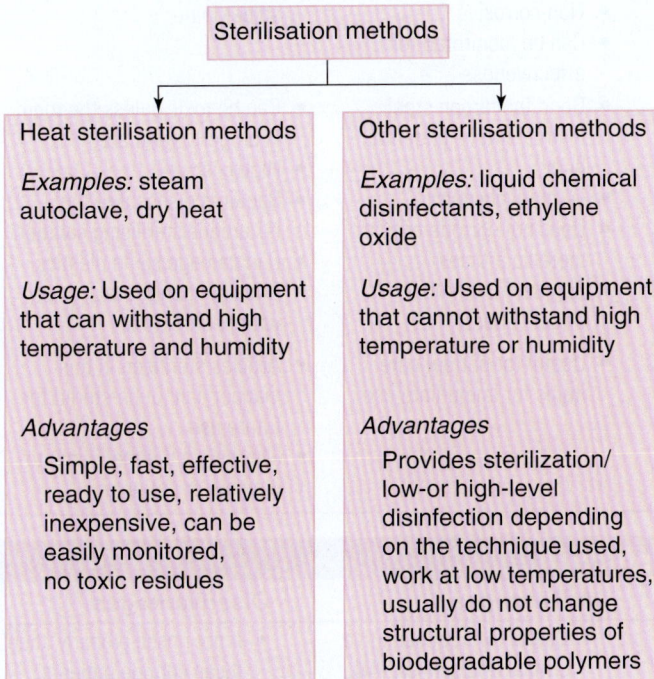

Flowchart 6.1 Types of Sterilisation Methods.

STERILISATION METHODS

All the sterilisation methods are classified in Flowchart 6.1. The different modalities of heat sterilisation, their conditions, efficacy testing, advantages and disadvantages are shown in Table 6.3 and those of other methods of sterilisation are shown in Table 6.4.

Autoclave

It is a physical method of sterilisation, which works on principle of moist heat. Moist heat denatures proteins by coagulating them. This kills the microbes. Autoclaves operate at producing steam with high temperature under high pressure, which kills microorganisms and spores. A basic autoclave (Fig. 6.1) is similar to a pressure cooker. Any material which comes in direct contact with steam at higher temperature for desired period of time gets sterilised. Desired temperature for autoclaving is at least 121°C for at least 30 min under at least 15 psi of pressure. Depending upon requirement, time and pressure may be increased to attain higher temperature.

Most Important Parameters of Autoclaving

- Time: 30 min or more
- Temperature: 121°C/132°C
- Steam to be circulated in the chamber at higher atmospheric pressure, that is 15 psi
- Pressure: 15 psi
- Vacuum is created initially. This displaces air and fills the vacuum with steam. The flow of steam is shown in Fig. 6.1
- Loading of items is very important. Items to be sterilised should come in direct contact with steam

Autoclaves are used to decontaminate desired material and inactivate medical waste. **Autoclave is unsuitable for sterilising heat sensitive objects.**

Autoclave Validation is Done By

Chemical Indicators. Chemical indicators are paper tapes with chemical markings on them. When exposed to 121°C, they change colour. Tape indicators are placed on the exterior of the waste load, and if exposed to 121°C in autoclave they change colour.

Biological Indicators. Biological indicators contain spores from *Bacillus stearothermophilus*. It does not get inactivated below 121.1°C exposed for 20 min. Autoclaves are validated regularly with biological indicators to ensure sterilisation.

Plasma Steriliser

Plasma steriliser is most appropriately used to sterilise expensive heat sensitive material. Plasma is recently described, unstable, fourth state of matter (solid, liquid, gas, and plasma) which is created when a gas is either heated sufficiently or exposed to a strong electromagnetic field. Plasma is an ionised gas, which produces ions (positively or negatively charged). Plasma sterilises by oxidation. Liquid hydrogen peroxide is heated up in order to turn it into gas. Once the hydrogen peroxide gas is heated to an even higher temperature, it turns into plasma. The high heat turns hydrogen peroxide into free radicals, which are highly unstable, and in order to stabilise, they latch on to the microorganisms in the load and destroy cellular components of microbes.

TABLE 6.3   Modalities of Heat Sterilisation/Disinfection

Modality	Conditions	Efficacy Testing	Advantages	Disadvantages
Autoclave (moist heat/ steam sterilisation)	121°C at 15 psi for at least 30 min	• Chemical indicator • Biological indicator (*Bacillus stearothermophilus* spore strips)	• Quick and easy • Economic and reliable • Loads are packed and then sterilised so can be stored in sterile state • Can be monitored for effectiveness	• May corrode and rust articles • May damage heat sensitive articles like plastic • May blunt some sharp instruments
Dry heat sterilisation: two types • Static air/oven-type steriliser • Forced air/rapid heat transfer steriliser	300°F (149°C)	• Chemical indicator • Biological indicator (*Bacillus subtilis* subsp *niger*)	• Can be used for articles that can be damaged by moist heat • Low operating costs • Non-corrosive • Can be monitored for effectiveness	• Prolonged process • High temperature is not suitable for certain articles like plastic, some metals, solder joint
Unsaturated chemical vapour sterilisation	Heating an alcohol solution with formaldehyde in a pressurised chamber	Chemical indicators	• Good for carbon steel instruments, e.g. dental burs • Is relatively quick • Does not corrode or rust metallic articles • Can be monitored for effectiveness	• Can be toxic unless the room is properly ventilated • High cost • Requires disposal of the hazardous sterilising solution • Can damage some plastics • Does not penetrate articles wrapped in fabric
Flash sterilisation (of unwrapped instruments)	Done with tabletop sterilisers according to conditions provided by manufacturer	• Mechanical monitors • Chemical indicators	• Useful for patient care items for immediate use	• Articles sterilised in this manner cannot be stored for future use • May cause thermal injury to workers or patients

TABLE 6.4   Other Modalities of Sterilisation/Disinfection

Modality	Conditions	Efficacy Testing	Advantages	Disadvantages
Liquid chemical germicides, e.g. glutaraldehyde, peracetic acid, hydrogen peroxide	• Complete immersion for at least 6–10 h • Followed by rinsing with sterile water to remove toxic residues	• Dilution test • Measuring basic bactericidal/ fungicidal/virucidal activity	• Useful for heat sensitive critical and semicritical instruments and devices • Provides high-level disinfection	• If not rinsed properly, then it can corrode patients' mucous membrane • Highly toxic for workers if not done using proper precautions
Alcohols as skin disinfectants	• 50%–80% ethyl alcohol • 70% isopropanol • 10–30 min	Disinfectant studies	• Low toxicity • Rapid action • Low residue • Non-corrosive • Can be used for skin disinfectant (antiseptic), surface decontamination, benchtop, cabinet wipedown	• Rapid evaporation limits contact time • Flammable • Eye irritant • May damage rubber, plastic, shellac • Ineffective against bacterial spores
Low-temperature sterilisation	Ethylene oxide gas (ETG) for 10–48 h	Biological indicator: *Bacillus subtilis* subsp *niger* spores	• Can be used for heat sensitive articles • Good for recycled prepacked articles	• Long sterilisation cycle • Potentially toxic to dental workers and patients • Stringent health safety requirements
Bead sterilisers	Glass beads or salt heat up to 220°C is used in heat transfer device. Articles are submerged into the beads and sterilised in 10 s	None	Useful for small instruments like endodontic files	Now obsolete

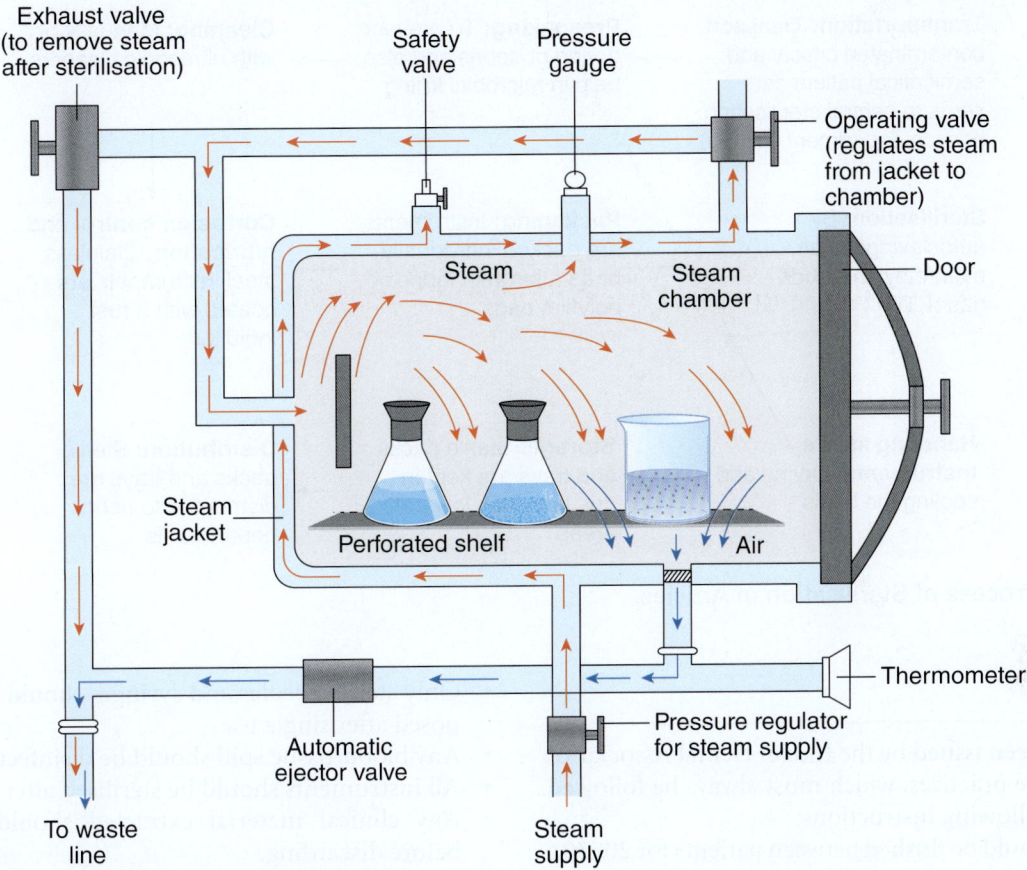

Fig. 6.1 Components and Working of an Autoclave.

Plasma sterilisers are used to sterilise endoscopes, heat sensitive appliances and instruments.

Advantages and Disadvantages of Hydrogen Peroxide Plasma

This method leaves no chemical residues, is safe for operators and environment and is quick. Disadvantages are that, it cannot sterilise powders, requires specific packaging of the load and is unable to take big loads.

There is a systematic process of sterilization of articles, which is highlighted in Flowchart 6.2.

DISINFECTION OF SURFACES

Contact surfaces such as light handles, switches, dental chairside computers, drawer handles, pens, telephone, door knobs etc. can get contaminated by direct spray of patient material or by practitioners' gloved hands. These can be protected from contamination by

- using barriers such as plastic wraps, bags, sheets etc. that should be replaced in between patients;
- cleaning and disinfecting the surface with low level (with HIV and HBV cidal claim) or intermediate level (with tuberculocidal claim) disinfectant; and
- general cleaning and disinfection of surfaces, dental unit surfaces and countertops should be done with gloved

hands daily at the end of the day with a liquid chemical germicide compatible with that equipment. Clean all the dental unit surfaces and counter tops daily (at least once at the end of the day) with a liquid chemical germicide (always check the compatibility of germicide with equipment).

Disinfection of Housekeeping Surfaces

Floors, walls and sinks usually do not pose a risk for disease transmission and can be cleaned only with a detergent and water. However, when these surfaces become contaminated with blood or other potentially infectious material, prompt removal and surface disinfection with chemicals such as sodium hypochlorite (1:100 dilution or approximately ¼ cup of 5.25% household chlorine bleach to 1 gallon of water) becomes essential.

Disinfection of Dental Prosthesis and Impressions

Before sending to the dental laboratory, impressions, gypsum casts, dental prosthesis, wax rims etc. should be thoroughly rinsed under gentle running water and then disinfected by immersing it in 0.5%–1% sodium hypochlorite solution for a short time. Other chemicals that can be used are glutaraldehyde, halogenated phenols or povidone-iodine.

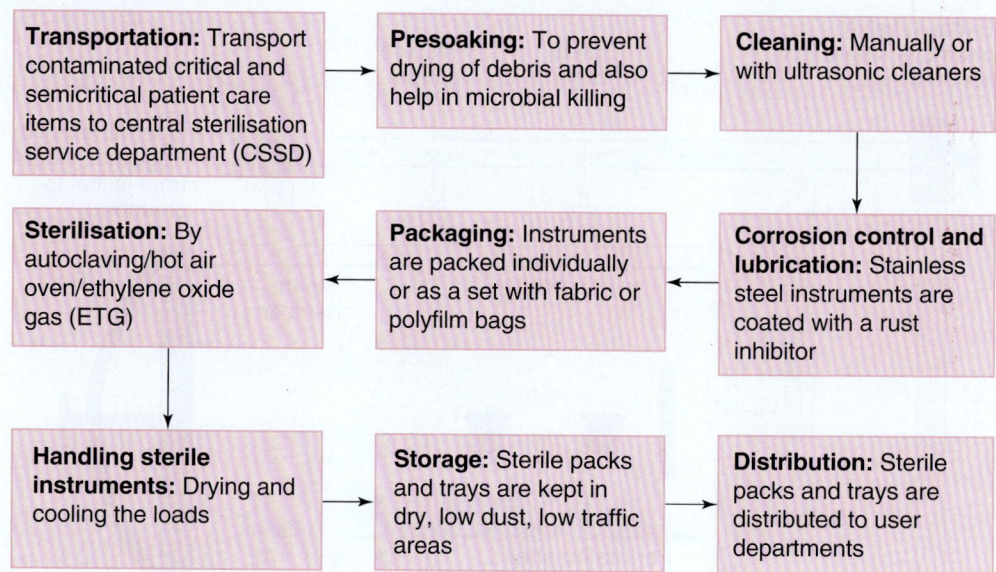

Flowchart 6.2 Process of Sterilisation of Articles.

CHECKLIST

A checklist has been issued by the Indian Dental Association (IDA) for routine practices, which must always be followed. It includes the following instructions:

- All devices should be flushed between patients for 20–30 s.

- Only sterile needles and syringes should be used and disposed after single use.
- Any blood/tissue spill should be disinfected immediately.
- All instruments should be sterilised after every use.
- Any clinical material extracted should be disinfected before discarding.

SECTION II

Aetiological Agents

PART A: Bacterial Pathogens

Chapter 7: Introduction to Bacterial Pathogens

Chapter 8: General Principles of Laboratory Diagnosis of Bacterial Pathogens

Chapter 9: Gram-Positive Cocci

Chapter 10: Gram-Positive Bacilli

Chapter 11: Mycobacteria, Actinomycetes and Nocardia

Chapter 12: Gram-Negative Cocci

Chapter 13: Gram-Negative Rods: Enterobacteriaceae

Chapter 14: Gram-Negative Rods: Other than Enterobacteriaceae

Chapter 15: Spirochaetes

Chapter 16: Chlamydia

Chapter 17: Rickettsia

Chapter 18: Mycoplasma

PART B: Viral Pathogens

Chapter 19: Introduction to Viral Pathogens

Chapter 20: Herpesviruses

Chapter 21: Poxviruses

Chapter 22: Hepatitis Viruses

Chapter 23: Orthomyxoviruses and Paramyxoviruses

Chapter 24: Arboviruses

Chapter 25: Enteroviruses and Rhinoviruses

Chapter 26: Human Immunodeficiency Virus

Chapter 27: Rabies Virus

Chapter 28: Miscellaneous DNA Viruses

Chapter 29: Miscellaneous RNA Viruses

Chapter 30: Other Miscellaneous Viruses

PART C: Parasitic Pathogens

Chapter 31: Introduction to Parasites

Chapter 32: Protozoa

Chapter 33: Helminths

Chapter 34: Ectoparasites

PART D: Fungal Pathogens

Chapter 35: Introduction to Fungal Pathogens

Chapter 36: Fungi Causing Superficial and Cutaneous Infections

Chapter 37: Fungi Causing Subcutaneous Infections and Mycetoma

Chapter 38: Yeasts Causing Human Infections

Chapter 39: Dimorphic Fungi Causing Human Infections

Chapter 40: Opportunistic Fungal Infections and Miscellaneous Fungal Diseases

SECTION II

Aetiological Agents

7

Introduction to Bacterial Pathogens

Parul Jain

LEARNING OBJECTIVES

Introduction, classification, morphology, physiology and metabolism of bacteria

INTRODUCTION

Bacteria are the oldest forms of life known to us and are still the most abundant living forms. They inhabit diverse environments on earth's surface such as ocean depths, boiling hot springs as well as the human and animal gut. Bacteria have a complex relationship with humans. They are important for ecology and recycling of elements, for our digestion and even for converting milk into yogurt. But at the same time they are also the most familiar agents of diseases like pneumonia and meningitis. The main definitions related to bacteria are listed in Box 7.1.

The major differences between eukaryotes and prokaryotes of kingdom Protista are shown in Table 7.1. Prokaryotes are simpler cells that do not possess a nucleus or other membrane-bound organelle. They reproduce by binary fission. Eukaryotes are complex cells that possess a nucleus and other membrane-bound organelles. They divide by mitosis or meiosis.

CLASSIFICATION

Bacteria have been classified according to phenotypic characters such as shape, size, staining properties, spore formation and biochemical properties. But recently bacteria have been classified on the basis of genotypic characteristics such as 16S ribosomal RNA. Although phenotypic and genotypic characters usually collate, yet dichotomies are known to exist. Therefore, presently all classifications are based on genotypic characters and are presented in **Bergey's Manual of Determinative Bacteriology**.

BOX 7.1 Important Definitions

Bacteria are microscopic single-celled organisms with simple cell structure. They are prokaryotic microorganisms that do not contain chlorophyll. They have no nucleus or membrane-bound organelles and their genetic information is contained in a single loop of DNA. They do not show true branching except the higher bacteria or the actinomycetales.

Bacteriology: Study of bacteria.

Clone: A population derived from a single cell by binary fission is called clone. A single bacterial colony represents a clone.

Strain: A population of bacteria obtained from a single source, e.g. a patient, is called a strain.

TABLE 7.1 Differences Between Eukaryotes and Prokaryotes

Characteristics	Eukaryotes	Prokaryotes
Nucleus	Present. They have a nuclear membrane and nucleolus	Absent
Membrane-bound organelles	Present. Include mitochondria, lysosomes, golgi apparatus, endoplasmic reticulum etc.	Absent
Chromosomal DNA	Contain multiple linear chromosomes with histone proteins	Usually contain a single circular chromosome with few proteins
Cell wall	Usually absent. But fungal cell wall contains chitin	Peptidoglycan containing cell wall is present in Eubacteria
Ribosomes	80S	70S
Mode of cell division	Mitosis, meiosis or cytokinesis	Binary fission
Examples	Fungi, slime moulds, other algae, protozoa	Bacteria, blue-green algae

NOMENCLATURE

The scientific name of bacteria usually contains two words, the first word being the 'genus name' (begins with a capital letter) and the second word being 'species name' (begins with a small letter). The whole name is italicised, for example, *Staphylococcus aureus*.

MORPHOLOGY

Bacteria are microscopic organisms, that is, they cannot be seen with the naked eye (limit of resolution is 200 μm for naked eye examination). The medically important bacteria usually measure 3–5 μm in length and 0.2–1.5 μm in diameter.

Shape of Bacteria

Bacteria have the following main shapes (Fig. 7.1):

Cocci

Spherical-shaped bacteria, for example, *Staphylococcus*, *Streptococcus*.

Bacilli

Rod-shaped bacteria, for example, *Escherichia*, *Bacillus*.

Comma Shaped

Curved rods, for example, *Vibrio*.

Spiral

Spiral-shaped bacteria, for example, *Spirilla*, spirochetes like *Leptospira*.

Filamentous Bacteria

They are elongated and have branches, for example, *Actinomycetes*.

Pleomorphic

Multiple shaped, for example, anaerobic bacteria.

Bacterial Arrangements

Bacterial cells are arranged in different patterns determined by the plane of cell division and tendency of the daughter cells to remain attached to the mother cell even after division. The various arrangements of bacteria are shown in Fig. 7.2.

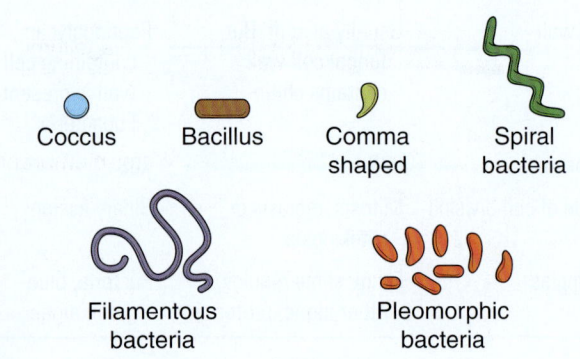

Fig. 7.1 Different Shapes of Bacteria.

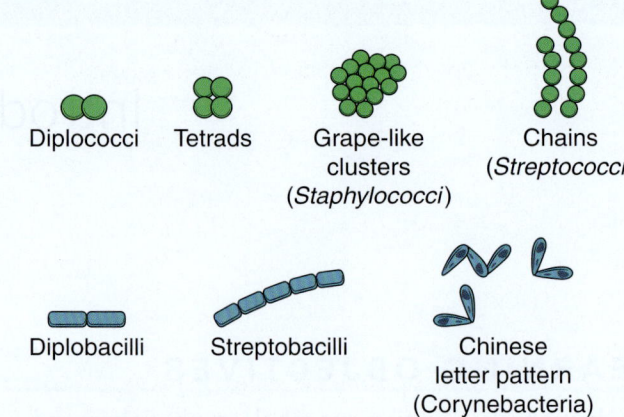

Fig. 7.2 Different Shapes and Arrangements of Bacteria.

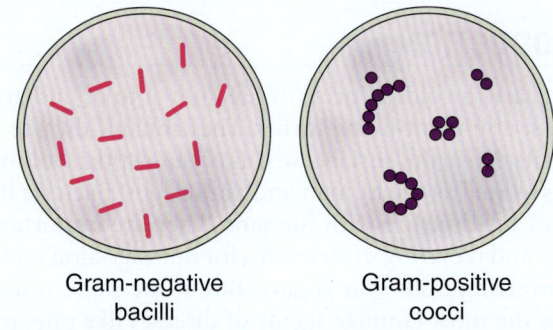

Fig. 7.3 Gram-Positive Cocci and Gram-Negative Rods.

Gram Reaction

On the basis of Gram's stain, bacteria are of two types (Fig. 7.3):

Gram-Positive

Bacteria that retain the colour of the primary stain, that is pararosaniline dye such as crystal violet, methyl violet or gentian violet, for example, *Staphylococcus*, *Streptococcus*, are called Gram-positive. They appear purple coloured in a Gram-stained slide.

Gram-Negative

Bacteria in which the primary stain gets decolourised and bacteria take the colour of the counterstain, such as safranine, are called Gram-negative, for example, *Escherichia*, *Klebsiella*. These bacteria appear pink coloured in a gram-stained slide.

BACTERIAL STRUCTURES

A bacterial cell has following structures:

Capsule or Slime Layer

When present it is the outermost layer surrounding a bacterial cell. It is an amorphous viscid secretion of some bacterial cells. When it is organised as a sharply demarcated structure, it is called capsule as seen in *Pneumococcus*. But when it is present as a loose undemarcated region it is called slime layer as seen in *Leuconostoc*. It is chemically composed of polysaccharide or polypeptide. Capsulated bacteria generally produce mucoid colonies.

Capsules can be demonstrated by (1) negative staining with India ink, (2) serological methods, for example, Quellung reaction as seen in *Pneumococcus*.

The capsule acts as a virulence factor of bacteria and protects it from phagocytosis and from the effect of lytic enzymes.

Cell Wall

It confers rigidity, ductility and shape to the bacterial cell, and is chemically composed of peptidoglycan layer formed by alternating units of N-acetyl glucosamine and N-acetyl muramic acid molecules cross-linked by peptide chains. The cell walls of **Gram-positive** and **Gram-negative** bacteria differ in their structure (Fig. 7.4). The Gram-positive bacteria have a thick (15–80 nm) peptidoglycan layer, and also contain lipid and teichoic acid. The Gram-negative bacteria have a thin (2–3 nm) peptidoglycan layer, and also contain lipopolysaccharide that has an endotoxic activity and O-antigen specificity. They also have an outer membrane containing proteins known as outer membrane protein (OMP) and transmembrane channels or porins through which small molecules can diffuse.

The cell wall can be demonstrated by mechanical rupture, electron microscopy, differential staining procedures, reaction with specific antibodies and microdissection.

Cytoplasmic Membrane

The plasma membrane measures 5–10 nm in thickness and forms the lining of cytoplasm. It is a semipermeable membrane that controls the passage of metabolites to and from the cytoplasm. It is a phospholipid bilayer arranged in 'Fluid mosaic model'.

Cytoplasm

It is a colloidal system containing both organic and inorganic solutes. It contains the following:

1. **Ribosomes**: It is a complex structure of 70S (sedimentation constant) and is composed of two subunits, 50S and 30S. They are held on strands of messenger-RNA known as polysomes or polyribosomes. They are the site of protein synthesis.
2. **Mesosomes**: These are convoluted structures formed by invagination of plasma membrane into the cytoplasm. They are the sites of respiration in bacterial cells.
3. **Cytoplasmic inclusions**: They are the sources of stored energy found in some bacteria, for example volutin granules (Babes-Ernst/metachromatic granules) in diphtheria bacilli, or polysaccharide, lipid, starch or glycogen granules.

Nucleus

It has no nuclear membrane or nucleolus. The genome consists of a single molecule of double-stranded circular DNA (haploid) not associated with histone protein. Extranuclear DNA/genetic elements may be present in some bacteria, known as plasmids. These plasmids may confer certain properties like antibiotic resistance or toxigenicity to bacteria, though they are not essential for survival.

Flagella

They are long, sinuous filaments that act as the organs of locomotion. A flagellum consists of three different regions, that is filament (hair like), hook (connects the filament outside the bacterial cell to the basal body inside the cell) and the basal body (Fig. 7.5). It is made up of flagellin protein and is antigenic in nature. Flagellar antibodies do not protect the bacteria, though they are used for serodiagnosis.

The different flagellar arrangements are shown in Table 7.2. Flagella can be demonstrated directly by dark ground microscopy, by staining with mordant dyes or by electron microscopy and indirectly by demonstrating motility.

Fimbriae

Fine hair-like appendages present on the surface of some Gram-negative bacilli are called fimbriae or pili. They are made up of self-aggregating monomers of pilin. They develop best in liquid cultures and tend to disappear on subculture

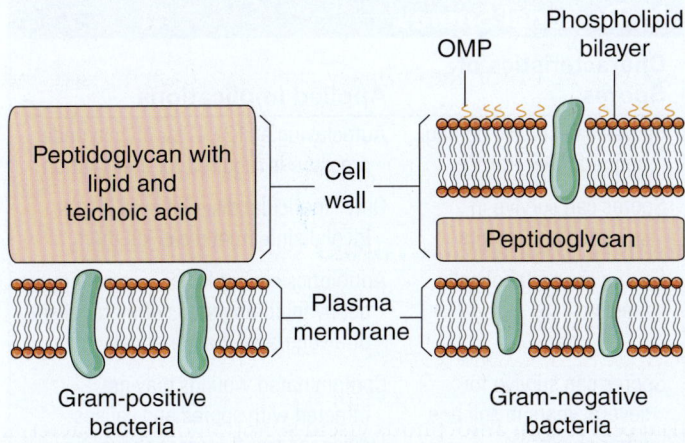

Fig. 7.4 Cell Walls of Gram-Positive and Gram-Negative Bacteria.

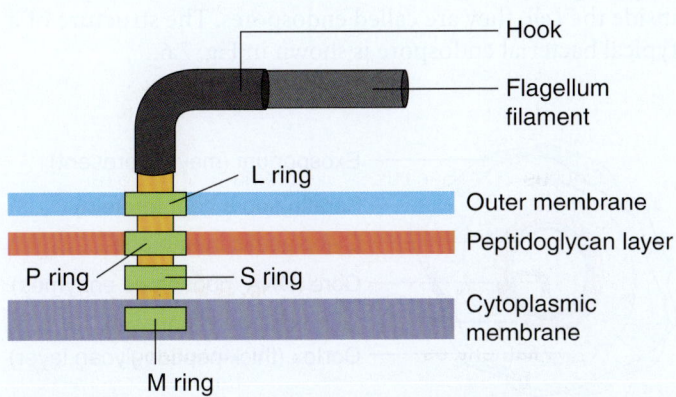

Fig. 7.5 Structure of Flagella in Gram-Negative Bacteria.

TABLE 7.2 Different Flagellar Arrangements

Bacterial Group	Flagellar Arrangement	Example
Atrichous	No flagella	*Klebsiella pneumoniae*
Monotrichous	One flagella	*Vibrio cholerae*
Lophotrichous	Tuft of flagella at one end	*Alcaligenes faecalis*
Amphitrichous	Flagella at both the polar ends	*Spirillum volutans*
Peritrichous	Flagella all over	*Escherichia*

TABLE 7.3 Shapes and Arrangements of Bacterial Spores

Position of Spores in bacilli	Shape of bacilli	Example
Central/equatorial and oval	Spindle shaped	*C. bifermentans*
Subterminal and oval	Club-shaped appearance	*C. perfringens*
Terminal and oval	Tennis racket appearance	*C. tertium*
Terminal and spherical	Drumstick appearance	*C. tetani*

on solid media. On the basis of their function pili are classified as common pili and sex pili. Their primary function is to adhere the bacteria firmly to a surface. They also agglutinate the RBCs of several animals such as fowls, guinea pigs, horses and so on.

They are demonstrated by haemagglutination. They are antigenic in nature and interfere with serological identification of bacteria, hence should be removed before serotyping.

Spore

Resistant dormant stage (i.e. no metabolic activity) of bacteria is called spore. They are formed when the environmental conditions become unfavourable to bacteria. They are highly resistant to heat, desiccation, radiation and chemicals. They are not a means of reproduction. A single bacterial cell forms one spore, which on availability of favourable environmental conditions (like water and nutrients) germinates to form a single vegetative form of bacteria. As these spores are present inside the cell, they are called endospores. The structure of a typical bacterial endospore is shown in Fig. 7.6.

Young spores are attached to the parent bacterial cell. Spores may occupy central, subterminal or terminal position in a bacterial cell and may be spherical or oval in shape (Table 7.3). They may be bulging (i.e. distend the bacillary body) or non-bulging (i.e. does not distend the bacillary body). The spores can be demonstrated in unstained preparations as refractile bodies. They appear as hollow unstained spaces in Gram's stain and can be stained by modified Ziehl–Neelsen stain. The characteristics of spores which require special attention in medical practices are listed in Table 7.4.

TABLE 7.4 Medical Practices Influenced by Bacterial Spores

Characteristics of Spores	Applied Implications
Spores can survive boiling temperature (100°C)	Autoclaving at 121°C for 15 min under pressure is required for sterilisation
Spores can survive in several disinfectants	Only 'sporicidal disinfectants' (high-level disinfectants) can kill them
Spores are metabolically inactive	Antibiotics cannot kill spores as they act by inhibiting steps of metabolism of a vegetative bacterial cell
Spores can survive for several years in soil and other objects	Contaminated wounds may get infected with spores and causes diseases and therefore should be immediately cared for

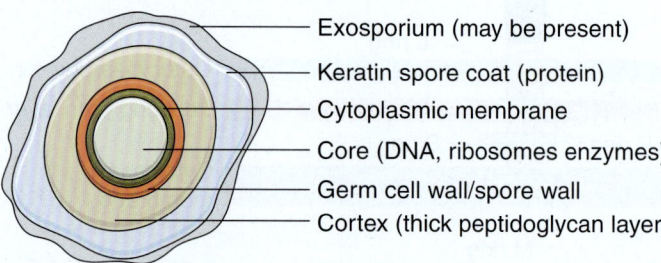

- Exosporium (may be present)
- Keratin spore coat (protein)
- Cytoplasmic membrane
- Core (DNA, ribosomes enzymes)
- Germ cell wall/spore wall
- Cortex (thick peptidoglycan layer)

Fig. 7.6 Structure of a Bacterial Endospore.

PHYSIOLOGY (GROWTH AND MULTIPLICATION) OF BACTERIA

Growth

It is a systematic increase in the chemical components of a cell.

Multiplication

It occurs after optimum growth of a cell has occurred and results in an increase in the number of bacteria. Most bacteria divide by binary fission in which a bacterial cell divides into two identical halves.

Generation Time

Time required for a bacterium to give rise to two daughter cells under optimum conditions. The generation time varies with bacterial species. It is about 20 min for *Escherichia coli*, 20 h for *Mycobacterium tuberculosis* and 20 days for *Mycobacterium leprae*. Theoretically, the multiplication can proceed logarithmically, but in reality the bacterial multiplication stops after few divisions due to depletion of nutrients or accumulation of toxic products.

Bacterial Growth Curve

When a small inoculum of bacteria is inoculated into a liquid culture medium, the population size increases following a classical pattern. A growth curve has following phases and can be obtained by plotting the number of bacteria in relation with time (Fig. 7.7).

Lag Phase

The multiplication of bacteria does not start immediately after inoculation. This time gap between inoculation and the initialising of multiplication is called the lag phase. This time is required for adaptation to the new environment during which the bacterial cell grows and accumulates all the necessary nutrients. This is the time when cells attain their maximum size. Its duration depends upon the bacterial species, culture medium, inoculum, incubation temperature and so on.

Log Phase/Exponential Phase

During this phase exponential increase in the number of bacteria occurs. The cells are small in size and stain uniformly. This phase is sensitive to antibiotics and other antibacterial agents.

Stationary Phase

After a log phase, increase in the number of bacteria ceases due to depletion of essential nutrients, change in pH and accumulation of toxic metabolic products. Bacterial cells start dying and the number of cells dying is just equal to the new cells formed by division. Hence, a state of equilibrium exists in this stage between the dying cells and the new cells. Cells are Gram variable and sporulation also occurs at this stage.

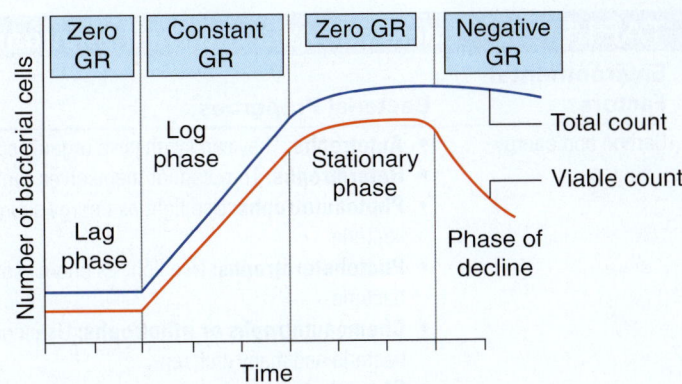

Fig. 7.7 Bacterial Growth Curve. GR = growth rate.

Decline Phase

The number of bacterial cells decreases due to death of bacterial cells. It occurs due to depletion of nutrients, toxic metabolite accumulation and autolytic enzymes. Involution forms are seen during this stage.

Continuous Culture

This can be obtained by continuously replenishing the culture media and removing bacterial population so that nutrients do not get depleted and toxic metabolites do not accumulate. This is done for industrial or research purposes.

Bacterial Types and Environmental Factors Influencing Bacterial Growth

The factors that affect bacterial growth are listed in Table 7.5.

Bacterial Counts

The bacterial counts can be done in terms of the following:

Cell Concentration

Number of cells per unit volume of culture

Total count. Total number of cells in a sample irrespective of whether they are living or dead. Total counts may be obtained by (1) comparing the opacity of a culture tube manually with standard opacity tubes or by measuring it with nephelometer, (2) direct counting methods including microscopic counts using a counting chamber or a hemocytometer.

Viable count. Number of living cells in a sample. It is obtained by dilution methods, for example 'presumptive coliform count' done in drinking water or by plating methods, for example 'Miles and Misra method'.

Biomass Concentration

Dry weight of cells per unit volume of culture.

TABLE 7.5   Bacterial Types and Their Environmental Requirements

Environmental Factors	Bacterial Properties
Carbon and energy sources	• **Autotrophs:** They can synthesise organic compounds by themselves using atmospheric CO_2 • **Heterotrophs:** They cannot themselves synthesise organic compounds and hence require them preformed • **Photoautotrophs:** Use light as energy source and CO_2 as carbon source, e.g. cyanobacteria, some purple and green bacteria • **Photoheterotrophs:** Use light as energy source and organic compounds as carbon source, e.g. some purple and green bacteria • **Chemoautotrophs or lithotrophs:** Use inorganic compounds as energy source and CO_2 as carbon source, e.g. few bacteria and many archaea • **Chemoheterotrophs:** Use organic compounds as energy and carbon source, e.g. most bacteria
Oxygen	• **Obligate aerobes:** Those requiring oxygen for growth, e.g. *Pseudomonas aeruginosa.* In these organisms oxygen acts as the terminal electron acceptor. Energy is provided by **oxidative phosphorylation**, that is when ADP (adenosine diphosphate) is converted to ATP (adenosine triphosphate) • **Obligate anaerobes:** Those can grow only in the absence of oxygen and are killed by oxygen, e.g. *Clostridium* spp. In these organisms molecules other than oxygen (e.g. nitrates or sulphates) act as the terminal electron acceptor. Energy is provided by **substrate-level phosphorylation**/fermentation, that is production of energy-rich phosphate bonds by introduction of organic phosphate into intermediate metabolites • **Facultative anaerobes:** Those requiring oxygen for growth but can grow without it. Most pathogenic bacteria lie in this category, e.g. *E. coli.* In these organisms oxygen or any other molecule acts as the terminal electron acceptor • **Microaerophilic bacteria:** Those requiring low oxygen and low carbon dioxide for growth, e.g. *Campylobacter jejuni*
Carbon dioxide	All bacteria require small amount of environmental CO_2 for growth (present in the atmosphere/endogenously produced) **Capnophilic bacteria:** Those requiring 5%–10% of CO_2 for growth, e.g. *H. influenzae*
Organic salts/growth factors/bacterial vitamins	• **Essential growth factors:** Which are absolutely essential for bacterial growth • **Accessory growth factors:** Which enhance bacterial growth but are not essential
Inorganic salts	Like sodium, potassium, magnesium, manganese, iron, calcium, phosphate, sulphate, cobalt etc. These need to be added to culture media for supporting bacterial growth
Temperature	• Optimum temperature (at which growth occurs best) for growth of pathogenic bacteria is 37°C. • **Psychrophiles:** That grow at cold temperatures, that is 0–20°C, e.g. most water and soil saprophytes • **Mesophiles:** That grow at moderate temperatures, that is 25–40°C, e.g. most pathogenic bacteria • **Thermophiles:** That grow at high temperatures, that is 55–80°C, e.g. *B. stearothermophilus* • **Thermal death point:** The lowest temperature that kills a bacterium under standard conditions in a given time is called thermal death time
pH/hydrogen ion concentration	• Optimum pH is 7.2–7.6 • **Acidophilic bacteria:** Grow at acidic pH, e.g. *Lactobacilli* • Bacteria tolerating alkaline pH, e.g. *V. cholerae* • Strong acids (e.g. HCl, H_2SO_4) and alkalis (e.g. NaOH) kill the bacteria
Light	• Bacteria grow well in dark • **Phototrophs:** Derive energy from sunlight • **Photochromogenic mycobacteria:** Produce pigment only when exposed to light • **Scotochromogenic mycobacteria:** Produce pigment only in the absence of light
Water	• Water constitutes 80% of the total weight of bacterial cells. Effect of drying varies with bacteria. • Bacteria can be highly sensitive to drying e.g. *Treponema pallidum* • Bacteria can survive in dry state for months, e.g. *Staphylococcus* • Spores can survive in dry state for decades • **Freeze drying/Lyophilisation:** Drying in vacuum in cold is the method used for preserving bacteria
Osmotic effect	• **Plasmolysis:** When a bacterial cell is suddenly exposed to a hypertonic solution, the protoplasm shrinks due to osmotic withdrawal of water from the cell • **Plasmoptysis:** When a bacterial cell is suddenly exposed to distilled water, the cell swells and finally ruptures due to excessive osmotic imbibition of water
Mechanical and sonic stress	Grinding, vigorous shaking with glass beads, or ultrasonic vibrations can disintegrate bacterial cells

BACTERIAL METABOLISM

It refers to the sum of all chemical reactions occurring within a living cell. It has two components (Fig. 7.8): (1) catabolism/energy generating or exergonic component and (2) anabolism/energy consuming or endergonic biosynthetic component. These reactions require the molecule **adenosine triphosphate** (ATP), which converts to adenosine diphosphate (ADP) releasing energy for biosynthetic reactions. Several catalysts, cofactors and enzymes (Box 7.2) play a role in bacterial metabolism.

ENERGY PRODUCTION

The two important aspects in energy production are mechanisms of ATP production and redox reactions.

Mechanisms of ATP Production

1. **Substrate-level phosphorylation:** Direct transfer of a phosphate group from an organic molecule to ADP produces ATP.
2. **Oxidative phosphorylation:** Electrons are transferred from organic compounds to electron carriers like NAD^+ and FAD. After passing through the electron transport chain, the electrons are accepted by O_2 molecules or other organic or inorganic molecules.
3. **Photophosphorylation:** It occurs in photosynthetic bacteria containing chlorophyll. Sugars are synthesised using CO_2, H_2O and light energy.

Redox Reactions

1. **Reduction:** Addition of one or more electrons to an atom or a molecule.
2. **Oxidation:** Removal of one or more electrons from an atom or a molecule.
3. Usually reduction and oxidation reactions occur together, called redox reactions.

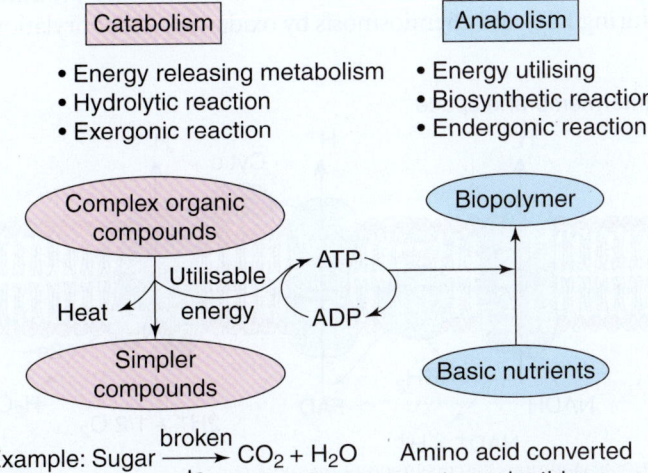

Fig. 7.8 Components of Bacterial Metabolism.

BOX 7.2 Important Terms in Reference to Bacterial Metabolism

Catalyst: Substance that enhances the speed of a chemical reaction without itself being permanently altered.
Substrate: Specific substances on which enzymes act.
Apoenzyme: Protein portion of an enzyme.
Cofactor: Non-protein portion of an enzyme, e.g. Mg, Fe, Zn, Ca and so on.
Coenzyme: Organic molecule acting as a cofactor, e.g. NAD^+ (nicotinamide adenine dinucleotide) and $NADP^+$ (nicotinamide adenine dinucleotide phosphate).
Holoenzyme: Whole active enzyme, that is apoenzyme + cofactor.

CATABOLISM/RESPIRATION

Catabolism involves biochemical activities that lead to breakdown of complex/high energy level substances to simpler/low energy level substances. Aerobic carbohydrate catabolism has the following steps.

Glycolysis

It is the most common pathway for glucose catabolism producing energy, reduced electron carriers and other precursors for cellular metabolism. This process occurs in the cell cytoplasm and breaks a glucose molecule (containing 6 carbon) into two pyruvate molecules (containing 3 carbon), which may be broken down further by aerobic or anaerobic respiration. It is also called **Embden–Meyerhof–Parnas (EMP) pathway** and consists of two phases: an initial energy investment/consuming phase wherein 2 ATP molecules are converted to 2 ADP molecules, and a later energy gaining phase wherein 4 ADP molecules are converted to 4 ATP molecules (Fig. 7.9) by substrate-level phosphorylation. Therefore, at the end of glycolysis there is a net gain of 2 ATP, 2 NADH and 2 pyruvate molecules. Glycolysis is not oxygen-dependent (can occur in the presence or absence of oxygen) and it does not use oxygen.

Alternatives of Glycolysis

1. **Entner–Doudoroff (ED) pathway:** In this pathway one glucose molecule is converted to two ethanol molecules and one ATP molecule is gained. For example, *Pseudomonas aeruginosa* and *Agrobacterium* use this pathway.
2. **Pentose phosphate pathway (PPP)/phosphogluconate pathway/hexose monophosphate (HMP) shunt:** In this pathway one glucose molecule is converted to 5 carbon sugars (including ribose-5-phosphate, a precursor for nucleotide synthesis) and 2 molecules of NADPH. This leads to a net gain of only one molecule of ATP for each molecule of glucose used. For example, *Escherichia Coli* and *Enterococcus faecalis* use this pathway.

Transition Reaction/Bridge Reaction

During this phase the pyruvate molecules (3-C) produced during glycolysis are decarboxylated to acetyl groups (2-C) by pyruvate dehydrogenase enzyme. These groups then

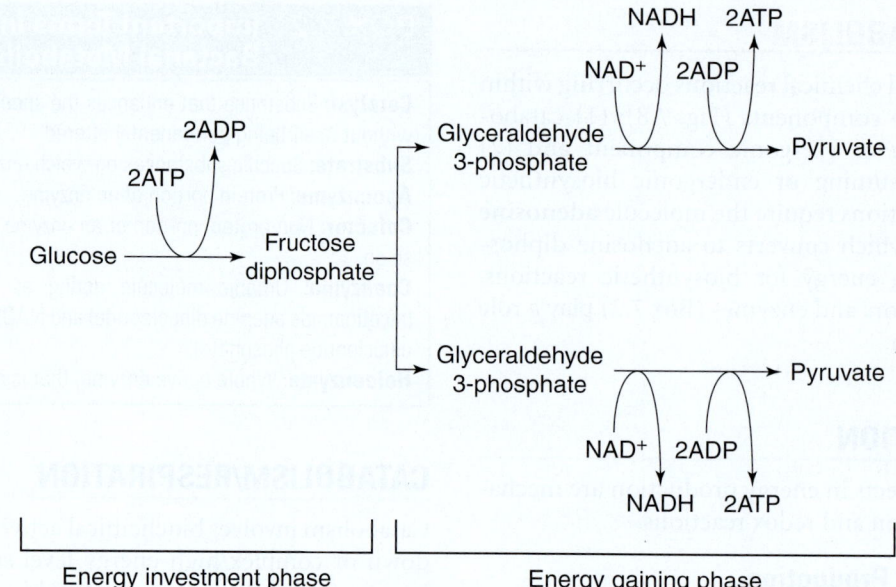

Fig. 7.9 Outline of EMP Pathway.

attach to a large carrier compound called coenzyme A (CoA) to form acetyl-CoA, the substrate for Krebs cycle. During this process NAD^+ is reduced to NADH. Therefore, the products of the transition phase are 2 molecules each of CO_2, NADH and acetyl-CoA.

Krebs Cycle/Citric Acid Cycle/Tricarboxylic Acid Cycle (TCA)

This is called cycle because oxaloacetate is the first reactant and the last product of this process. The acetyl group of acetyl-CoA enters the Krebs cycle, which is added to oxaloacetate (4-C) to produce citric acid (6-C). Total eight reactions occur in the Krebs cycle (outside the scope of this book). At the end of the cycle, from each acetyl-CoA molecule entering the cycle the products produced are 2 molecules of CO_2, 3 molecules of NADH, 1 molecule of $FADH_2$, 1 molecule of ATP or other high-energy nucleotide molecule by substrate-level phosphorylation.

As a glucose molecule produces 2 molecules of pyruvate, all the above numbers are doubled. The summary of products formed during the three steps of respiration is shown in Table 7.6.

Electron Transport Chain (ETS) and Oxidative Phosphorylation (OP)

The ETS is a series of proteins and organic molecules found on the bacterial cytoplasmic membrane. A series of oxidation-reduction electron transfer reactions occur resulting in release of energy. This energy is then used for formation of ATP by a process called chemiosmosis.

NADH and $FADH_2$ produced during above cycles transfer their electrons to molecules near the beginning of the transport chain (Fig. 7.10). The electrons move from a higher energy to a lower energy level releasing energy. Some energy is used

TABLE 7.6   Summary of Products Formed in Aerobic Respiratory Cycle

	Glycolysis	Transition Reaction	Krebs Cycle	Total
CO_2	0	2	4	6
ATP	2	0	2	4
NADH	2	2	6	10
$FADH_2$	0	0	2	2

to pump H^+ from matrix to the intermembrane space, thus creating an electrochemical gradient. The electrons are transferred to molecular oxygen that combines with H^+ to form water. The protons flow down into the matrix along their electrochemical gradient through an enzyme ATP synthase that synthesises ATP. A total of 34 ATP molecules are produced during ETC and chemiosmosis by oxidative phosphorylation.

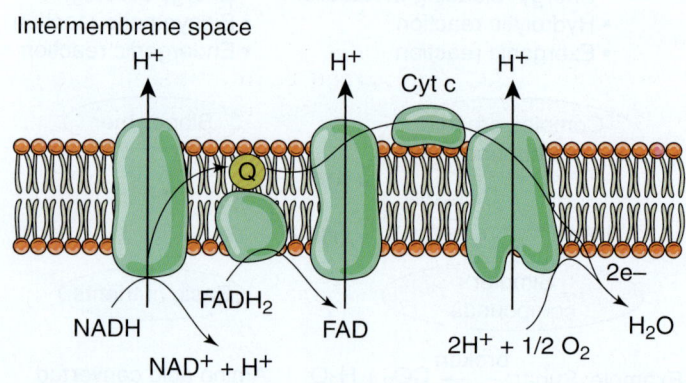

Fig. 7.10 Electron Transport Chain.

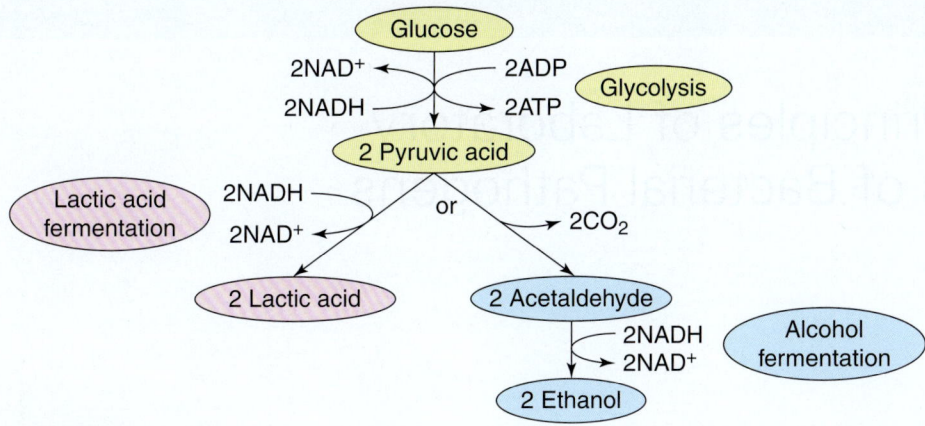

Fig. 7.11 Types of Fermentation.

ANAEROBIC CELLULAR RESPIRATION

It is similar to aerobic respiration in that the electrons extracted from a molecule pass through an ETC, thereby releasing ATP. But here the terminal electron acceptor is an inorganic compound rather than O_2. Some bacteria like *Desulfovibrio* and *Archaea* use SO_4^{2-} as the final electron acceptor, producing H_2S. Some bacteria like *Pseudomonas* and *Bacillus* have the enzyme nitrate reductase due to which they can use nitrate (NO_3^-) as the terminal electron acceptor reducing it to nitrite ion and finally to nitrous oxide or nitrogen gas. Some bacteria like methanogens use carbonate as the terminal electron acceptor producing methane as a by-product.

The amount of ATP produced depends on the organism and the pathway but is always less than that produced in aerobic respiration and therefore anaerobes tend to grow slower than the aerobes.

FERMENTATION

Some organisms lack an ETS. Therefore, they cannot use oxygen as the terminal electron acceptor; instead they use organic carbon (commonly pyruvate). In the absence of ETS they should have an alternative mechanism to oxidise NADH to NAD^+ for carrying out normal metabolic pathways like glycolysis. This alternate pathway is fermentation, which is an anaerobic reaction and does not require oxygen. As a result of fermentation many reduced metabolic by-products are formed like ethanol, acetate, lactate and butyrate. Fermentative organisms have a great industrial importance and are used to make different types of food products, pharmaceuticals and diagnostic kits. Most pathogens are facultative anaerobes, that can undergo aerobic respiration in the presence of oxygen and fermentation in its absence.

Fermentation is of two main types: alcoholic and lactic acid fermentation (Fig. 7.11). Lactic acid fermentation can be homolactic, producing only lactic acid as the fermentation

product, for example, *Lactobacillus* or heterolactic, producing a mixture of lactic acid, ethanol and/or acetic acid and CO_2, for example *Leuconostoc*. Lactic acid is also produced by normal microbiota of human body, thereby lowering pH of certain body parts and preventing the growth of pathogens for example in gastrointestinal tract and vagina. Alcohol fermentation is characterised by decarboxylation of pyruvic acid to acetaldehyde, which is then reduced to ethanol. This type of fermentation is used for producing alcoholic beverages and biofuels.

The comparison of effectiveness of the three types of heterotrophic pathways in terms of number of ATP molecules produced is shown in Table 7.7.

TABLE 7.7   Comparison of Total Number of ATP Molecules Produced by Different Metabolic Pathways

Type of Metabolism	Final Electron Acceptor	Pathway	Maximum ATP Produced
Aerobic respiration	O_2	Glycolysis	2
		Krebs cycle	2
		ETS and OP	34
		Total	38
Anaerobic respiration	Inorganic compounds (SO_4^{2-}, NO_3^-)	Glycolysis	2
		Krebs cycle	2
		ETS and SP	<32
		Total	5–36
Fermentation	Organic compound (pyruvate)	Glycolysis	2
		Fermentation	0
		Total	2

OP = oxidative phosphorylation; SP = substrate level phosphorylation.

8

General Principles of Laboratory Diagnosis of Bacterial Pathogens

Amita Jain

LEARNING OBJECTIVES

- Common methods for diagnosis of bacterial infections:
  - Rapid tests
  - Smear examination
- Culture of bacterial pathogens
- Serological assays
- Nucleic acid amplification tests

INTRODUCTION

Clinical manifestations of bacterial infection are non-specific, hence they may require the use of supportive bedside or confirmatory laboratory tests to confirm or exclude the diagnosis. The laboratory diagnosis of bacterial infections is usually done by direct or indirect demonstration of aetiological agents in clinical specimen. Clinical microbiology laboratories demonstrate pathogens and determine the antimicrobial resistance in bacterial pathogens. Detection of pathogenic agents is done usually by using one of the following tests (Flowchart 8.1):

- Microscopic visualisation of pathogens in clinical specimen.
- Culture of microorganisms in artificial culture media/cell line/animal models.
- Identification of bacteria based on phenotypic and biochemical characteristics.
- Detection of bacterial antigens and/or specific antibodies.
- Detection of bacterial nucleic acid for diagnosis, quantification and identification.
- Typing of bacterial strains.

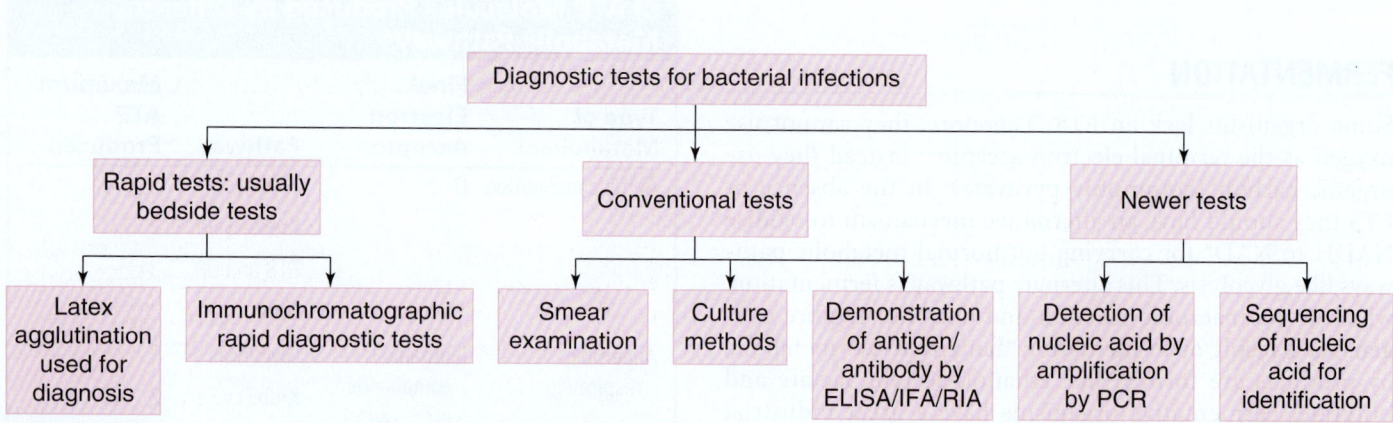

Flowchart 8.1 Tests for Diagnosis of Bacterial Infections.

APPROACH TO A CASE OF BACTERIAL INFECTION

Clinical Diagnostics

Full clinical evaluation with identification of localising signs and symptoms is important to make a clinical diagnosis of bacterial infection.

The laboratory tests, as shown in Flowchart 8.1, are discussed below.

Rapid Bedside Tests

Some of the examples which provide rapid bedside diagnosis are listed in Box 8.1. These tests are rapid and useful for bedside diagnosis. These tests do not require big laboratory infrastructure and usually have high sensitivity. However, for confirmation of a clinical diagnosis a more specific laboratory test is usually advocated.

Laboratory Tests

Laboratory tests which are done to diagnose bacterial diseases can be divided as the following:

Tests to Confirm Inflammation

1. **Peripheral white cell count (WCC):** is a non-specific indicator of infection, usually elevated in bacterial infections.
2. **C-reactive protein (CRP)** is a non-specific test for inflammation. It is positive in the presence of inflammation; however, it is not specific enough to differentiate bacterial from nonbacterial causes.
3. **Procalcitonin (PCT)** is more specific for bacterial infection; however, it is very costly.

Demonstration of Bacterial Pathogens in Stained/Unstained Smears by Microscopy

The microscopic examination of specimens provides useful diagnostic information. Staining techniques provide clarity in examination.

1. **Wet mount** is the simplest method. Wet mounts can be examined stained as well as unstained.
2. **Gram's stain** differentiates between Gram-positive organisms with thick peptidoglycan cell walls (retain the basic dye within bacterial cell) and Gram-negative organisms with thin peptidoglycan cell walls (basic dye gets dissolved in acetone and stain is washed off). Based on Gram's stain

TABLE 8.1   Common Special Stains Used in Bacteriology

Bacteria/Bacterial Part	Stain
Mycobacteria	Acid-fast stain/auramine–rhodamine stain
Treponema/Leptospira	Silver impregnation stains
Bacterial capsule	India ink preparation
Bacterial flagella	Leifson flagella stain
Granules in corynebacteria	Albert's stain
Spores in bacteria	Schaeffer–Fulton method

and cellular morphology, organisms are classified into groups, for example Gram-positive organisms in cluster or Gram-negative rods.

3. **Special stains** such as acid-fast stain, auramine–rhodamine combination fluorescent dye technique, Albert's stain, capsular stains, flagellar stains, silver impregnation stains etc. are done to demonstrate bacteria or their cellular structures which are difficult to demonstrate with Gram's stain (Table 8.1).
4. **Immunofluorescent stains** can be done by two methods:
   a. *Direct* immunofluorescent antibody technique uses antibody specific against some bacterial antigens and tagged with a fluorescent compound (e.g. fluorescein).
   b. *Indirect* immunofluorescent antibody technique is a two-step process. First, an unlabelled antibody specific against bacterial antigen (primary antibody) binds that specific antigen. Second, a fluorescein-labelled polyclonal antibody directed against primary antibody (secondary antibody) is used for staining. This method is more sensitive and specific than direct method.

Culture of Bacterial Pathogens in Artificial Culture Media

Culture-based tests are most important standards in bacteriology. Cultures are used in clinical practice to diagnose bacterial infection and decide on antibacterial therapy. As culture results are usually available no sooner than 48 h, mostly empirical antibiotics are given after collection of samples. Methods for culturing bacteria can be conventional, which usually takes longer time, or automated, which are usually faster. To culture bacterial pathogens, an appropriate sample must be inoculated into suitable culture media. Results often depend on the collection and transport process, time of sample collection, choice of test, choice of media, culture conditions and methods used.

Antimicrobial Susceptibility Testing (AST) of Bacteria

It is one of the most important responsibilities of the clinical microbiology laboratory. Results are used for patient care and for monitoring of infection control practices. Results are reported as qualitative results, that is

BOX 8.1   Commonly Used Bedside Tests

The urine dipstick test for diagnosis of urinary tract infection by detection of nitrites released from leucocytes.

Wet film microscopy, e.g. looking at number of pus cells in CSF/urine to suggest diagnosis of pyogenic meningitis/UTI, respectively.

Latex agglutination test, e.g. demonstration of pneumococcal antigen in CSF for diagnosing pneumococcal meningitis.

Immunochromatographic assay, e.g. lateral flow assay for detection of pneumococcal antigen in urine sample of cases of pneumococcal invasive disease.

categorised as susceptible, resistant or intermediate, or as quantitative results, that is *minimal inhibitory concentration* (MIC) of drug. AST is usually done by placing paper disks impregnated with antibiotics on a culture plate inoculated with the bacterial strain under test (**Kirby-Bauer or disk/agar diffusion method**). Results are read as measurement of the zones of growth inhibition after incubation. Another method is **breakpoint method**, which is based on use of broth cultures containing a set concentration of antibiotic. Test strain of bacteria is inoculated into a series of broth cultures (or agar plates) with increasing concentrations of antibiotic. The lowest concentration of antibiotic that inhibits visual microbial growth is known as the *minimal inhibitory concentration* (MIC). A newer version of disk/agar diffusion method uses a quantitative diffusion gradient of antibiotic on absorbent strip along its length (**epsilometer test [E-test]**). When the strip is placed on the surface of an agar inoculated with test strain, antibiotic diffuses into the medium, and bacterial growth is inhibited. This test also reports MIC.

Identifying an Antigen

Antigen testing is sometimes preferred over culture in situations when organism under consideration is:

1. Too fastidious to grow
2. Too dangerous to grow
3. May take too long to grow
4. Bacteria are not cultivable, by the time patient reports in the hospital

Antigen tests may have low sensitivity compared with culture. They are also unable to provide antimicrobial susceptibility results. Several ELISA platforms are available for the purpose. These are usually rapid and inexpensive methods for identifying organisms, extracellular toxins and bacterial antigens. Such assays may be performed directly on clinical samples or on culture isolates.

Demonstration of Antibodies Against Pathogen

Serological tests are sometimes used to diagnose infections caused by bacteria that are difficult to culture or take too long to culture. Serum or body fluids of host may have antibodies against the pathogen. These antibodies can be detected in platforms like ELISA and make a basis of diagnosis. However, these tests may show high cross-reactivity among pathogens and may have low specificity. Moreover, antibodies usually appear a week or so, after infection; hence, these tests may be false-negative in initial phase of illness. As seen in endemic areas, individuals may have pre-existing antibodies against pathogen, as a result of previous exposure. This may result in false-positive test reaction.

Nucleic Acid-Based Tests

Techniques for the detection and quantitation of specific DNA and RNA base sequences of culture isolates for identification and in clinical specimens for diagnosis are most often used. These tests are used for (1) detection and quantification of pathogens in clinical specimens, (2) identification of

---

> **BOX 8.2   Advantages of Nucleic Acid-Based Tests**
>
> - Rapid
> - Generally specific and highly sensitive
> - Multiplexing, that is testing for more than one organism in same assay is possible
> - Both qualitative and qualitative assays possible
> - Can be useful for diagnosis and for monitoring response to treatment

organisms (usually bacteria), (3) strain relatedness testing and (4) predicting the sensitivity of organisms to chemotherapeutic agents. These methods detect organism-specific DNA or RNA sequences extracted from the microorganism, with certain advantages mentioned in Box 8.2.

Tests using unamplified DNA: There are probes available which can identify microbial DNA in a sample, even when unamplified. Sample should have high bacterial load. Bacterial isolates grown in culture can also be identified with probes.

Test using nucleic acid amplification techniques (NAAT): NAAT is a technique where a very small amount of DNA or RNA present in a clinical sample (theoretically a single copy) is amplified several fold to make it detectable using a detection system. Culture may be avoided. NAAT is useful for organisms that are fastidious to culture or not cultivable, takes too long to grow or is too risky to grow. There are many platforms present. Few are listed in Box 8.3.

Amplification methods are highly sensitive and false-positive results may occur due to contamination. Test needs a special attention. Laboratories need special designing to avoid contamination. Every test step is quality controlled and needs technical expertise.

False-negative results sometimes occur. To avoid false-negative results, right sample should be collected at right time in right manner. For example, swabs with wooden shafts or cotton tips should be avoided. Sample should reach laboratory as soon as possible. Minus 20°C (transport on dry ice) is best temperature to transport if delay is unavoidable.

> **BOX 8.3   Platforms Used for NAAT**
>
> - PCR
> - Reverse transcriptase-PCR (RT-PCR)
> - Real-time PCR
> - Nested PCR
> - Multiplex PCR
> - Strand displacement amplification
> - Signal amplification (e.g. branched DNA assays, hybrid capture)
> - Post-amplification analysis (e.g. sequencing of the amplified product, microarray analysis and melting curve analysis, as is done in real-time PCR)

Typing Methods

These methods are used for identification and typing of bacterial isolates.

1. **Phenotypic typing** characters such as colony size, shape, colour, culture properties, odour, microscopic appearance and so on may suggest the identity of pathogen. It may be confusing at times.

2. **Biotyping**: It is still the most common method in most of the conventional laboratories. Production of specific enzymes, ability to metabolise specific sugars or proteins or the production of certain metabolites make the basis of biotyping. Normally, these methods take long, but currently some rapid manual and automated methods have become available. These systems are based on growing these isolates on multiple substrates, and the reaction pattern is compared with known patterns for various bacterial species.

3. **Serotyping** of bacteria is done by the presence or absence of surface antigens using antisera.

4. **Gas–liquid chromatography** is used to detect metabolic end products of bacterial fermentations. This method is mostly used for identifying anaerobic bacteria. Both automated and manual platforms are available.

5. **Nucleic acid-based typing**: We can test a sample for the presence of a bacteria or virus by looking at its DNA, which is unique for every organism. DNA and RNA base sequences of isolates or bacterial strains in clinical specimens can be detected for the identification of these organisms by several methods.

Methods for nucleic acid-based typing

   a. Pulsed-field gel electrophoresis is the gold standard for bacterial strain typing. It uses restriction enzymes that recognise unique sequences of nucleotides to cut bacterial DNA at specific points, resulting in large DNA fragments. These fragments are separated by gel electrophoresis and then are visualised. Similar band patterns (i.e. differences in ≤3 bands) suggest that different bacterial isolates are closely related, or clonal.

   b. DNA sequencing of single or multiple genes.

   c. PCR-based amplification of repetitive DNA sequences in the bacterial chromosome.

6. **Phage tying**: Bacteriophage (phage) is a virus that infects bacteria. Phages are bacteria specific, meaning thereby that each type of Phage only attacks a particular type of bacteria. This method is most commonly used for strain typing of staphylococci.

9

Gram-Positive Cocci

Parul Jain

LEARNING OBJECTIVES

- Definition, epidemiology, cultural characteristics, identification, antibiotic resistance, pathogenesis, diseases and treatment:
  - *Staphylococcus* spp.

- *Streptococcus* spp.
- Gram-positive anaerobic cocci

INTRODUCTION

Gram-positive cocci (GPC) are spherical bacteria (cocci) that appear purple after Gram staining (Gram-positive) and are arranged in pairs, chains or clusters. Medically, important

GPC are *Staphylococcus* spp., *Streptococcus* spp., *Enterococcus* spp. and anaerobic cocci. The algorithmic approach to differential diagnosis of GPC is shown in Flowchart. 9.1.

The salient differences between GPC are listed in Table 9.1.

STAPHYLOCOCCUS SPECIES

DEFINITION

Staphylococci are GPC arranged in clusters (Fig. 9.1). They are non-spore forming, non-motile, usually catalase-positive, not capsulated and facultative anaerobic bacteria.

EPIDEMIOLOGY

Staphylococci are seen worldwide and colonise skin and mucosa. Nasal carriage in healthy individuals varies from 10% to 40%. They can also be recovered from axilla and perineum and have also been reported from oral cavity.

CULTURAL CHARACTERISTICS

Staphylococci can grow on nutrient agar, blood agar, MacConkey agar and liquid media. After 24 h of incubation on nutrient agar, they produce **large, circular, smooth, convex, opaque, easily emulsifiable, yellow (*Staphylococcus aureus*) or white colonies**. On sheep or horse blood agar, *S. aureus* usually **produces haemolysis**, especially when incubated in the presence of CO_2. The colonies are small and pink on MacConkey agar, and they produce a uniform turbidity in liquid medium.

IDENTIFICATION

They are **catalase-positive, MR- and VP-positive,** and reduce nitrates to nitrites. On the basis of coagulase test, *Staphylococcus* species are divided into (1) **coagulase-positive species,**

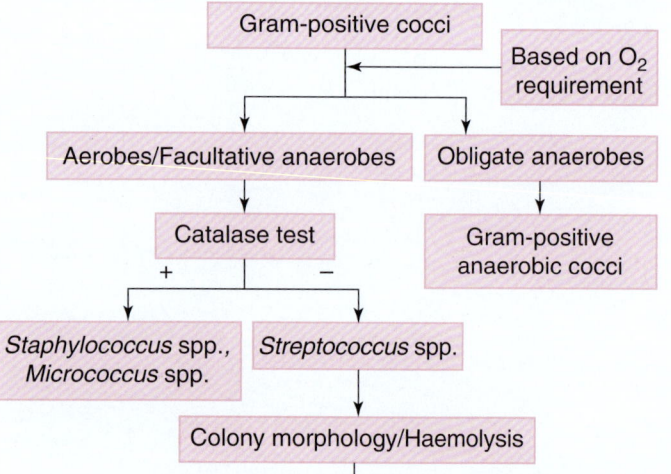
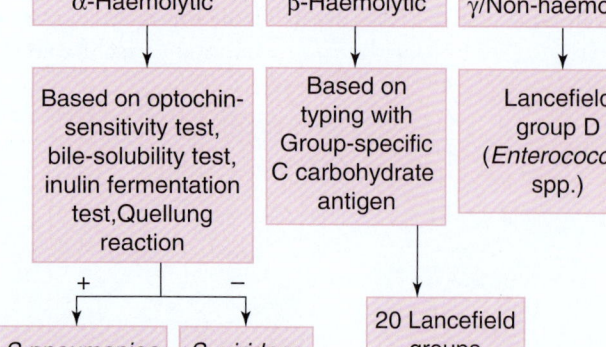

Flowchart 9.1 Algorithm for Differential Diagnosis of Gram-Positive Cocci.

TABLE 9.1	**Salient Features of Common Medically Important Gram-Positive Cocci**			
	Staphylococcus	*Streptococcus*	*Enterococcus*	**Anaerobic Cocci**
Morphology (Fig. 9.1)	Cocci are about 1 µm in size, arranged in pairs, tetrads, short chains and characteristic grape-like clusters	Cocci are 0.5–1 µm in size, arranged in chains	Oval in shape, seen as single cells, pairs, short chains	Cocci or coccobacilli in pairs, short chains, clumps or clusters
O₂ Requirement	Facultative anaerobes	Facultative anaerobes	Facultative anaerobes	Obligate anaerobes
Medically important species	*S. aureus, S. lugdunensis, S. epidermidis, S. haemolyticus, S. saprophyticus*	*S. pyogenes, S. mutans, S. sobrinus, S. pneumoniae*	*E. faecalis, E. faecium*	*Anaerococcus prevotii, Peptoniphilus asaccharolyticus, Finegoldia magna, Peptostreptococcus anaerobius, Parvimonas micra*
Diseases caused	Skin and soft tissue infections, bloodstream infections, infectious endocarditis, meningitis, pericarditis, pulmonary infections, osteomyelitis, septic arthritis, septic bursitis and pyomyositis, oral infections	Suppurative complications, endocarditis, dental caries, otitis media, bacteraemia	Bacteraemia and endocarditis, UTI, meningitis, intra-abdominal and pelvic infections, neonatal infections, skin and soft tissue infections	Odontogenic infections, oral infections, infections of abdominal cavity, bone, skin and soft tissue infections, bacteraemia, pleural empyema
Antimicrobial resistance	*S. aureus* has developed resistance to almost all classes of drugs. It is essential to identify methicillin-resistant *Staphylococcus aureus* (MRSA)	Though usually sensitive to penicillins but resistance is being increasingly reported	Enterococci are intrinsically resistant to several drugs. Of the drugs available for treatment, vancomycin-resistant enterococci (VRE) and enterococci with high-level resistance (HLR) to aminoglycosides pose clinical problems	GPAC possess variable resistance to penicillins, metronidazole, clindamycin, tetracycline and erythromycin

HLR = high-level resistance; SBA = 5% sheep blood agar; UTI = urinary tract infection; VRE = vancomycin-resistant enterococci; GPAC = Gram-positive anaerobic cocci.

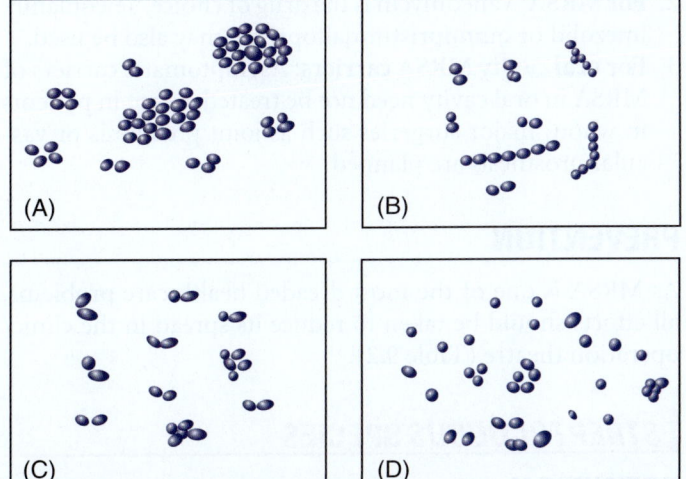

Fig. 9.1 Morphology of (A) *Staphylococci* spp., (B) *Streptococcus* spp., (C) *Enterococcus* spp. and (D) Gram-positive anaerobic cocci (diagrammatic representation).

for example, *S. aureus,* and (2) coagulase-negative *Staphylococci* (CONS), for example, *Staphylococcus epidermidis, Staphylococcus saprophyticus, Staphylococcus lugdunensis* and *Staphylococcus haemolyticus.*

Molecular Methods Available for Identification

Several molecular methods are available for rapid identification (results within few hours), antibiotic resistance determination and strain typing. The methods include molecular typing, pulsed field gel electrophoresis (PFGE), multilocus sequence typing (MLST) and so on.

PATHOGENESIS

S. aureus has several pathogenic factors in its armamentarium. They include the following:

1. Cellular structures like capsule, peptidoglycan layer and surface adhesins.
2. Secreted enzymes and toxins like haemolysins, Panton–Valentine toxin, enterotoxins, staphylococcal toxic shock

syndrome toxin (TSST-1) and exfoliative toxin. Enterotoxins and TSST-1 are superantigens (cause massive non-specific stimulation of the immune system).

ANTIBIOTIC RESISTANCE

Resistance to β-Lactams

Most of the clinical isolates of *S. aureus* are resistant to penicillin group of drugs. *S. aureus* exhibits two main mechanisms of resistance to β-lactam drugs:

1. By producing penicillinase (β-lactamase) enzyme that hydrolyses penicillin and other penicillinase-susceptible compounds into inactive penicilloic acid. It is an inducible enzyme encoded by a plasmid gene.
2. By producing altered penicillin-binding protein (PBP2A) that has reduced ability to bind the β-lactam drugs. The *mecA* gene responsible for producing PBP2A is present on chromosome. It confers high-level intrinsic resistance to all β-lactams, including penicillins, penicillin derivatives, cephalosporins and carbapenems. Such strains are called methicillin-resistant *S. aureus* (MRSA), which can be hospital acquired (HA-MRSA) or community acquired (CA-MRSA). In laboratory, cefoxitin is used as a surrogate marker for detecting MRSA. The other strains that are sensitive to β-lactams are called methicillin-sensitive *S. aureus* (MSSA).

DISEASES CAUSED

S. aureus can cause pyogenic and toxin-mediated diseases.

Pyogenic Diseases

1. *Systemic infections*: Skin and soft tissue (e.g. folliculitis, carbuncle, furuncle, impetigo), respiratory (e.g. pharyngitis, tonsillitis, empyema), CNS (e.g. meningitis, abscess), endovascular (bacteraemia, septicaemia, endocarditis) and musculoskeletal (e.g. osteomyelitis, arthritis) infections.
2. *Oral infections*: Angular cheilitis, jaw cysts, acute dentoalveolar infections, oral mucosal lesions, denture-induced stomatitis and so on.

Toxin-mediated Diseases

Staphylococcal food poisoning, toxic shock syndrome and staphylococcal scalded skin syndrome.

Diseases of Oral Cavity

MRSA has been found in samples from implant infections, acute parotitis, acute mucositis, dental abscesses, dental stomatitis and angular cheilitis. Cross infection has also been reported from dentist to patient in cases with dental abscesses.

Oral Cavity as a Source of Systemic Infections

Staphylococcus spp. are also carried in healthy oral cavity (with or without nasal cavity) that act as a source of infection for systemic diseases, particularly in immunocompromised

individuals, for example, studies have reported endocarditis by *S. lugdunensis* after tooth extraction and bacteraemia by *S. epidermidis* and *Streptococcus oralis* in children receiving treatment for leukaemia. MRSA has also been detected on oral dentures, which may serve as a source of infection to others or as a reservoir for colonising other body sites.

LABORATORY DIAGNOSIS

The laboratory diagnosis is shown in Flowchart 9.2.

TREATMENT

1. **For MSSA**: The antibiotics of choice are penicillinase-resistant penicillins, for example, flucloxacillin and dicloxacillin. In case of penicillin hypersensitivity, clindamycin, lincomycin or erythromycin may be used.
2. **For MRSA**: Vancomycin is the drug of choice. Teicoplanin, linezolid or quinupristin/dalfopristin may also be used.
3. **For oral cavity MRSA carriers**: Asymptomatic carriers of MRSA in oral cavity need not be treated except in patients in whom major surgeries such as joint prosthesis or vascular prosthesis are planned.

PREVENTION

As MRSA is one of the most dreaded health care problems, all efforts should be taken to reduce its spread in the clinic/operation theatre (Table 9.2).

STREPTOCOCCUS SPECIES

DEFINITION

Streptococci are GPC arranged in pairs or chains (Fig. 9.1). They are catalase-negative, non-motile, non-spore forming,

---

Sample collection: representative samples are collected, e.g., swabs or aspirates in abscess, sputum or throat swab in respiratory infections

↓

Direct smear from sample: GPC in clusters

↓

Culture on blood agar and nutrient agar: characteristic colonies

↓

Identification by Gram stain and biochemical tests

↓

Antibiotic susceptibility test including cefoxitin disc test for MRSA

Flowchart 9.2 Laboratory Approach for Diagnosis of Staphylococcal Infections.

TABLE 9.2 Sources and Preventive Methods for Staphylococcal Dental Infections

Source of Infection	Possible Prevention
Aerosols generated by processes such as ultrasonic scaling, use of air–water syringe, abrasive devices, or by high-speed rotation	• Using high-volume evacuators • Pre-procedural rinse with chlorhexidine mouthwash • Use of rubber dam • Reducing number of small devices on table surfaces to a minimum • Keeping the instrument drawers closed
Cross contamination via indirect contact	• Wearing proper personal protective equipment • Hand hygiene • Efficient surface disinfection • Changing towels, washcloths etc. in between patients and washing soiled linen in hot water with bleach followed by drying in hot dryer
In case of oral cavity or nasopharyngeal carriers	• Treatment with local mupirocin • Heat sterilisation and daily disinfection of dentures during the course of treatment

nutritionally fastidious and facultative anaerobic bacteria. Some strains are capsulated.

CLASSIFICATION

1. **On the basis of haemolysis produced on blood agar**: *Streptococci* are divided into; (1) alpha (α)-haemolytic, causing partial haemolysis resulting in greenish discolouration around the colonies, for example, *Streptococcus pneumoniae* and viridans *Streptococcus*; (2) beta (β)-haemolytic, causing complete haemolysis resulting in complete clearing around the colonies, for example, *Streptococcus pyogenes* and (3) gamma (γ)-haemolytic, causing no haemolysis, for example, *Enterococci*.

The oral viridans *Streptococci* can be divided into five groups (Table 9.3).

TABLE 9.3 Predominant Oral Viridans Streptococci

Oral Viridans Streptococci Group	Predominant Species
Mutans group	S. mutans, S. sobrinus
Salivarius group	S. salivarius
Anginosus group	S. anginosus, S. intermedius
Sanguinis group	S. sanguinis, S. gordonii
Mitis group	S. mitis, S. oralis

The *Enterococci*/faecal *Streptococci* have been reclassified as a different genus *Enterococcus* and contain several species such as *Enterococcus faecalis*, *Enterococcus faecium* and *Enterococcus durans*.

2. **Lancefield grouping** of β-haemolytic *Streptococci* is done on the basis of serological reaction with carbohydrate (C) antigen on cell wall. Twenty Lancefield groups are known of which group A (*S. pyogenes*), group B (*Streptococcus agalactiae*), group D (*Streptococcus bovis*) and some others are clinically important.

EPIDEMIOLOGY

S. pyogenes

It is carried in the nasopharynx and skin of up to 20% people. The infection can occur at any age, though it is most common in school-aged children. Winter season and overcrowding facilitate its transmission.

Viridans Streptococci

They are normal inhabitants/commensals of mouth, pharynx and intestine. Species like *Streptococcus gordonii*, *Streptococcus sanguinis* and *Streptococcus oligofermentans* are early colonisers of oral cavity, while *Streptococcus mutans* and *Streptococcus sobrinus* are late colonisers. They are usually harmless, but may cause disease under special circumstances like an increase in dietary carbohydrates, particularly sucrose, which alters the whole microenvironment of mouth. *S. mutans* is the most widely studied species.

S. pneumoniae

It is the most common cause of pneumonia worldwide. The bacteria are carried in the nasopharynx of people (5%–10% in healthy adults, 20%–40% in healthy children) and spreads via droplet infection. Risk factors for acquiring pneumococcal disease are ages <2 years or >65 years, asplenia, cigarette smoking and chronic heart disease. On the basis of capsular polysaccharide, 93 different serotypes are known of which types 6B, 9V, 14, 19F and 23F are most common in young children and type 1 is most virulent.

Enterococcus spp.

They are normal inhabitants of intestinal and genital tract and transiently colonise the oral cavity. Virulent, antibiotic-resistant *E. faecalis* strains may be present in oral cavity that may act as a critical reservoir.

CULTURAL CHARACTERISTICS

Streptococcus spp. are exacting in its nutritional requirements and cannot grow on nutrient media. On blood agar, they form colonies 1–2 mm in size, white to grey in colour, circular, surrounded by a zone of α-, β- or γ-haemolysis. Growth and haemolysis are enhanced by 5%–10% CO_2. Colonies may be mucoid because of the presence of capsule. *Enterococcus* spp. produce tiny pink colonies on MacConkey medium, and are usually non-haemolytic on blood agar.

IDENTIFICATION

S. pyogenes

It produces β haemolysis on blood agar, is Lancefield group A positive, is sensitive to 0.04 U of bacitracin, gives positive reaction for pyrrolidonyl-beta-naphthylamide (PYR) hydrolysis and does not ferment ribose sugar.

S. mutans

It may produce α, β or γ haemolysis, is VP-positive, hydrolyses esculin and produces acid from most sugars like mannitol, sorbitol, lactose, trehalose, inulin and raffinose.

S. pneumoniae

It produces α haemolysis on blood agar, is bile soluble, ferments inulin and is sensitive to optochin. It also differs from other *Streptococci* in being lanceolate diplococci, capsulated and in producing dome shaped colonies initially that later become draughtsman colonies (umbilicated, due to autolysis of bacterial cells in the center of colony).

Enterococcus spp.

Some unique features include growth in the presence of 6.5% sodium chloride, 40% bile, high alkalinity (at pH 9.6), high temperature (up to 45°C); ability to ferment mannitol, sucrose and sorbitol; and to give positive bile esculin test.

PATHOGENESIS AND VIRULENCE FACTORS

The pathogenesis and virulence factors of important *Streptococcal* species are discussed below.

S. pyogenes

It has several somatic constituents and extracellular products that play a role in pathogenesis (Box 9.1).

BOX 9.1   Virulence Factors Found in *S. pyogenes*

Somatic constituents
- Capsule (made of hyaluronic acid)
- M protein present in the cell membrane
- Serum opacity factor
- Cell wall adhesins

Extracellular products: It is because of these extracellular enzymes that *S. pyogenes* causes spreading infections in contrast to staphylococcal infections that are usually contained in a space.
- Haemolysins, which are of two types: Streptolysin O (oxygen labile toxin causing haemolysis in anaerobic cultures or in subsurface colonies) and streptolysin S (oxygen stable toxin causing haemolysis on blood agar plate surface colonies)
- Deoxyribonucleases (DNases or DNA degrading enzymes)
- Hyaluronidase (degrades connective tissue hyaluronic acid)
- Streptokinase (causes clot dissolution)
- Streptococcal exotoxin B (protease)
- Streptococcal pyrogenic exotoxins (causing necrotising fasciitis or streptococcal toxic shock syndrome)

Viridans Streptococci in Dental Plaque and Caries

In the formation of dental plaques, salivary molecules containing glycoproteins, mucins and so on adsorb to the tooth enamel. This provides a medium for biofilm formation by oral bacteria primarily *S. sanguis* and *Actinomyces viscosus*. Other bacterial species present in mouth and pharynx, such as *S. mutans* and *S. sobrinus*, attach to these biofilms resulting in bacterial growth on tooth surface or the dental plaques. *S. mutans* and *S. sobrinus* are acidogenic and produce extracellular polysaccharides (EPS) in the presence of sugars like sucrose, fructose and glucose. These EPS play a vital role in cariogenicity by *S. mutans*. Thus, caries is principally a disbalance of oral bacterial ecology and is polymicrobial in nature, predominated by *S. mutans* and certain lactobacilli.

Pathogenic mechanisms of *S. mutans* are as follows:
- Produces lactic acid as a part of its metabolism.
- Produces bacteriocins called mutacins that are capable of (1) inhibiting the growth of other competitive bacteria, and (2) causing leakage of cell contents of other bacteria by targeting membrane of the susceptible species.
- Produces water-insoluble glucans that help it to bind to tooth surfaces.

S. mutans can be transmitted by the following:
- Vertical transmission: From mother to child shortly after birth.
- Horizontal transmission: By frequent close contact, for example day care centre, within family and so on.

S. pneumoniae

The capsular polysaccharide, pneumolysin (a membrane-damaging toxin) and autolysins contribute to virulence of *S. pneumoniae*.

Enterococcus spp.

They have several virulence factors related to dentine adhesion and biofilm formation, including secretory metalloprotease gelatinase E gene, collagen adhesion protein, aggregation substance, cytolysin and so on. Some strains also produce a capsule that has anti-phagocytic properties.

ANTIBIOTIC RESISTANCE

S. pyogenes

Penicillin resistance has not been reported till now, though resistance to azithromycin and clarithromycin occurs in some communities.

S. mutans

There are reports stating resistance of *S. mutans* to quaternary ammonium monomers (QAMs) that are incorporated in dental resins.

S. pneumoniae

An increasing resistance of *S. pneumoniae* to β-lactam antibiotics is being reported, which has been attributed to its ability to acquire genetic material from other coexisting bacteria. Resistance has also been reported to macrolides, clindamycin, trimethoprim–sulphamethoxazole, doxycycline and quinolones.

Enterococcus spp.

They are intrinsically resistant to cephalosporins, clindamycin (drug frequently used in severe oral infections) and exhibit low-level resistance to aminoglycosides. Therefore, enterococcal infections are never treated by cephalosporins or alone with aminoglycosides. Some strains also exhibit high-level resistance (HLR) to aminoglycosides. Recently, vancomycin-resistant enterococci (VRE) have appeared that pose a significant health care problem. Strains from oral cavity can be resistant to conventional dental treatment regimens including tetracycline and erythromycin.

DISEASES CAUSED

S. pyogenes

The diseases caused include streptococcal pharyngitis, scarlet fever, suppurative complications like peritonsillar cellulitis or abscess, mastoiditis, acute sinusitis, otitis media and so on. Non-suppurative complications of *S. pyogenes* infection include acute rheumatic fever and acute post-streptococcal glomerulonephritis.

Viridans Streptococci

They are usually harmless residents of oral cavity. But species like *S. gordonii*, *S. sanguinis* and *S. oligofermentans* may cause endocarditis when disseminated through blood stream. Viridans *Streptococci* have also been implicated in bacteraemia, pneumonia or meningitis. *S. mutans* has a well-established role in caries or periodontal disease. It may also cause pyogenic infections at other sites such as mouth, heart, joints, skin and muscle. Persistent *S. mutans* colonies in oral cavity at a young age can result in lifelong dental problems and also its role has been suggested in older age problems like high blood pressure, multiple sclerosis, diabetes or heart disease.

S. pneumoniae

It can cause mild respiratory tract infections like otitis media and sinusitis to severe diseases like meningitis, pneumonia, septicaemia and acute exacerbation of chronic obstructive pulmonary diseases.

Enterococcus spp.

They can cause several systemic infections such as urinary tract infections, endocarditis, septicaemia, wound infection, biliary tract infection and intra-abdominal abscess. VREs causing nosocomial infections are of considerable concern and can cause life-threatening infections. *E. faecalis* has also been implicated in several oral diseases such as posttreatment apical periodontitis, endodontic infections, peri-implantitis, caries etc. It has often been seen to result in

---

Sample collection: representative samples are collected, e.g. throat or pus swab, sputum, CSF

↓

Direct smear from sample: GPC in pairs or short chains. For pneumococcus, Gram-positive diplococci can be seen intracellularly and extracellularly

↓

Culture on blood agar and chocolate agar: characteristic colonies appear after incubation at 37°C under 5%–10% CO_2.

↓

Identification by Gram staining and biochemical tests

↓

Antibiotic susceptibility test

Flowchart 9.3 Laboratory Diagnosis of *Streptococci*.

---

failure of endodontic treatment because (1) of its ability to cause persistent infections and form biofilms in treated and untreated root canals, and (2) high resistance to medications used in endodontic treatment.

LABORATORY DIAGNOSIS

The laboratory diagnosis is shown in Flowchart 9.3.

Other tests: For *S. pneumoniae*, other tests are also used, for example blood culture, and antigen detection in blood, urine or CSF.

TREATMENT

- The treatment of choice for streptococcal infections is penicillin or amoxycillin. In patients with penicillin allergy, first- and second-generation cephalosporins (cephalexin or cefadroxil), clindamycin, azithromycin or clarithromycin can be used.
- *For early childhood caries and other dental diseases*, treatment is to fill each cavity and educate families regarding dental hygiene and proper dietary practices.
- The choice of treatment for enterococcal infections is a combination therapy with penicillin and aminoglycosides. HLR enterococci and VRE are not the usual problems associated with oral enterococci, but if present can be treated, respectively, with vancomycin and linezolid or quinupristin/dalfopristin.

Prophylaxis: Infections with oral *Streptococci* can be prevented by the following:

- By maintaining proper oral health.
- The risk of having postsurgical enterococcal infections can be minimised by observing adequate asepsis, use of irrigants, properly sterilised instruments, chlorhexidine mouthwashes and other disinfectants.

GRAM-POSITIVE ANAEROBIC COCCI

DEFINITION

Gram-positive anaerobic cocci (GPAC) are a heterogeneous group of obligate anaerobic, non-spore forming cocci or coccobacilli, occurring in pairs, tetrads, short chains or clumps or irregular masses. The individual coccus varies from 0.3 to 2 mm depending on the species.

CLASSIFICATION

GPAC has been divided into various genera: (1) *Anaerococcus* (type species *A. prevotii*), (2) *Peptoniphilus* (type species *P. asaccharolyticus*), (3) *Finegoldia* (type species *F. magna*), (4) *Peptostreptococcus* (type species *P. anaerobius*), (5) *Parvimonas* (only species *P. micra*) and (6) *Ruminococcus*.

EPIDEMIOLOGY

GPAC constitute a major portion of the normal microbiota colonising the skin and mucosa of oral cavity, upper respiratory tract, female genitourinary tract and the gastrointestinal tract.

CULTURAL CHARACTERISTICS

GPAC are usually cultured on fastidious anaerobic agar or on anaerobically incubated blood agar usually for 48 h but may take up to 7 days. They are difficult to isolate because of their oxygen sensitivity, slow growth and their general association with polymicrobial infections.

IDENTIFICATION

It is based on morphological appearance, Gram-stain appearance and sensitivity to metronidazole. Certain biochemical tests like inhibition by sodium polyanethol sulphonate (SPS), nitrate reduction, urease production, positive indole test, pigment production and proteolytic enzyme profile analysis can be done for species or genus identification. Gas–liquid chromatography-based volatile fatty acid detection and carbohydrate fermentation can also be used for such identification. Today, several commercial biochemical assays are available in the market for rapid detection and identification of GPAC. MALDI-TOF MS (matrix-assisted laser desorption/ionisation time-of-flight mass spectrometry) can be used to rapidly identify the GPAC isolated from patients' specimens.

Molecular Identification

GPAC can be identified by molecular techniques such as PCR, multiplex PCR, 16S rRNA gene sequencing and pyrosequencing.

PATHOGENESIS AND VIRULENCE FACTORS

Several virulence factors have been identified in GPAC, though several still remain unidentified. The most well studied are protein L, PAB, SufA, FAF and enzymes like collagenase, gelatinase, hippurate hydrolase and so on.

ANTIBIOTIC RESISTANCE

GPAC possess variable resistance to penicillins, metronidazole, clindamycin, tetracycline and erythromycin.

DISEASES CAUSED

GPAC have been known to cause infections of abdominal cavity, bone, skin and soft tissue infections; chronic wound infections; bacteraemia; pleural empyema and leg ulcers.

Oral Infections

Species like *Peptostreptococcus* and *Parvimonas* have been associated with oral infections; *Atopobium parvulum* has been associated with infections of dental implants and other odontogenic infections and with oral malodour.

LABORATORY DIAGNOSIS

The laboratory diagnosis is shown in Flowchart 9.4.

TREATMENT

GPAC can be treated with penicillins, cephalosporins, carbapenems, chloramphenicol, β-lactam/β-lactamase inhibitor combinations and metronidazole. This shall be supplemented with drainage, debridement of necrotic tissue, or both.

Specimen collection: pus or fluid aspirate in anaerobic transport medium is the sample of choice. Swabs are unacceptable. Sample sent immediately to the laboratory in a capped syringe after expulsion of air can also be accepted

↓

Gram stain and culture on Brucella, Columbia, Schaedler or any other anaerobic agar base supplemented with vitamin K, 5% sheep blood and hemin

↓

Identification by Gram staining and other methods

↓

Antibiotic susceptibility test

Flowchart 9.4 Laboratory Diagnosis of *Gram-Positive Anaerobic Cocci*.

Gram-Positive Bacilli

Parul Jain

INTRODUCTION

Gram-positive bacilli (GPB) are rod-shaped bacteria that appear purple after Gram staining (Gram-positive). They can be aerobes/facultative anaerobes or obligate anaerobes.

AEROBIC GRAM-POSITIVE BACILLI

The approach to differentiate aerobic bacilli is outlined in Flowchart 10.1.

CORYNEBACTERIUM DIPHTHERIAE

INTRODUCTION

C. diphtheriae (Greek: 'korynee' or club and 'diphthera' or leather hide) is an aerobic non-motile, non-sporulating, nonencapsulated Gram-positive bacillus that has a characteristic V-shaped, L-shaped or 'Chinese letter pattern' morphology (Fig. 10.1). It has four biovars on the basis of colony characteristics, biochemical and haemolytic reactions: gravis, mitis, intermedius and belfanti.

Other *Corynebacteria*, such as *Corynebacterium ulcerans* and *Corynebacterium pseudotuberculosis*, can also at times cause diphtheria-like illness.

EPIDEMIOLOGY

Human-to-human transmission occurs via droplet secretions or via direct contact with respiratory secretions or exudates. Humans are the only reservoir of these bacteria. Owing to widespread vaccinations the incidence of diphtheria has decreased in several parts of the world, but it still remains high in Southeast Asia, South America and Africa.

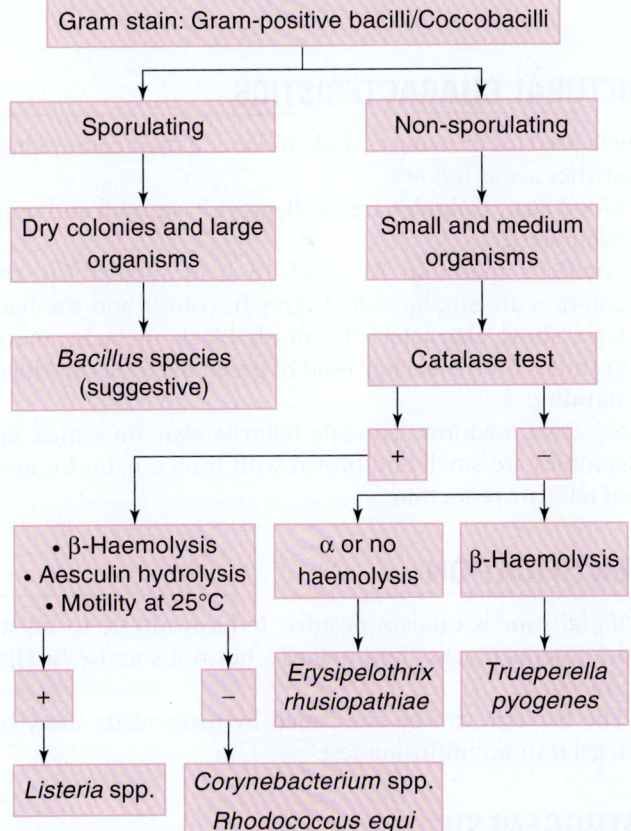

Flowchart 10.1 Algorithm for Diagnosis of Aerobic GPB.

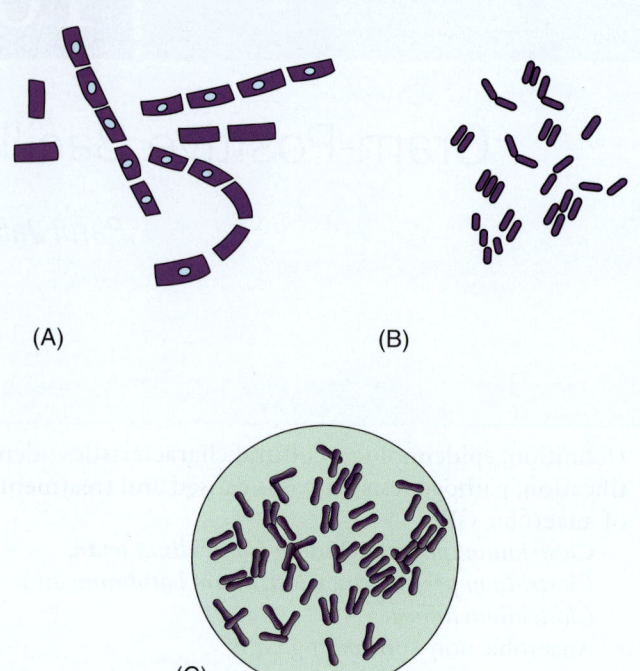

(A)   (B)

(C)

Fig. 10.1 Morphology of GPB. (A) *Bacillus* spp., (B) *L. monocytogenes* and (C) *C. diphtheriae.*

CULTURAL CHARACTERISTICS

Media used for growing *C. diphtheriae* and their colony characteristics are as follows:

1. Blood agar: Colonies are small, greyish, smooth and non-haemolytic.
2. Loeffler's medium: After 12–18 h of incubation, the colonies are small, whitish-grey in colour and the bacteria show characteristic bluish black metachromatic granules on the background of green bacteria on Albert staining.
3. Selective medium: Tinsdale tellurite agar on which the colonies are small and brown with brown halos because of tellurite reduction.

IDENTIFICATION

C. diphtheriae is catalase positive, reduces nitrate to nitrite and ferments glucose and maltose, but not sucrose in Hiss serum water.

The bacteria can be confirmed by toxigenicity assay by Elek gel immunodiffusion test.

PATHOGENESIS

C. diphtheriae produces its pathogenic effect by an exo-toxin. The production of exotoxin is determined by the presence of lysogenic β-phage that carries the tox+ gene for producing toxin. This toxin forms a leather-like adherent grey-brown pseudomembrane that may be restricted to tonsils, pharynx or nose or may extend to involve the tracheobronchial tree.

CLINICAL FEATURES

1. **Asymptomatic carriage**: It is common in endemic regions. Carriers act as an important source of infection for others.
2. **Respiratory tract diphtheria**: Infection may be limited to anterior nasal, faucial regions when patient presents with low-grade fever, sore throat, malaise and dirty grey colour membrane on tonsils, oropharynx and nasopharynx. The infection might have spread to larynx and tracheobronchial tree when child develops hoarseness of voice, dyspnoea, stridor, etc. Later due to effect of toxin, patient may develop cardiotoxicity, neurotoxicity, renal failure, hypotension or pneumonia.
3. **Cutaneous diphtheria**: It is common in tropical climates. Skin gets inflamed with pain, redness and swelling. Ulcers covered by a gray membrane may also develop in this form.

LABORATORY DIAGNOSIS

Presumptive diagnosis of diphtheria is clinical. The suggestive symptoms include fever with painful tonsillitis associated with a membrane, or adenopathy and cervical swelling. The laboratory diagnosis is shown in Flowchart 10.2 However, since diphtheria may be confused clinically with conditions such as Vincent's angina, severe streptococcal throat infection or glandular fever, laboratory diagnosis becomes imperative.

TREATMENT

Diphtheria antitoxin and antibiotics, such as penicillin and erythromycin, are the mainstay of treatment. Mortality rate is 30%–40% in patients with systemic disease.

Sample required: Throat swab, bits of membrane or submembrane swabs transported immediately to lab

Perform following stains on smears prepared from swabs:
1st: Gram stain (shows Chinese letter pattern)
2nd: Albert's stain (may show metachromatic/volutin/Babes–Ernst granules)

Laboratory diagnosis: Chinese letter pattern on Gram stain, characteristic colonies on Tinsdale medium, metachromatic granules with methylene blue stain

Final diagnosis: Biochemical tests, demonstration of toxin production by in vitro Elek precipitation test or molecular tests

Flowchart 10.2 Laboratory Diagnosis of Diphtheria.

TABLE 10.1	Formulations of Diphtheria and Tetanus Vaccines		
Type of Vaccine	Components	Age Group	Schedule
DT	Diphtheria toxoid (6.7–25 Lf), tetanus toxoid (5 –7.5 Lf)	<7 years	• Used in children where pertussis vaccine is contraindicated • 3 doses are given at an interval of 4–8 weeks; 4th dose is given 6–12 months after the 3rd dose. Booster is given at 4–6 years
DTaP	Diphtheria toxoid, tetanus toxoid, acellular pertussis	<7 years	Total 5 doses for primary immunisation at below-mentioned age: • one dose each at 2 months • 4 months • 6 months • 15–8 months • 4–6 years
Tdap	Diphtheria toxoid (<2Lf), tetanus toxoid, acellular pertussis	>7 years (adolescents and adults)	• One dose at 11–12 years or as soon as possible • One dose must be given to all health care workers and pregnant women
Td	Diphtheria toxoid (<2 Lf), tetanus toxoid (5–7.5 Lf)	>7 years (adolescents and adults)	• Booster dose should be given every 10 years • May be taken earlier after a severe dirty burn or wound

aP = Acellular pertussis; *d* = diphtheria for adults; *Lf* = limits of flocculation; *p* = lower concentration of pertussis.

PREVENTION

Diphtheria can be prevented by vaccination. Four types of vaccines are available as shown in Table 10.1.

For children <7 years old: Diphtheria and tetanus (DT) vaccines and diphtheria, tetanus and pertussis (DTaP) vaccines

For older children and adults: Tetanus, diphtheria and pertussis (Tdap) vaccines and tetanus and diphtheria (Td) vaccines.

In India as per Universal Immunization Programme, three doses of DPT are given to all children at 6, 10 and 14 weeks of age followed by boosters at 16–24 months and 5–6 years of age.

OTHER *CORYNEFORM* BACTERIA

Other *coryneform* bacteria that have morphology similar to *C. diphtheriae* include genera *Corynebacterium, Arcanobacterium, Microbacterium, Rothia, Turicella, Dermabacter* and *Arthrobacter*. These are widely distributed in nature and are normal inhabitants of soil and water. They are also found as commensals of skin and mucous membrane. The *coryneform* bacteria are pleomorphic, non-spore forming, variable in size and appearance (coccoid to bacillary forms) and are arranged in 'Chinese letter pattern'.

Coryneform bacteria are often found as contaminants in the clinical samples because of their widespread presence. The conditions where these can be considered significant are mentioned in Box 10.1. They can be cultured on blood agar plate, thioglycollate broth and blood culture media. Species identification can be done on tryptic soy agar (with and without Tween 80). Identification is done based on catalase test, nitrate reduction test, urease test and glucose, maltose, sucrose, mannitol and xylose hydrolysing patterns.

BOX 10.1 Conditions When Coryneform Bacteria are Considered Significant

1. Cultured from sterile sites (e.g. blood or CSF)
2. On Gram stain of clinical sample, significant number of bacteria are visible
3. Obtained in pure culture
4. Large colony count on culture

C. ulcerans and *C. pseudotuberculosis* primarily cause diseases in animals (bovine mastitis and caseous lymphadenitis in sheep, respectively) but may cause zoonotic infections in humans. *C. ulcerans* causes exudative pharyngitis (indistinguishable from diphtheria) and *C. pseudotuberculosis* causes granulomatous lymphadenitis. *Corynebacterium jeikeium* is an important cause of nosocomial infections. *Corynebacterium urealyticum* causes urinary tract infection in debilitated and immunosuppressed people. *Arcanobacterium haemolyticum* causes pharyngitis in humans with illness varying from mild disease to those like diphtheria. *Rhodococcus equi* may cause pneumonia as well as extrapulmonary infections.

Treatment should always be based on antibiotic susceptibility data though macrolides, penicillins, cephalosporins, tetracyclines or aminoglycosides can be used for empirical treatment.

LISTERIA MONOCYTOGENES

INTRODUCTION

L. monocytogenes is a facultatively anaerobic, non-sporulating Gram-positive bacillus (Fig. 10.1B) that exhibits a characteristic tumbling motility at 25°C. *L. monocytogenes* is the only pathogenic species; others, such as *Listeria welshimeri*, *Listeria innocua* and *Listeria grayi* are non-pathogenic. Some cases caused by *Listeria seeligeri* and *Listeria ivanovii* have been reported.

EPIDEMIOLOGY

L. monocytogenes is acquired by ingestion along with contaminated raw vegetables, raw milk, cheese and meat, etc. Occasionally, it can also be transmitted from mother to child transplacentally, or by cross infection in neonatal nurseries.

CLINICAL MANIFESTATIONS

Infection may result in listeriosis especially in immuno-compromised persons, such as persons with AIDS or cancer, elderly, pregnant women, foetus or neonates. Usually it causes flu-like disease but may cause severe complications, such as stillbirths, neonatal septicaemia, bacteraemia, CNS infections, such as meningitis and encephalitis, and endocarditis.

LABORATORY DIAGNOSIS

Diagnosis may be made by culture on selective media, such as Oxford agar, PALCAM agar, MOX agar and LBM agar and confirming by haemolysis, Gram stain, motility, Christie–Atkins–Munch-Petersen (CAMP) test, sugar fermentation tests or immunoprecipitation tests. Diagnosis can also be done by molecular methods, such as real time PCR or DNA hybridisation.

TREATMENT AND PREVENTION

Treatment is based on antimicrobial susceptibility tests but ampicillin or trimethoprim–sulphamethoxazole may be initiated empirically. The disease may be prevented by cooking food from animal sources thoroughly, washing raw vegetables thoroughly before cooking and avoiding consumption of raw milk or foods made from raw milk.

ERYSIPELOTHRIX RHUSIOPATHIAE

E. rhusiopathiae is a thin, pleomorphic, non-sporulating GPB arranged singly, in pairs or in groups.

CULTURAL CHARACTERISTIC

On blood agar, it produces small, α-haemolytic colonies with a smooth shining surface. It is catalase, oxidase, indole, MR and VP negative but hydrogen sulphide test positive.

CLINICAL FEATURES

The diseases caused include (1) erysipeloid: localised subacute cellulitis of skin, (2) diffuse cutaneous eruption with systemic symptoms and (3) bacteraemia associated with endocarditis.

TREATMENT

For treatment, penicillin is the drug of choice, though ampicillin, ceftriaxone and daptomycin may also be used.

BACILLUS SPP.

INTRODUCTION

Bacillus species is ubiquitous and is found in soil, lake and deep water. They are Gram-positive aerobic bacilli that form endospores and can tolerate extremes of temperature and moisture. This hardiness under extreme conditions is exploited to test efficiency of heat sterilisation cycle (spores of *Bacillus stearothermophilus*) and fumigation procedures (spores of *Bacillus subtilis*). *Bacillus species* is a common contaminant and often gives false positive results in laboratories. Important species are B. *anthracis* (causing anthrax) and B. *cereus* (causing food poisoning).

BACILLUS ANTHRACIS

INTRODUCTION

B. *anthracis* (Greek: 'anthrakis' means coal) usually causes animal diseases and is used as an important agent of bioterrorism. These are catalase positive, non-motile, large (3–8 μm in size) GPB that form a prominent capsule made of poly-D-γ-glutamic acid (PGA). They readily form spores on exposure to oxygen that are central or paracentral and do not cause swelling of the bacilli (Fig. 10.1A). These spores are not found in infected tissues.

EPIDEMIOLOGY

Anthrax disease has a worldwide distribution and is found in domestic and wild animals. Animals acquire infection while grazing in fields or by bites of flies that have fed on contaminated carcasses. Human cases occur on exposure to infected animals or contaminated animal products, such as wool, hair, bone meat, hides and horns.

CULTURAL CHARACTERISTICS

On blood agar, B. *anthracis* forms non-haemolytic, grey white, 'Medusa head appearance' colonies, that are tenacious. On gelatin liquefaction test by stab technique, maximum liquefaction appears on the surface than at the bottom, which is known as 'inverted fir tree appearance'.

IDENTIFICATION

B. *anthracis* can be differentiated from other *Bacillus species* by the presence of a prominent capsule, catalase positivity and non-motility. A pink capsule is visible around a blue cell after staining with polychrome methylene blue (McFadyean's reaction). Definitive identification can be done by lysis by γ phage, capsular detection by fluorescent antibody and PCR detection of toxin genes.

PATHOGENESIS

The main virulence factors of *B. anthracis* are PGA capsule and two exotoxins. The toxin genes produce three components: protective antigen (PA), oedema factor (EF) and lethal factor (LF), which are inactive individually but combine to produce oedema toxin (PA and EF) and lethal toxin (PA and LF) that cause pathogenic lesions.

CLINICAL FEATURES

B. anthracis causes anthrax, which is of three primary forms depending on the route of exposure:

1. **Cutaneous:** Occurs in 95% cases (hide porter's skin). Anthrax spores get inoculated in the skin by contaminated animal hide usually on exposed areas, following which a trivial trauma leads to the development of a small pruritic papule. This later turns into a characteristic black eschar with extensive subcutaneous oedema.
2. **Gastrointestinal:** May be associated with bacteraemia, sepsis and seeding of other sites and has a high mortality if left untreated.
3. **Inhalational:** Extremely rare (wool sorter's disease) and occurrence of this form should raise the possibility of bioterrorism.

LABORATORY DIAGNOSIS

Clinical hints: Clinical suggestions are Presence of a painless lesion, oedema out of proportion to the size of lesion and GPB reported in Gram stain of vesicular fluid or swab. Flowchart for laboratory diagnosis of *B. anthracis* is shown in Flowchart 10.3.

Serology: ELISA for detecting anti-PA antibodies is available.

TREATMENT

Ciprofloxacin or doxycycline is the treatment of choice. The treatment should be taken for 60 days (maximum time taken for the anthrax spores in human body to get activated is 60 days).

BIOTERRORISM

B. anthracis is one of the most common agents used for biological attack or bioterrorism (bioterrorism is defined as deliberate release of bacteria, viruses or other agents that can harm or kill people, livestock or crops).

PREVENTION

A vaccine available for protection against anthrax bacilli is the anthrax vaccine adsorbed (AVA). This vaccine is approved for use by FDA in two situations: before exposure to people with possible occupational exposure and after possible exposure for post-event emergency use in people aged 18 through 65.

B. CEREUS

B. cereus can cause toxin-mediated food poisoning within 24 h of consuming foods, such as fried rice, meat, vegetables and sauces. It can be of two types: diarrhoeal and emetic, both of which are self-limiting and usually resolve within 24 h. Rarely *Bacillus species* can cause bacteraemia, meningitis, pneumonia and ophthalmitis.

B. cereus can be detected by culturing food, vomitus or diarrhoeal fluid. Presence of more than 10^5 or more bacteria/g of stool is diagnostic.

Food poisoning can be prevented by cooking food adequately and consuming it immediately or storing in refrigerator. Treatment is usually not required but serious cases may be treated with vancomycin or clindamycin along with an aminoglycoside.

ANAEROBIC GRAM-POSITIVE BACILLI

INTRODUCTION

Most important GPB implicated in dental diseases are the anaerobic bacteria, such as *Clostridium, Actinomyces, Lactobacillus, Propionibacterium, Bifidobacterium* and *Eubacteria*. The differential diagnosis of anaerobic GPB is shown in Flowchart 10.4. These are obligate anaerobes, Gram-positive

Preferred sample: vesicle fluid/ulcer swab/punch biopsy in cutaneous anthrax; sputum/blood/pleural effusion in inhalational anthrax; stool, blood, rectal swab or ascitic fluid in gastrointestinal anthrax

↓

Gram stain: Gram-positive bacilli

↓

Definitive diagnosis by culture/-PCR/silver staining and immunohistology

Flowchart 10.3 Laboratory Diagnosis of *B. anthracis*.

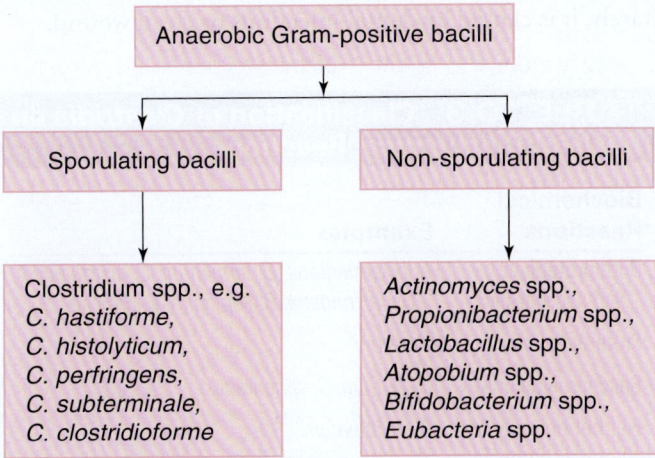

Flowchart 10.4 Differential Diagnosis of Anaerobic GPB Causing Dental Diseases.

TABLE 10.2 Morphology of Different Species of *Clostridium bacilli*

Position of Spores in Bacilli	Shape of Bacilli	Structure	Examples
Central/equatorial and oval	Spindle-shaped appearance		*C. bifermentans*
Subterminal and oval	Club-shaped appearance		*C. perfringens*
Terminal and oval	Tennis racket appearance		*C. tertium*
Terminal and spherical	Drumstick appearance		*C. tetani*

(but may be Gram variable or even negative) bacilli that may be sporulating (*Clostridium* spp.) or non-sporulating (non-sporulating Gram-positive anaerobic bacilli (GPAB)).

Clostridium spp. (kloster: means spindle) have a swollen appearance because their spores are wider than bacillary bodies. The different positions of spores and shapes of bacilli are mentioned in Table 10.2.

Clostridia of medical importance are also classified on the basis of biochemical reactions as in Table 10.3.

CLOSTRIDIUM TETANI

INTRODUCTION

Clostridium tetani is an obligate anaerobic Gram-positive bacillus with terminal and spherical spores (drumstick appearance) that are extremely stable and can cause disease indefinitely. It is the causative agent of tetanus, a disease characterised by persistent tonic spasm, with sudden short exacerbations, particularly of jaw and neck muscles (trismus, lockjaw).

EPIDEMIOLOGY

C. tetani is found in soil, dust and intestines of humans and animals and is usually acquired when spores enter the body through cuts or wounds caused by contaminated articles. Rarely, it is caused by dental infections or clean wound.

TABLE 10.3 Classification of *Clostridia* on Basis of Biochemical Reactions

Biochemical Reactions	Examples
Both proteolytic and saccharolytic	*C. bifermentans, C. botulinum, C. histolyticum, C. perfringens, C. septicum, C. difficile*
Proteolytic only	*C. tetani*
Saccharolytic only	*C. tertium, C. sphenoides*
Neither proteolytic nor saccharolytic	*C. cochleatum*

CULTURAL CHARACTERISTICS

C. tetani grows on ordinary media under strict anaerobic conditions, producing a swarming growth on the surface of the medium. Growth is improved by blood and serum. In gelatin stab cultures, a fir tree growth appears with slow liquefaction.

IDENTIFICATION

C. tetani is feebly proteolytic but not saccharolytic. It is indole positive, MR and VP negative, H_2S negative and does not reduce nitrates.

PATHOGENESIS

C. tetani produces a toxin 'tetanospasmin' that can cause local failure of neuromuscular transmission leading to muscular rigidity and spasms. It also produces tetanolysin with unknown significance.

CLINICAL FEATURES

C. tetani causes tetanus. Manifestations of the disease include difficulty in swallowing (hydrophobia), lockjaw (trismus), fever, elevated blood pressure and other muscle spasms triggered by sensory stimuli, stiffness and pain. There are four clinical types of tetanus: generalised (most commonly seen), localised, cephalic and neonatal (due to infection of umbilical stump).

LABORATORY DIAGNOSIS

Tetanus is usually diagnosed clinically and laboratory diagnosis is not required. However, it should be differentiated from strychnine poisoning. Lab diagnosis can be made by microscopy, culture or in vitro and in vivo (in mouse) identification and toxigenicity testing.

TREATMENT

Antibiotics (penicillin or metronidazole), tetanus antitoxin and a full course of immunisation are used for treatment.

PROPHYLAXIS

Different vaccines are available for active protection against tetanus for children and adults, (Table 10.1) as follows:

For children <7 years old: Diphtheria and tetanus (DT) vaccines and diphtheria, tetanus and pertussis (DTaP) vaccines.

For older children and adults: Tetanus, diphtheria and pertussis (Tdap) vaccines and tetanus and diphtheria (Td) vaccines.

These vaccines are given in three doses at an interval of 0, 1 and 6 months followed by two boosters in the primary vaccination and then a booster after every 10 years for lifelong protection.

For passive vaccination tetanus antitoxin (ATS) is given to the persons at risk with infected wounds, wounds contaminated with mud or soil and wounds with devitalised tissue. Usually a combination of active and passive immunisation is given to persons at risk.

CLOSTRIDIUM PERFRINGENS

INTRODUCTION

Clostridium perfringens is an anaerobic Gram-positive bacillus but can tolerate oxygen. It is pleomorphic, filamentous and has rounded ends and parallel sides. It is usually found singly, in chains or small bundles, and is capsulated and nonmotile. It forms subterminal and oval spores giving a club-shaped appearance. It is the most common cause of gas gangrene (others are *C. septicum, Clostridium sordellii, Clostridium novyi, C. bifermentans* and *C. histolyticum*).

EPIDEMIOLOGY

C. perfringens is found in normal intestinal and vaginal microflora. It is usually acquired by a traumatic injury that breaches the skin followed by contamination with clostridial spores present in soil or faeces.

CULTURAL CHARACTERISTICS

On blood agar, the colonies grow rapidly (within 12–16 h), are small, yellowish to grey in colour, opaque and show target haemolysis (narrow zone of complete haemolysis surrounded by a wider zone of incomplete haemolysis). On Robertson's cooked meat medium, good growth occurs and meat is turned pink but is not digested (predominantly saccharolytic).

IDENTIFICATION

Biochemical reaction profile of *C. perfringens* includes fermentation of glucose, maltose, lactose and sucrose with the production of acid and gas; indole positive; MR positive; VP negative; abundant H_2S production and nitrate reduction. Strains of *C. perfringens* are divided into five types (A to E) on the basis of toxins produced.

PATHOGENESIS

C. perfringens produce several toxins. The major toxins produced are α (alpha) toxin (lecithinase), β (beta) toxin (necrotoxic), ε (epsilon) toxin (haemorrhagic), ι (iota) toxin (lethal), CPE enterotoxins (cytopathic), δ (delta) toxin (haemolysin), κ (kappa) toxin (collagenase), μ (mu) toxin (hyaluronidase) and ν (nu) toxin (DNase).

CLINICAL FEATURES

C. perfringens can cause the following:

1. Gas gangrene/clostridial myonecrosis: The disease usually begins within 24–72 h after crushing traumatic injury or surgery. The disease develops as a deep tissue infection presenting as increasing pain, tenderness and oedema of the affected part and signs of toxaemia. Crepts develop in tissue because of gas accumulation. If left untreated, profound toxaemia and finally death occurs due to circulatory failure.
2. Food poisoning: This is almost always caused by *C. perfringens* type A. Consumption of at least 10^8 viable enterotoxin-producing cells is required for food poisoning, which most commonly occurs on consumption of improperly cooked meat and meat-containing products. After an incubation period of 6–24 h, self-limiting watery diarrhoea, vomiting, abdominal cramps and fever develop.
3. Other infections: Bacteraemia, abdominal infections, biliary tract infections and infections of the female genital tract.

LABORATORY DIAGNOSIS

1. **Gas gangrene:** Sample required is fluids or exudates from the affected tissues. On Gram staining, large numbers of regularly shaped GPB without spores are seen along with few polymorphonucleocytes. On culture, typical colonies are seen on blood agar.

   On egg yolk agar medium, α toxin can be detected by white precipitate around the colonies due to lecithinase production. Neutralisation of this precipitate by *C. perfringens* antitoxin gives the presumptive evidence of the organism and is called **Nagler reaction**. Reverse CAMP test can also be used. PCR methods are also available.
2. **Food poisoning:** For presumptive diagnosis, quantitative culture for *C. perfringens* is done from stool of patient and from incriminated food. A colony count of more than 10^6 cfu/g is considered significant. Other modalities of diagnosing are detection of the enterotoxin by cytopathic toxin assay, latex agglutination test or an enzyme immunoassay.

TREATMENT

Immediate debriding surgery is the mainstay of treatment. Antibiotics (penicillin, metronidazole and gentamicin) are also given.

CLOSTRIDIUM BOTULINUM

DEFINITION

Clostridium botulinum is a large, obligatory anaerobic GPB that forms a subterminal spore, is non-capsulated and motile by peritrichate flagella. The species is divided into four groups (I–IV) on the basis of toxin produced (A–G). It causes **botulism** (a paralytic disease of different clinical types: food-borne botulism, wound botulism, infant botulism, adult intestinal toxaemia botulism and iatrogenic botulism).

EPIDEMIOLOGY

C. botulinum spores are found globally in soil, animal manure and marine sediments. Spores are highly resistant and can tolerate heat (at 100°C for several hours) and radiation.

CULTURAL CHARACTERISTICS

On ordinary media, *C. botulinum* forms large, irregular and semitransparent surface colonies.

PATHOGENESIS

C. botulinum produces a powerful heat-labile and low-potency neurotoxin, the botulinum toxin. The toxin after absorption (from intestines in case of food-borne botulism or from wound) reaches neuromuscular junction and other peripheral cholinergic synapses where it prevents the release of the neurotransmitter acetylcholine. Thus the motor system is paralysed.

CLINICAL FEATURES

The predominant clinical types of botulism are as follows:
1. **Food-borne botulism:** Usually occurs within 12–36 h of ingestion of preformed toxin in foods especially canned foods, such as fish, salmon or vegetables. Common symptoms include nausea, dry mouth, diarrhoea, dysphagia, ocular paresis and difficulty in breathing. Acute bilateral cranial neuropathies along with symmetric descending paralysis/paresis are the classic presentation.
2. **Wound botulism:** Occurs after wound infection with *C. botulinum*. Symptoms are similar to food-borne botulism except the gastrointestinal manifestations.
3. **Infant botulism and adult intestinal toxaemia botulism:** Acquired by ingestion of spores in contrast to the other forms that are due to ingestion of preformed toxin. Toxin is produced within the infant's/adult's intestine after germination of spores. The disease is manifested as poor feeding, hypotonia, weak cry, airway obstruction, etc. in infants.
4. **Iatrogenic botulism:** Occurs due to overdose of botulinum toxin for cosmetic or medical applications.

LABORATORY DIAGNOSIS

Diagnosis can be made by two modalities:
1. Detection of *C. botulinum*: The sample can be serum, faeces of patient and the implicated food. Demonstration of GPB in direct smears followed by isolation of *C. botulinum* on anaerobic culture is required for diagnosis.
2. Detection of toxin: Samples are serum, gastric fluid, faeces or food samples. Toxin may be detected by mouse bioassay or ELISA. Toxin gene may be detected by PCR.

TREATMENT

Supportive treatments, such as maintenance of airway and agents that improve acetylcholine release, are given. Equine antitoxin therapy is the mainstay of treatment. Antibiotics (metronidazole, penicillin G, aminoglycosides and tetracyclines) are also given. Debridement of the affected part is done in case of wound botulism.

PREVENTION

Prevention is possible by proper food preparation and handling. Packed foods that appear to bulge should not be consumed. At present, no vaccine is approved by FDA, but several vaccines, such as DNA-based, viral vector-based and recombinant protein-based vaccines are under development.

CLOSTRIDIUM DIFFICILE

C. difficile is a long, slender, Gram-positive bacillus, which may appear Gram-negative, with large, oval and subterminal spores. It is non-haemolytic, saccharolytic and weakly proteolytic. It has now become an important cause of hospital-acquired infection (HAI) especially in persons on antibiotics. *C. difficile* produces two toxins: an enterotoxin (TcdA) and a cytotoxin (TcdB) that causes cytotoxicity of the intestinal epithelial cells.

CLINICAL FEATURES

C. difficile causes *C. difficile*-associated disease (CDAD) which is associated with considerable antibiotic use and is prevalent in medicine and dentistry. The manifestations of CDAD range from self-limiting diarrhoeal illness to pseudomembranous enterocolitis.

LABORATORY DIAGNOSIS

Laboratory Diagnosis can be made by culture and non-culture methods. Culture methods include inoculation of stool sample on a selective agar media (e.g. CCFA or cycloserine, cefoxitin, fructose and egg yolk agar), which is incubated anaerobically. The colonies of *C. difficile* are flat, 4–6 mm in diameter and grey in colour with 'barnyard' odour. Gram stain from these colonies shows slender Gram-positive rods with subterminal spores.

Non-culture methods include detection of *C. difficile* toxin in stool samples by enzyme immunoassay or by PCR. Other tests that can be used are glutamate dehydrogenase (GDH) test and cell cytotoxicity assay.

Drug of choice for CDAD is vancomycin. However, fidaxomicin (Dificid) or metronidazole can also be used. Infection by *C. difficile* can be prevented by implementing appropriate infection control measures and by adhering to prescribing appropriate antibiotics.

ANAEROBIC NON-SPORE-FORMING GRAM-POSITIVE BACILLI

Non-sporulating GPAB are found as commensals of the digestive tract, urogenital tract and skin. Medically important genera include *Eubacterium, Mobiluncus, Propionibacterium, Bifidobacterium, Actinomyces* and *Lactobacillus*.

Members of genus **Eubacterium** are part of normal mouth and intestinal flora, but sometimes can cause periodontitis. **Propionibacterium** is a skin commensal that often contaminates improperly collected blood and CSF cultures and may cause acne. **Lactobacillus** is found in adult vagina (Döder-

lein's bacilli), mouth and intestines. **Mobiluncus** causes bacterial vaginosis (a polymicrobial infection with a malodorous vaginal discharge).

Non-sporulating GPAB can cause dental abscess in which the usual presentation is pain, swelling and erythema around the affected tooth, fever, tenderness to palpation and extra- and intraoral swelling. Rarely complication may occur when infection extends to the neck and chest leading to deep neck abscesses and necrotising mediastinal abscesses, respectively.

Culture methods are not generally used for non-sporulating GPAB since they are difficult to grow on conventional laboratory media. Overall anaerobes are difficult to culture; the process is expensive or misleading because of growth of normal flora. For dental abscess, GPAB can be detected and identified by molecular techniques, such as PCR, DNA hybridisation and 16S rRNA gene sequencing.

Treatment of dental abscess includes surgical drainage/debridement along with antibiotics, such as penicillins and cephalosporins. In patients with penicillin allergy, clarithromycin or azithromycin, metronidazole and clindamycin may be used. It may be prevented by maintaining oral hygiene.

11

Mycobacteria, Actinomycetes and Nocardia

Amita Jain

LEARNING OBJECTIVES

- History of tuberculosis
- Classification of mycobacteria
- Morphology and culture characteristics of *Mycobacterium tuberculosis*
- Disease pathogenesis and drug resistance of tuberculosis

- Atypical mycobacteria
- *Mycobacterium leprae*
- Actinomycetes
- Nocardia
- Difference in mycobacteria, nocardia and actinomycetes

MYCOBACTERIA

INTRODUCTION

Mycobacterium tuberculosis causes a disease 'tuberculosis' (TB) also known as *kshay rog* (wasting disease) or *raj rog* (king of diseases), which is known to affect human for centuries. Tuberculosis was also known as 'consumption', 'wasting disease' and 'white plague'.

HISTORICAL LANDMARKS

TB has affected humans for centuries. Till mid-1800, it was believed that TB was hereditary. Until 1940s and 1950s, there was no cure for TB. For many people, TB was a killer disease. Important landmarks in history of tuberculosis are listed in Box 11.1 in chronological order.

CLASSIFICATION

Mycobacteria have ubiquitous presence of more than130 species, of which, only few species cause most infections. Tuberculosis is caused by *M. tuberculosis*, sometimes called tubercle bacilli (TB). Some mycobacteria are called tuberculous mycobacteria because they cause TB, for example, *M. tuberculosis*, *Mycobacterium bovis* and *Mycobacterium africanum*. Other mycobacteria are called non-tuberculous mycobacteria (NTM)/mycobacteria other than tuberculosis (MOTT) because they do not cause TB, for example, *Mycobacterium avium* complex.

Preliminary classification of mycobacteria is done by growth properties and colonial morphology: In 1954, Runyon and Timpe gave first MOTT classification (Table 11.1). Other names for MOTT are: atypical mycobacteria, environmental mycobacteria and opportunistic mycobacteria. There are >100 MOTT species. MOTT has variable pathogenicity and geographic distribution. They usually cause diseases in immunocompromised hosts.

> ### BOX 11.1 Important Landmarks in History of Tuberculosis
>
> - **2400 BC:** Tubercular lesion in spine of Egyptian mummy was noted
> - **460 BC:** Hippocrates identifies 'consumption' as 'most widespread disease of the time' and 'is always fatal'
> - **1679:** Physician Sylvius described the lung pathology as seen in TB
> - **1702:** The fact that TB is of infectious nature was accepted
> - **1720:** Physician Marten wrote 'wonderfully minute living creatures' as cause of consumption
> - **1854:** Brehmer opened first sanatorium
> - **1865:** A French surgeon, Jean-Antoine Villemin proved that TB was contagious and demonstrated human to cow to rabbit transmission of TB
> - **1882:** A German scientist Robert Koch isolated *M. tuberculosis, the causative organism* in pure culture
> - **1895:** BCG vaccine was developed by Calmette and Guerin
> - Until 1950s, many people with TB were sent to sanatoria, where they followed a prescribed routine every day. No specific treatment was given. No one knows whether sanatoriums really helped.
> - **1943:** Streptomycin discovered by Selman Waksman
> - Following drugs were subsequently discovered:
>   - **1949:** *p*-aminosalicylic acid
>   - **1952:** Isoniazid
>   - **1954:** Pyrazinamide
>   - **1955:** Cycloserine
>   - **1962:** Ethambutol
>   - **1963:** Rifampin
>   - **1970–90:** Aminoglycosides (kanamycin and amikacin)
>   - Quinolones (ofloxacin and ciprofloxacin)

MYCOBACTERIUM TUBERCULOSIS

MORPHOLOGICAL CHARACTERISTICS

- Thin, straight rods, about 0.4 × 3 μm in size
- These cannot be classified as either Gram-positive or Gram-negative (difficult to stain with Gram stain)

TABLE 11.1 Classification of Mycobacteria

Mycobacterial Species	Subgroups	Growth Characteristics
M. tuberculosis complex		Slow-growing, no pigmentation
NTM; MOTT	• Runyon Group I	• Slow-growing, yellow pigment (+) in light
	• Runyon Group II	• Slow-growing, yellow pigment (+) in dark
	• Runyon Group III	• Slow-growing, non-pigmented
	• Runyon Group IV	• Rapidly growing
M. leprae		Non-cultivable

MOTT, Mycobacteria other than tuberculosis; NTM, Non-tuberculous mycobacteria.

- **Ziehl–Neelsen (ZN) technique** of staining is employed for identification of acid-fast bacteria
- Acid-fast, alcohol-fast (due to presence of excessive mycolic acid in cell wall)
- Once stained by basic dyes, they cannot be decolourised by alcohol, regardless of treatment with iodine
- These can be demonstrated by yellow-orange fluorescence after staining with fluorochrome stains (e.g. auramine, rhodamine)

CELL WALL

Cell wall of mycobacteria (Fig. 11.1) is rich in lipids. It is responsible for many characteristic morphological properties (acid-fastness, resistant to disinfectants and antibiotics, antigenicity, slow growth, clumping) and forms 60% of bacterial dry weight.

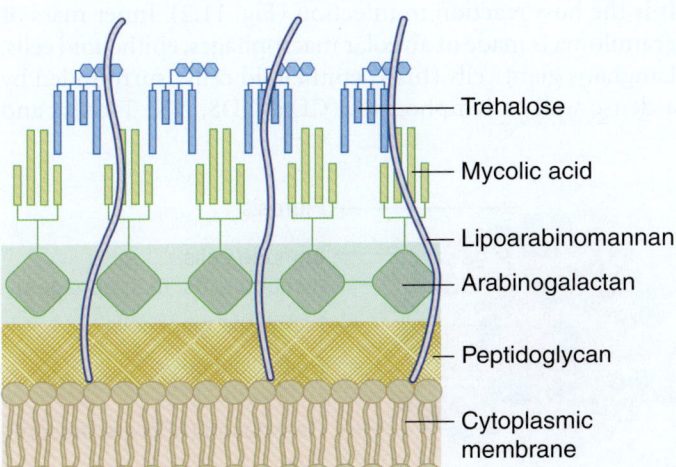

Fig. 11.1 Cell Wall of Mycobacteria.

Labels: Trehalose, Mycolic acid, Lipoarabinomannan, Arabinogalactan, Peptidoglycan, Cytoplasmic membrane

BOX 11.2 Composition of LJ Medium

- LJ medium is light green, opaque and opalescent
- For preparing media the following ingredients are added
- Malachite green: gives colour to media and makes it specific for *M. tuberculosis*
- Glycerol: enhances growth of *M. tuberculosis*
- Asparagine: source of nitrogen
- Coagulated eggs: enrichment and solidifying agent
- Mineral salt solution
  - Potassium di-hydrogen phosphate
  - Magnesium sulphate
  - Sodium citrate
- For cultivation of *M. bovis*, sodium pyruvate is added in place of glycerol

Mycolic acid present in bacterial is resistant to common laboratory stain. Once stained, it cannot be decolourised with acid solutions causing acid-fastness.

Polypeptides of mycobacteria present in cell wall are mainly transport proteins and porins.

Purified protein derivatives (PPD) are part of cell wall which induce cell-mediated immunity.

CULTURAL CHARACTERISTICS

Mycobacteria are strict aerobe. The Lowenstein–Jensen medium, more commonly known as LJ medium (Box 11.2) is the commonly used solid media for growing mycobacteria. Other media are also available for growing *M. tuberculosis*, which includes several liquid media (e.g. Middlebrook medium) available from commercial sources.

Colonies on LJ media are light brown/straw-coloured, granular (sometimes called 'buff, rough and tough'). Time for formation of colonies is more than 4 weeks due to the slow doubling time of *M. tuberculosis* compared with other bacteria (15–20 h in comparison to 20 min for *E. coli*).

GROWTH CHARACTERISTICS

- Mycobacteria are obligate aerobes
- They derive energy from the oxidation of simple carbon compounds
- Increased CO_2 tension enhances growth
- Growth rate is much slower than that of most bacteria
- Saprophytic mycobacteria may grow more rapidly, at lower temperature, that is 22–33°C. They may be more pigmented and less acid-fast than pathogenic mycobacteria

TUBERCULOSIS, THE DISEASE

Some important definitions in reference to drug resistance and tuberculosis are given in Box 11.3.

Epidemiology

As per World Health Organization's 2015 estimates, estimated 10.4 million new (incident) TB cases, of which 1.2

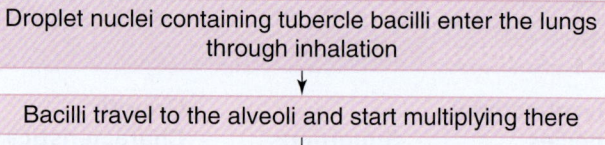

Droplet nuclei containing tubercle bacilli enter the lungs through inhalation
↓
Bacilli travel to the alveoli and start multiplying there
↓
Some tubercle bacilli may enter blood stream through which a haematogenous spread occurs within whole body
↓
Bacilli may get settled anywhere in body, including lungs, kidneys, brain or bone
↓
Within 2–10 weeks, immune system produces macrophages that surround tubercle bacilli
↓
Macrophages contain bacilli under control (TB infection)
↓
If immune system cannot keep bacilli under control, the bacilli begin to multiply rapidly
↓
TB disease (granuloma formation)

Flowchart 11.1 Pathogenesis of Tuberculosis.

million (11%) are HIV positive are present worldwide. Total 60% of the new cases are found in India, Indonesia, China, Nigeria, Pakistan and South Africa. New cases of multidrug-resistant TB (MDR-TB) were 480,000. Total 1.4 million TB deaths occurred in 2015. An additional 0.4 million deaths resulted in HIV positives due to TB.

Clinical Presentations

1. **Pulmonary TB:** About 85% of TB cases present as pulmonary TB, which is the most infectious form of TB. Patients mostly have cough and an abnormal chest X-ray.
2. **Extra-pulmonary TB (EPTB):** This occurs in tissues other than lungs. Larynx, lymph nodes, pleura, brain, kidneys, bones, joints are common sites. EPTB is commoner in HIV-positive individuals than HIV-negative individuals. This form of TB is mostly non-infectious.
3. **Miliary TB:** This is rare but serious form of infection. Mycobacteria cause bacteraemia, and may get settled at any site to cause more generalised form of disease. Chest X-ray has appearance of millet seeds scattered throughout; hence the name miliary TB is given.

Transmission

TB is air borne. When a person with pulmonary TB coughs or sneezes, droplet nuclei around 1–5 microns large, are expelled into the air. Droplet nuclei contain *M. tuberculosis* and can remain suspended in the air for several hours in a still environment. If another person inhales air that contains these droplet nuclei, transmission may occur.

The probability that TB will be transmitted depends on three factors:

1. How contagious is the TB patient?
   a. AFB excreted in sputum or not: Patients with AFB-positive sputum are highly contagious.
   b. Larger droplets settle on the floor and cannot remain suspended in the air. Hence, medium-sized droplets

are highly infectious. Moreover, larger droplets if inhaled get lodged in airway. Smaller particles cannot hold bacilli.

2. In what kind of environment does the exposure occur?
   a. If the environment is ventilated (about 8–10 air changes in an hour) the bacilli are cleared from the air rapidly. Stagnant air is favourable for transmission of infection.
3. How long did the exposure last?
   a. If the person lasts longer in the room, for example, living or working together, chances of exposure are more.

Pathogenesis

It is explained in Flowchart 11.1.

Granuloma

It is the host reaction to infection (Fig. 11.2). Inner mass of granuloma is made of alveolar macrophages, epithelioid cells, Langhans giant cells (fused epithelioid cells) surrounded by a dense wall of lymphocytes (CD4, CD8, NK, T cells) and

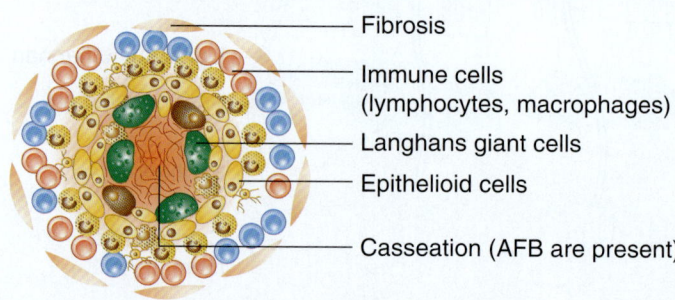

- Fibrosis
- Immune cells (lymphocytes, macrophages)
- Langhans giant cells
- Epithelioid cells
- Casseation (AFB are present)

Fig. 11.2 Tubercular Granuloma.

TABLE 11.2 Differences in TB Infection and Disease		
Characteristics	**TB Infection**	**TB Disease (in Lungs)**
Tubercle bacilli present in the body	Yes	Yes
Tuberculin skin test reaction	Positive	Positive
Chest X-ray	Usually normal	Usually abnormal
Sputum smears and cultures	Negative	Positive
Clinical symptoms	No symptoms	Cough, fever weight loss
Infectious	Not infectious	Often infectious
Case of TB disease	Not a case of TB	A case of TB
Need treatment	No	Yes

macrophages. Finally, a fibrin layer develops around granuloma (fibrosis), which 'walls off' the lesion.

Typical progression in pulmonary TB involves casseation, calcification and cavity formation.

Progression

All the people who get **exposed** to TB may not get sick with it. Some of them **get infection**, meaning thereby that TB bacilli settle in host, multiply there and get controlled by host defence. Persons who get infection may or may not develop disease. Only a small fraction of TB infections translate in **TB disease**. People with risk factors are more prone to develop disease. The difference in TB infection and disease are listed in Table 11.2 and the risk factors for development of TB disease are listed in Table 11.3.

Anti-Tubercular Drugs

Anti-tubercular drugs as first- and second-line drugs and their mechanism of action are listed in Fig. 11.3 and Table 11.4.

TABLE 11.3 Risk Factors and Their Association With Risk of Developing TB Disease	
Risk Factors	**Risk of TB Disease (Times)**
AIDS	170
HIV infection	113
Recent TB infection (within past 2 years)	15
Certain medical conditions[a]	3–16

[a]For example, diabetes, certain types of cancer, immunosuppressive therapy, silicosis, severe kidney disease, certain intestinal conditions and low body weight ($\geq$10% below ideal). acquired immunodeficiency syndrome.

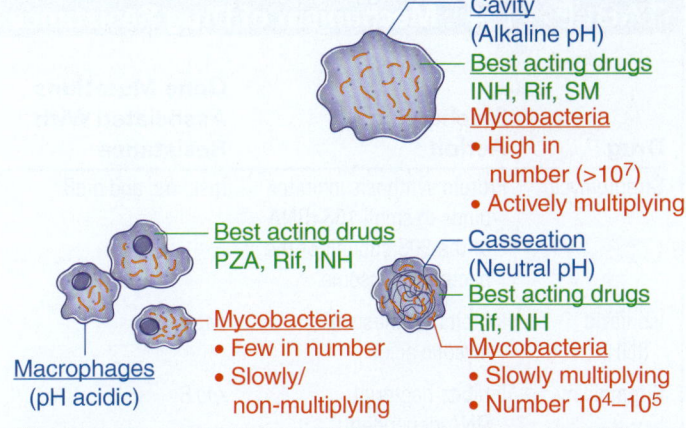

Fig. 11.3 Best Effective Anti-Tubercular Drugs at Different Sites.

- **First-line drugs** are used in initial treatment of TB. These are isoniazid, rifampin, streptomycin, ethambutol and pyrazinamide.
- **Second-line drugs** such as PAS, ethionamide, amikacin, kanamycin, capreomycin, cycloserine, ciprofloxacin, levofloxacin and clofazimine are used to treat drug-resistant tuberculosis.

Mechanism of Drug Resistance

Drug resistance in mycobacteria is caused by chromosomal mutations. Mutations occur **spontaneously**. Estimated frequency of INH and rifampicin resistance is 5×10^{-6} for and 3.1×10^{-8}, respectively. Risk of a double spontaneous mutation for INH and rifampicin together is extremely low i.e. one in 9×10^{-14} mutations (translates in emergence of MDR *M. tuberculosis*). Drug-resistant mutants are selected as a result of ineffective treatment, hence, drug resistant TB is a man-made phenomenon.

Diagnosis

Modalities available for laboratory diagnosis of TB (Table 11.5) can help us with the following:
- **Diagnosis of disease**: Decision on active/latent disease is also important, more so in countries where tuberculosis is endemic as this decision will help clinician decide whether to treat or not.
- **Detection of drug resistance**: This will help in choice of treatment and predicting response to therapy.
- **Monitoring response to treatment**: Since treatment for tuberculosis is long, mid-term evaluation of response to treatment is desired.

  Samples required: Sample should be collected from representative site, for example:
- Pulmonary tuberculosis: sputum/gastric aspirate/bronchoalveolar lavage
- Extra-pulmonary tuberculosis: CSF/other body fluids, tissue biopsy, pus, needle aspirates etc.
- Blood/serum are not acceptable samples for TB diagnosis except in case of military tuberculosis.

TABLE 11.4 Mechanism of Drug Resistance to First-Line Anti-Tubercular Drugs

Drug	Mechanism of Action	Gene Mutations Associated With Resistance	Action on Actively Multiplying Extracellular Mycobacteria	Action on Slowly or Non-Replicating Mycobacteria	Action on Acid pH
Streptomycin	Protein synthesis inhibitor, binds to small 16S rRNA of the 30S subunit of the bacterial ribosome	rpsL, rrs, and gidB	+++	−	−
Isoniazid (INH)	Inhibits synthesis of mycolic acids	kat G, INA	++	+	+
Rifampicin	Inhibits bacterial DNA-dependent RNA synthesis thus preventing synthesis of proteins	rpo B	++	+	+
Pyrazinamide	Inhibits the growth of M. tuberculosis, a prodrug to Pyrazinoic acid	pncA	±	++	++
Ethambutol	Inhibition of arabinosyl transferase, cell wall synthesis inhibition	embCAB operon	±	±	±

TABLE 11.5 Tests for Diagnosis of Tuberculosis

Testing Modality	Methods	Turnaround Time (TAT)	Sensitivity/Detection Limit (Bacilli/mL)
Smear examination for AFB	• ZN staining • AR (fluorescent) staining	2 h	1000
Culture	• Liquid media-based automated methods like MGIT 960 • solid media-based manual method (e.g. LJ media)	2 days–8 weeks	100
NAAT/(PCR)	• Conventional NAAT • Real time PCR (in-house/commercial)	2 days	10
Gene-Xpert (@Cepheid)	Semi-automated method, additionally detects resistance to rifampicin	4 h	10

AFB, Acid-fast bacilli; AR = auramine–rhodamine; MGIT 960 = mycobacterial growth indicator Tube 960; LJ = Lowenstein–Jensen; NAAT = nucleic acid amplification test; PCR = polymerase chain reaction; TAT = turnaround time; ZN = Ziehl–Neelsen.

Available Diagnostic Tests

These tests are listed in Box 11.4.

Commonly used methods in a tuberculosis lab are:

1. Smear examination: It is the oldest, inexpensive, rapid but low sensitivity and 'most widely used' test. Sensitivity has been improved with the use of LED fluorescent microscopy (Fig. 11.4).

What does AFB smear tells us?
- IF positive, it is TB or MOTT infection
- If negative, no infection or infection but less than 10^4 AFB/mL in the specimen
- A 10% culture-positive specimen may be smear negative

2. Nucleic acid amplification tests (NAAT): Many home brew and commercial systems are in use currently but quality controls and system checks may not be in place, hence, variation within laboratories is commonly seen. There are no clear guidelines on use and interpretation of NAAT. Hence, manual NAAT assays have no clear recommendation for clinical diagnosis.

3. Immunological methods (ELISA): Detection of antibodies to species-specific/disease-specific *MTB* antigens is **not recommended** for clinical diagnosis in endemic countries like India. This was the **first ever negative recommendation from WHO.**

4. Mycobacterial culture and drug susceptibility testing (DST): Uses, advantages and disadvantages of culture are mentioned in Box 11.5.

LJ culture: Positivity time is high usually 6–8 weeks. It is labour-intensive and technically challenging. Colony

BOX 11.4 Available Tests for Diagnosis and Management of TB

For Establishing Diagnosis (Turnaround Time) (Sensitivity)
- Smear examination for acid-fast bacilli (AFB)[a] (turnaround time 2 h) (detection limit 1000 bacilli/mL)
  Methods in use: Ziehl–Neelsen (ZN) staining/auramine–rhodamine (AR) staining
- Culture[a] (Turnaround time 2 days–8 weeks) (detection limit 100 bacilli/mL)
  Methods in use: Liquid media–based automated methods like MGIT 960/solid media-based manual method (e.g. Lowenstein–Jensen (LJ) media)
- Nucleic acid amplification test (NAAT)/polymerase chain reaction (PCR) (turnaround time 2 days) (detection limit 10 bacilli/mL)
  Conventional NAAT/real time PCR (in-house/commercial)
- Gene-Xpert (@Cepheid) (turnaround time 4 h) (detection limit 10 bacilli/mL)
  Semi-automated method, additionally detects resistance to rifampicin

Other Methods Which Support Diagnosis
1. Fine needle aspiration cytology
2. Radiology: X-ray and magnetic resonance imaging (MRI)
3. Histology
4. Adenine deaminase detection
5. Lysozyme detection

Methods Which Diagnose Latent Disease, Hence Are Not Recommended for Diagnostic Use
1. Enzyme-linked immunosorbent assay (ELISA) for detection of anti-tubercular antibodies
2. Tuberculin test
3. Interferon gamma assay

Available Tests for Detection of Drug-Resistant Tuberculosis
- Culture for *M. tuberculosis* and drug susceptibility testing
  - **Solid culture methods (LJ media): 1% proportion (turnaround time 42 days)[b]**
  - **Liquid culture automated methods (e.g. MGIT) (turnaround time 10 days)[a]**
- Some liquid media-based methods not commonly used:
  - **Microscopic observation of drug susceptibility (MODS)**
  - **Nitrate reduction assay (NRA)**
  - **Resazurin microtitre assay (REMA)**
- Molecular (non-culture) methods: test can be applied directly on samples. No need to grow *M. tuberculosis*
  - **Line probe assay (turnaround time 2 days)[a]**
  - **TB-Xpert (turnaround time 4 h)[a]**
  - **Sequencing**

[a](Differentiation of *M. tuberculosis* and MOTT will be required once test is positive!)
[b](World Health Organization (WHO) Approved Tools)

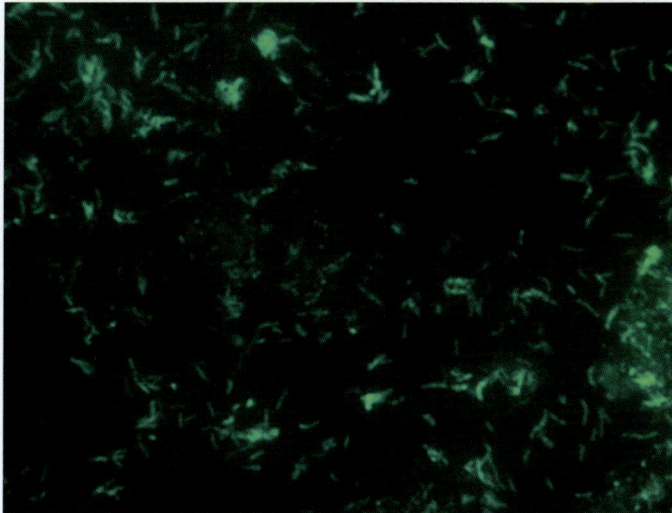

Fig. 11.4 AFB-Positive Smear by Auramine–Rhodamine Stain.

BOX 11.5 Uses, Advantages and Disadvantages of Culture

Uses
1. Drug resistance surveillance
2. Diagnosis: Especially helpful in following group of patients:
   a. Diagnostic confirmation of smear-negative TB cases
   b. HIV-positive cases
   c. Children
   d. Extra-pulmonary tuberculosis
3. Research

Advantages
1. More sensitive and specific than microscopy
2. Essential step before performing DST
3. Diagnosis of microscopy negative cases
4. Follow-up of patients on anti-tubercular treatment to study the response to treatment

Disadvantages
1. Takes long time to get the results
2. Requires more materials/equipment, so expensive
3. Complex procedure, require skilled technicians
4. Prerequisite is good referral system with safe/rapid specimen transport mechanism
Poor accessibility, technical complexity, long TAT, and high cost, offset advantage of higher sensitivity of culture

morphology is characteristic and is easy to identify (Fig. 11.5).

Advantages and disadvantages of liquid culture over solid media: Liquid culture, such as Mycobacterial Growth Indicator Tube 960 (MGIT 960) has short turnaround time (TAT) and higher sensitivity (by 10%). However, high media

and set up cost, relatively high contamination rate, poor accessibility and biosafety-related issues are major disadvantages, which need to be addressed. Many more commercial automated and manual methods are available. However, each one of them has similar limitations.

5. **Line probe assay**: It is a family of novel DNA strip-based tests which uses PCR and reverse hybridisation methods. This is designed to identify *M. tuberculosis* complex and simultaneously detect mutations associated with drug

Fig. 11.5 LJ Culture Media With Colonies of *Mycobacterium tuberculosis*.

resistance. The majority of studies had shown sensitivity of 95% or greater and nearly 100% specificity.

6. **Gene–Xpert**: It has the following properties:
- Fully automated molecular test for detection of tuberculosis and R_{if} resistance
- Uses heminested real-time polymerase chain reaction (PCR) assay
- Amplifies an MTB specific sequence of the *rpoB gene*
- Uses a set of molecular beacon probe
- Detects mutations in rifampin-resistance determining region (RRDR)
- Integrates sample processing and NAAT in a disposable plastic cartridge
- Cartridge contains reagents required for bacterial lysis, nucleic acid extraction, amplification and amplicon detection
- Uses a manual step to add bactericidal buffer to sample and transferring a predetermined volume to MTB/RIF cartridge, which is then inserted into GeneXpert device
- Results are available within 2 h
- Allows a relatively unskilled health care worker to diagnose tuberculosis and detect resistance to a key antibiotic within 90 min

Advantages of Gene–Xpert over conventional nucleic acid amplification tests are as follows:
- Simple to perform
- Minimal training needed
- Not prone to cross-contamination,
- Requires minimal biosafety facilities
- High sensitivity in smear-negative tuberculosis (relevant in patients with HIV infection)

Limitations are high cost, testing only for rifampicin resistance, detection of relatively small number of mutations and inability to indicate which patients are 'sputum smear AFB positive' for reporting purposes, cannot be used for infection-control intervention and treatment monitoring.

Treatment

Therapy requires at least two effective drugs concurrently. Within 2 weeks patient responds to appropriate treatment. Continue treatment for at least 3–6 months after the sputum becomes negative. Minimum length of therapy is 6–9 months. Never use single drug to treat TB and never add a single drug to a failing regimen. Refer to recommendations of Revised National Tuberculosis Control Program (RNTCP) for management of tuberculosis.

Control

The most effective method to control spread of TB in a population is to effectively treat sputum smear-positive cases (source of infection). As per the WHO, the strategy that is most successful for TB control is directly observed treatment short course (DOTS), which has sustained political and financial commitment.

ATYPICAL MYCOBACTERIA

INTRODUCTION

Atypical or non-tubercular mycobacteria are found in water, soil, food and animals. They do not spread from person to person. Disease caused is not a notifiable disease. They are relatively resistant to chlorination and ozonisation and outbreak in hospitals are common. HIV and dialysis patients are commonly affected. They are classified as shown in Flowchart 11.2. Box 11.6 lists the common pathogens of human.

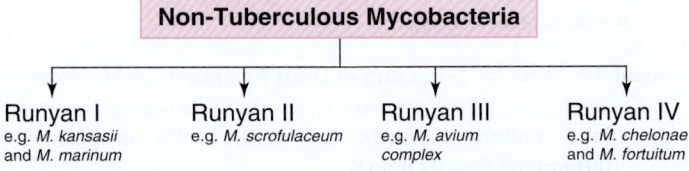

Flowchart 11.2 Non-Tuberculous Mycobacteria (Runyon Classification).

BOX 11.6	**Runyon Classification**

Slow Growers (Runyon Class I, II and III)
Growth time: 1–2 weeks in broth and 2–4 weeks in solid media.
Examples:
- *M. avium*
- *Mycobacterium kansasii*
- *Mycobacterium scrofulaceum*
- *Mycobacterium ulcerans*
- *Mycobacterium xenopi*
- *Mycobacterium gordonae*

Rapid Growers (Runyon Class IV)
Growth time: 1–3 days in broth and <1 week in solid media.
Examples:
- *Mycobacterium abscessus*
- *Mycobacterium chelonae*
- *Mycobacterium fortuitum*

EPIDEMIOLOGY

Many people harbour atypical mycobacteria in their respiratory secretions without any evidence of disease. Person-to-person transmission of atypical mycobacteria does *not* occur, with the exception of organisms causing skin lesion. Most infections arise from environmental exposure to organisms in infected water, soil, dust or aerosols. In some individuals, disease involving the lungs, skin or lymph nodes may result; may also infect open wounds. Immunocompetent patients usually present with localised skin and soft tissue infections only. Individuals with respiratory disease do not need to be isolated as do **not** readily infect others. Preventive treatment of close contacts is not indicated.

Commonly reported infections with NTM are: *M. avium intracellular complex* (MAC) 40%, rapidly growing NTM 10%, unknown 15%, *M. gordonae* 2.5%, *M. kansasii* 2.5% and *M. xenopi* 1%.

COMMON CLINICAL SYNDROMES

Lymphadenopathy
Chronic pulmonary disease
Skin and soft tissue infections (often associated with trauma or a foreign body) sometimes with extension to bone and joint
Disseminated disease

ASSOCIATED RISK FACTORS

Immunosuppression (HIV, medications)
Ageing
No BCG vaccination
Cystic fibrosis
Fibronodular bronchiectasis

PATHOGENESIS

Mortality is rare. NTM can cause morbidity, especially when not diagnosed and not treated effectively. Often cutaneous infection can resolve on its own.

Common aetiological agents of pulmonary disease caused due to NTM:

M. avium complex (MAC)
M. kansasii
M. abscessus
M. xenopi

Usually adults are affected. Symptoms of cough, sputum production and weight loss are common. Isolation of same species at least twice from sputum or isolation, even once, from broncho-alveolar lavage/biopsy tissue/specimen from sterile site is diagnostic. Distribution of isolates varies regionally.

LAB DIAGNOSIS

Identification and sensitivity may be required at times for better management of case. Colony morphology, growth time, biochemical characteristics and molecular assays are helpful in identification. DNA probes for MAC, *M. kansasii* and *M. gordonae* are available.

TREATMENT

First ensure that the isolate is clinically significant. This can be done by repeated culturing of specimen from the same site. Isolation of the same organism from the same site, more than once may be significant.

In most cases a course of antibiotics, such as rifampicin, ethambutol, isoniazid, minocycline, ciprofloxacin, clarithromycin, azithromycin and cotrimoxazole is necessary. Treatment should usually combine at least two drugs of proven efficacy. Treatment of rapidly growing mycobacteria should be guided by in vitro susceptibilities. In some cases, successful treatment requires aggressive debridement of all infected subcutaneous tissues and skin. In general, 6–12 months is required following negative cultures.

Contact follow-up is not necessary since NTM are not transmitted from person to person.

LEPROSY

INTRODUCTION

In India it is also called **kustharoga** and was attributed to curse from God. Gerhard Henrik Armauer Hansen was a physician, who identified *M. leprae* as the cause of leprosy in 1873; hence the name Hansen's disease. Mainly peripheral nerves are affected. Over the past 20 years, >14 million cases of leprosy are cured.

CAUSATIVE ORGANISM

M. leprae resembles *M. tuberculosis* morphologically. Features specific to *M. leprae* are:
- AFB with parallel sides and rounded ends (Fig. 11.6)
- Present in large numbers in lepromatous leprosy
- Globular masses of bacilli are present within the lepra cells known as globi

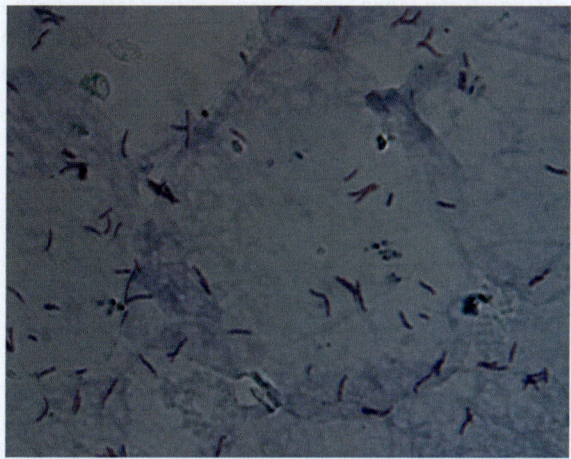

Fig. 11.6 Ziehl–Neelsen Staining of Lepra Bacilli.

- Often arranged like palisade, known as bundles of cigars arrangement

CULTURE CHARACTERISTICS

M. leprae cannot be cultured in artificial culture media. However, it can grow on foot pad of mice and nine banded armadillos. It is an obligate intracellular organism which grows with in nerve cells. Bacilli prefers lower temperature, hence prefers to localise in nerve cells in skin and other cooler areas of host body.

EPIDEMIOLOGY

Leprosy is a social disease with associated stigma. The diagnosis, treatment and the rehabilitation are all problematic. Transmission is either through droplet infection or contact transmission. Source of infection are multibacillary cases, actually all patients with 'active leprosy' are infectious. Portal of exit from infected individual is nose and ulcerated skin. Attack rate is 4.4%–12% among household contacts. Incubation period averages 3–5 years; however, symptoms may take as long as 20 years to appear.

As per WHO report from 138 countries, the prevalence of leprosy at the end of 2015 was 0.2 cases per 10,000 people. The transmission is still not stopped because new cases keep occurring in 14 countries including India.

LEPROSY, THE DISEASE

It occurs in polar forms. Leper means scaly. Individuals suffering from disease are referred as Lepers. Cardinal signs of leprosy are: hypopigmented patches, partial or total loss of cutaneous sensations in affected areas, presence of thickened nerves and presence of AFB in skin or nasal smears. Suspect signs in skin are: Nodules on ear lobes, smooth, shiny and oily skin, loss of eyebrows and eyelashes, multiple patches and swelling or oedema or blisters on hands and feet. Based on severity, the disease is called either paucibacillary or multibacillary.

Paucibacillary leprosy: 1–5 skin patches with definite sensory loss or any one nerve trunk affected by leprosy.

Multibacillary leprosy: 6 or more skin patches with definite sensory loss or 2 or more nerve trunks affected by leprosy or 5 skin patches and 1 nerve trunk which means 6 lesions.

PATHOGENESIS

Pathogenesis is shown in Flowchart 11.3.

CLASSIFICATION OF LEPROSY

There are many systems to classify leprosy (Box 11.7).

LABORATORY DIAGNOSIS

Diagnosis is easy in lepromatous leprosy and difficult in tuberculoid leprosy.

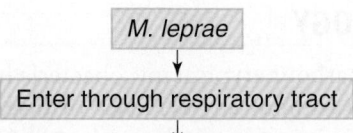

M. leprae
↓
Enter through respiratory tract
↓
Localise in Schwann cell in cutaneous and peripheral nerves
↓
If host cell-mediated immunity (CMI) is good: Either no lesions/skin lesion followed by healing/paucibacillary leprosy (tuberculoid leprosy)
↓
If CMI is weak: Multibacillary leprosy; involvement of skin, nerves, eyes, testes, kidney, voluntary and smooth muscles, reticuloendothelial cells, vascular endothelium (lepromatous leprosy)

Flowchart 11.3 Pathogenesis of Leprosy.

BOX 11.7   Classification of Leprosy

Indian classification
- Indeterminate
- Tuberculoid
- Borderline
- Lepromatous
- Pure neuritic

Madrid classification
- Indeterminate
- Tuberculoid
- Borderline
- Lepromatous

Ridley Jopling classification
- Tuberculoid (TT)
- Borderline tuberculoid (BT)
- Borderline (BB)
- Borderline lepromatous (BL)
- Lepromatous (LL)

Clinical classification
- Multibacillary (MB)
- Paucibacillary (PB)

Bacterial Smear Examination: Demonstration of AFB

Sites for AFB demonstration are:
- Skin smear from lesion, ear lobules, median side of eye brows (5–6 sites)
- Nasal smear from mucosa
- Skin biopsy

Skin smears: It is obtained by slit and scrap method.

Nasal smears: It can be prepared from early morning mucus material. It is used for assessing patient's infectivity.

Grading of smear positivity is done as following using **Ridley's algorithmic scale** (at least 100 fields are examined):

Number of Bacilli/Number of Field	Interpretation
No bacilli seen	Negative
1–10 bacilli/100 oil immersion fields	1+
1–10 bacilli/10 oil immersion fields	2+
1–10 bacilli/each oil immersion field	3+
10–100 bacilli/each oil immersion field	4+
100–1000 bacilli/each oil immersion field	5+
More than 1000 bacilli/each oil immersion field	6+

Bacteriological Scale (BI)

BI is an objective way of monitoring benefit of treatment. It indicates number of bacilli in smears and is calculated by adding up the index from each site examined and divided by total number of slides examined.

Morphological Scale (MI)

The percentage of solid staining bacilli in a stained smear is MI. MI is calculated after examining 200-pink stained free standing bacilli. It helps to signal drug resistance. Only viable leprosy bacillus stains with carbol-fuchsin as solid acid-fast rods. Dead bacilli stain irregularly. Indicator is valuable to measure response to treatment.

Foot Pad Culture

It is 10 times more sensitive than smear examination and is used for detecting drug resistance, for evaluating potency of drugs and for detecting viability of bacilli during treatment.

Immunological Assays

- Lepromin test
- Lymphocyte transformation test
- FLA-ABS test
- Monoclonal antibodies
- ELISA

Lepromin Test

It is done to classify the stage of leprosy. Intradermal injection of lepromin, such as Dharmendra antigen or Mitsuda antigen is given on the forearm of patient. In tuberculoid leprosy, there is a positive delayed reaction at the injection site, while in lepromatous leprosy, there is no reaction.

Early reaction (known as Fernandez reaction):
An inflammatory response develops within 24–48 h and disappears after 3–4 days. Test is called positive, if the diameter of red area is more than 10 mm at the end of 48 h.Positive reaction indicates that a person is previously sensitised by exposure to or infection by leprosy.

Late Reaction (known as Mitsuda reaction): This reaction becomes apparent in 7–10 days and reaches its maximum in 3 or 4 weeks. Test is read at 21 days and at the end, if nodule is more than 5 mm in diameter, reaction is said to be positive. The nodule may even ulcerate and heal with scarring.

Histamine Test

Reliable method for detecting early stage nerve damage due to leprosy. Total 0.1 mL of 1:1000 solution of histamine is injected into hypopigmented patch. In normal person, it gives rise to wheel surrounded by erythematous flare. In leprosy flare response is lost.

TREATMENT

- Medical measures
- Social support
- Programme management
- Evaluation

Drug Regimen

For paucibacillary leprosy, dapsone and rifampicin is given for 6 months in daily doses. Adults get 100 mg dapsone and 600 mg rifampicin daily while children get half the dose. Multibacillilary cases get clofazimine additionally; 50 mg daily and 300 mg once in month, under supervision. These patients get treatment for 12 months. People with single pausibacillary lesion get a onetime dose of rifampicin 600 mg, plus ofloxacine 400 mg plus minocycline 100 mg. Children get half the dose.

Prophylactic treatment is given for at least 3 years or until the index case in each household becomes negative. Acedapsone, one intra muscular injection every 10 weeks also can be given.

ACTINOMYCETES

INTRODUCTION

Actinomycetes commonly presents as subacute-to-chronic infection. Typical characteristics are: suppurative and granulomatous inflammation and formation of multiple abscesses and sinus tracts that may discharge sulphur granules. Lesion spreads to surrounding tissues. Common clinical forms of disease are: cervicofacial actinomycosis (i.e. lumpy jaw), thoracic actinomycosis and abdominal actinomycosis. In women, pelvic actinomycosis is possible. More than 30 species of *Actinomyces* have been described. *Actinomyces israelii* is the most prevalent species isolated in human infections (Box 11.8).

MORPHOLOGICAL CHARACTERISTICS

Actinomycetes are Gram-positive filamentous bacteria, do not form spores and are not acid-fast. They grow slowly in anaerobic-to-microaerophilic conditions. Colonies have characteristic molar tooth appearance.

EPIDEMIOLOGY

Actinomycosis occurs worldwide in people of all ages. Persons with low socioeconomic status, young to middle-aged adults with poor dental hygiene often get infected. Treatment with antibiotics shows high cure rate. Mortality is rare.

PATHOGENESIS

Actinomyces are prominent commensals of the human oropharynx. They are also found in lower gastrointestinal tract and female genital tract. Actinomycosis is often an endogenous infection. Break in the integrity of mucous membranes and devitalized tissue, may facilitate invasion in deeper body structures.

Actinomycosis is a polymicrobial infection. Isolates from infection sites may be 5–10 bacterial species, in number. These are called companion bacteria. These bacteria release some kind of toxin, which causes either tissue damage or inhibit host response, thus helping in establishing infection. Most important companion bacteria is *Actinobacillus* and *Actinomycetemcomitans*. Others are as follows:

- *Peptostreptococcus*
- *Prevotella*
- *Fusobacterium*
- *Bacteroides*
- *Staphylococcus*
- *Streptococcus* species
- Enterobacteriaceae

Tissue response is suppurative, granulomatous, inflammatory response, followed by fibrosis. Infection invades surrounding tissues or organs. Draining sinus tracts that discharge purulent material containing yellow (i.e. sulphur) granules are common. Hematogeneous dissemination to distant organs may occur in any stage of infection. Lymphatic dissemination is unusual.

Cervicofacial Actinomycosis

Cervicofacial actinomycosis is the most common manifestation of actinomycosis. Of total cases, around 50%–70% cases present with cervicofacial manifestations. Cases commonly have history of oral surgery or poor dental hygiene. Commonly it is characterised by soft-tissue swelling of the perimandibular area and invasion of cranium.

Thoracic Actinomycosis

Thoracic actinomycosis Accounts for nearly 15%–20% of cases. Aspiration of oropharyngeal secretions containing actinomycetes is usual mechanism of infection. Common presentation is pulmonary infiltrate or mass, involving pleura, pericardium and chest wall.

Pelvic and Abdominal Actinomycosis

Accounts for nearly 10%–20% of reported cases. Patients with recent or remote bowel surgery may have this presentation. Mostly Ileocecal region is involved and classical presentation is that of a slowly growing tumour. Actinomycosis of the pelvis is most commonly associated with intrauterine contraceptive devices (IUCDs).

LABORATORY DIAGNOSIS

Specimen

Acceptable specimen: Pus or sulphur granules from draining sinuses, deep needle aspirate, biopsy specimens
Unacceptable specimen: Swabs, sputum, and urine specimens

Gram-Stained Smear

Gram-stained smear may demonstrate the presence of beaded, branched, Gram-positive filamentous rods. These are not acid-fast, hence ZN stain is not needed.

Cultures

Cultures should be placed immediately under anaerobic conditions and incubated for 48 h or longer (may require 2–3 weeks). Antimicrobial susceptibility testing is not indicated because of their predictable antibiogram.

Histopathology

Suppurative and granulomatous inflammatory reactions, connective tissue proliferation and the presence of sulphur granules are common features. Coexisting with them are companion bacteria.

Sulphur granules are almost pathognomonic. These are large (0.1–1 mm in diameter) enough to be seen with the naked eye as yellowish particles in pus. Microscopically, granules are Gram-positive cauliflower like structures seen at 10× magnification. At 100× magnification is, a clump of filamentous actinomycete microcolonies is seen. These granules are often surrounded by polymorphonuclear neutrophils.

TREATMENT

Penicillin G is the drug of choice. Incision and drainage of abscesses, excision of sinus tracts and recalcitrant fibrotic lesions, decompression of closed-space infections may be needed. Most of other antibacterials have no activity against actinomycetes.

NOCARDIA SPECIES

INTRODUCTION

Nocardia is found worldwide in soil and dust. These are weakly Gram-positive, acid-fast, filamentous bacteria, usually found in soil and dust. Human disease was first described by Eppinger in 1890, after bovine disease was described by Nocard in 1888.

MORPHOLOGY

Delicate filamentous Gram-positive branching rods that appear similar to *Actinomyces* species. *Nocardia* has mycolic acid content in the cell wall (acid-fast) and grow under aerobic conditions, whereas *Actinomyces* is not acid-fast and grow under anaerobic conditions. It has unique capability to disseminate to any organ, particularly the central nervous system, and to relapse or progress despite appropriate therapy.

COMMON SPECIES

The genus *Nocardia* includes more than 80 species, at least 33 of which cause disease in humans. Common species pathogenic to human are as follows:
- *Nocardia brasiliensis*
- *Nocardia asteroides*
- *Nocardia pseudobrasiliensis*
- *Nocardia otitidis-caviarum* (formerly *Nocardia caviae*)
- *Nocardia farcinica*
- *Nocardia nova*
- *Nocardia transvalensis*
- *Nocardia carnea*
- *Nocardia elegans*
- *Nocardia paucivorans*
- *Nocardia puris*
- *Nocardia takedensis*
- *Nocardia abscessus*
- *Nocardia africana*

EPIDEMIOLOGY

Nocardia species are not members of normal human flora. They are found worldwide in soil, decaying vegetable matter, and aquatic environments. They can become airborne. Inhalation of the organism is considered to be the most common mode of entry. Others can be: ingestion of contaminated food via the gastrointestinal tract, direct inoculation of the organism in skin (cutaneous disease) and nosocomial transmission.

> **BOX 11.9 Important Pathogenic Mechanisms of Nocardia**
>
> - Bacteria in log-phase are resistant to phagocytosis
> - Inhibition of phagosome–lysosome fusion occurs post phagocytosis
> - Production of a bacterial cell surface-associated superoxide dismutase
> - Possibly increased production of catalase

PATHOGENESIS

Nocardia spp. possess multiple mechanisms to overcome the immune response of the host (Box 11.9). Cell-mediated immunity is crucial in containing *Nocardia* spp. infection. Conditions that have been associated with nocardiosis include immunocompromised status, glucocorticoid therapy, malignancy, organ transplant recipients, HIV infection, diabetes mellitus, alcoholism etc.

LESIONS PRODUCED

Acute, subacute, or chronic suppurative infection with pronounced tendency to remission and exacerbation are usually produced by Nocardia infection. Lesions may be localised or disseminated. Cutaneous or lymphocutaneous infections occur after contamination of an abrasion. Disseminated and fulminant disease mainly occurs in immunocompromised hosts.

TREATMENT

Trimethoprim–sulphamethoxazone (TMP–SMX) is the choice of treatment for long. However, since the resistance is frequently seen alternative choices are third generation cephalosporin, amikacin, imipenem, minocycline etc. Mycetoma may also require surgical intervention.

DIFFERENTIATION IN MYCOBACTERIUM, ACTINOMYCETES AND NOCARDIA

At times it may be difficult to differentiate these three organisms morphologically. Table 11.6 lists the common differentiating features.

TABLE 11.6 Characteristics of Mycobacterium, Actinomycetes and Nocardia

Characteristics	Mycobacterium	Actinomycetes	Nocardia
Acid-fastness	Yes	No	Yes
Gram-positive	Weak	Yes	Variable
Branching	No	Yes	Yes
Aerobic	Strict	Anaerobic, microaerophilic	Aerobic
Pathognomonic diagnostic finding	Beaded acid-fast curved rods; tuberculoma	Sulphur granules, Sunray granuloma	Branching beaded acid-fast rods
Most common disease	Tuberculosis	Lumpy jaw	Mycetoma
Treatment	Anti-tubercular drugs	Penicillin G	Trimethoprim–sulphamethoxazone

12

Gram-Negative Cocci

Parul Jain

LEARNING OBJECTIVES

- Definition, epidemiology, cultural characteristics, identification, pathogenesis, diseases caused and treatment of GNC, including
  - *Neisseria* species
- *Moraxella catarrhalis, Kingella, Acinetobacter* species
- Anaerobic GNC

INTRODUCTION

Medically, most important aerobic Gram-negative cocci (GNC) belong to the family Neisseriaceae which includes the genera *Neisseria* (*N. meningitidis, N. gonorrhoeae*), *Moraxella, Kingella* and *Acinetobacter* (*A. calcoaceticus–A. baumannii* complex) and anaerobic GNC including *Veillonella* spp. The salient differential characteristics of important aerobic GNC are listed in (Table 12.1).

NEISSERIA SPECIES

INTRODUCTION

Neisseria species are aerobic Gram-negative diplococci with fastidious growth requirements. These are catalase and oxidase positive, oxidise carbohydrates by production of acid only and reduce nitrates to nitrites.

Most *Neisseria* species, for example, *N. sicca* and *N. subflava*, are non-pathogenic and are found as normal commensals of mouth and nasopharynx. However, there have been reports of septicaemia, meningitis, abscess etc. due to these species, especially in people with deficient immune responses. *N. meningitidis* and *N. gonorrhoeae* are important pathogenic species and are described below.

NEISSERIA MENINGITIDIS

INTRODUCTION

N. meningitidis is a Gram-negative diplococcus with flat adjacent sides (Fig. 12.1A). On the basis of serogrouping with capsular polysaccharide antigen, *N. meningitidis* can be divided into 13 serogroups, that is A, B, C, D, E, H, I, K, L, X, Y, Z, W-135 of which 8 serogroups (A, B, C, L, X, Y, Z, W-135) mostly cause human disease. Group A usually causes epidemics, group C causes localised outbreaks and group B causes both epidemics and outbreaks.

EPIDEMIOLOGY

N. meningitidis occurs in carrier state in nasopharynx of otherwise healthy individuals (10% adults). Human-to-human transmission occurs through respiratory route or through fomite. Meningitis most commonly occurs in children aged 3 months and 5 years. Epidemics usually occur in people living in crowded conditions. Frequent epidemics of meningitis have occurred in the 'meningitis belt of Africa' extending from Ethiopia to Senegal.

TABLE 12.1	Differential Characteristics of Aerobic GNC		
Properties	Moraxella catarrhalis	Neisseria gonorrhoeae	Neisseria meningitidis
Sugar fermentation	None	Glucose only	Both glucose and maltose
Capsule	None	None	Present and is an important virulence factor
Nitrate reduction	Yes	No	No

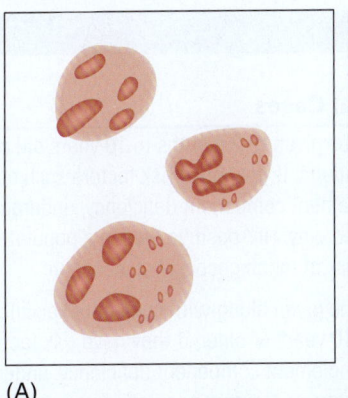 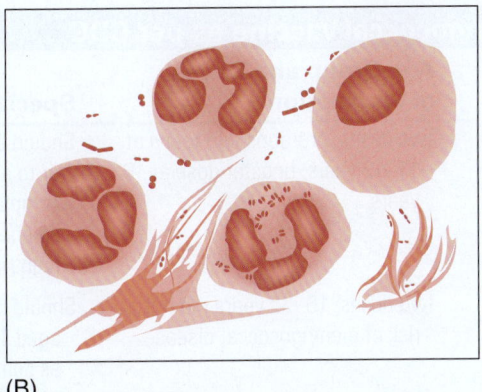 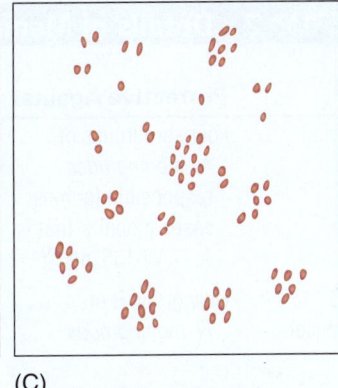

(A)   (B)   (C)

Fig. 12.1 Morphology of Common GNC. (A) Intracellular *N. meningitidis* in CSF (Gram-negative). (B) Intra- and extracellular *N. gonorrhoeae* in urethral smears (Gram-negative). (C) Gram stain of *Acinetobacter baumannii*.

CULTURAL CHARACTERISTICS

N. meningitidis is nutritionally demanding and does not grow on ordinary media. It grows on blood agar, trypticase soy agar, Mueller Hinton agar and supplemented chocolate agar at 35–37°C in the presence of 5%–10% carbon dioxide. The colonies are small, round, transparent, non-pigmented, non-haemolytic, convex and mucoid. Selective medium most commonly used is modified Thayer-Martin medium (supplemented with vancomycin, colistin and nystatin).

IDENTIFICATION

It metabolises glucose and maltose with the production of acid only, but fails to metabolise sucrose or lactose. It has a rapid autolytic rate. It is oxidase and catalase positive but indole and H_2S negative.

PATHOGENESIS

The virulence factors of *N. meningitidis* include (1) outer membrane lipooligosaccharide that acts as an endotoxin; (2) polysaccharide capsule that prevents phagocytosis; (3) fimbriae, pili and surface-exposed proteins Opa and Opc that help in bacterial adherence to nasopharyngeal epithelial cells.

CLINICAL FEATURES

The classic manifestation of meningococcal disease is a **triad of haemorrhagic rash, meningismus and impaired consciousness** (also found in pneumococcal meningitis). The symptoms range from transient fever and bacteraemia to meningococcaemia (**petechial rashes** appear), meningitis and fulminant meningoencephalitis with death occurring within hours of onset of symptoms. Other severe complications include **Waterhouse–Friderichsen syndrome** (massive, bilateral haemorrhage of adrenal glands).

LABORATORY DIAGNOSIS

N. meningitidis can be diagnosed by its isolation from sterile sites. Blood and CSF are most useful samples. Others are skin scrapings from petechial rashes and sterile body fluids such as synovial, pleural or pericardial fluid. The diagnosis is made on the basis of Gram stain appearance (Gram-positive intracellular diplococci), culture and identification followed by Antibiotic Susceptibility Assay.

Meningococcal antigens can be detected in supernant of CSF by latex agglutination or by ELISA. Diagnosis can be made even in cases treated with antibiotics.

Real-time PCR may also be used for diagnosis. Its advantages over culture are availability of results within 24 h and no effect of prior antibiotic administration.

TREATMENT

Penicillin or ceftriaxone is the treatment of choice. In case of penicillin allergy, chloramphenicol may be used.

PREVENTION

Rifampin, ciprofloxacin and ceftriaxone may be used for household chemoprophylaxis. For immunoprophylaxis, at present four types of meningococcal vaccines are available of which two are conjugate vaccines (quadrivalent vaccines that provide protection against serogroups A, C, W and Y) and two are recombinant protein vaccines (monovalent vaccine against serogroup B). The details of the two types of vaccines are mentioned in Table 12.2.

NEISSERIA GONORRHOEAE

INTRODUCTION

N. gonorrhoeae is a non-motile, non-sporulating, Gram-negative diplococcus with flattened adjacent sides found

TABLE 12.2   Details of Meningococcal Vaccine as per CDC

	Protective Against	Age Group and Doses of Vaccination	Special Cases
Meningococcal conjugate vaccines	Four serogroups of *N. meningitides* responsible for most cases globally, that is, A, C, W-135, and Y	Two doses: preteens and teens at 11–12 years, booster dose at 16 years	Should be given to 2 months to 10 years old children or to adults, if they have risk factors such as complement component deficiency, undergone splenectomy, HIV positive, live in a population with high risk of meningococcal disease, etc.
Serogroup B meningococcal vaccine	Only group B of *N. meningitides*	Two doses: 16–23 years old not at risk of meningococcal disease	Should be given along with conjugate vaccine to people aged 10 years or older, if they have risk factors such as complement component deficiency, undergone splenectomy, live in a population with high risk of group B meningococcal disease, etc.

both intracellularly and extracellularly (Fig. 12.1B). The bacteria is transmitted by sexual contact or perinatally and causes a disease called **gonorrhoea**.

CULTURAL CHARACTERISTICS

It grows on blood agar and selective media (Thayer Martin agar, New York City agar) at 35–37°C in the presence of 5%–10% carbon dioxide. The colonies are small, transparent, non-pigmented, non-haemolytic, convex and mucoid.

IDENTIFICATION

It metabolises glucose but not maltose, sucrose or lactose with the production of acid only and reduces nitrates to nitrites.

PATHOGENESIS

N. gonorrhoeae is always pathogenic. The bacteria infects urogenital mucosa (cervix, urethra or rectum), or the oropharyngeal mucosa where it interacts with various cells including polymorphonuclear leucocytes and replicates inside them. Type IV pili, a surface antigen, and mechanisms such as gene shuffling or gene conversion play a role in pathogenesis of *N. gonorrhoeae*.

CLINICAL FEATURES

The clinical features may range as follows:
1. *Asymptomatic infection*: More common in women than in men.
2. *Symptomatic uncomplicated gonorrhoea*: In women common symptoms are odourless mucopurulent vaginal discharge or bleeding. In men they are mucopurulent penile discharge and dysuria.
3. *Complicated gonorrhoea*: Gonococci can ascend to the upper genital tract causing cervicitis, endometriosis or pelvic inflammatory disease (PID) in women and epidid-

ymitis in men. If untreated PID may lead to scarring of fallopian tubes, ectopic pregnancy or infertility.
4. *Disseminated gonococcal infection* (**DGI**): It occurs due to *N. gonorrhoeae* bacteraemia. Septic polyarthritis and dermatitis are the main features.
5. *Neonatal infections:* Ophthalmia neonatorum or gonococcal conjunctivitis is the most common manifestation. Others are sepsis, arthritis, meningitis, vaginitis or urethritis.

LABORATORY DIAGNOSIS

Culture and nucleic acid amplification test (NAAT) are available for diagnosing *N. gonorrhoeae*. As the disease gonorrhoea has medicolegal aspects, Center of Disease Control, USA, has defined three categories of diagnosis. The diagnosis of gonorrhoea is **suggestive** when there is presence of mucopurulent endocervical or urethral exudate on physical examination and a history of sexual exposure to a *N. gonorrhoeae*-infected person.

Diagnosis if **presumptive** when on microscopic examination of urethral exudate from men or endocervical secretions from women, typical Gram-negative intracellular diplococci are visible *or* on culture of these samples on a selective culture medium colonies are obtained that have a typical colonial morphology, oxidase positive, and have a typical Gram-negative morphology *or N. gonorrhoeae* is detected by a non-culture laboratory test.

The diagnosis is **definitive** when *N. gonorrhoeae* is isolated on a selective culture medium from urethra, endocervix, throat, rectum or other sites of exposure with typical colonial and Gram stain morphology, positive oxidase reaction, *and* is confirmed by biochemical, enzymatic, serologic or nucleic acid testing techniques.

TREATMENT

Monotherapy with Benzathine penicillin G, cefixime or ceftriaxone was the treatment of choice for gonorrhoea. But looking at the increasing resistance of *N. gonorrhoeae* to antimicrobials, CDC recommends combination therapy with ceftriaxone (preferred)/cefixime AND azithromycin/doxycycline. In case of penicillin allergy, monotherapy with azithromycin may be given.

PREVENTION

Screening of sexually active people, use of condoms and microbicides may help prevent gonococcal infection.

NON-GONOCOCCAL URETHRITIS

Reiter's syndrome is a triad of urethritis, conjunctivitis and arthritis caused due to *N. gonorrhoeae* and other organisms such as *Chlamydia trachomatis*, *Ureaplasma urealyticum* and *Mycoplasma hominis*, rarely by herpes virus and cytomegalovirus. Urethritis may also be caused by *Trichomonas vaginalis*, *Candida albicans* and *Gardnerella vaginalis*.

MORAXELLA CATARRHALIS

M. catarrhalis (also known as *Neisseria catarrhalis* or *Branhamella catarrhalis*) is a Gram-negative, oxidase-positive, catalase-positive, DNase-positive, non-motile, aerobic diplococcus. Its colony morphology resembles *Neisseria*. It was thought to be normal flora of oropharynx and nasopharynx for decades. But it is now known to cause otitis media, pneumonia, sinusitis and bacteraemia. For treatment, amoxicilline–clavulanate, tetracyclins and oral cephalosporins may be used.

Chronic angular blepharoconjunctivitis is the classical infection caused by *Moraxella lacunata*. Occasionally, it can cause serious systemic disease. Penicillin is usually used for treatment, but β-lactamase-producing strains are now found.

KINGELLA SPECIES

Kingella species are short Gram-negative (sometimes may appear Gram-positive) cocci or coccobacilli that grow on blood and chocolate agar but not on Mac Conkey agar;

are oxidase positive, catalase negative, urease negative and indole negative; and produce acid from glucose and maltose. It may be found as a commensal in respiratory tract of healthy individuals, but it has now been identified to cause disease. *Kingella kingae* is the most common pathogenic species identified, though other species *K. indologenes*, *K. denitrificans* and *K. oralis* are also known. Invasive infections by *K. kingae* most commonly occur in children less than 4 years of age and include skeletal infections, endocarditis and bacteraemia. For treatment, cefotaxime or ceftriaxone is the drugs of choice for *Kingella* species and other HACEK organisms.

ACINETOBACTER SPECIES

Acinetobacter species are non-motile, colourless, non-fermenting, non-nitrate reducing, aerobic, encapsulated, indole negative, catalase positive, Gram-negative cocci or coccobacilli. The species have been grouped as *A. calcoaceticus–A. baumannii* complex. It can cause community acquired as well as hospital acquired infections, including bacteraemia, respiratory tract infections, genitourinary infections, intracranial infections or soft tissue infection. Its isolation from clinical samples is of doubtful significance because of its widespread distribution in nature and its colonising ability of healthy or damaged tissues, including skin, pharynx, gastrointestinal tract, urethra, conjunctiva and vagina. However, repeated isolation of the same organism from clinical samples should be considered to be pathogenic. Nosocomial species are resistant to several antibiotics; however, fluoroquinolones, tigecycline, imipenem, meropenem, polymyxin B and colistin may be used for treatment.

ANAEROBIC GRAM-NEGATIVE COCCI

Anaerobic GNC include *Veillonella*, *Megasphaera* (commensals in mouth, upper respiratory tract, gastrointestinal and vaginal flora), *Acidaminococcus* and *Anaeroglobus*. They are usually arranged in pairs, but can also be seen as single cells, masses or chains. Rarely, *Veillonella* species may cause serious infections, including meningitis, osteomyelitis, bacteraemia and endocarditis.

13

Gram-Negative Rods: Enterobacteriaceae

Parul Jain

LEARNING OBJECTIVES

- Definition, classification, epidemiology, cultural characteristics, identification, pathogenesis, diseases caused and treatment of members of family Enterobacteriaceae:
  - *Escherichia coli*
  - *Citrobacter*
  - *Edwardsiella*
  - *Klebsiella*
- *Enterobacter*
- *Hafnia*
- *Serratia*
- *Salmonella*
- *Shigella*
- *Proteus, Providencia* and *Morganella* (Tribe Proteeae)
- *Yersinia*
- Miscellaneous members of Enterobaceriaceae

INTRODUCTION

The family Enterobacteriaceae contains a large number of genera that are genetically and biochemically related to each other. The members of **family Enterobacteriaceae** are facultatively anaerobic, non-sporing, non-acid-fast, catalase positive, oxidase negative and Gram-negative bacilli that grow readily on ordinary media, ferment glucose with the production of acid only or acid and gas and reduce nitrates to nitrites. The motile species have peritrichous flagella. Members of the family are classified on the basis of lactose fermentation. Examples of **lactose fermenters** are *Escherichia* and *Klebsiella* and of **non-lactose fermenters** are *Salmonella*, *Shigella* and *Proteus*.

The family Enterobacteriaceae has been divided into four tribes:

1. Tribe I: Escherichiae (Genera: *Escherichia, Edwardsiella, Citrobacter, Salmonella, Shigella*)
2. Tribe II: Klebsielleae (Genera: *Klebsiella, Enterobacter, Hafnia, Serratia*)
3. Tribe III: Proteae (Genera: *Proteus, Morganella, Providencia*)
4. Tribe IV: Yersiniae (Genus: *Yersinia*)
5. Tribe V: Erwinieae (Genera: *Erwinia*)

The broad identification of members of Enterobacteriaceae is shown in Flowchart 13.1.

ANTIGENIC STRUCTURE

Most members of Enterobacteriaceae have following major antigens:

1. **O antigen**: These are heat-stable oligosaccharides attached to lipopolysaccharide core of the outer membrane, which forms the basis for serogroup classification. In *Escherichia coli* alone, there are more than 170 different serogroups.

2. **H antigen**: These are heat-labile flagellar antigens present in motile bacteria. *Salmonella* has biphasic H antigen whereas other members have monophasic H antigen.

3. **K antigen**: These are acidic polysaccharide antigens present in the capsule, which surrounds the O antigen.

4. **F antigen**: These are heat-labile surface antigens present in Fimbriae.

ANTIBIOTIC RESISTANCE

Resistance to antibiotics is increasingly been noticed among members of the family Enterobacteriaceae. This resistance is mainly because of mutations in chromosomal genes and mobile resistance genes, such as plasmids. Carriage of several resistance genes on a single plasmid makes the bacterium multidrug resistant.

The significant resistant mechanisms are production of **extended-spectrum β-lactamases (ESBLs) and carbapenem-resistant Enterobacteriaceae (CRE)**. ESBLs are a group of enzymes that break down antibiotics of the class penicillin and cephalosporin and render them ineffective. These can be inhibited by clavulanic acid, sulbactam or tazobactam. The genes responsible for ESBLs production are transmissible and can be transmitted horizontally and vertically. The ESBLs are classified on the basis of two schemes: Ambler molecular classification (four classes on the basis of protein homology of enzymes) and Bush–Jacoby–Medeiros classification (on the basis of functional properties of enzymes).

CREs produce enzymes that break down carbapenems and make them ineffective. *Klebsiella pneumoniae* carbapenemase (KPC) and New Delhi metallo-beta-lactamase (NDM) are some of CREs. Besides the carbapenems (imipenem, meropenem, doripenem etc.), some CRE bacteria exhibit resistance to most available antibiotics, thereby complicating treatment.

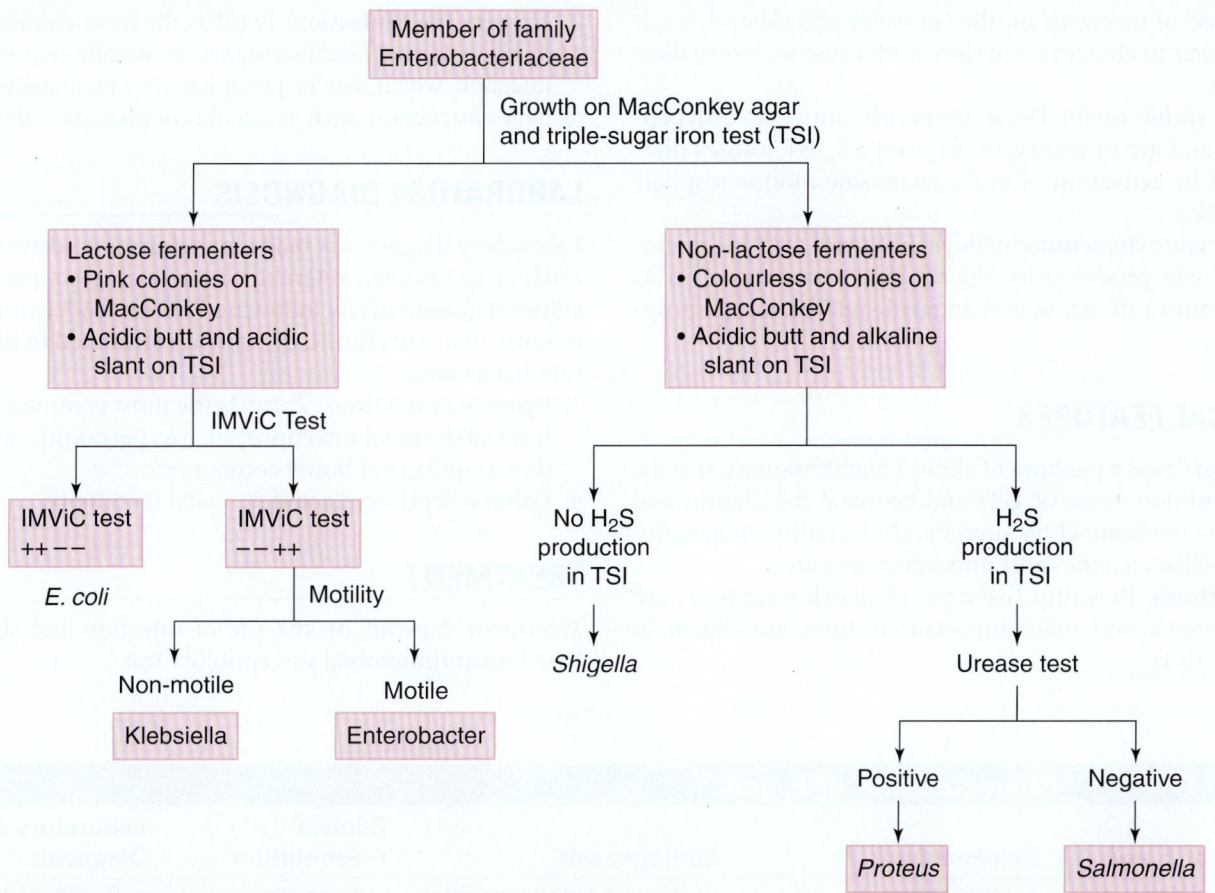

Flowchart 13.1 Presumptive Identification of Enterobacteriaceae.

ESCHERICHIA COLI

INTRODUCTION

The genus *Escherichia* is named after the scientist Escherich who first described this organism in neonatal faecal flora. The type species is *E. coli*, which is the most common bacteria found in human gut and is most common pathogen of this family. It is further divided into several serotypes and biotypes. Other species of less significance are *Escherichia fergusonii*, *Escherichia hermannii* and *Escherichia vulneris* that sometimes may cause disease.

DEFINITION

E. coli is non-sporing, capsulated, motile by peritrichate flagella, Gram-negative bacilli measuring 1–3 × 0.4–0.7 μm arranged singly or in pairs. Though it is found as the normal commensal of the intestine, it causes several diseases, such as urinary tract infection, diarrhoea, nosocomial pneumonia, cholecystitis, osteomyelitis, arthritis etc.

CULTURAL CHARACTERISTICS

E. coli grows on ordinary media at temperatures ranging from 10°C to 40°C. The colonies obtained are large,

greyish white, smooth, moist, opaque and easily emulsifiable. On blood agar, colonies may show haemolysis and on MacConkey agar colonies are pink, due to lactose fermentation.

IDENTIFICATION

E. coli is indole positive, methyl red (MR) positive, Voges–Proskauer (VP) negative and citrate negative. It ferments sugars (except sucrose) with the production of acid and gas, does not liquefy gelatin, is urea negative and does not grow on KCN medium.

VIRULENCE FACTORS AND PATHOGENESIS

The virulence factors include (1) **surface O antigens** that protect from phagocytosis and complement activation, and have endotoxin activity, (2) **fimbriae**, which help in initial attachment and colonisation and (3) toxins, which are of two types: **haemolysins and enterotoxins**. Enterotoxins are further of three types:

1. *Heat-labile toxin*: It has one active subunit A and five binding subunits B. The B subunit binds the toxin to the intestinal epithelial ganglioside receptor following which the A subunit increases the production of cyclic adenosine 5' monophosphate (cAMP). This leads to diarrhoea

because of increased outflow of water and electrolytes. It is similar to cholera toxin since both cause secretory diarrhoea.

2. **Heat stable toxin**: These are poorly antigenic polypeptides and are of two types: ST_A and ST_B. ST_A causes diarrhoea by activation of cyclic guanosine monophosphate (cGMP).

3. **Verocytotoxin/verotoxin/Shiga-like toxin** (SLT): It resembles toxin produced by *Shigella dysenteriae* type 1 in its mechanism of action and antigenic and biological properties.

CLINICAL FEATURES

E. coli can cause a plethora of clinical manifestations. It is the most common cause of UTI and neonatal meningitis, and can cause nosocomial pneumonia, cholecystitis, cholangitis, osteomyelitis etc.; the most important ones are;

1. **Diarrhoea**: Presently, five types of diarrhoeagenic *E. coli* are known and their important features are shown in Table 13.1.

2. **Urinary tract infection**: *E. coli* is the most common cause of UTI. Normal faecal serotypes are usually responsible for infection, which may be precipitated by pregnancy and urinary obstruction, such as calculus or prostatic enlargement.

LABORATORY DIAGNOSIS

Laboratory diagnosis is made by culturing midstream clean catch urine sample, preferably first morning sample. Urine is cultured quantitatively following Kass method, colony count of more than 10^5 cfu/mL of urine is considered to be significant bacteriuria.

3. **Pyogenic infections**: *E. coli* is the most common cause of intra-abdominal infections, such as peritonitis, infections due to spillage of bowel contents etc.

4. **Others**: Septicaemia and neonatal meningitis.

TREATMENT

Treatment depends on the site of infection and should be based on antimicrobial susceptibility test.

TABLE 13.1	Characteristics of *E. coli*			
Types of *E. coli*	**Epidemiology**	**Pathogenesis**	**Clinical Presentation**	**Laboratory Diagnosis**
Enteropathogenic *E. coli* (EPEC)	• Leading cause of nosocomial and community-acquired neonatal diarrhoea in developing countries • Transmitted from person to person	By attaching and effacing effect on the intestinal epithelial cells. It carries bundle-forming pili (BFP), which are responsible for adherence	Vomiting and severe acute diarrhoea	• By PCR or DNA probes • Tissue culture
Enterotoxigenic *E. coli* (ETEC)	• Most common cause of childhood diarrhoea in the developing countries • Causes traveller's diarrhoea • Infection is transmitted through contaminated food or water	Main pathogenic mechanism is mucosal adherence and toxin-mediated fluid secretion by heat-stable and heat-labile toxins	Range from asymptomatic carriage to severe cholera-like illness	Toxin detection by PCR or DNA probes
Enteroinvasive *E. coli* (EIEC)[a]	Transmitted via contaminated food	Cellular invasion similar to *Shigella*, intracellular motility and cell-to-cell spread	Watery diarrhoea or dysentry	PCR or DNA probes
Enterohaemorrhagic *E. coli* (EHEC), Shiga-toxigenic *E. coli* (STEC), Verotoxigenic *E. coli* (VTEC) (Typical serotype O157: H7)	• Spreads by contaminated food, water and from person to person • Important cause of bloody diarrhoea in developed countries	• Targets vascular endothelial cells causing capillary microangiopathy	• Watery diarrhoea • Fatal haemorrhagic colitis • Haemolytic uremic syndrome (HUS)	• Growth on Sorbitol MacConkey agar • Immunoassay for Shiga toxin • PCR or DNA probes
Enteroaggregative *E. coli* (EAEC)	• Unknown mode of transmission • Persistent diarrhoea in developing countries	Fimbriae-mediated aggregative adherence	Mucoid persistent diarrhoea	• Tissue culture assay • PCR

[a]Enteroinvasive *E. coli* (EIEC): It is similar to *Shigella* strains in several respects. Both are non-motile, non-lactose fermenters and lysine decarboxylase negative. EIEC strains can be differentiated from *Shigella* as it ferments glucose and xylose. Both can cause watery diarrhoea, dysentery, fever, severe abdominal pain and stools mixed with blood and mucus.

CITROBACTER

Members of the genus *Citrobacter* can utilise citrate as their sole carbon source, grow in KCN medium, produce H_2S and are late lactose fermenters. Medically important species include *Citrobacter koseri*, *Citrobacter freundii* (gives typical biochemical reactions) and *Citrobacter amalonaticus*. They are antigenically similar to *Salmonella* species.

Citrobacter can be cultured from urine and respiratory tract, which more commonly represents colonisation instead of symptomatic infection. It can also cause intra-abdominal infections, soft tissue infections and osteomyelitis. It has also been incriminated in outbreaks of neonatal meningitis.

Citrobacter can be resistant to several antibiotics due to plasmid encoded resistance gene like inducible ampC gene that determines resistance to ampicillin and first generation cephalosporins.

EDWARDSIELLA

Members of the genus *Edwardsiella* are small, motile, non-capsulated, weakly fermentative Gram-negative bacilli that are relatively biochemically inert. It ferments only glucose and maltose, forms indole and H_2S, utilises citrate and decarboxylates lysine and ornithine. It grows on ordinary media to produce small colonies after 24 h.

The type species *Edwardsiella tarda* (tarda refers to weak fermentation of sugars) is a cause of haemorrhagic septicaemia in fish but recently has been associated with human gastrointestinal and extra-intestinal infections, such as wound infection, meningitis and septicaemia. It has also been cultured from normal and diarrhoeal stools. Transmission appears to occur through infected water or consumption of fish.

KLEBSIELLA

INTRODUCTION

Klebsiella species is normally found as commensals in human intestine and faeces but can also be found in health care settings. The bacteria does not spread through air but is transmitted in health care settings by person-to-person contact, by contamination of the environment or through fomites (e.g. through ventilators).

DEFINITION

Klebsiella is a non-sporing, capsulated, non-motile, Gram-negative short and plump bacilli measuring $1–2 \times 0.5–0.8$ μm. Genus *Klebsiella* has three medically important species: *K. pneumoniae* (type species), *K. oxytoca*, and *K. granulomatis*. *K. pneumoniae* has two subspecies: *K. pneumoniae* subspecies *ozaenae* and *K. pneumoniae* subspecies *rhinoscleromatis*.

CULTURAL CHARACTERISTICS

K. pneumoniae grows on ordinary media at temperatures ranging from 10°C to 40°C. It ferments lactose, produces highly mucoid colonies on plates and is non-motile.

IDENTIFICATION

K. pneumoniae is indole negative, methyl red (MR) negative, VP positive and citrate positive (IMViC--++). It ferments sugars with the production of acid and abundant gas and forms urease.

PATHOGENESIS

The main virulence factor of *K. pneumoniae* is a polysaccharide capsule that inhibits phagocytosis and has more than 70 antigenic types. It also produces fimbrial types (including type 1 pili) that help in adherence to host cells.

CLINICAL FEATURES

K. pneumoniae causes urinary tract infections, liver abscess or pneumonia in otherwise healthy people. Pneumonia generally occurs in middle aged or older persons with alcoholism, chronic bronchopulmonary diseases or diabetes mellitus. It predominantly affects upper lobes with the production of 'currant jelly' sputum thereby causing hemoptysis. It is also responsible for several infections among debilitated or hospitalised people, such as wound infections, biliary tract infections, peritonitis and meningitis. It is the second most common cause of Gram-negative bacteraemia (after *E. coli*).

K. pneumoniae subspecies *rhinoscleromatis* causes rhinosleroma, a chronic granulomatous disease of the respiratory tract.

K. pneumoniae subspecies *ozaenae* has been thought to be associated with ozaena (chronic atrophic rhinitis).

K. oxytoca can cause several nosocomial infections and can also cause haemorrhagic colitis occurring after antibiotic use.

K. granulomatis causes chronic ulcerative genital disease.

LABORATORY DIAGNOSIS

Laboratory diagnosis can be done by culturing appropriate clinical sample and identifying it by biochemical reactions.

TREATMENT

K. pneumoniae has developed resistance to several antibiotics by producing enzymes, such as ESBLs and recently KPC. Treatment depends on site of infection and should always be based on antimicrobial susceptibility test.

ENTEROBACTER

Members of the genera *Enterobacter* are motile, lactose fermenting, and produce mucoid colonies. These are MR and indole negative but VP and citrate positive. There are two medically important species: *E. cloacae* and *E. aerogenes.*

It is usually found in intestine but can cause infections in people on antibiotics and in intensive care units. Infection in hospitalised patients usually arises from endogenous intestinal flora but can also be transmitted from person to person. The clinical syndromes include pneumonia, UTIs, infections of prosthetic devices, meningitis and wound infections.

HAFNIA

The only species of this genus is *Hafnia alvei*, which is a normal commensal in human and animal intestines. It is motile, non-lactose fermenting, indole and MR negative, VP and citrate positive.

Most infections are known in patients with underlying diseases, such as malignancies. As this is also a commensal, isolation of *Halvei alvei* from a sample should be interpreted with caution.

SERRATIA

The only medically important species of this genus is *Serratia marcescens*. It is motile, pleomorphic and produces a pink or red non-diffusible pigment called **prodigiosin** at room temperature that changes its colour with ageing of colonies. It is widely found in the environment in water, soil and food but is not a part of human intestinal flora.

Infection is acquired exogenously and gives rise to wide range of nosocomial infections (UTIs, respiratory tract and wound infections, meningitis, endocarditis) and infections in intravenous drug users. *S. marcescens* produces an inducible, chromosomal AmpC β-lactamase (similar to *Enterobacter*), which make it resistant to ampicillin and first-generation cephalosporins. It may also develop resistance to other drugs, such as fluoroquinolones, trimethoprim–sulphamethoxazole, carbapenems and aminoglycosides.

SALMONELLA

INTRODUCTION

The genus *Salmonella* was named after a pathologist Salmon, who first isolated *Salmonella choleraesuis*. The genus is usually found in intestines of animals and can infect humans causing enteric fever, gastroenteritis, septicaemia etc.

DEFINITION

Salmonellae are Gram-negative, non-spore forming, non-lactose fermenting, facultatively anaerobic bacilli, measuring $2-3 \times 0.4-0.6$ μm in size that produce acid on sugar fermentation, reduce nitrates to nitrites and are motile by peritrichous flagella (except *Salmonella gallinarum-pullorum*, which is non-motile).

On the basis of disease caused, Salmonellae may be divided into two groups: typhoidal or the enteric fever group comprising of typhoid and paratyphoid bacilli and non-typhoidal or the food poisoning group comprising of other Salmonellae.

EPIDEMIOLOGY

The typhoidal strains, *Salmonella typhi* and *Salmonella paratyphi*, have only human hosts and are therefore transmitted by faecal contamination of water, ice, foods and drinks, or by consumption of raw fruits and vegetables fertilised with sewage. Enteric fever is endemic in underdeveloped countries, such as African and Asian countries but also in countries of Europe, Central and South America and the Middle East. In India, *S. typhi* and *S. paratyphi* A are endemic (in a proportion 10:1) and *S. paratyphi* B and C are rare. Infants, preschool- and school-aged children are most commonly affected in endemic regions.

The non-typhoidal Salmonella (NTS) have several animal reservoirs and transmission may occur through consumption of inadequately cooked food, its improper storage or post cooking contact with contaminated raw ingredients, or contact with animals or their environment. The most commonly associated food products include animal products, for example eggs, poultry and dairy products. Despite improvements in hygiene, the incidence of NTS continues to increase in both developed and underdeveloped countries.

CULTURAL CHARACTERISTICS

Salmonellae grow readily on simple media to produce large colonies (2–3 mm in diameter), which are circular, low convex, smooth and translucent. On MacConkey agar, they produce non-lactose fermenting colonies. On selective media, such as **deoxycholate citrate agar (DCA), xylose-lysine-deoxycholate (XLD) agar** or **Hektoen agar** colonies grow with black heads due to H_2S production.

On **Wilson and Blair bismuth sulphite medium,** jet-black colonies appear with a metallic sheen due to H_2S production. *S. paratyphi* A and others produce green colonies owing to lack of H_2S production.

Using enrichment broth, such as tetrathionate and selenite-F, may enhance the recovery of *Salmonella.*

IDENTIFICATION

Salmonellae ferment sugars including glucose, mannitol and maltose, forming acid and gas except *S. typhi* that does not produce gas on sugar fermentation. Lactose, sucrose and salicin are not fermented, indole negative, MR positive, VP negative, urease negative and citrate positive (except *S. typhi* and a few other Salmonellae that do not grow on Simmons' citrate medium). They produce H_2S except by *S. paratyphi* A, *S. choleraesuis* and some others.

On triple-sugar iron (TSI) agar and lysine-iron agar (LIA) slants, the characteristic reaction of Salmonellae is acid butt, alkaline slant, with H_2S and gas.

ANTIGENIC STRUCTURE

Salmonellae are mainly classified on the basis of following antigens:

1. *Flagellar (H) antigen*: H antigen, a heat-labile protein antigen is present on flagella. It is highly immunogenic and can occur in one of the two phases.
2. *Somatic (O) antigen*: O antigen is a phospholipid–protein–polysaccharide complex, which is an integral component of cell wall.
3. *Surface (Vi) antigen*: Polysaccharide heat-labile antigen that acts as virulence factor by inhibiting phagocytosis and is poorly immunogenic. When present, it masks O antigen and inhibits agglutination with O antiserum.

ANTIGENIC VARIATION

Following antigenic variations are seen in *Salmonella*:

H-O Variation

A temporary loss of flagella occurs when Salmonellae are grown on a media containing phenol, which later reappear on subculturing on media without phenol. Flagella may also be lost by mutation, for example, 901-1 strain (non-motile variant of *S. typhi*). But some flagellated cells also occur in such cultures that can be separated by culturing in Craigie's tube (Fig. 13.1).

Phase Variation

The H antigen of most Salmonellae occurs either in Phase 1 (specific) or in Phase 2 (non-specific or group). Phase 1 antigens are designated by alphabets (a, b, c etc. till z followed by z1, z2 etc.). Phase 2 antigens are referred by numbers (1, 2 etc.). A culture contains flagellar antigens of both phases but usually one of them predominates. A culture in Phase 1 may be converted to Phase 2 by subculturing in Craigie's tube containing Phase 1 antiserum and vice versa.

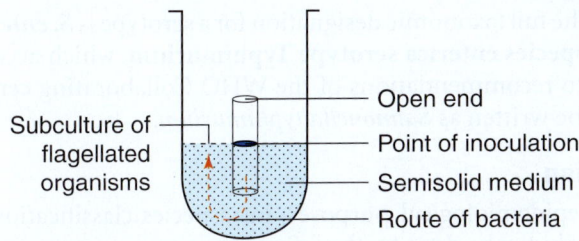

Fig. 13.1 Craigie's Tube for Separating H Forms.

V-W Variation

Forms of Salmonellae covered with Vi antigen surface layer that are agglutinable with Vi antiserum but not with O antiserum are called V form. Others, which are agglutinable with O antiserum but not with Vi antiserum, are the W form.

S-R Variation

Conversion from smooth to rough forms occurs due to mutations on repeated subcultures. The colonies become large, rough, irregular and there is loss of virulence.

NOMENCLATURE AND CLASSIFICATION

The genus *Salmonella* has two species: *Salmonella enterica* that further is divided into six subspecies (I. *S. enterica* subspecies *enterica*; II, *S. enterica* subspecies *salamae*, IIIa, *S. enterica* subspecies *arizonae*, IIIb, *S. enterica* subspecies *diarizonae*, IV, *S. enterica* subspecies *houtenae* and V, *S. enterica* subspecies *indica*) and *Salmonella bongori* (earlier species V). The name of the serotype is usually based on the location where it was first isolated. Nearly all serotypes of *S. enterica* subspecies *enterica* are pathogenic for man. These subspecies are further serotyped into more than 2500 serotypes/serovars on the basis of somatic polysaccharide (O) antigen, flagellar (H) antigen and capsular (Vi) antigen according to **Kauffmann–White scheme**.

According to this scheme, on the basis of O antigen (1, 2, 3 etc.), the Salmonellae are classified into serogroups (A, B, D etc and after Z, 51–67). These serogroups on the basis of phase 1 and 2 flagellar antigens are further classified into serotypes (Table 13.2).

TABLE 13.2	Kauffmann–White Scheme			Antigen H	
Serogroups	**Serotypes**	**Antigen O**		**Phase I**	**Phase II**
A	*S. paratyphi A*	1, 2, 12		A	—
B	*S. paratyphi B*	1, 4, 5, 12		B	1.2
	S. typhimurium	1, 4, 5, 12		I	1.2
C1	*S. paratyphi C*	6, 7 (Vi)		C	1.5
C2	*S. muenchen*	6, 8		D	1.2
D	*S. typhi*	9, 12 (Vi)		D	—
	S. enteritidis	1, 9, 12		g,m	—
E1	*S. anatum*	3 B		e,h	1.6

The full taxonomic designation for a serotype is **S. enterica subspecies enterica serotype Typhimurium**, which according to recommendations of the WHO Collaborating centre, can be written as *Salmonella typhimurium*.

Typing

For epidemiological purposes, intraspecies classification of *S. typhi* can be done by the following:

1. **Bacteriophage typing:** In India bacteriophage typing is done at National Salmonella Phage Typing Center at the Lady Hardinge Medical College, New Delhi
2. **Antibiotic susceptibility pattern**
3. **Molecular methods,** such as multilocus enzyme electrophoresis (MLEE), pulsed field gel electrophoresis (PFGE) and rapid amplified polymorphic DNA analysis (RAPD) are quite popular.

CLINICAL MANIFESTATIONS

Organisms belonging to *S. enterica* subspecies *enterica* produce different clinical manifestations in humans, including enteric fever, gastroenteritis, bacteraemia, localised infections and chronic carrier state.

Enteric Fever

It is also known as typhoid fever, it is a severe systemic illness caused by *S. Typhi*. *S. typhi* derives its name from the Greek word 'typhos' meaning an ethereal smoke or cloud that was believed to cause the disease and madness associated with later stages of the disease. A similar but less severe illness, paratyphoid fever, is caused by *S. paratyphi* A and B (uncommonly *paratyphi* C). The incubation period is usually 7–14 days but may range from 5 to 21 days. Untreated, the disease is fatal in 25% cases. The disease predominantly occurs in areas with poor sanitation and lack of clean drinking water.

Infection is usually acquired through ingestion of contaminated food and water. In the intestine the bacteria penetrate lamina propria and submucosa where they are phagocytosed by neutrophils and macrophages. Within the cells, the bacteria enter the mesenteric lymph nodes and then the bloodstream. It is during this bacteraemia that the bacteria get seeded in the liver, gall bladder, spleen, bone marrow, lungs etc. In the intestine, the bacteria involve Payer's patches and the lymphoid follicles that become inflamed, necrosed and finally slough off forming characteristic typhoid ulcers.

Typhoid disease has a gradual onset with fever, headache, nausea, loss of appetite, abdominal discomfort and constipation or diarrhoea. Fever typically has a step-ladder pattern and is associated with bradycardia and toxaemia. Hepatosplenomegaly can occur. Rose spots may also appear. Overall the symptoms are non-specific.

Complications that may occur include intestinal perforation, haemorrhage and circulatory collapse. Others are meningitis, neuritis, neuropsychiatric symptoms, Guillain–Barre syndrome or delirium.

TABLE 13.3 Week-Wise Test Modalities Preferred for Diagnosing Typhoid Fever

Week of illness	Test Modality	Sensitivity
First Week	• Blood culture • Bone marrow culture • Stool culture	• 90% • 55–90% • 30%
Second Week	• Widal test • Rapid diagnostic tests for detecting specific antibodies (e.g. typhidot, Tubex, Test-it Typhoid)	• 47–77% • 69–84%
Third Week	• Blood culture • Stool culture	• 50% • 60% in children, 27% in adults

Laboratory Diagnosis

The **gold standard for diagnosis** of typhoid and paratyphoid fever is blood culture, the yield of which varies from 90% during the first week of infection to 50% during the third week. Diagnosis can also be made by culturing bone marrow (80% sensitivity), stool (30% positivity during the first week of infection that may increase during the third week in untreated cases), urine, rose spots, bile, intestinal or gastric secretions. Multiple cultures are usually required. The ideal modality to diagnose typhoid fever by which diagnosis can be made in more than 90% patients, is to culture blood, bone marrow and stool samples (Table 13.3). Stool samples are directly plated on selective media (DCA/XLD/Wilson Blair media). Identification is done by biochemical tests and serotyping is done by agglutination with O and H antisera. Unusual serotypes can be sent to National Salmonella Reference Centre at Central Research Institute, Kasauli or those of animal origin to Indian Veterinary Research Institute, Izatnagar.

Serologic tests, such as Widal test become positive during the second week of infection but are not usually recommended because of high rates of false positivity and false negativity. PCR may be used for diagnosis but is not widely available.

Carriers: Patients who continue to have positive stool cultures for at least 1 year after clinical recovery are called carriers. Approximately 10% of cases may become carrier. The bacilli persist in gall bladder or kidneys and are shed intermittently. These individuals shed large number of bacilli in faeces (faecal carriers) or urine (urinary carriers) and act as a source of infection for others by contaminating food and water. Carrier state also occurs in livestock animals and is responsible for food-borne outbreaks. It is important to identify and treat carriers for the control of disease outbreaks. An infamous incidence was 'Typhoid Mary', a cook in New York who caused seven food-borne outbreaks over 15 years. These carriers can be identified by isolation of the bacteria from stool or from bile. Tracing of carriers in cities can be done by sewer swab technique or by filtration of sewage through Millipore membranes and then plating on selective media.

Treatment and Antibiotic Resistance

The first-line antimicrobials for the treatment of typhoid and paratyphoid fevers were ampicillin, chloramphenicol and co-trimoxazole. With the development of resistance to these first-line antibiotics, ciprofloxacin was used as the drug of choice in the 1990s. But because of selection pressures exerted by irrational use of ciprofloxacin in humans and poultry, quinolone resistance appeared and presently, quinolone-resistant strains (designated as nalidixic acid-resistant *S. typhi* or NARST, since nalidixic acid is used to screen fluoroquinolone resistance) have become endemic in the Indian subcontinent. Also it has been documented that ciprofloxacin cannot be used as a marker for the other quinolones (gatifloxacin, levofloxacin and ofloxacin) and therefore all fluoroquinolones should be tested individually. So now third- and fourth-generation cephalosporins, azithromycin, tigecycline and carbapenems are used for treatment of typhoid fever depending upon the susceptibility patterns.

Prophylaxis

WHO recommends the use of any of the following three typhoid vaccines to control endemic typhoid fever and for outbreak control:

1. **Typhoid conjugate vaccine (TCV):** An injectable vaccine consisting of Vi polysaccharide linked to tetanus toxin protein. It can be used in children from 6 months of age to adults up to 45 years old. It is preferred because of better immunological properties, can be used in younger children and provides a longer duration of protection.
2. **Vi-PS vaccine:** An injectable unconjugated polysaccharide vaccine based on the purified Vi antigen. It can be used in persons aged 2 or more.
3. **Ty21a (typhoral) vaccine:** An oral live attenuated vaccine to be used in persons above 6 years of age.

Besides vaccination, the WHO advocates health education, improvements of water quality and sanitation, training of health care professionals in management of typhoid fever as other measures to control the disease.

Gastroenteritis

A self-limiting acute gastroenteritis or food poisoning is caused by non-typhoidal *Salmonella*-like *S. enteritidis*, *S. choleraesuis*, *S. Newport*, *S. anatum*. Contaminated animal products, such as poultry, meat, milk or milk products and water transmit the disease. Symptoms like nausea, vomiting and diarrhoea occur within 6–48 h after ingestion of food. Diarrhoea is usually moderate, self-limited and resolves within 3–7 days but sometimes may be 'cholera-like' (watery and large volume) or 'dysentery-like' (small volume stools with tenesmus).

Isolating *Salmonella* from food or from faeces makes the diagnosis. Treatment is usually symptomatic. Antimicrobial therapy is not recommended because it may increase the duration of carriage. However, it may be given in serious invasive cases. The disease may be controlled by prevention of food contamination at all levels (infection in bird or animal to contamination of the cooked food). The disease is more common in developed countries in contrast to typhoid fever, which is a disease of underdeveloped nations.

Septicaemia

Occurs in up to 8% of patients with non-typhoidal *Salmonella* infection, particularly *S. choleraesuis* and *S. Dublin* infection and in persons at extremes of age or those with immunocompromised conditions. Focal infections, such as osteomyelitis, pneumonia, meningitis may be associated with septicaemia. The case fatality rate may be high.

Isolating *Salmonella* from blood or from pus from the suppurative lesion makes the diagnosis.

| SHIGELLA

INTRODUCTION

Shigella species causes **dysentery**, a clinical condition characterised by frequent passage of stool with blood and mucous often associated with tenesmus and painful defecation. Dysentery can be either bacillary or amoebic. **Bacillary dysentery** is caused by *Shigella* species, enteroinvasive *E. coli*, *Campylobacter* spp., *Vibrio parahemolyticus* etc. Amoebic dysentery is caused by the parasite *Entamoeba coli* (see Chapter 32).

DEFINITION

Shigella spp. are Gram-negative bacilli, members of the family *Enterobacteriaceae*, tribe *Escherichieae* and genus *Shigella*. They are non-motile, non-sporing and non-capsulated.

EPIDEMIOLOGY

Shigellosis is endemic in developing countries possibly because of lack of clean drinking water, poor hygiene, overcrowding and malnutrition. Outbreaks have been reported in crowded or unhygienic environments like prisons or asylums. Incidence is highest during summers and rainy season. Children under 5 years of age are most commonly affected. In HIV infected people, *Shigella* can cause severe disease.

Transmission occurs by faeco–oral route, which may be through contaminated fingers, fomites, food or water. In tropical countries, flies may play an important role in transmission of the disease. Transmission also occurs through sexual route in men who have sex with men.

S. dysenteriae usually causes outbreaks and epidemics of dysentery whereas *Shiglla flexneri* and *Shigella sonnei* cause endemic disease in developing and developed countries respectively. *Shigella boydii* has a restricted geographical distribution and is limited to Southeast Asian countries. In India temporal shifts have been observed in prevalent serogroups of shigellae. Overall *S. flexneri* has been the predominant species followed either by *S. dysenteriae* or *S. sonnei*. *S. boydii* is the least common serotype.

TABLE 13.4   Typical Biochemical Reactions of *Shigella* and Their Exceptions

Typical Reaction of *Shigella* spp.	Exception
Catalase positive	*S. dysenteriae* type 1: catalase negative
Anaerogenic, that is ferments glucose with the production of acid only	*S. flexneri* type 1: biotypes Newcastle and Manchester; *S. boydii* types 13 and 14, produce gas
Ferment mannitol	*S. dysenteriae*: does not ferment mannitol
Does not ferment lactose and sucrose	*S. sonnei*: late lactose fermenter

CULTURAL CHARACTERISTICS

Shigellae are aerobes and facultative anaerobes that grow on ordinary media. Selective media, such as **MacConkey agar** (general purpose medium), **Xylose-lysine-deoxycholate (XLD) agar** and **Eosin-methylene blue (EMB) agar** (more selective media) are generally used for isolation. After overnight incubation at 37°C, the colonies are small (2 mm in diameter), circular, smooth, convex, lactose non-fermenting (except *S. sonnei*, which is a late lactose fermenter and produce pale pink colonies). On **deoxycholate citrate agar (DCA)** and XLD, the colonies do not have a black centre (in contrast to *Salmonella* that produces red colonies with a black centre). It does not grow on Wilson and Blair bismuth sulphite medium (a selective media for Salmonellae).

Use of enrichment broth like GN broth or Selenite broth may enhance the recovery of Shigellae.

IDENTIFICATION

Shigellae are MR positive, VP negative, citrate and urease negative, and do not produce H$_2$S. Serotypes of *Shigella* show several deviations from the typical biochemical reactions (Table 13.4).

On TSI agar and lysine-iron agar (LIA) slants, the characteristic reaction of *Shigella* is acid butt, alkaline slant and no gas or H$_2$S.

ANTIGENIC STRUCTURE

Shigellae have a simple antigenic structure compared to the complex antigenic structure of Salmonellae. They possess major and minor somatic 'O' antigens and fimbrial antigens. Some strains may also possess K antigens. There is considerable antigenic sharing between *E. coli* and *Shigella*.

CLASSIFICATION

The genus *Shigella* comprises four species and several serotypes based on biochemical reactions (mannitol fermentation) and antigenic properties.
1. *S. dysenteriae* (subgroup A): has 17 serotypes. *S. dysenteriae* type 1 produces Shiga toxin.
2. *S. flexneri* (subgroup B): has six classical serotypes (1–6) and several subserotypes (1a, 1b, 2a, 2b, 3a, 3b, 3c, 4a, 4b, 5a, 5b)

3. *S. boydii* (subgroup C): 15 serotypes have been identified
4. *S. sonnei* (subgroup D): has a single serotype but two forms (phase I and II) and several colicin types. It was named after a Danish bacteriologist, Carl Olaf Sonne.

PATHOGENICITY

A bacillary load of 10–100 is enough to cause disease. It is because of such low minimum infectious dose that *Shigella* has a high secondary attack rate and can cause large outbreaks. The reason may be that virulent Shigellae can survive the low pH of gastric juice.

Mucosal Invasion

Virulent *Shigella* and enteroinvasive *E. coli* invade the intestinal mucosa superficially (rarely beyond the mucosa). Thus, despite hyperpyrexia and toxaemia, blood cultures are rarely positive for *Shigella*. The organism binds to M cells of the colon and rectum, invade the lamina propria, multiply intracellularly and spread to neighbouring cells by an actin-dependent process causing severe inflammation, finally destroying the colonic mucosa.

Toxigenicity

S. dysenteriae type 1 produces an exotoxin called Shiga toxin. It has three toxic properties: neurotoxicity (can cause paralysis and death in mice and rabbits), enterotoxicity (can cause fluid accumulation in ligated rabbit ileal loop), and cytotoxicity (can cause cytopathic changes in Vero cell culture). Though, the main virulence factor is the ability to penetrate, the toxin may be responsible for watery diarrhoea during the initial two days of illness. It may also be important in causation of haemorrhagic colitis and haemolytic uremic syndrome.

CLINICAL MANIFESTATIONS

Shigella causes Shigellosis, a disease spectrum varying from mild diarrhoea to bacillary dysentery. Dysentery has a short incubation period (usually 48 h). The main symptoms are frequent passage of loose, scanty stools, with blood and mucus, accompanied with abdominal cramps and straining at stools. Fever and vomiting may be present. Some patients (especially people at extremes of age) may experience significant dehydration because of excessive fluid loss from diarrhoea and vomiting. Complications that may occur include arthritis, conjunctivitis, parotitis, intussusception

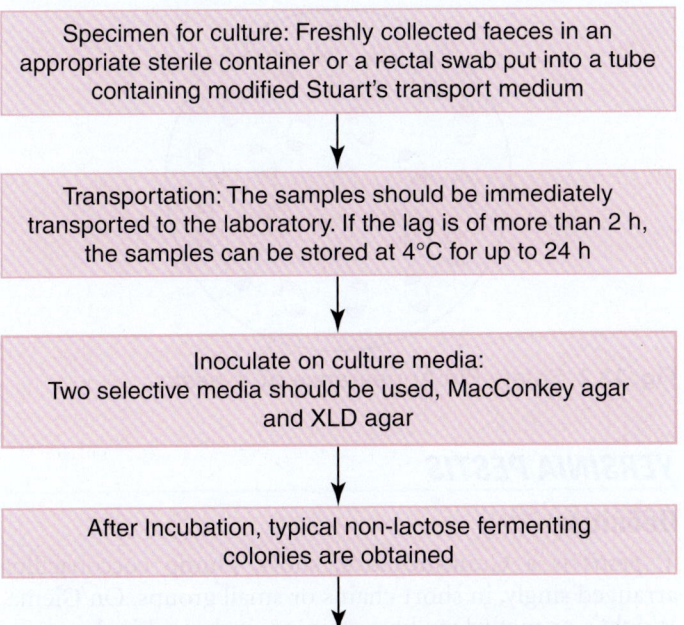

Specimen for culture: Freshly collected faeces in an appropriate sterile container or a rectal swab put into a tube containing modified Stuart's transport medium

↓

Transportation: The samples should be immediately transported to the laboratory. If the lag is of more than 2 h, the samples can be stored at 4°C for up to 24 h

↓

Inoculate on culture media:
Two selective media should be used, MacConkey agar and XLD agar

↓

After Incubation, typical non-lactose fermenting colonies are obtained

↓

Identification by biochemical reactions and by slide agglutination reaction using polyvalent and monovalent sera

Flowchart 13.2 Laboratory Diagnosis of *Shigella*.

(in children) and haemolytic uremic syndrome (HUS), all of which are more commonly seen in infection with *S. dysenteriae* type 1.

LABORATORY DIAGNOSIS

The recovery of *Shigella* from stool is easier during the early stages of the disease but decreases during the later stages because of decrease in viable bacterial counts. The identification of *Shigella* is shown in Flowchart 13.2.

TREATMENT

As per the WHO guidelines, fluoroquinolones are the first-line drugs for treatment, β-lactams and cephalosporins are the second-line drugs. Azithromycin may be used in regions with high ciprofloxacin resistance. The advantages of appropriate antibiotic therapy are clinical improvement within 48 h, shorter duration of symptoms, decreased risk of complications and death and decreased risk of secondary transmission.

CONTROL

Measures that may control shigellosis include the following:
1. Safe water supply
2. Effective sewage disposal system
3. Insecticides during peak seasons for decreasing vector population
4. Health education regarding hand washing, appropriate storage of prepared food and environmental hygiene
5. Breast feeding for infants

PROTEUS, PROVIDENCIA AND *MORGANELLA* (TRIBE PROTEEAE)

INTRODUCTION

Proteus, *Providencia* and *Morganella* are related members of the tribe Proteeae. They are non-lactose fermenters, motile, pleomorphic, produce phenylalanine deaminase (PAD), have a strong urease activity (except some *Providencia* strains), MR positive and VP negative. They are normal inhabitants of the intestine and are opportunistic pathogens. The name 'Proteus' is derived from the Greek God 'Proteus' who was capable of changing form.

The characteristics of tribe Proteeae are as follows:
1. They produce the enzyme phenylalanine deaminase that converts phenylalanine to phenyl pyruvic acid (PPA reaction)
2. They have a strong urease activity
3. Degrade tyrosine
4. Do not acidify lactose, dulcitol or malonate or decarboxylate arginine or lysine

PROTEUS

Members of the genus *Proteus* are saprophytes and are found in dead and decaying matter, sewage and faeces. They are also prevalent in hospital environment and can inhabit the skin of patients and health care workers.

They possess O and H antigens, which share antigenic homogeneity with rickettsia and hence are used for producing antigens for the Weil Felix test used for diagnosis of rickettsial infections. OX2, OX19 and OXK are the three non-motile strains of *Proteus* that are used for the test.

They show swarming motility on culture media. Swarming on the culture plate is swim out of bacteria from the site of inoculation in successive waves so that concentric circles of thin filmy layer appear in the surface moisture film. Swarming occurs because of hyperflagellation of the greatly elongated bacteria. This phenomenon is important in medical bacteriology because when *Proteus* is present in mixed cultures from an infection site, it obscures the other organisms present on the culture plate. Therefore, in laboratory swarming is usually inhibited by various methods such as increasing agar concentration (6%), incorporation of chloral hydrate, sodium azide, alcohol, detergent or other surface-acting agents, or by use of MacConkey agar. The cultures also have a characteristic fishy or seminal smell.

The genus *Proteus* contains several species, but the two medically important species are *Proteus mirabilis* and *Proteus vulgaris*. Both produce urease and H_2S but can be differentiated by indole test. *P. mirabilis* is indole negative and *P. vulgaris* is indole positive.

Proteus species are common causes of urinary tract infections, especially in patients with indwelling catheters or those with abnormalities of the urinary tract. They are also an important cause of bloodstream infections secondary to

TABLE 13.5   Biochemicals Differentiating Members of Family Proteae

Biochemical Test	Proteus	Morganella	Providencia
Hydrolysis of gelatin at 22°C	+	–	–
Urease	+	+	±
Lipase	+	–	–
Fermentation of adonitol	–	–	±
H$_2$S production	+	±	–
Swarming	+	–	–
Fermentation of D-mannose	–	+	+
Indole	±	+	+

UTI. In addition, they can also cause miscellaneous nosocomial infections.

PROVIDENCIA

The genus *Providencia* can be differentiated form *Proteus* and *Morganella* as it is citrate positive and can ferment D-mannitol (Table 13.5). The most common species of this genus is *Providencia stuartii* but *Providencia rettgeri* and *Providencia alcalifaciens* can also be obtained from clinical specimens. These pathogens cause nosocomial infections and are often resistant to several antibiotics. Therefore therapy should be guided by antibiotic susceptibility assay. They are also resistant to commonly used disinfectants and hence are important pathogens in burn units.

MORGANELLA

Morganella morganii is the only species of the genus *Morganella* and is occasionally isolated form nosocomial infections. It does not produce swarming on culture plates and is citrate negative.

YERSINIA

INTRODUCTION

The genus *Yersinia* is a member of the family Enterobacteriaceae that includes several species causing infections primarily in rodents, birds, pigs and other animals. The pathogens are incidentally transmitted to humans who are not a part of the natural disease cycle. Three species are medically important: *Yersinia pestis* (causes plague), *Yersinia enterocolitica* and *Yersinia pseudotuberculosis* (causes gastroenteritis and systemic disease).

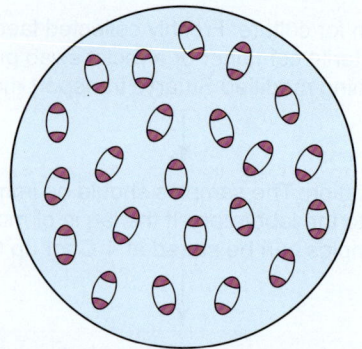

Fig. 13.2 Safety Pin Appearance of *Y. pestis*.

YERSINIA PESTIS

Definition

Y. pestis is a Gram-negative, short, plump coccobacillus arranged singly, in short chains or small groups. On Giema, Wright's or methylene blue staining, it shows bipolar staining or safety pin appearance (Fig. 13.2). It is non-sporing, non-motile, non-acid fast and capsulated.

History of Plague

Plague is an ancient human disease and has caused some of the most devastating epidemics in history. In the 14th century, it caused the disease 'Black death' when as much as one third of Europe's population died. The second and third huge pandemics arose in Asia in the late 19th and early 20th centuries that eventually spread globally causing millions of deaths.

In India, in 1994, outbreaks of bubonic and pneumonic plague occurred in district Beed, Maharashtra and in district Surat, Gujarat, respectively. Later, cases of pneumonic plague were reported from Delhi, Varanasi, Karnataka and other states. In 2002 and 2004 outbreaks of plague were reported from Himachal Pradesh and Uttarkashi, Uttarakhand, respectively. As per the National Centre for Disease Control (NCDC), there are four sylvatic foci in India: The tri-junction of south India i.e. Karnataka, Tamil Nadu and Andhra Pradesh, Beed belt in Maharashtra and Uttarakhand and Rohru in Himachal Pradesh.

Epidemiology

Y. pestis is found on all continents except Australia. Its foci are dynamic that change according to climate, landscape, rodent population migration and other factors. Plague is transmitted to humans via wild rodent fleas, cannibalism or by contaminated food. Two natural cycles of plague exist in nature:

1. **Wild or sylvatic plague:** This occurs in wild animals and rodents independent of humans. Human cases occur when they come in contact with wild rats while working, hunting or camping.
2. **Urban or domestic plague:** Associated with rodents (genus *Rattus*) living intimately with humans and is transmitted by bite of rat fleas *Xenopsyllacheopis*. They have a potential for causing epidemics.

Cultural Characteristics

Y. pestis grows aerobically on most culture media like blood agar, MacConkey agar, nutrient broth, brain heart infusion broth. After 24–48 h of incubation at 35°C, small, delicate colonies are formed on solid media (non-lactose fermenting colonies on MacConkey agar). In broth with oil or ghee added on the top, a characteristic **stalactite growth** appears that hangs down into the surface from the top.

Identification

Y. pestis ferments glucose, maltose and mannitol with the production of acid only but does not ferment lactose or sucrose. It is indole, citrate and urease negative and MR, catalase and esculin positive. On TSI agar, it produces alkaline slant and an acid butt.

Pathogenesis

Fleas acquire infection by feeding on a bacteraemic host. The bacteria colonise the flea midgut, replicate and block the proventriculus. In order to survive, these blocked fleas feed aggressively on another host regurgitating bacteria into the bite thereby transmitting infection. The bacteria in the human body are killed by polymorphonuclear cells but a few bacteria may escape killing in the mononuclear cells. The bacteria multiply intracellularly and are carried via lymphatics to the regional lymph nodes where they create bubo by an intense inflammatory reaction. Bacteraemia occurs that may progress to sepsis, pneumonia, and necrotic, haemorrhagic lesions in various organs. The body releases several inflammatory mediators that may lead to DIC (disseminated intravascular coagulation), organ failure and irreversible shock. Bleeding diathesis results in purpuric lesions that later become dark purple and finally slough (black death).

Clinical Manifestations

Y. pestis causes plague/black death, the clinical features of which vary depending on the route of exposure. The most common type is bubonic plague, followed by septicaemic plague and then pneumonic plague.
1. **Bubonic plague:** After an incubation period of 2–7 days, the patient presents with sudden onset fever, chills, headache, and regional lymphadenitis usually in groin, axilla or neck. The glands enlarge, are extremely tender and suppurate.
2. **Pneumonic plague:** It is of two types—primary (occurs by droplet infection) and secondary (occurs as a result of haematogenous spread of bacteria from bubo). It is usually complicated by sepsis.
3. **Septicaemic plague:** It is usually a complication of the other types of plague but sometimes may occur primarily.

Laboratory Diagnosis

The laboratory diagnosis of plague is shown in Flowchart 13.3.

In case of negative cultures but strong clinical suspicion, serology may be done. The first serum sample should

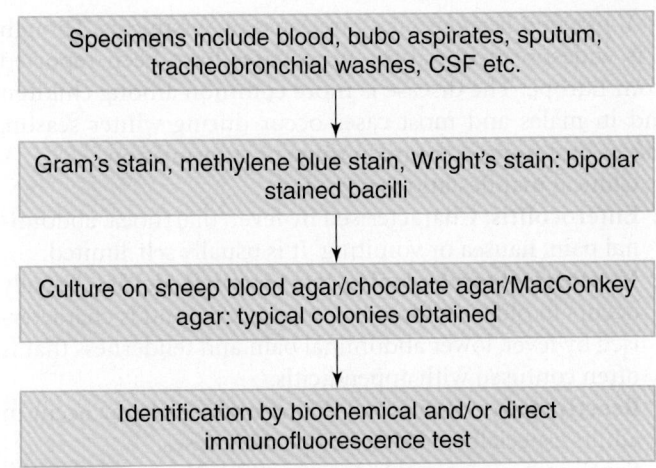

Flowchart 13.3 Laboratory Diagnosis of Plague.

be taken early in illness and 4–6 weeks later, a convalescent sample should be collected.

PCR may be done on clinical samples.

Treatment

The first-line treatment for plague is gentamicin and fluoroquinolones for 10–14 days.

Prophylaxis

Postexposure prophylaxis should be given in persons who came in close contact with a case of pneumonic plague or with infected body fluid or tissues. Doxycycline or ciprofloxacin is recommended for 7 days.

No vaccine is available for plague currently but few are under development. The other measures that can be taken to reduce the incidence of plague are as follows:
1. Reducing rodent habitat around the surroundings
2. Wearing gloves while handling potentially infected substances
3. Using repellents, such as DEET and permethrin

YERSINIA ENTEROCOLITICA AND *YERSINIA PSEUDOTUBERCULOSIS*

These are pleomorphic, Gram-negative bacilli that are non-lactose fermenting, urease positive organisms that can grow over a wide range of temperatures and in different media. These are motile at 25°C but not at 37°C. Colonies grow slowly and appear after 48 h of incubation. *Y. enterocolitica* is classified into 6 biotypes and more than 60 serotypes. *Y. pseudotuberculosis* has six serotypes and four subtypes.

Y. enterocolitica is widely distributed throughout the world and can be isolated from water, foods and a wide range of animals. All ages are susceptible to infection but children are more commonly affected. Transmission of infection occurs by ingestion of contaminated food like raw or undercooked pork, unpasteurised milk or milk products, contaminated water or by direct contact with infected patients or animals. Since, it can grow at 4°C, refrigerated meats can be a source of infection.

Y. pseudotuberculosis is a rare cause of Yersiniosis. Though it is widespread in nature, most cases have been reported from Europe. The disease is more common among children and in males and most cases occur during winter season. Faeco–oral route is the main route of transmission.

Clinical manifestations include:

1. **Enterocolitis**: Characterised by fever, diarrhoea, abdominal pain, nausea or vomiting. It is usually self-limited.
2. **Mesentric lymphadenitis or terminal ileitis**: Usually occurs in older children and adolescents and is characterised by fever, lower abdominal pain and tenderness that is often confused with appendicitis.
3. **Reactive polyarthritis, erythema nodosum** may occur in adults especially in HLA-B27 positive cases.
4. **Septicaemia** may rarely be caused by *Y. enterocolitica* in immunocompromised persons or in elderly.

The diagnosis can be made by isolating *Yersinia* from stool, mesenteric lymph nodes, peritoneal fluid or from blood in case of septicaemia. If Yersinia is clinically suspected laboratory should be specifically informed and specimens from non-sterile sites should also be cultured on a selective medium such as cefsulodin-irgasan-novobiocin (CIN) agar. The plate should be incubated at 25°C to facilitate growth of *Y. enterocolitica* over other organisms. The colonies can be identified by biochemical or molecular tests.

Treatment is usually not required as the infection is self-limited. However in severe cases aminoglycosides, fluoroquinolones, third-generation cephalosporins, trimethoprim–sulphamethoxazole may be used.

MISCELLANEOUS MEMBERS OF ENTEROBACTERIACEAE

Plesiomonas shigelloides (previously classified with vibrios because of oxidase positivity and polar flagella) is rarely isolated from patients with gastroenteritis. It is usually associated with raw shellfish consumption.

Ewingella is a rare cause of nosocomial peritonitis and septicaemia. *Erwinia* (now called *Pantoea agglomerans*) are plant pathogens and has occasionally been isolated from hospitalised patients with respiratory or urinary tract infections.

14

Gram-Negative Rods: Other than Enterobacteriaceae

Parul Jain

LEARNING OBJECTIVES

- Definition, epidemiology, cultural characteristics, identification, pathogenesis, diseases caused and treatment of Gram-negative rods other than Enterobacteriaceae:
  - *Pasteurellaceae*
  - *Vibrionaceae*
- *Pseudomonadaceae*
- Miscellaneous Gram-negative Bacilli
- Anaerobic Gram-negative rods

INTRODUCTION

There are several Gram-negative bacilli other than members of the family Enterobacteriaceae. The medically important ones can be classified on the basis of growth characteristics, biochemical reactions and morphology (Flowchart 14.1).

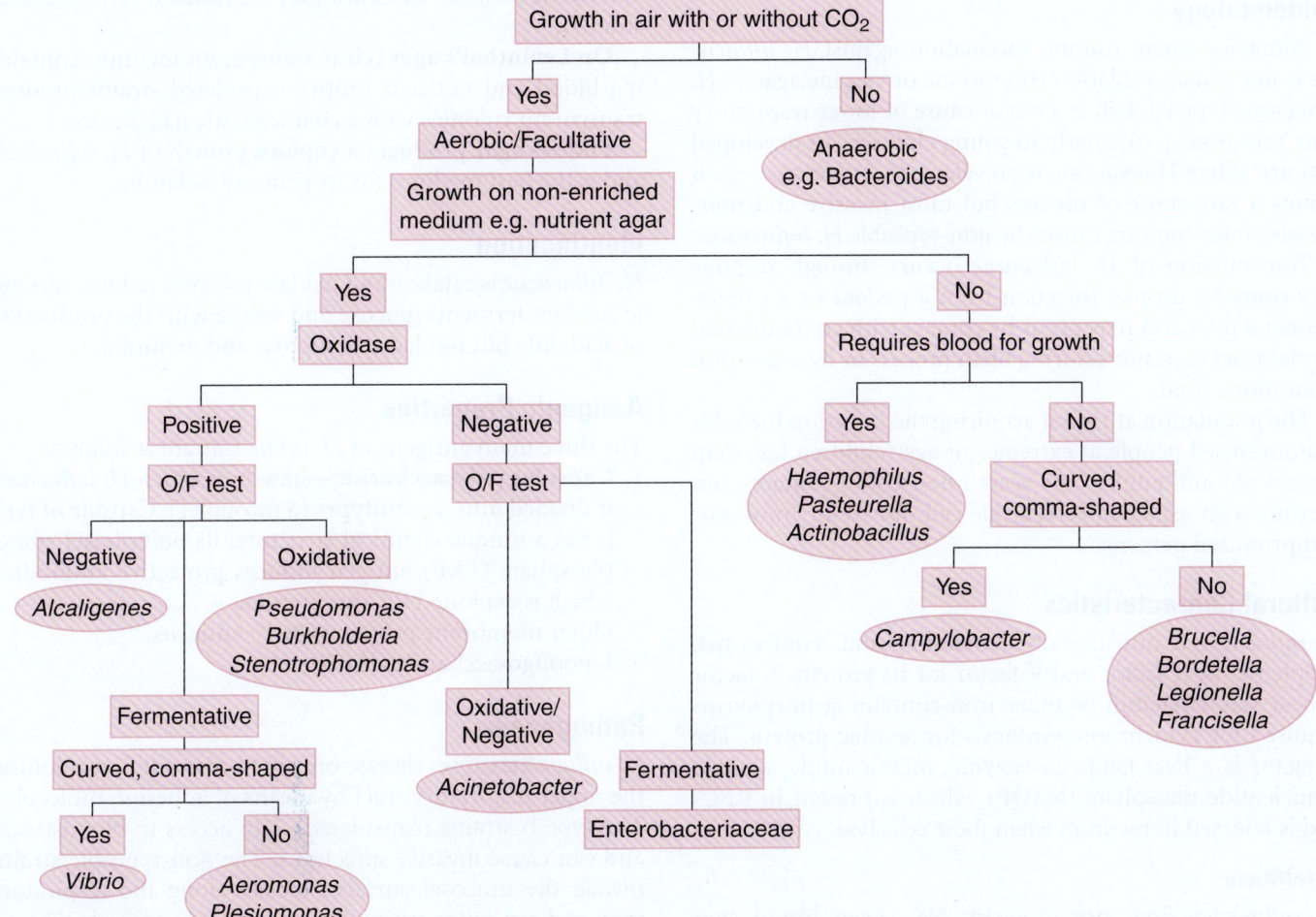

Flowchart 14.1 Classification of Gram-Negative Bacilli Other Than Enterobacteriaceae.

PASTEURELLACEAE

Members of the family *Pasteurellaceae* include commensals and potential pathogens but survive poorly in the external environment. The most important members of this family include three genera, that is, *Haemophilus*, *Pasteurella* and *Actinobacillus*.

The genus *Haemophilus* contains small, non-motile, non-spore-forming, pleomorphic, oxidase-positive, Gram-negative coccobacilli that are pathogenic to humans and animals and are found principally in the upper respiratory tract. The main pathogens in this genus are *H. influenzae*, *H. parainfluenzae*, *H. aegypticus*, *H. ducreyi*, *H. haemolyticus*, and *H. aphrophilus*.

HAEMOPHILUS INFLUENZAE

Introduction

H. influenzae is a pleomorphic Gram-negative coccoba-cillus that may be encapsulated or unencapsulated. The encapsulated ones can be typed with antisera, but not the unencapsulated ones. On the basis of capsular polysaccharides, the encapsulated strains are divided into six serotypes (a through f). Most invasive infections are caused by the 'b' serotype, that is, *H. influenzae* type b (Hib).

Epidemiology

In countries where routine vaccination against *H. influenzae* is not widely available (Hib vaccine or vaccine against *H. influenzae* type b), Hib is a major cause of lower respiratory tract infections particularly in young children. In developed nations, where Hib vaccine is prevalent, *H. influenzae* type b is now a rare cause of disease, but most invasive and non-invasive infections are caused by non-typeable *H. influenzae*.

Transmission of *H. influenzae* occurs through respiratory route by droplet infection from a patient or a carrier. Neonatal infection may occur by contact with contaminated genital tract secretions during birth process or by aspiration of amniotic fluid.

The population at risk of acquiring the infection includes unimmunised people at extremes of age (children less than 5 years old and adults >65 years), household contacts and persons with asplenia, HIV, sickle cell disease or immunocompromised persons.

Cultural Characteristics

H. influenzae is nutritionally demanding and requires two supplements, X factor and V factor for its growth. X factor is heat stable haemin or other iron-containing porphyrins required for cytochrome synthesis for aerobic growth. The V factor is a heat-labile co-enzyme, nicotinamide adenine dinucleotide phosphate (NADP), which is present in RBCs and is released in medium when these cells lyse.

Satellitism

H. influenzae does not grow on 5% sheep blood agar because it lacks factor V, though it contains factor X. But

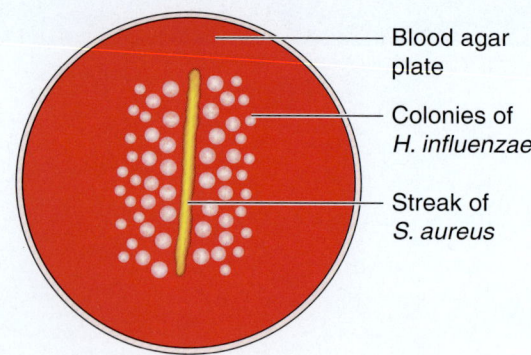

Fig. 14.1 Satellitism as Shown by *H. influenzae*.

when *Staphylococcus aureus* is streaked across a plate inoculated with *H. influenzae*, it grows around the colonies of *S. aureus*, which provide factor V in the culture medium. Size of colonies decreases as they go farther from the *S. aureus* streak. This phenomenon is known as satellitism (Fig. 14.1).

Some strains of *H. influenzae* require 5%–10% CO_2 for optimal growth.

On chocolate agar at 37°C, colonies of *H. influenzae* are circular, initially small (about 0.8 mm in size) that enlarge to 1.5 mm on prolonged incubation (48 h), transparent and dome shaped. Colonies may be mucoid (encapsulated strains).

On Levinthal's agar (clear transparent medium containing blood and nutrient broth), capsulated strains produce transparent colonies with a characteristic iridescence.

Fildes's agar produces a copious growth of *H. influenzae* and is the best medium for its primary isolation.

Identification

H. influenzae is catalase and oxidase positive, reduces nitrates to nitrites, ferments glucose and xylose with the production of acid only but not lactose, sucrose and mannitol.

Antigenic Properties

The three main antigens of *H. influenzae* are as follows:
1. **Capsular polysaccharide**—based on which *H. influenzae* is divided into six serotypes (a through f). Capsule of type b has a unique chemical structure; its polyribosyl ribitol phosphate (PRP) antigen induces protective antibodies, which is exploited for vaccination.
2. Outer membrane protein (OMP) antigens.
3. Lipooligosaccharides (LOS).

Pathogenesis

H. influenzae causes disease only in humans. It first colonises the upper respiratory tract by means of adhesion molecules. The type b strains (capsulated) gain access to bloodstream and can cause invasive infections. The non-typeable strains invade the mucosal surfaces locally along the respiratory tract and can cause otitis media, sinusitis or exacerbations of chronic obstructive pulmonary diseases.

The non-typeable strains can also form biofilms in the middle ear. **Biofilms** are community of bacteria enmeshed in a matrix and attached to a surface that imparts more stability and antibiotic resistance to bacteria.

Clinical Manifestations

H. influenzae can affect many organ systems. Most of the clinical manifestations listed below are caused by Hib, however non-b H. influenzae can occasionally cause similar manifestations. Non-typeable *H. influenzae* usually causes otitis media and bronchitis, but can also cause invasive disease.

1. **Meningitis** is the most serious manifestation. This occurs most commonly in young children and cannot be clinically differentiated from other forms of purulent meningitis. The case fatality rate lies between 3% and 5% in children or higher in adults >65 years of age. Permanent hearing loss or other neurological sequelae may occur in up to 20% survivors.
2. **Epiglottitis (croup)** usually occurs in children aged 2–7 years and is characterised by sudden onset of fever, sore throat, and dyspnea that progress to dysphagia and airway obstruction. Blood cultures are positive in this condition.
3. **Pneumonia** usually occurs in infants and young children. It is insidious in onset.
4. **Bacteraemia** may occur without any local disease especially in young children. The child has a high fever and raised neutrophil counts. Early diagnosis and treatment is essential.
5. **Suppurative lesions**—otitis media, cellulitis, septic arthritis, purulent pericarditis, endocarditis and osteomyelitis are other manifestations caused by *H. influenzae*.

Laboratory Diagnosis

The laboratory diagnosis of *H. influenzae* is shown in Flowchart 14.2.

Treatment

The clinical conditions caused by *H. influenzae* shall be rapidly treated, which otherwise can get fatal rapidly. Cefotaxime or ceftriaxone for 10 days is the treatment of choice.

Prophylaxis

For immunoprophylaxis, polysaccharide conjugate vaccines are available, which are made by joining Hib capsular PRP to a protein carrier (process referred to as conjugation). Three monovalent conjugate vaccines and one combination vaccine (Hib-PRP combined with DTaP/IPV vaccines) are available in the market. A 3-dose series (2-dose series for combination vaccine) should be given to infants 2–6 months of age followed by a booster at 12–15 years of age. Hib vaccine should also be given to persons at risk of invasive disease, that is, people with splenectomy, sickle cell disease or leukaemia.

Chemoprophylaxis with rifampicin should be given to all household contacts.

Specimen: CSF, blood or sputum (depending on the site of infection)
Transport: Samples should immediately be transported to the laboratory. These should never be refrigerated as low temperature kills *H. influenzae*
Direct microscopy: Pleomorphic Gram-negative bacilli in CSF is indicative of *H. influenzae* infection
Antigen detection: By latex particle agglutination test (commercial kits are available) for capsular polysaccharide antigen detection in CSF
Culture: The specimens should be cultured on chocolate agar/blood agar with a streak of *S. aureus*/Levinthal's medium/Filde's agar. The cultures should be incubated at 37°C in presence of 5%–10% CO_2
Identification by biochemicals and typing by slide agglutination

Flowchart 14.2 Laboratory Diagnosis of *H. influenzae*.

HAEMOPHILUS INFLUENZAE BIOGROUP AEGYPTIUS

Formerly known as *H. aegyptius*, it is distributed worldwide and causes conjunctivitis. It also caused Brazilian purpuric fever. An outbreak occurred in Brazil of a fulminant systemic illness characterised by high fever, vomiting and abdominal pain followed by purpura and vascular collapse in children. The disease was associated with a high mortality. This biogroup is unique in that in spite of being unencapsulated, it can cause invasive infections.

HAEMOPHILUS DUCREYI

Ducrey first demonstrated the bacterium in chancroid lesions, an infection with genital ulcer and inguinal lymphadenitis. It exclusively affects human beings. It is highly prevalent in developing countries and causes small outbreaks in developed countries. Infection is transmitted heterosexually and males are more commonly affected. It is also associated with drug abuse.

Infection causes **chancroid/soft sore** characterised by genital ulceration. The lesion is tender, well circumscribed, non-indurated and has ragged edges. It is often associated with inguinal lymphadenopathy. Natural infection does not confer immunity, and patients who have chancroid may have repeated infections.

The main differential diagnosis include other causes of genital ulcers like primary syphilis, genital herpes, donovanosis, lymphogranuloma venereum etc. (refer to Chapter 41).

Diagnosis is made by Gram stain of a swab of the ulcer that may show predominantly Gram-negative coccobacilli.

For isolation, a swab from genital ulcer or an aspirate of suppurative lymph nodes should be cultured on supplemented media like media containing freshly clotted rabbit blood. The organism is difficult to grow and diagnosis is mostly based upon a molecular test like PCR.

The treatment of choice is azithromycin. Ceftriaxone, ciprofloxacin or erythromycin can also be used for therapy. Recent contacts of patients with chancroid (within 10 days prior to onset of symptoms) should also be treated.

OTHER SPECIES

Other *Haemophilus* species like *H. parainfluenzae*, *H. paraphrophilus*, *H. parahaemolyticus* and *H. haemolyticus* can cause human infections. They are commensals of the upper respiratory tract of humans. They have different growth requirements, and this property along with others such as better growth in the presence of CO_2, catalase positivity and haemolysis can be exploited for differentiating the species. *H. parainfluenzae*, *H. paraphrophilus* and *H. parahaemolyticus* need only factor V for growth, but *H. haemolyticus* and *H. aphrophilus* require factor X and V or X only.

HACEK GROUP OF BACTERIA

Haemophilus species are members of the HACEK group of bacteria (*Haemophilus* species, *Aggregatibacter actinomycetemcomitans*, *Cardiobacterium* species, *Eikenella* species and *Kingella* species) that can cause endocarditis and are slow growing. Although earlier prolonged incubation was advocated for these organisms, it is not of much use in automated blood cultures. Treatment should be guided by antibiotic susceptibility assay.

PASTEURELLA

Pasteurella species are the causative agents of haemorrhagic septicemia and pneumonia in animals and rarely may infect humans. *Pasteurella* species are facultatively anaerobic, non-motile, Gram-negative coccobacilli that are oxidase, catalase and indole positive and produce acid from sucrose. They have fastidious growth requirements and can grow on blood and chocolate agar producing smooth, iridescent and mucoid (encapsulated strains) colonies. They cannot grow on MacConkey agar. The most common species affecting human is *P. multocida*.

P. multocida is found as a commensal in upper respiratory tract and gastrointestinal tract of animals such as dogs, cats, cattle and sheep. Humans acquire infection mainly from bite of dogs or cats, but infection can also occur after other animal exposures or even without any known animal contact.

The most common clinical manifestations include skin and soft tissue infections (at the site of animal bite or scratch), bacteraemia and pneumonia. Other manifestations include osteomyelitis, meningitis, respiratory tract infections, endocarditis or appendicitis.

Penicillin is the drug of choice. In case of penicillin allergy, fluoroquinolone with metronidazole or clindamycin can be used.

ACTINOBACILLUS

Actinobacillus spp. are facultatively anaerobic, non-motile, non-sporeforming Gram-negative coccobacilli that stain irregularly (Morse-code appearance). They usually colonize animals and are pathogenic to them but can cause opportunistic infections in humans. A few human cases of soft tissue infections originating from contact or bite of animals (horse, pigs, cattle or sheep) have been reported.

VIBRIONACEAE

Members of the family *Vibrionaceae* are widely distributed in the environment and can cause infections in both humans and animals. It includes genera *Vibrio* and *Aeromonas*.

The most important species of the genus *Vibrio* is *V. cholerae* that causes cholera, a dreaded diarrhoeal disease known since ancient times. Seven pandemics of cholera have occurred in the modern era. The first six were caused by *V. cholerae* O1 classic biotype between 1817 and 1923 and originated from the Indian subcontinent. The seventh is caused by *V. cholerae* O1 El Tor biotype that appeared for the first time in 1961 and originated from Indonesia.

VIBRIO CHOLERAE

Introduction

V. cholerae is a short and curved (comma-shaped) Gram-negative bacillus, about $1–3 \times 0.5–0.8$ µm in size (Fig. 14.2) that is actively motile with a single-sheathed polar flagellum. The motility is typically described as 'darting motility' and vibrios when seen in acute cholera stool, suggesting a 'swarm of gnats'.

Classification (Flowchart 14.3)

V. cholerae is classified into serogroups O1 and non-O1 on the basis serological reactions with somatic O antigen. Gardner and Venkatraman first described this classification in 1935. Currently, 206 'O' serogroups are known of which O1

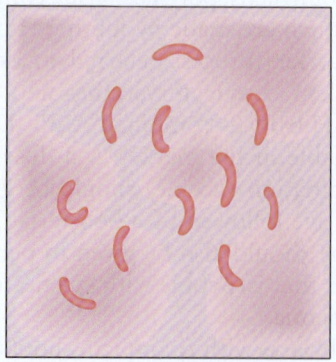

Fig. 14.2 Morphology of *V. cholerae*.

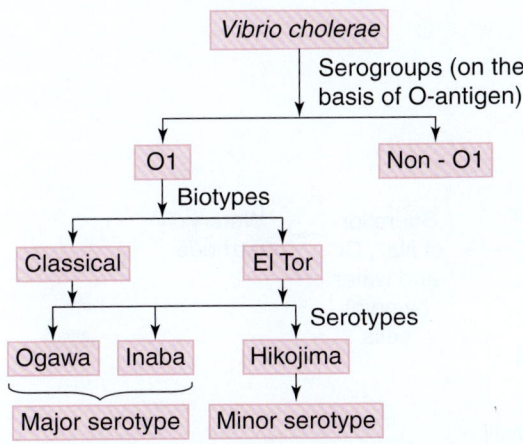

Flowchart 14.3 Classification of *V. cholerae*.

and O139 can cause diarrhoea and have pandemic potential. The O1 serogroup has two biotypes, classical and El Tor. The differences between them are listed in Table 14.1.

These biotypes are further divided into two major serotypes (**Ogawa** and **Inaba**) and a minor serotype (**Hikojima**). The Ogawa serotype expresses A and B antigens, Inaba strain expresses A and C antigens and Hikojima strain expresses all the three antigens. All three serotypes have similar clinical spectra and are important for epidemiological purposes.

Epidemiology

Cholera has unique epidemiological features and can exist in sporadic, endemic, epidemic or pandemic forms. The largest population at risk of cholera lives in the Southeast Asia region including India and Bangladesh. These countries have all the risk factors for cholera outbreaks like poverty, high population density, poor sanitation and extreme climatic conditions such as frequent floods.

V. cholerae (both O1and non-O1 strains) are found in aquatic environments. Humans acquire infections through contaminated water sources or food. Once humans are infected they shed bacteria into the environment, which in turn further contaminate water and food. The secondary attack rate of the disease is high and therefore outbreaks or epidemics occur rapidly in susceptible populations.

Epidemics are more common in summer season and in rainy season, especially following floods. Some host factors like O blood group, infection by *Helicobacter pylori* and breast-feeding also play a role in transmission of cholera.

Cultural Characteristics

V. cholerae is aerobic and grows optimally at a temperature range of 16–40°C (ideal temperature 37°C). Growth is favoured by alkaline pH (8.2, range 6.4–9.6) and in the presence of NaCl (0.5%–1%).

On nutrient agar, the colonies are circular, moist and translucent with a distinctive odour. It shows a bluish tinge in transmitted light. On MacConkey agar, the colonies are colourless, which on prolonged incubation develop a reddish colour (due to late lactose fermentation). On blood agar, a zone of haemolysis surrounds the colonies on prolonged incubation. The special media used for *V. cholerae* are as follows:

1. **Transport medium:** Venkataraman Ramakrishnan (VR) medium, Cary-Blair medium and autoclaved seawater are used as transport media.
2. **Enrichment medium:** Alkaline peptone water or Monsur's taurocholate tellurite peptone water can be used as transport or enrichment media.
3. **Selective agar:** TCBS (thiosulphate citrate bile salts sucrose agar), GTTA (gelatin taurocholate trypticase tellurite agar) and BSA (alkaline bile salt agar) are most commonly used. On TCBS agar, *V. cholerae* produces yellow colonies due to sucrose fermentation.

Identification

V. cholerae ferments sugars like glucose, sucrose, mannitol, maltose and sucrose but not lactose (may be late lactose fermenter), or arabinose, with the production of acid only. It is oxidase and catalase positive but MR and urease negative. It decarboxylates lysine and ornithine but not arginine. Specific identifying tests are as follows:

1. **String test:** When a drop of sodium deoxycholate (0.5%) is mixed with a loopful of colonies, the suspension becomes mucoid and forms a string when loop is withdrawn.
2. **Cholera red reaction:** *V. cholerae* is indole positive and reduces nitrates to nitrites. Therefore, it forms red coloured nitroso-indole when concentrated sulphuric acid is added to a 24-h peptone water culture.

Pathogenesis

V. cholerae enters the human body through contaminated food or water. The infectious dose of bacteria varies with the vehicle (ranging from 10^3–10^6 when water is the vehicle to 10^2–10^4 when food is the vehicle). Gastric acid kills vibrios, and hence people with achlorhydria (use of antacids, chronic gastritis, gastrectomy) are more susceptible to severe disease. The incubation period is 12–72 h.

V. cholerae produces an **enterotoxin** that is encoded by the gene ToxR. The toxin has two A subunits and five B subunits.

TABLE 14.1 Difference Between Classical and El Tor Vibrio		
Test	**Classical Cholera**	**El Tor**
Haemolysis of sheep RBC	−	+
VP test	−	+
Haemagglutination of chick erythrocyte	−	+
Susceptibility to polymyxin B	+	−
Susceptibility to Mukherjee phage IV	+	−
Susceptibility to Mukherjee phage V	−	+

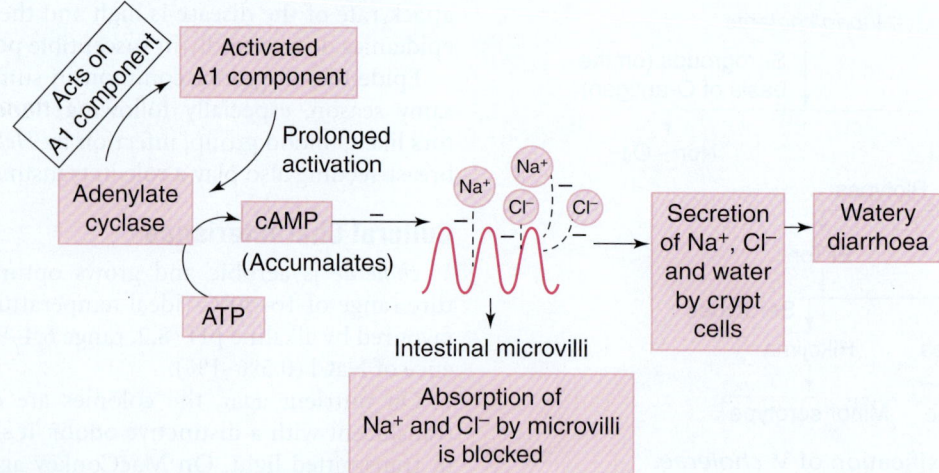

Flowchart 14.4 Mechanism of Action of *V. cholerae* Enterotoxin.

The B (binding) subunit attaches the toxin molecule to the GM1 ganglioside receptors located on the intestinal epithelial cells. The A (active) subunit has two units, A1 and A2, of which A1 on activation by adenylate cyclase causes accumulation of cyclic adenosine monophosphate (cAMP). This cAMP blocks the absorption of sodium and chloride leading to copious amount of water and electrolytes in small intestine and consequent watery diarrhoea (Flowchart 14.4).

Clinical Manifestations

The characteristic of cholera is abrupt onset watery diarrhoea with dehydration varying from mild to severe and life-threatening. Diarrhoea is profuse, painless, without strain, watery containing flecks of mucous and is accompanied with copious vomiting. The diarrhoea stool is called '**rice water stool**' because it resembles water in which rice has been washed. It has a characteristic sweetish odour.

Symptoms of dehydration manifest with progress in diarrhoea such as muscular cramps, oliguria, sunken eyes and so on. Complications that may occur include pulmonary oedema, renal failure and cardiac arrhythmias.

Disease caused by El Tor vibrios is generally milder than that caused by classical vibrios.

Laboratory Diagnosis

Laboratory diagnosis can be made by the following:
1. **Sample collection:** Freshly collected stool sample during the acute stage of disease and prior to administration of antibiotics is the sample of choice. Rectal swabs may also be collected. Unacceptable samples are stool collected from pans or vomitus.
2. **Sample transport:** Stool samples should ideally be immediately plated on the culture plates and inoculated plates should be sent to the laboratory. If this is not possible, the sample may be transported in a transport media such as VR medium or Cary Blair medium at 4°C or in enrichment media such as **alkaline peptone water** or **Monsur's medium**.
3. **Direct microscopy of stool sample:** Darkfield microscopy of a fresh stool sample shows the characteristic '**Darting**

motility' of *V. cholerae*. The diagnosis is confirmed by blocking this movement by using serotype-specific antisera.
4. **Culture:** For definitive diagnosis, stool culture is required. Culture should be done on selective medium like TCBS or GTTA and on a non-selective medium like MacConkey agar.
5. **Identification:** Done by biochemical reactions. For phage typing samples may be sent to National Institute of Cholera and Enteric Disease (NICED) at Kolkata. Serotyping is done by slide agglutination test.
6. **Molecular tests:** PCR and NASBA (nucleic acid sequence-based amplification) may be used for detecting vibrios in stool and environmental samples.
7. **Field testing:** A rapid, highly sensitive and specific immunochromatographic dipstick test for fresh stools is available.

Treatment

The aim of cholera treatment is to restore promptly and adequately the fluid losses caused by diarrhoea and vomiting. This is done orally or by intravenous administration of fluids according to the condition of the patient. Antibiotics have a secondary role in treatment. Doxycycline is the treatment of choice followed by ciprofloxacin. But recently drug resistance has increased particularly in India and Bangladesh. In resistant cases Azithromycin has shown promising results.

Prevention

The most effective way to prevent cholera is by provision of safe drinking water and improvement of environmental sanitation. Following vaccines are available for protection against cholera, but these provide incomplete protection.
1. **Live oral cholera vaccine:** Vaxchora (lyophilised CVD 103 HgR) is a single-dose vaccine recently approved for adults (18–64 years old) travelling to an area of active cholera transmission.
2. **Oral inactivated vaccines:** WHO has approved the following two types:
   a. Killed oral O1 with whole cell with B subunit.
   b. Killed oral O1 and O139.

HALOPHILIC VIBRIOS

Vibrios that require a high concentration of salt for their growth are called halophilic vibrios. These vibrios are ubiquitous in seawater and marine life. The medically important species include *V. parahemolyticus*, *V. vulnificus* and *V. alginolyticus*.

Vibrio parahemolyticus

V. parahemolyticus was first isolated in an outbreak of food poisoning in Japan, and has now been identified as a major cause of acute diarrhoeal illness in several other countries. It is often associated with consumption of raw or undercooked shellfish or oysters.

It resembles *V. cholerae* in morphology. The difference is that *V. parahemolyticus* is capsulated, shows bipolar staining and is pleomorphic. It requires NaCl for its growth and up to 8% NaCl can be added to a media. The selective medium used is TCBS agar on which it produces green opaque colonies in contrast to *V. cholerae* that produces yellow colonies. All the clinical strains isolated are **Kanagawa reaction** positive (β-haemolysis of human erythrocytes) due to production of a heat stable haemolysin. Final identification is done by biochemical tests include catalase, oxidase, nitrate, indole and citrate positive. It ferments sugars like glucose, maltose, mannitol and mannose but not sucrose, lactose or inositol with the production of acid only.

Clinical manifestations include gastroenteritis, ranging from mild watery diarrhoea to frank dysentery-like syndrome. The illness usually begins within 24–72 h following ingestion of contaminated seafood and may be associated with abdominal pain, diarrhoea, fever and vomiting. Dehydration of mild-to-moderate severity occurs, which recovers rapidly. *V. parahemolyticus* can also cause wound infections and septicaemia.

The disease is self-limited and treatment is not usually required. Antibiotics like tetracycline or quinolone may be given to persons with diarrhoea lasting longer than 5 days. Avoiding consumption of raw or inadequately cooked seafood can help prevent the disease.

Vibrio alginolyticus

V. alginolyticus can cause cellulitis and acute otitis media or externa. The disease can occur in healthy people with local trauma who come in contact with seawater. It differs from *V. parahemolyticus* in that it can tolerate higher salt concentration (up to 10%), is VP positive and ferments sucrose. Sea fish is an important carrier of the pathogen.

Vibrio vulnificus

V. vulnificus is a member of normal marine flora and is commonly found in oysters. It is VP negative, ferments lactose but not sucrose. It is the most virulent non-cholera vibrio. It is associated with severe soft tissue infection and septicemia.

V. vulnificus can cause infection in two ways:
1. By contamination of a superficial wound by warm seawater, it can cause intense cellulitis and ulcer formation.

2. On ingestion: It can invade the bloodstream causing bacteraemia. This most frequently occurs in immunocompromised persons and is fatal in >50% patients.

Cellulitis and wound infection respond to antibiotic therapy if initiated timely. Response to antibiotics is not good in bacteraemia in immunocompromised persons. The only effective means of prevention is proper cooking of seafood.

AEROMONAS

Aeromonas spp. are members of the family *Vibrionaceae* along with *Vibrio* spp. They are oxidase-positive, Gram-negative bacilli that are motile by polar flagella. In humans, *Aeromonas hydrophila* has been isolated from cases of diarrhoea and some pyogenic lesions. They resemble *Vibrio* spp., but can be differentiated by some biochemical tests. *Plesiomonas shigelloides* was earlier a member of the family *Vibrionaceae* but has now been re-categorized to the family *Enterobacteriaceae* owing to its molecular relatedness to the latter. It has been associated with some outbreaks of diarrhea following consumption of contaminated water and oysters.

PSEDOMONADACEAE

The family *Pseudomonadaceae* includes genera *Pseudomonas*, *Burkholderia* and *Stenotrophomonas*. These are large, aerobic non-sporing, Gram-negative bacilli. They are ubiquitous and are found in moist environments, soil and water. The main pathogenic species of the genus *Pseudomonas* is *Pseudomonas aeruginosa*. Members of the genus *Burkholderia* were earlier classified as *Pseudomonas*. It has several species, but only three are medically important: *B. cepacia*, *B. pseudomallei* (causing melioidosis) and *B. mallei* (causing equine glanders).

PSEUDOMONAS AERUGINOSA

Definition

P. aeruginosa is a Gram-negative bacilllus measuring 1–3 μm × 0.5–1 μm in size that is oxidase and catalase positive, and is a non-fermenter, actively motile by polar flagella. The term *aeruginosa* is derived from the green-blue hue within the colonies seen in several clinical isolates. It is non-capsulated, but many strains, particularly those isolated from respiratory samples of patients with cystic fibrosis, have a mucoid slime layer. This slime is extracellular polysaccharides composed of alginate polymers in which the microcolonies of the bacillus are enmeshed.

Epidemiology

P. aeruginosa is an important nosocomial pathogen because it can grow in a variety of environments at a wide range of temperature, and has minimal nutritional requirements. In the hospital it can colonise moist surfaces of patients' skin (axilla and perineum), toilets, showers, ventilators and so on. Patients undergoing mechanical ventilation, chemotherapy, surgery, antibiotic therapy or patients with burns usually acquire the infection. The infection usually spreads by hands of health care workers or by contaminated equipment.

In community, infection is mainly acquired after using swimming pools, whirlpools and other water recreational sources when patients acquire otitis externa. Ulcerative keratitis may occur in contact lens users.

Cultural Characteristics

P. aeruginosa is strictly aerobic and can grow over a wide range of temperature (5–42°C). On nutrient agar it forms typical large, opaque colonies with a metallic sheen and often has a slimy appearance. Most strains produce a characteristic fruity or grapelike smell. It produces several different morphological types of colonies like dwarf, coliform and mucoid. It produces non-lactose fermenting colonies on MacConkey agar and haemolytic colonies on blood agar. The selective media used is **cetrimide agar** (detergent containing media that inhibits the growth of several bacteria).

P. aeruginosa produces following characteristic pigments on laboratory media:
1. **Pyocyanin:** Blue coloured, phenazine pigment, soluble in water and chloroform.
2. **Pyoverdin/Fluorescein:** Greenish-yellow coloured, diffusible, soluble in water but not in chloroform.
3. **Pyorubin:** Red coloured
4. **Pyomelanin:** Brown coloured.
   The role of these pigments in pathogenesis is not known.

Identification

P. aeruginosa is strongly oxidase positive, catalase positive, and is a non-fermenter. It oxidatively utilises sugars such as glucose, fructose and xylose using nitrate as a terminal electron acceptor and produces only acid. It reduces nitrate to nitrogen gas. On triple sugar iron agar, it produces an alkaline slant and alkaline butt and no H_2S. It is positive for Simmon's citrate and L-arginine dihydrolase.

For epidemiological purposes, the strains can be typed by serotyping, bacteriocin typing, bacteriophage typing and molecular methods such as pulse field gel electrophoresis.

Pathogenesis

The outcome of infection by *P. aeruginosa* is a complex interplay between the host immune response and bacterial virulence factors. The virulence factors include the following:
1. Toxins including exotoxins, endotoxins, enterotoxin
2. Structural products like cell wall lipopolysaccharide, pili, flagella
3. Enzymes including proteases, phospholipases, elastases, haemolysins
4. Exopolysaccharides, slime formation and biofilm formation also help in pathogenesis
5. Pyocyanin
   The host innate response plays a major role in protecting against *P. aeruginosa* infections. The host factors that make a person susceptible to infection include neutropenia, breach in the anatomic barrier (skin and mucosal surfaces), for example, by burn, use of indwelling devices such as intravenous or urinary catheters, or endotracheal tubes.

Clinical Manifestations

P. aeruginosa can cause infection at any site of the body and can colonise any damaged site. It can cause bacteraemia; respiratory tract infections like acute pneumonia, chronic respiratory tract infections especially in persons with cystic fibrosis; ear infections varying from mild swimmer's disease to malignant otitis externa/necrotising otitis externa; urinary tract infections in catheterised patients; eye infections varying from keratitis in contact lens wearers to endophthalmitis; skin and soft tissue infections especially in burn patients; and rarely central nervous system infections and bone and joint infections.

Laboratory Diagnosis

On routine laboratory media, diagnosis can easily be made based upon pigment production and the characteristic smell. However, about 10% strains are non-pigmented that can be identified by prompt oxidase reaction and other biochemical reactions (see above).

Treatment and Antimicrobial Resistance

P. aeruginosa infection needs careful treatment with antibiotics. It is intrinsically resistant to several antibiotics. The antibiotics with antipseudomonal activity include penicillins (carboxypenicillins and ureidopenicillins), cephalosporins, monobactams, quinolones, aminoglycosides and carbapenems. But as it has developed resistance to several antibiotics, combination therapy (a beta-lactam with an aminoglycoside or a carbapenem with a quinolone) is now preferred over monotherapy. Therapy should always be guided by antibiotic susceptibility test.

Prevention

In hospitals the most effective way to prevent *P. aeruginosa* infections is to religiously follow infection control practices with special attention on hand washing and environmental cleaning. Proper maintenance of pools or water bodies, care of contact lenses and its solutions can help prevent community acquired infections.

STENOTROPHOMONAS

S. maltophilia is the only species in the genus *Stenotrophomonas* (formerly called *Pseudomonas/Xanthomonas*). They are motile, free living saprophytes, glucose non-fermenting Gram-negative bacilli. They can be found on several environmental sources such as tap water, salads, solutions and so on, and grow on most routine media used in a laboratory producing pale yellow or greyish colonies that emit ammonia-like odour. They are catalase positive but oxidase negative and acidify glucose, maltose, lactose and sucrose, and mostly cause opportunistic infections like bloodstream infections and pneumonia that have a high fatality rate especially when not timely and appropriately treated. Treatment should be directed by antimicrobial susceptibility test results but empirically a combination of cotrimoxazole and ticarcillin-clavulanate therapy can be started along with removal of necrotic tissue or foreign material.

BURKHOLDERIA CEPACIA

B. cepacia, formerly called Pseudomonas cepacia, is a free living, motile, non-fermentative, oxidase positive, aerobic, Gram-negative bacillus. It is classified as genomovar I of the B. cepacia complex (Bcc), which contains 10 genomovars/genomic species. It is distributed ubiquitously in the environment particularly aquatic environment. It can colonise fluids (e.g. intravenous fluids, irrigation solutions) used in the hospital. It can grow in media containing antibiotics or disinfectants, for example, B. cepacia selective agar, P. cepacia agar and the oxidation-fermentation polymyxin bacitracin lactose agar on prolonged incubation (up to 72 h). The bacteria have low virulence and rarely infect healthy humans. However, it is an opportunistic pathogen capable of causing pneumonia and bacteraemia, especially in persons with predisposing conditions like cystic fibrosis and chronic granulomatous disease. It is resistant to several antibiotics but can be treated with trimethoprim-sulphamethoxazole, chloramphenicol and minocycline.

BURKHOLDERIA PSEUDOMALLEI

B. pseudomallei, formerly called Pseudomonas pseudomallei, is a motile, oxidase positive, aerobic, Gram-negative bacillus. It shows a 'safety pin' appearance on staining and can be easily isolated on routine laboratory medium. It is found in soil and surface water in endemic areas like Southeast Asia and Australia. Transmission of infection to humans and animals occurs by direct contact with the contaminated source by inoculation, inhalation or by ingestion.

It causes **Melioidosis (or Whitmore's disease)** especially in patients with risk factors such as diabetes, liver disease, cancer or other immunocompromised conditions. The incubation period varies from days to several years. The disease can be either acute or chronic.

Acute Infection

The acute disease can be localised infection at the site of inoculation, pulmonary infection ranging from mild bronchitis to severe pneumonia, bloodstream infection or disseminated melioidosis with abscess formation in various organs of the body like bones, viscera, brain, lymph nodes and so on.

Chronic Infection

Disseminated melioidosis with multiple caseous foci may occur in chronic infection.

Diagnosis in laboratory is made by the following:
1. Staining: **'Safety pin' appearance**
2. Culture: B. pseudomallei can be isolated from blood, urine, skin lesions or abscess
3. Serological diagnosis by detecting specific antibodies in blood

Treatment usually requires intravenous therapy with ceftazidime or meropenem for 10–14 days followed by oral treatment with trimethoprim-sulphamethoxazole or amoxicillin-clavulanic acid.

In areas where the disease in endemic, it can be prevented by avoiding contact with soil and standing water like wearing boots during agricultural work. The organism can be a potential agent of bioterrorism.

BURKHOLDERIA MALLEI

B. mallei is a small, oxidase-positive, aerobic Gram-negative bacillus. It is non-motile and needs an equine host for survival in contrast to B. pseudomallei, which is motile and can survive in the environment without a host. Its potential for zoonotic transmission is also greater than that of B. pseudomallei. It not only causes glanders, which is primarily a disease affecting equine such as horses, donkeys and mules, but can also affect humans and other animals. Glanders disease can be acute or chronic.

Acute Disease

On respiratory inoculation of the bacteria, fever, mucopurulent nasal discharge, lobar or bronchopneumonia and septicemia with dissemination to several organs can occur. On percutaneous inoculation, local skin nodules and regional lymphadenopathy may occur.

Chronic Disease

Chronic pneumonia or lymphatic tract nodules can occur depending on the mode of inoculation, which can later lead to a disseminated disease. Central nervous system may also be involved.

Laboratory diagnosis is made by isolating B. mallei on routine laboratory media from blood, sputum, urine or skin lesions. But laboratory cultures are highly infectious and should be processed with caution.

It is primarily an occupational disease and people who handle infected animals or their specimens like veterinarians, laboratorians, abattoir workers and butchers are at risk. The disease can be treated with antibiotics such as sulphadiazine, tetracyclines, ciprofloxacin, imipenem and so on.

MISCELLANEOUS GRAM-NEGATIVE BACILLI

LEGIONELLA PNEUMOPHILA

Introduction

Legionella species cause Legionellosis that includes **Legionnaires' disease** (acute pneumonia) and **Pontiac fever** (febrile, non-pneumonic illness). Legionnaires' disease was first identified when an outbreak of pneumonia occurred in the members of American Legion convention in Philadelphia in 1976. Several species of the genus Legionella have been described but 20 can infect humans; the most important one is L. pneumophila.

L. pneumophila is a small, obligatory aerobic, Gram-negative bacillus with fastidious growth requirements that uses proteins as the energy source. It can grow at wide temperature range (20–42°C). It contains at least 16 different serogroups.

Epidemiology

L. pneumophila is widely distributed in aquatic environment like lakes, streams, coastal oceans and hot springs at temperatures varying from 5 to 50°C. It is a facultative intracellular parasite of free-living amoebae and protozoa and can survive and multiply within them.

Outbreaks of Legionnaires' disease usually occur in buildings like hotels, hospitals and so on, where the bacterium may grow and spread in human-made water systems like air conditioning plants, hot tubs, water tanks or decorative fountains. Transmission of infection occurs by inhalation of aerosols created. Healthy people usually do not develop the disease. The host risk factors for contracting the disease are age >50 years, smoking, chronic lung disease, terminal illnesses and immunocompromised conditions. Human-to-human transmission does not occur.

Most cases occur during summers and early fall, though cases may occur round the year.

Cultural Characteristics

Legionella species are nutritionally exacting and require complex media such as **BCYE medium** (buffered charcoal, yeast extract agar) with L-cysteine and antibiotic supplements. Growth is slow and occurs best at 35°C. The colonies have a typical opal-like appearance and take 2–5 days (rarely up to 14 days) after inoculation to grow.

Identification

Legionella species can be identified by typical colony morphology, Gram stain appearance and growth dependence on L-cystein. These are thin, pleomorphic Gram-negative bacilli, measuring 2–20 μm. After growth on agar surface, long filamentous forms may develop. Basic fuchsin must be used as the counterstain as the bacterium stains poorly with safranin. Molecular techniques may also be used.

Pathogenesis

Factors required for disease causation:
1. Virulent strains in an environmental site
2. Aerosolisation of bacteria by sources such as air conditioning plants
3. Proper environmental conditions that favour survival and growth of bacteria
4. Susceptible host

Virulence factors of *L. pneumophila* include type IV pili, flagella, major OMP and heat shock protein 60 that help in bacterial adherence to epithelial cells and macrophages or their invasion. The lipopolysaccharide has some endotoxic activity.

L. pneumophila serogroup 1 is the most common causative organism of Legionnaires' disease.

TABLE 14.2   Differences Between Legionnaires' Disease and Pontiac Fever

	Legionnaires' Disease	Pontiac Fever
Clinical features	Fever, cough, myalgia	A flu-like milder illness
Pneumonia	Yes	No
Pathogenesis	Replication of organism	An inflammatory response to endotoxin
Incubation period	2–10 days after exposure	24–72 h after exposure
Percent people contracting the disease on exposure	<5%	>90%
Treatment and outcome	Antibiotics are required and hospitalisation is common. Case fatality rate is high (10%)	The disease is self-limited and only supportive care is required. Hospitalisation is not common and case fatality is rare
Isolation of organism	Can be done	Not demonstrated

Clinical Manifestations

The two common manifestations of disease caused by *L. pneumophila* are Legionnaires' disease and Pontiac fever; the differences among them are listed in Table 14.2.

Laboratory Diagnosis

Laboratory diagnosis is done by any of the following modalities:
1. **Culture:** Isolation of *L. pneumophila* from sputum, bronchial aspirate, lung biopsy or pleural fluid on BCYE medium.
2. **Urinary antigen test:** Detects only *L. pneumophila* serogroup 1. The negative test does not rule out disease caused by other species and subgroups of *Legionella*. The test is also used in epidemiological studies.
3. **Direct Fluorescent Antibody test:** In lung biopsy specimens.
4. **PCR.**
5. **ELISA:** For detecting serum antibodies in acute and convalescent sera.

Treatment

For Legionnaires' disease macrolides, ciprofloxacin or tetracyclines may be used. Antibiotics are not required for Pontiac fever.

Prevention

The disease may be prevented by minimising growth of Legionella in water systems of the building by regular maintenance and disinfection. Timely identification and case reporting are also important.

BORDETELLA PERTUSSIS

Introduction

The genus *Bordetella* was named after Bordet who along with Gengou discovered the bacilli causing whooping cough in 1900 and in 1906 developed a culture medium for isolating the organism. The genus *Bordetella* has 10 known species of which *B. pertussis* and *B. parapertussis* are most common.

B. pertussis is a fastidious, non-motile, non-sporing, capsulated, Gram-negative coccobacillus. It is the causative agent of 'pertussis' or whooping cough. The bacilli have a **thumb print appearance** in smears prepared from cultures that is arranged in loose clumps with clear spaces in between (Fig. 14.3).

Epidemiology

Pertussis remains a problem in the developing and developed world despite vaccination. With the introduction of whole cell pertussis vaccine in 1940s, the disease rates fell drastically. But since then the incidence rates have increased and now disease outbreaks occur every 3–5 years. Most cases occur in unimmunised children younger than 1 year. But in the recent years the incidence of the disease has also increased in adolescents and young adults, which often go undiagnosed as pertussis is generally thought of as a paediatric disease. Immunity provided by pertussis vaccine or by natural infection is not lifelong and diminishes over time. These factors contribute to endemicity of infection and periodic outbreaks.

The organism spreads by droplet infection and is highly contagious among close contacts. Carriers are an important source of infection. Carrier state commonly occurs in adults, but transient nasopharyngeal carriage also occurs in immunised children. Incubation period is generally 7–10 days but may extend up to 21 days.

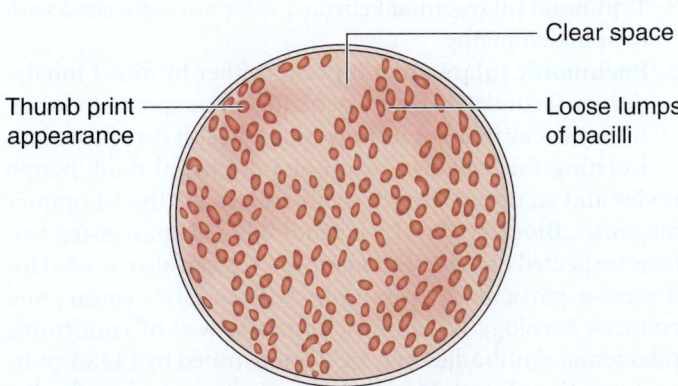

Fig. 14.3 Morphology of *B. pertussis*.

Thumb print appearance — Loose lumps of bacilli — Clear space

Cultural Characteristics

B. pertussis is an obligate aerobe and is highly fastidious and slow growing. It is inhibited by the components commonly found in culture media like fatty acids, metal ions and so on. Isolation requires addition of blood, starch or charcoal to the medium. Bordet Gengou glycerine potato agar (BG) or Regan Lowe charcoal medium (RL) are generally used in the laboratory for growth of *B. pertussis*.

After 48–72 h, colonies are small, smooth, opaque, greyish white, refractive, dome shaped, shining and have a 'bisected pearl' or 'mercury drop' appearance. Confluent growth has an 'aluminium paint' appearance.

Identification

B. pertussis is biochemically inert and does not ferment sugars. It is negative for citrate and urease and does not reduce nitrates.

Pathogenesis

B. pertussis produces several antigenic and biologically active products, including pertussis toxin (PT), filamentous haemagglutinin (FHA), adenylate cyclase, pertactin and tracheal cytotoxin. The steps and the virulence factors involved in developing pertussis are listed in Flowchart 14.5.

Clinical Manifestations

There are three clinical stages of pertussis, that is, catarrhal, paroxysmal and convalescent.

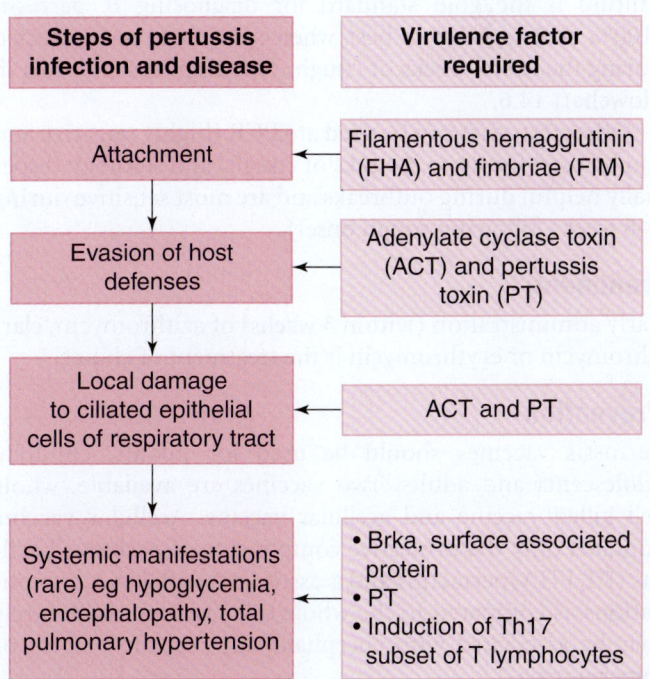

Flowchart 14.5 Steps and Virulence Factors Associated With Development of Pertussis.

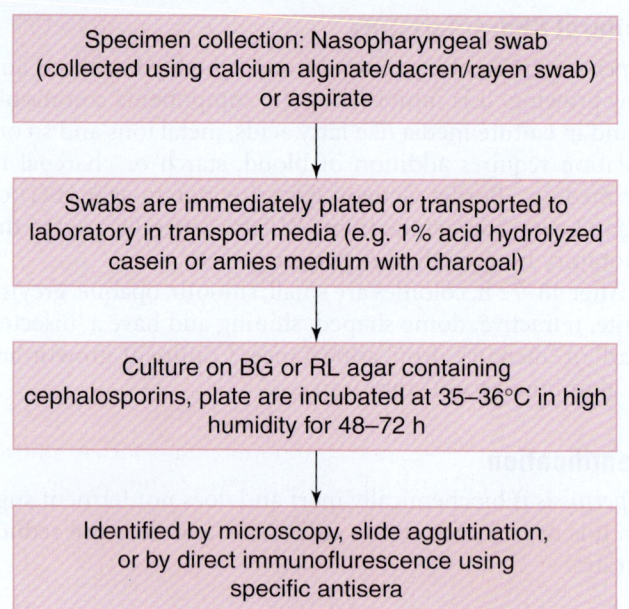

Specimen collection: Nasopharyngeal swab (collected using calcium alginate/dacren/rayen swab) or aspirate

↓

Swabs are immediately plated or transported to laboratory in transport media (e.g. 1% acid hydrolyzed casein or amies medium with charcoal)

↓

Culture on BG or RL agar containing cephalosporins, plate are incubated at 35–36°C in high humidity for 48–72 h

↓

Identified by microscopy, slide agglutination, or by direct immunoflurescence using specific antisera

Flowchart 14.6 Laboratory Diagnosis of *B. pertussis*.

1. **Catarrhal** stage lasts for 7–10 days and is characterised by coryza, low-grade fever and occasional mild cough. This is the most infective stage but can be arrested by antibiotics.
2. **Paroxysmal** stage lasts for 1–6 weeks and is characterised by paroxysms of numerous, rapid cough and long inspiratory effort with a characteristic 'whoop' at the end of paroxysms. Cyanosis, vomiting and exhaustion follow this.
3. **Convalescent** stage lasts for 1–7 days during which frequency and severity of cough gradually decrease.

Laboratory Diagnosis

Culture is the gold standard for diagnosing *B. pertussis* (100% specific) but is best when specimens are collected during the first 2 weeks of cough. The diagnosis is shown in Flowchart 14.6.

Other tests that can be used are PCR (highly sensitive and can be used for up to 4 weeks of cough) and serology (especially helpful during outbreaks and are most sensitive during 2–8 weeks following cough onset).

Treatment

Early administration (within 3 weeks) of azithromycin, clarithromycin or erythromycin is the treatment of choice.

Prevention

Pertussis vaccines should be used for infants, children, adolescents and adults. Two vaccines are available, whole cell killed vaccine and acellular vaccine. Acellular vaccine contains only the protective components of pertussis bacillus (PT, FHA, pertactin) and is associated with lesser complications as compared to the whole cell vaccine, which rarely may be associated with encephalopathy, seizures, shock or hyperpyrexia.

These are generally given as combined vaccines (along with diphtheria and tetanus) (see Chapter 10). As per CDC,

5 doses of DTaP is given to children between 2 months to 6 years of age followed by 1 dose of Tdap at 11–12 years.

BORDETELLA PARAPERTUSSIS

It causes a mild disease and is not protected by the pertussis vaccine. It can be differentiated from *B. pertussis* by its non-fastidious growth characteristics and production of a diffusible brown pigment after 2 days of growth.

FRANCISELLA TULARENSIS

F. tularensis is a small, aerobic, catalase-positive, pleomorphic, Gram-negative coccobacillus primarily affecting animals and rarely humans. It causes tularaemia, a disease responsible for significant mortality and morbidity in spite of availability of effective antibiotics. *F. tularensis* has regained importance because of its use as a biological weapon.

F. tularensis has fastidious growth requirements and requires cysteine or cysine for growth. It therefore does not grow on majority of solid media used routinely in laboratory. It can, however, be grown on modified Thayer Martin medium, thioglycolate broth, chocolate agar and so on. It should be suspected when small, slow growing colonies are obtained on chocolate agar and not on blood agar. Growth occurs best at 35°C, under 5% CO_2 and requires 2–5 days to appear.

Strains of *F. tularensis* have been divided into several **biotypes** based on their virulence and epidemiology. Highly virulent strains are found in North America and low virulence strains are also found in Europe and Asia.

The incubation period is usually 3–5 days, but may range from 1 to 21 days. The **clinical manifestations** depend on the portal of entry of organism. They are as follows:

1. **Ulceroglandular tularaemia:** Most common presentation and is associated with tick bites and animal contacts.
2. **Glandular tularaemia:** Patient has tender regional lymphadenopathy but without a cutaneous lesion.
3. **Oculoglandular tularaemia:** It occurs rarely when organisms enter through conjunctiva from contaminated fingers or splashes.
4. **Pharyngeal tularaemia:** When the site of invasion is oropharynx.
5. **Typhoidal tularaemia:** Febrile disease not associated with lymphadenopathy.
6. **Pneumonic tularaemia:** It occurs either by direct inhalation or secondarily from haematogenous spread to lungs. It may occur in laboratory workers, health care providers.

Isolating the organism from blood, pleural fluid, lymph nodes and so on on proper media can make the laboratory diagnosis. **Biosafety level 3** is required while processing isolates suspected of being *F. tularensis*. PCR can also be used for diagnosis particularly in samples with negative smears and cultures. Serology is the most common way of confirming tularaemia. Antibodies may be demonstrated by ELISA or by agglutination. Fourfold or greater rise in titre of antibodies between acute and convalescent sera confirms the diagnosis.

TABLE 14.3 Characteristics of Medically Important *Bartonella* spp.

Bartonella spp.	Vector	Geographical Distribution	Non-Human Vertebrate Reservoir	Disease Caused
B. bacilliformis	Sandfly vector (genus *Lutzomyia*)	Geographically limited to Andes mountains	None	Oroya fever/Carrion's disease, verruga peruana
B. quintana/Rochalimaea quintana	Human body louse (*Pediculus humanus*)	Distributed globally	None	Trench fever
B. henselae	Cat flea (*Ctenocephalides felis*)	Globally endemic	Cats	Bacillary angiomatosis, cat scratch disease

Treatment is by antibiotics like streptomycin or gentamicin. The disease can be prevented by avoiding exposure to the organism by taking measures such as (1) treatment of community water supply with standard chlorination, (2) using chemical tick repellents and (3) practicing universal precautions for contaminated secretions. Vaccines against *F. tularensis* are ineffective.

BARTONELLA

Members of the genus *Bartonella* are genetically closely related to *Brucella* and *Agrobacterium* and distantly related to Rickettsia. *Bartonella* species primarily infect animals. Humans are incidentally affected when transmission occurs via arthropod vectors or direct inoculation. The medically important species and their characteristics are shown in Table 14.3. *B. quintana* was earlier classified with Rickettsia and was called *R. quintana*. Later because it was found to be different from Rickettsia, it was classified as Rochalimaea and later on the basis of taxonomical shift it was named as *B. quintana*.

Bartonella species are curved Gram-negative bacilli, measuring 0.25–0.5 μm by 1–3 μm in size, non-acid-fast, and stain with silver impregnation techniques (e.g. Steiner, Warthin–Starry) in tissue sections. The different diseases caused by medically important Bartonella species are listed in Table 14.3.

Oroya fever caused by *B. bacilliformis* presents as fever and progressive anaemia due to bacterial invasion of erythrocytes. It has a high mortality in untreated cases. Verruga peruana is a late sequel in survivors. Trench fever, caused by *B. quintana*, is self-limited febrile disease characterised by periodic febrile paroxysms, each lasting approximately 5 days. Cat scratch disease, caused by *B. henselae*, is a febrile illness with lymphadenopathy that occurs after a cat scratch. Bacillary angiomatosis is a disease in which vascular nodules or tumours appear on skin and mucosa.

Laboratory diagnosis of *Bartonella* species is most commonly done by isolating the bacteria from blood and tissue, which should be processed as soon as possible. In case of delay the samples should be frozen. Use of lytic blood culture systems increases the yield. The samples are plated on freshly prepared blood and chocolate agar plates containing rabbit heart blood. Incubation for at least 7 days is required at 35–37°C, under 5%–10% CO_2 and more than 40% humidity. Two morphologic types of colonies are obtained in the same culture: cauliflower or verrucous type and smaller, circular type. Presumptive identification is done by slow growth, small curved Gram-negative bacilli, oxidase and catalase negativity. Definite identification is done by cellular fatty acid composition by gas–liquid chromatography, PCR or by serologic testing.

Treatment of choice is chloramphenicol, though ciprofloxacin, tetracyclines or trimethoprim-sulphamethoxazole can also be used. Avoiding contact with arthropod vectors or cat may prevent the disease.

BRUCELLA

Introduction

Brucella species causes **brucellosis (Malta fever/undulant fever/Mediterranean fever)**, a zoonotic disease that may be transmitted to human beings. The organism is a potential bioterrorism agent due to its ability to undergo aerosolisation. Bruce first isolated *B. melitensis* in 1886 from victims of Malta fever on the island of Malta.

Brucellae are small, aerobic, non-sporing, non-motile, non-capsulated, Gram-negative coccobacilli that are catalase positive, but give variable reactions for oxidase, urease and H_2S production. The medically important species are *B. abortus*, *B. melitensis* and *B. suis*. Their important characteristics are shown in Table 14.4.

TABLE 14.4 Principal Species of *Brucellae* Affecting Humans

	B. melitensis	*B. abortus*	*B. suis*
Principal host	Goats and sheep	Cattle	Swine
Pathogenicity	Highest	Moderate	High
Mode of transmission	By consumption of unpasteurised dairy products	Contact with vaginal discharges and birth products	Usually transmitted to hunters who come in contact with infected wild pigs

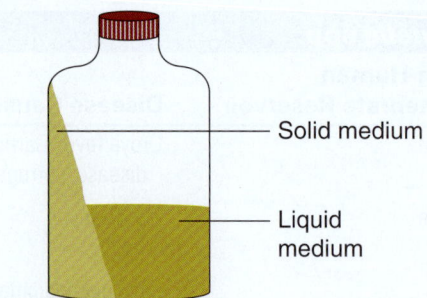

Fig. 14.4 Castaneda Medium.

Epidemiology

Brucellosis is a worldwide disease, but is mainly found in Asia, Africa, the Caribbean, Middle East, parts of America and Europe. The disease can be transmitted by the following:
1. **Ingestion** of bacteria through undercooked meat or unpasteurised/raw dairy products.
2. **Inhalation** of dried material from animal origin.
3. **Contact** through skin wounds or mucous membranes.

Occupations with increased risk of exposure to brucellosis include butchers, veterinarians, hunters, meat packers and laboratory workers.

Cultural Characteristics

Brucellae are aerobic, though some require 5%–10% CO_2 for growth. They can be grown on serum dextrose agar, serum potato infusion agar, trypticase soy agar or tryptose agar. The colonies obtained are small, moist, transluscent and shining.

Blood sample is usually cultured on Castaneda medium (Fig. 14.4), a biphasic blood culture bottle containing trypticase soy broth and solid medium in the same bottle. This method is useful as for isolation of brucellae repeated subcultures are required on solid media every 3–5 days. For subculture the bottle is tilted so that liquid media flows on the surface of solid media. Hence, repeated manipulation of the media is not required. This minimises the chance of contamination and also exposure to the laboratory workers.

Identification

Brucellae are catalase positive, citrate negative; indole, MR and VP tests negative. They give variable reactions for oxidase, urease and H_2S production.

Pathogenesis

The outcome of disease after infection with brucellae is determined by several factors such as immune and nutritional status of the host, size of inoculum and route of transmission.

Brucella is facultative intracellular pathogen that affects the reticuloendothelial system. This intracellular existence makes them less susceptible to antibiotics and circulating antibodies.

Host Immunity

Immunity to brucellosis is mainly cell mediated. Activated macrophages, Th1 type of T-helper cell response, tumour necrosis factors α and γ, and interleukins 1 and 12 mediate the protective response.

The antibodies also appear in response to infection; IgM antibodies appear within the first week of infection followed by IgG antibodies, which appear by the end of second week. These antibodies decline slowly with recovery after 2–3 years. Persistent detection of IgG antibodies suggests relapse or chronic infection.

Clinical Manifestations

The initial symptoms of brucellosis can be acute or insidious in onset and are non-specific including fever, sweat, malaise, fatigue, headache, anorexia, joint or back pain. These symptoms appear 2–4 weeks after inoculation. In untreated patients, symptoms that may persist are recurrent or undulant fever, arthritis, endocarditis, neurologic symptoms, swelling of the testicle, chronic fatigue and depression.

The disease may relapse within 3–6 months after discontinuing therapy. Chronic brucellosis is usually caused by persistent foci of infection in tissues such as bone, liver, spleen or other organs.

Laboratory Diagnosis

As the symptoms of brucellosis are non-specific, laboratory diagnosis is essential. The following laboratory methods are used:
1. **Bacterial culture:** *Brucella* is most commonly isolated from blood cultures using Castaneda medium, which usually requires 6–8 weeks for producing results. However, automated blood cultures can provide results within 5–7 days.

   Other samples that can be used for isolation are bone marrow, CSF, wounds, pus or joint fluid.

   Proper precautions should be taken while working with *Brucella* because it has a low infectious dose and can aerosolise. No wonder, it is the **most commonly reported laboratory-associated bacterial infection**. Precautions that should be taken to minimise the risk of an accidental laboratory exposure include handling of organism by an experienced person, all work to be performed under **BSL-3** conditions and informing the laboratory about the possible organism in case of clinical suspicion.

2. **Serology:** It can be performed in addition to bacterial culture. Antibodies to *B. abortus*, *B. melitensis* and *B. suis* can be detected using **serum/tube agglutination test (SAT)** or its modification, **Brucella microagglutination test (BMAT)**. Two serum samples are required for serology, the first during acute illness (within 7 days of disease onset) and the second after 2–4 weeks (convalescent sample). A fourfold or greater rise in antibodies is diagnostic of brucellosis.

3. **Detection in animals:** The laboratory methods similar to those used in humans can be used. In addition, immunofluorescence on pathological specimens or **Rose Bengal card test** can be done.

4. **Detection in milk:** Milk ring test is done. A blue ring forms when stained *Brucella* antigen is mixed with milk containing specific antibodies.

Treatment

Treatment of choice is a combination therapy with doxycycline and gentamicin.

Prevention

Avoiding consumption of undercooked meat and unpasteurised dairy products (like milk, cheese or ice cream) prevents the disease. Animal handlers should use proper personal protective equipment like rubber gloves, goggles and gowns while handing animals.

Vaccines are available for immunising animals that include **B. abortus RB51** and **B. abortus S19** for cattle and **B. melitensis Rev-1** for sheep and goats. But precautions should be taken while handling these vaccines as they can also cause infection in humans.

CAMPYLOBACTER

Introduction

The term *Campylobacter* is derived from a Greek word 'campylos', meaning curved. It was identified as a major human pathogen in 1970s and later the related genera *Arcobacter* and *Helicobacter* were identified. It causes Campylobacteriosis. Medically important *Campylobacter* spp. cause two types of diseases:
1. Enteric: *C. jejuni*, *C. coli*.
2. Extraintestinal: *C. fetus*, *C. conciscus*.

Campylobacter is a slender, spirally curved (comma shaped or S shaped), Gram-negative bacillus measuring 0.2–0.5 μm × 0.5–5 μm. They are non-sporing, motile with a single-sheathed polar flagellum, strongly oxidase positive but do not ferment carbohydrates.

Epidemiology

Campylobacteriosis is a zoonosis found globally. Campylobacters live as commensals in the gastrointestinal tract of animals like cattle, dog, sheep, cats, rodents and so on. Infection is transmitted to humans by the following:
1. Consuming contaminated food (meat, poultry, unpasteurised milk) and water.
2. Direct contact with infected animals especially household pets with diarrhoea.
3. Person to person by faeco-oral transmission.

In developed nations, it is the major cause of diarrhoeal illness. In developing countries *C. jejuni* is endemic and infection is more common in children less than 5 years of age but is often asymptomatic.

Cultural Characteristics and Identification

Campylobacter spp. are microaerophilic, that is require 5%–10% oxygen for growth. They grow at 37°C, but *C. jejuni* grows best at 42°C. The selective media used for isolation are **Skirrow's, Butzler's** and **Campy-BAP media.** Filtration techniques can enhance their growth. After 24–48 h of incubation, the colonies are grey, non-haemolytic, moist and flat. Suspicious colonies are confirmed by Gram stain, **'darting'**

motility and other biochemical tests. *C. jejuni* is positive for oxidase test, catalase test, nitrate reduction test and hippurate hydrolysis test. Molecular tests such as PCR can also be used for identification.

Pathogenesis

The factors determining the pathogenesis include infecting dose of the organism, pathogen virulence and host immunity. The incubation period varies from 1 to 7 days. It colonises the small intestine and causes tissue injury in jejunum, ileum and colon, thereby causing diarrhoea. It sometimes invades the mesenteric lymph nodes to cause portal bacteraemia. In a normal host, this bacteraemia is transient but in an immunocompromised host, sustained bacteraemia and sepsis may occur.

The important virulence factors are flagella, lipopolysaccharide endotoxins in the outer membrane and toxins such as enterotoxin and cytotoxin.

Clinical Manifestations

1. **Enteric:** The most common presentation is acute enteritis, which is usually self-limited. Fever, headache, myalgia and so on occur along with diarrhoea, which may range in severity from loose stools to massive watery or grossly bloody stools. Gastroenteritis is also caused by the related genus *Arcobacteria* (*A. butzleri* and *A. cryaerophila*).
2. **Extra-intestinal:** Disease is a systemic illness with bacteraemia, meningitis, abscesses and vascular infections. The disease may be fatal in immunocompromised host.

Laboratory Diagnosis

Laboratory diagnosis can be made by the following methods:
1. **Specimen:** Freshly passed stool sample (within 2 h of passage)
2. **Direct microscopy of stool:** By darkfield or phase contrast microscopy that shows darting motility of *Campylobacter* spp.
3. **Gram stain of stool:** Gram-negative vibroid forms are suggestive. RBCs and neutrophils are also visible.
4. **Culture:** On selective media under specific conditions and identification by biochemical methods (see above).
5. **Direct methods from stool** that are still investigational: PCR, FISH (fluorescence in situ hybridisation).

Treatment

For enteric illness, fluid and electrolyte replacement is the standard treatment. For invasive diseases, early antibiotic therapy is required with erythromycin, tetracycline, aminoglycosides and so on.

HELICOBACTER

Introduction

Barry Marshall and Robin Warren (Australian physicians) first isolated *H. pylori* (earlier known as *Campylobacter pylori*) from gastric mucosa of patients with peptic ulcer disease. For this discovery, they were awarded the Nobel Prize in Medicine.

H. pylori is microaerophilic, small measuring 0.5–1 μm × 2.5–4 μm, Gram-negative spiral rod that is motile by multiple polar sheathed flagella (lophotrichous). The motility is typically described as '**rapid corkscrew motion**'. It produces abundant urease.

Epidemiology

H. pylori affects only humans and monkeys. It has been isolated from people in all parts of the world. The prevalence is higher in developing countries than in the developed world. Human gastric mucosa is the only habitat of *H. pylori*. The mechanism of transmission seems to be oral–oral or faecal–oral and conditions such as poverty, overcrowding and poor hygiene favour transmission.

H. heilmanii, a larger spiral bacterium, is occasionally found in humans and in cats and dogs.

Cultural Characteristics and Identification

Generally, a combination of two media is used for isolation: (1) a selective medium such as Skirrow's medium and (2) a non-selective medium such as chocolate agar. The plates are incubated at 35–37°C for 2–5 days in the presence of 5%–10% oxygen (microaerophilic conditions). Suspicious colonies may be identified as *H. pylori* by comma or S-shaped organisms on Gram stain, typical motility, and catalase, oxidase and strongly urease positive tests.

Pathogenesis and Clinical Manifestations

H. pylori usually causes asymptomatic colonisation of the gastric mucosa usually in the gastric antrum. The infection is strictly confined to the mucosa and does not invade it. The infection may be transient, but generally persists for years or decades. Tissue and serologic response develops in all persistently infected persons. This tissue inflammation may result in either of the following responses:

1. **Atrophic gastritis:** That may lead to gastric adenocarcinoma
2. **Hyperacidity:** That may cause duodenal ulceration
3. **Antigenic stimulation:** Leading to B-cell lymphoma or mucosa-associated lymphoid tissue (MALT) lymphomas
4. **Non-ulcer dyspepsia**, reflux oesophagitis and its sequelae

Laboratory Diagnosis

The diagnosis can be made by the following tests, each of which has a sensitivity and specificity of >95%:

Invasive Tests

Endoscopy and biopsy. Specimen obtained is subjected to
1. **Microscopy and culture** (see above). The organisms may also be seen on histological examination with silver stains, Giemsa, acridine orange stain, immunofluorescence and so on.
2. **Rapid urease test.**

Non-invasive Tests

1. **Serologic analysis:** Specific IgG or IgA antibodies can be detected.

2. **Urea breath tests:** A person drinks carbon ($_{13}$C or $_{14}$C) labelled urea solution, which is detected in breath. It can also be used for monitoring response to eradication therapy.
3. **Faecal antigen analysis:** It can be used to detect *H. pylori* antigen in stool and also for monitoring response to treatment.

Treatment

It is indicated only for symptomatic disease. The standard regime is a combination of bismuth salt, amoxicillin or tetracycline and metronidazole for 14 days. Clarithromycin and a proton pump inhibitor like omeprazole can also be used as an alternative.

ANAEROBIC GRAM-NEGATIVE RODS

Medically important anaerobic Gram-negative rods (GNAR) include *Bacteroides*, *Porphyromonas*, *Prevotella* and *Fusobacterium* and less commonly *Bilophila* and *Leptotrichia*. They are colonisers of oropharynyx, gastrointestinal tractand urogenital tract of humans. Usually they live as commensals but can cause opportunistic infections.

BACTEROIDES

They are non-sporing, obligately anaerobic rod-shaped bacilli, which may appear pleomorphic with straight rods, branching forms or coccobacilli. The genus *Bacteroides* contains more than 30 species, but *B. fragilis* is clinically most important. *Bacteroides* are the most commonly isolated anaerobes from clinical specimens.

On blood agar *B. fragilis* forms small (2–3 mm), smooth, circular, white or grey, shiny colonies. It shows a good growth in brain heart infusion agar when incubated in the presence of 10% CO_2 in an anaerobic environment. It can grow on bile esculin agar owing to its bile tolerance. It is highly resistant to antibiotics including vancomycin, colistin and kanamycin, but is susceptible to metronidazole, clindamycin and chloramphenicol. It is beta-lactamase positive.

PREVOTELLA

Earlier classified as *Bacteroides*, these are pigmented and saccharolytic anaerobic Gram-negative bacilli. The genus includes more than 20 species, but *P. melaninogenica*, *P. buccalis* and *P. denticola* are clinically most important. They are pleomorphic and may vary from short Gram-negative rods to coccobacilli.

Prevotella forms small (1–2 mm), circular, convex, shiny grey colonies on blood agar that are pigmented (brown or black coloured) and may show a dark red fluorescence upon exposure to a Wood's Lamp. It is resistant to antibiotics including vancomycin, colistin and kanamycin.

PORPHYROMONAS

Previously classified as *Bacteroides*, these are pigmented and asaccharolytic anaerobic Gram-negative bacilli. The important species are *P. gingivalis* and *P. endodontalis*. These are shorter rods than *Prevotella* and may appear as coccobacilli. They form grey-black colonies on blood agar. They are sensitive to vancomycin but resistant to colistin.

FUSOBACTERIUM

It is filamentous (long, thin or spindle-shaped bacilli with pointed ends) Gram-negative rods. It is pleomorphic on Gram stain and morphology varies from coccoid to rod-shaped forms. On blood agar it forms small pinpoint colonies that may umblicate to form 'fried egg colonies' after 3–5 days of incubation. It is sensitive to kanamycin and colistin but resistant to vancomycin. It is indole positive, bile sensitive and metabolises threonine to proprionate.

Virulence Factors of GNAR

The virulence factors include capsular polysaccharides, lipopolysaccharide endotoxins, fimbriae, pili and exotoxins.

Clinical Manifestations of GNAR

The GNARs can infect any system of the body. They can cause bacteraemia, skeletal infections, skin and soft tissue infections, intra-abdominal infections, CNS infections, infections of the oropharynx, cardiovascular system and urogenital systems. The suggestive features of infections caused by GNAR are shown in Box 14.1.

Laboratory Diagnosis of GNAR

The most important requirements for correct laboratory diagnosis of GNAR are as follows:

1. **Correct specimen:** Specimens that are collected with syringe (aspirated pus), tissue or biopsy samples, urine collected by suprapubic aspiration etc. are the appropriate specimens. Any specimen that has been in contact with air or collected from an open cavity (e.g. sputum, lochia, voided urine) is not acceptable. As far as possible swabs should be avoided.
2. **Proper collection of specimens:** The specimen should be collected properly avoiding contact with normal microbial flora.

BOX 14.1	**Suggestive Features of GNAR Infections**

- Usually follow some precipitating factor like surgery, trauma, presence of foreign bodies and so on.
- Predisposing factors like diabetes, malignancy, immunocompromised conditions are usually present.
- Systemic infections are often polymicrobial.
- Most infections arise endogenously from the colonised sites.
- The suggestive features include proximity to a mucosal site; pus which is putrid, with a pervasive, nauseating odour; tissue necrosis; and gas in tissues.

3. **Transportation:** Samples must be immediately transported to the laboratory either in sealed vials gassed out with carbon dioxide or in transport media like Robertson cooked meat medium (RCM), or pre-reduced anaerobic sterilised (PRAS) transport medium. If swabs must be collected they should be transported in Stuart's transport medium or RCM.
4. **Gram's stain:** Gram's stain usually shows pleomorphic organisms along with numerous pus cells.
5. **Culture:** Several anaerobic media are available but freshly prepared blood agar with neomycin, haemin, yeast extract and vitamin K are usually used. The plates are incubated in an anaerobic environment containing 10% CO_2, which can easily be done by Gaspak system.
6. **Gas–liquid chromatography:** When used along with the biochemical tests, it results in accurate diagnosis of anaerobic bacteria. But the disadvantages are that it is time-consuming, expensive and is not widely available in many clinical laboratories.

Treatment

The mainstay of treatment is surgery along with antibiotic therapy with penicillin, metronidazole and clindamycin. As *Bacteroides* spp. produces beta-lactamase, it can be treated with amoxicillin-clavulanic acid combinations or with carbapenems.

Spirochaetes

Amita Jain

INTRODUCTION

Spirochaetes are long (5–500 μm in length), spiral, motile and flexible bacteria, which are motile by cork screw movement. They have complex machinery for movement, that is axial or endoflagella, varying in number, woven around the cell protoplast in a helical manner and situated between outer membrane and cell wall that is periplasmic space (Fig. 15.1). These are polar flagella. Cell wall is like that of Gram-negative bacteria. As they are very thin, it may not be possible to visualise them by Gram stain. They are best visualised either by silver impregnation methods or in wet mount using phase contrast. They divide by transverse fission. Some of these bacteria are human pathogens, while others are either saprophytic or commensals (Flowchart 15. 1). In this chapter we will discuss only about human pathogens.

SPIROCHAETES OF ORAL CAVITY

Borrelia buccalis and several treponemes, which are morphologically like pathogenic ones, are present in oral cavity. Treponemes present in oral cavity are listed in Box 15.1. These are usually harmless. Most of them are strict anaerobes. However, in case there is some damage in oral mucosa, the saprophytic spirochaetes of the mouth, **together with anaerobic fusiform bacilli**, that is fusobacteria, multiply in large numbers, resulting in **odontogenic infections** and gangrenous stomatitis.

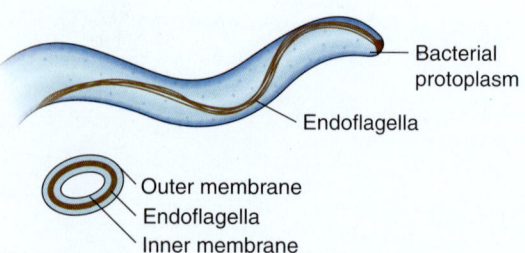

Fig. 15.1 Arrangement of Flagella in Spirochaetes.

TREPONEMA

INTRODUCTION

Genus *Treponema* contains both pathogenic and non-pathogenic species. Treponemes are about 0.2 μm in width and 5–15 μm in length, thin, with tapering ends, spiral (around 10 spirals present), actively motile, rotating around their endoflagella. They are not easily visible in Gram stain, but immunofluorescent stain, darkfield illumination and silver impregnation methods of staining are helpful.

Pathogenic *Treponema* includes the following:
- *T. pallidum* subspecies *pallidum*: Causative organism of syphilis

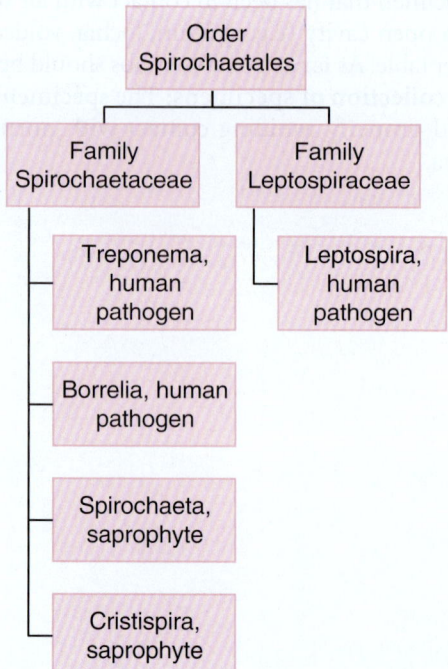

Flowchart 15.1 Classification of Spirochaetes.

BOX 15.1 *Treponema* spp. Found in Oral Cavity

They are usually harmless. At times they are supposedly associated with periodontitis and implant periarthritis. Species commonly detected in the oral cavity are as follows:

Treponema denticola,
Treponema scaliodontum,
Treponema macrodentium,
Treponema oralis,
Treponema intermedia,
Treponema maltophilum,
Treponema socranskii,
Treponema vincentii, and so on.

- *T. pallidum* subspecies *pertenue*: Causative organism of yaws
- *T. pallidum* subspecies *endemicum*: Causative organism of bejel (also known as endemic syphilis)
- *T. carateum*: Causative organism of pinta
  Non-pathogenic *Treponema*

T. PALLIDUM SUBSPECIES PALLIDUM

INTRODUCTION

T. pallidum causes syphilis. Morphology is described earlier. It is microaerophilic and is not cultivable in vitro. However, non-pathogenic treponemes like Reiter's strain, which is a saprophyte, can be cultured anaerobically in vitro. They share antigen with *T. pallidum*. *T. pallidum* does not survive outside body for long time. It is a fragile organism. Drying, temperature more than 42°C, antiseptics and disinfectants, soap and contact with oxygen can easily kill *Treponema*. However, in whole blood it can survive for 24 h, kept at 4°C. Storing the blood in refrigerator for more than 3 days makes it safe for transfusion, so far as syphilis is concerned. All the strains are penicillin sensitive.

As *T. pallidum* is difficult to grow, antigenic constitution of Treponemal antigens are poorly studied. There is cross-reactivity among pathogenic and non-pathogenic species of *Treponema*.

Some of the antigens which could be characterised are as follows:
- **Outer membrane proteins** are lipoproteins with lipid moiety at their terminal end. These lipid moieties do not allow access of antibodies to the proteins, hence are protective in nature.
- **Hyaluronidase** is present which breaks down the hyaluronic acid present in tissue, thus enhancing its invasive property.
- **Endoflagella** contain three core proteins resembling bacterial flagellin proteins. In addition, they have a sheath protein.
- **Cardiolipin** is an important component of the treponemal antigens.

Infection with *T. pallidum* induces antibody production in host body which is found in serum. These antibodies form the basis for the serologic diagnosis of syphilis. At least three types of antibodies are detected in an infected human:
1. **Reagin antibodies:** These are non-specific antibodies, which react with cardiolipin. These antibodies are picked by non-specific serological tests for syphilis (STS) like Venereal Disease Research Laboratory (VDRL) test. Cardiolipin has a lipid hapten, which is also found in body tissues like cardiac tissue.
2. **Group antigen-specific antibodies:** These antigens are present in all the pathogenic and non-pathogenic *Treponema* and are more specific than cardiolipin.
3. **Species antigen-specific antibodies:** These are specific to *T. pallidum* and form the basis of specific tests for syphilis.

SYPHILIS

Syphilis is a sexually transmitted disease caused by *T. pallidum*, occurring only in humans. The pathogen enters through intact mucous membrane or abraded skin and mucous membrane. It multiplies slowly at local site, spreads to local lymph node and disseminates through haematogenous route. Incubation period is 2–10 weeks after which a papule develops at the site of inoculation, which is painless and remains for a very short duration; hence, can be missed by patient. This papule finally forms an ulcer known as **hard chancre**.

Primary Syphilis

Hard chancre is clean, painless, with indurated edges and hard base. Induration is due to accumulation of lymphocytes and plasma cells. It is usually found on genitals or occasionally it is intrarectal, perianal, on nipples or fingers. Chancre is covered with thick transparent exudate, which is rich in treponemes. It heals within 40–50 days even without treatment. Regional lymph nodes are enlarged, rubbery, painless and discrete. This stage is called **primary syphilis**.

Secondary Syphilis

About 2–10 weeks after the primary ulcer heals, the **secondary** lesions appear, which subside spontaneously without treatment. Red maculopapular rash anywhere on the body, including the hands and feet, and moist condylomas in the anogenital region, axillas and mouth are typical features. Complications like syphilitic meningitis, chorioretinitis, hepatitis, nephritis or periostitis may occur rarely.

Tertiary Syphilis

Only 1/3 of cases progress to tertiary stage. Of remaining 2/3, half heal spontaneously and the remaining half develop latent disease. Tertiary syphilis is characterised by granulomatous lesions known as **gummas** in skin, bones and liver. Complications are cardiac and neurological, for example aortitis, aortic aneurysm, aortic valve insufficiency, meningovascular syphilis, paresis, tabes and so on. These lesions are also known as **late tertiary syphilis**.

These lesions have either no or very occasional treponemes. Most of these occur due to host immune response.

Syphilis patients are most infectious in first 5 years of illness. Primary and secondary lesions are full of spirochaetes, hence are highly infectious. Cases with tertiary and latent syphilis are not infectious.

Congenital Syphilis

Syphilis can be transmitted from mother to foetus through the placenta. Most often transmission occurs in 10th–15th weeks of gestation. Mothers with recent syphilis, that is in first 2 years of disease, more often transmit the infection to foetus than those with older lesions. Stillbirth, miscarriage and abortions are usual. Some live births do occur. Interstitial keratitis, Hutchinson's teeth, saddle nose, periostitis and central nervous system anomalies are classical symptoms of congenital syphilis, which manifest in children after 2 years of age. Adequate treatment of the mother during pregnancy prevents congenital syphilis. In congenital infection, the child makes IgM antitreponemal antibody. Antibodies are detectable in newborn. In case these are maternal antibodies they fade after few weeks.

If syphilis is acquired non-sexually as an occupational hazard to scientists/doctors/nurses, course of illness remains same as in venereal syphilis.

Laboratory Diagnosis of Syphilis

Specimen

- For demonstration of treponemes: Tissue fluid expressed from hard chancre and maculopapular rash of secondary syphilis
- For serological assays: Blood/serum/CSF

Demonstration of Treponemes

Demonstration of treponemes is done by the following methods:

1. **Darkfield examination:** Spirochaetes appear as bright spiral motile thin bacteria.
2. **Immunofluorescence:** Stained with a fluorescein-labelled anti-*Treponema* serum spirochaetes are examined by fluorescent microscope.
3. **Silver impregnation methods:** Staining with Fontana's or Levidity method allows examination of treponemes. Slides can be preserved for long-term storage.

Serologic Tests for Syphilis

1. **Non-specific tests** utilise **non-treponemal antigens** for detection of **reagin antibodies**. Reagin contains both IgM and IgG class of antibodies directed against the cardiolipin–cholesterol–lecithin-mixed antigen. Cardiolipin is the lipid antigen extracted from beef heart with lecithin and cholesterol. It is diphosphatidylglycerol. Lecithin and cholesterol enhance antigen–antibody reaction. Cardiolipin reacts with **reagin** antibodies. Tests become positive after about 2 weeks of infection. Secondary and tertiary stage show strongly positive test. Test becomes negative after 6 months of successful treatment. Quantitation of antibodies is possible if test is done on diluted samples. Titers can be used

to assess the response to treatment. The titre value also helps in ruling out the non-specific test positives from true positives. Reagin does not cross blood–brain barrier, but is formed in nervous system in cases of neurosyphilis. Hence, CSF can also be a sample for testing and monitoring neurosyphilis. These tests are economical and can be done in field as they do not require much equipment. Commonly used tests are as follows:

a. Venereal disease research laboratory test (VDRL) is a flocculation test. Test results are read microscopically. Serum needs inactivation by heating at 56°C for 30 min.
b. Rapid plasma reagin (RPR) works on same principle as VDRL, except it has coloured particles added to it, so that reaction is enhanced and can be read without microscope with unaided eyes. No serum treatment is required.
c. Toluidine red unheated serum test (TRUST) works exactly like RPR tests.

Biological false-positive: tests for 'reagins' may also be reactive in non-syphilitic conditions, for example,

a. systemic lupus erythematosus,
b. polyarteritis nodosa,
c. rheumatic disorders,
d. malaria,
e. leprosy,
f. measles,
g. infectious mononucleosis, and
h. vaccinations.

2. **Treponemal antibody tests** are specific serological tests for syphilis which use *T. pallidum* for testing the antibodies in patient's serum. These are more specific tests.

a. **Fluorescent treponemal antibody (FTA-ABS) test** is a specific and sensitive test, which uses killed *T. pallidum* fixed on a glass slide to detect specific antibodies in patient's serum. Fluorescence label-tagged antihuman gamma globulin are used as detection system. Test becomes positive in early syphilis, and is positive in secondary and tertiary syphilis. Test does not turn negative following treatment. IgM FTA in the blood of newborns should be demonstrated to prove in utero infection.

b. **_T. pallidum_-particle agglutination (TP-PA) test:** Inert particles are coated with *T. pallidum* subspecies *pallidum* antigens which is used as antigen for detection of anti-treponemal antibodies. The presence of antibodies in patient's sera agglutinates these particles and makes a mat which can be visualised by unaided eyes. These tests are specific and sensitive. Following two formats are commercially available:

i. *T. pallidum* haemagglutination (**TPHA**)
ii. Microhaemagglutination for *T. pallidum* (**MHA-TP**)

Treatment

Penicillin is the treatment of choice. A single injection of benzathine penicillin G intramuscularly is good enough to treat a case suffering from less than a year. Three intramuscular injections of benzathine penicillin given at weekly intervals are good enough to take care of cases with longer duration, neurosyphilis and latent syphilis. Patients sensitive to penicillin can be given other antibiotics, for example, tetracycline or erythromycin. Few hours following treatment, toxic products from dying spirochaetes cause typical **Jarisch–Herxheimer reaction**. It is a condition that resembles bacterial sepsis manifested as fever, chills, hypotension, tachycardia, exacerbation of skin lesions, etc. and is usually self limited. Patient, who is immunosuppressed, for example, co-infected with HIV, may show relapse following treatment.

Epidemiology

Syphilis is sexually acquired, except at times mother to foetus and experimentally transmission can occur. Persons treated in past may get re-infected. Untreated patients with late or latent syphilis may become non-infectious. There is no vaccine available. Early and effective treatment of all cases and contacts and safe sex with condoms are preventive measures. Every case of syphilis should be investigated for other sexually transmitted diseases as well.

DISEASES CAUSED BY *TREPONEMA* OTHER THAN *T. PALLIDUM*

These treponemes are related to *T. pallidum*. Antibodies against them cross-react with syphilis. These organisms are non-cultivable. These diseases are *not* sexually transmitted, but need direct contact for transmission.

- **Yaws**—caused by *T. pallidum* subspecies *pertenue*. It is endemic in tropical countries. Children are commonly affected. Transmission is person-to-person. To begin with an ulcerating papule, it occurs usually on extremities, which heals with scar. Bones may be involved. Other organs are usually not involved. Disease development and diagnosis resembles non-venereal form of syphilis. Penicillin is the drug of choice.
- **Bejel**—caused by *T. pallidum* subspecies *endemicum*. Such cases are seen in certain parts of world, for example, Africa, Middle East and Southeast Asia. Children are mainly affected. Transmission is person-to-person. Lesions are highly infectious. Penicillin is the drug of choice.
- **Pinta**—caused by *T. carateum*. *It is endemic* in Mexico, Central and South America, Philippines. All age groups can be affected. Transmission occurs by direct contact. Occasionally flies can transmit infection. Non-ulcerating papule on exposed areas of body is the first lesion to occur, which later on heals with a hyperpigmented scaly scar. Diagnosis and treatment is like that of syphilis.

LEPTOSPIRA

INTRODUCTION

Genus *Leptospira* is classified in species which are pathogenic like *Leptospira interrogans* and non-pathogenic species like *Leptospira biflexa*. *L. interrogans* causes Leptospirosis (Weil's disease), a zoonotic disease. *Leptospira*, both interrogans and biflexa, has several serovars. These serovars are grouped

---

BOX 15.2	**Common Serovars of *Leptospira interrogans***

1. Autumnalis
2. Ballum
3. Bovis
4. **Canicola**
5. Grippotyphosa
6. Hebdomadis
7. **Icterohaemorrhagiae**
8. Mitis
9. Pomona

together based on diseases caused by them and are serologically related to each other, showing cross-reactivity. Serological variation in these groups is the basis of the classification of *Leptospira* species. It also determines the specificity of the human immune response to leptospirae. The serogroups of *L. interrogans* are pathogenic and divided into many groups. Some are listed in Box 15.2, of which Icterohaemorrhagiae and Canicola cause human infections. These are named as *L. interrogans* serovar Icterohaemorrhagiae. *L. biflexa* and other serogroups are non-pathogenic. Leptospirae are also divided based on DNA relatedness; however, for clinical utility only serovars are used.

The lipopolysaccharide in outer envelope of *Leptospira* varies from strain to strain, and is used to differentiate members within serogroups.

MORPHOLOGY AND GROWTH CHARACTERISTICS

Leptospirae are thin, 5–15 µm long and tightly coiled spiral bacilli. One end is hooked. They are highly motile. The methods of demonstration are same as used for *Treponema*. They can be grown under aerobic conditions at 28–30°C. Semisolid media containing serum like **Fletcher's** and **Stuart's** are good; however, **Ellinghausen–McCullough–Johnson–Harris (EMJH)** liquid medium with fatty acids and **bovine serum albumin** is the most commonly used media at present. They do not make well-formed colonies, instead diffuse growth is seen at the top of the semisolid media, where oxygen tension is high. Addition of neomycin or 5-fluorouracil makes media selective for *Leptospira*. *Leptospira* survives in alkaline media.

PATHOGENESIS (Flowchart 15.2)

Leptospirosis or Weil's disease is a zoonotic disease. Human gets infection after coming in contact with contaminated water. Water contamination occurs with the excreta of animal hosts, that is rats, mice, wild rodents, dogs and so on. Animals are asymptomatic carriers which excrete *Leptospira* in urine. Leptospirae survive in water for many weeks as it can survive at alkaline pH. Infection occurs through breech in skin or mucous membrane. Ingestion usually does not cause infection. Incubation period is 1–2 weeks. The illness is **often biphasic**. Once *Leptospira* reach bloodstream, they can infect parenchymatous organs like liver, brain, heart,

```
┌─────────────────────────────────────────────────────────────┐
│         Infected animals, urine contaminate water bodies      │
└─────────────────────────────────────────────────────────────┘
                              ↓
┌─────────────────────────────────────────────────────────────┐
│  Infection: With water contaminated with leptospira/through   │
│                   breech in skin/aerosols                      │
└─────────────────────────────────────────────────────────────┘
                              ↓
┌─────────────────────────────────────────────────────────────┐
│       Incubation period: 1–3 weeks, patient asymptomatic      │
└─────────────────────────────────────────────────────────────┘
                              ↓
┌─────────────────────────────────────────────────────────────┐
│  First week of illness: Acute phase, patient has fever. Other │
│  symptoms may or may not be present. Leptospira present in    │
│  blood, blood culture positive, antibodies low (non-detectable),│
│  CSF culture may or may not be positive, urine culture negative│
│  (treatment with doxycycline is useful)                       │
└─────────────────────────────────────────────────────────────┘
                              ↓
┌─────────────────────────────────────────────────────────────┐
│  Second week of illness: Convalescent phase, fever subsides   │
│  and then recur (biphasic). Jaundice, hepatitis, encephalitis,│
│  nephritis, starts appearing. Leptospira disappears from blood.│
│  Blood culture negative, antibodies present in high level, CSF │
│  culture positive, urine culture positive                     │
└─────────────────────────────────────────────────────────────┘
                              ↓
┌─────────────────────────────────────────────────────────────┐
│  Third week of illness: Untreated patient: Weil's disease.    │
│  Fever high. Jaundice, hepatitis, encephalitis, nephritis,    │
│  present. Patient gets critically ill. Blood culture negative,│
│  antibodies present in high level, CSF culture positive, urine │
│  culture positive. Treatment may not work                     │
└─────────────────────────────────────────────────────────────┘
                              ↓
┌─────────────────────────────────────────────────────────────┐
│  Fourth week of illness: Weil's disease. Fever high. Jaundice,│
│  hepatitis, encephalitis, nephritis, present. Patient         │
│  critically ill. Blood culture negative, antibodies present in │
│  high level, CSF culture positive, urine culture positive.    │
│  Treatment may not work                                       │
└─────────────────────────────────────────────────────────────┘
                              ↓
┌─────────────────────────────────────────────────────────────┐
│  Some patients die, others become convalescent shedder or     │
│  reservoir. Antibodies provide serovar-specific immunity      │
└─────────────────────────────────────────────────────────────┘
```

Flowchart 15.2 Development of Weil's Disease.

lungs and kidneys. Bleeding and necrosis of parenchyma in infected organs causes manifestations like jaundice, fever and liver and kidney dysfunction. Patients show recovery before a second serious phase happens. Second phase manifests itself as multiorgan dysfunction that is 'aseptic meningitis', nephritis, hepatitis and skin, muscle and eye lesions. Antileptospiral IgM antibody titer is high. Subclinical infection can occur occasionally. Liver and kidney haemorrhage and biochemistry changes in patient's body signifying hepatic and kidney damage are marked. Antibodies are produced in large number, produce immunity but are serovar-specific.

LABORATORY DIAGNOSIS

Specimens

Specimens include whole blood, cerebrospinal fluid, urine, serum and tissues.

Microscopic Examination

Darkfield examination of wet mount made from centrifuged deposit of urine or fresh blood and immunofluorescent staining of stained smears can demonstrate *Leptospira*.

Serology

- **ELISA:** Many commercial platforms are available for demonstration of anti-leptospiral antibodies by immunoassay. This is most commonly used.
- **Microscopic agglutination of live organisms (MAT):** This is the gold standard, most sensitive test, done by only reference laboratories, as it can be dangerous due to use of

live bacteria. Usually antibodies are very high in concentration (1:10,000 or higher).
- **Indirect haemagglutination of red blood cells with adsorbed leptospirae:** Rarely used.

Culture

Whole fresh blood, urine, CSF or tissue is inoculated in specific EMJH liquid medium with fatty acids and bovine serum albumin or in semisolid media like Fletcher's medium. It is slow growing. Culture takes time around 2–8 weeks. Not routinely used in laboratories for culture.

Molecular Tests

The presence of bacterial DNA can be demonstrated in whole fresh blood, urine, CSF or tissue.

Animal Inoculation

It is a sensitive test though is hazardous. Not routinely used for diagnosis.

Treatment

Treatment includes oral or injectable antibiotics doxycycline, ampicillin or amoxicillin.

BORRELIA

INTRODUCTION

Borrelia has 52 known species; of them about 12 species cause diseases. Many species occur as commensals in oral cavity and genitals. *Borrelia burgdorferi* causes Lyme disease, which is transmitted by ticks (*Ixodes dammini* and related Ixodes ticks) or by louse (*Pediculus humanus corporis*). *Borrelia recurrentis* causes relapsing fever and is transmitted by the human body louse (*P. humanus corporis*). Vincent's angina is a polymicrobial disease, in which *B. vincentii* plays an important pathogenic role.

MORPHOLOGY

Borreliae are irregular spiral organisms measuring 10–30 μm in length and 0.3–0.7 μm in width (larger than *Treponema* and *Leptospira*). They are loosely twisted; the space between two twists is 2–4 μm. They move both by rotation and twisting and are very flexible organisms. Giemsa's stain and Wright's stain are used to demonstrate them. They are Gram-negative.

CULTURE CHARACTERISTICS

They grow with difficulty. They can be grown in liquid culture media containing serous fluids in a microaerophilic environment at 28–30°C. They grow on chorioallantoic membrane of chick embryo. They lose their pathogenic potential when passaged through animals.

RELAPSING FEVER

Relapsing fever is mostly caused by *B. recurrentis* and occurs in epidemic, endemic and sporadic forms. It is transmitted by ticks or louse.

- **Epidemic form** is transmitted by louse and is transmitted from human to human. There is no animal reservoir. This is seen where people are bound to live in poor hygienic conditions like war camps or jails. Borreliae are present in haemolymphatic system of lice and are not transmitted by lice bite or its excreta. Only crushing of lice on human skin releases the borreliae, which gets inoculated through small cuts or abrasions on skin. High fatality is seen.
- **Endemic form** is transmitted by ticks, and is geographically restricted. It is endemic in many parts of the world. It is a milder disease than epidemic form, but relapses are many. Normally, life cycle runs between rodents or other mammals and ticks. Human is an accidental host. Borreliae are present in infected tick and are shed in saliva and excreta of tick. It is transmitted by tick bite or its excreta. *Ornithodorus* and *Argas* species of ticks spread infection in India.

Pathology and Clinical Manifestations

Incubation period varies from 2 to 10 days, after which sudden rise of high fever occurs. Fever lasts for 3–5 days and then declines. After 4–10 days fever again reappears. Following infection with borreliae antibodies develops and level gets very high. The borreliae are known to undergo genetic variation in the host. This is due to recombination occurring between several plasmids present in it and its linear chromosome DNA. Due to recombination, a new antigenic variant appears in the host, in a single episode of infection. The antigenic changes are strong enough, which allow antibodies to select and let only antigenic variant strain survive. The relapsing course of the disease is due to the multiplication of antigenic variant strain. This new strain keeps on multiplying; patient runs fever till new antibodies are formed. This relapsing nature of the fever, against which the host then develops new antibodies, goes on for 3–10 relapses unless treated.

Borreliae are seen in large numbers in the parenchymatous organs like spleen and liver. Haemorrhage and necrosis occur in kidneys and the gastrointestinal tract. It can occasionally cause meningoencephalitis. During the febrile period, borreliae are present in the blood; during the afebrile periods, they disappear.

Laboratory Diagnosis

Blood collected during the rise in fever is used for demonstration of borreliae. For antibody demonstration, serum is used.

- **Microscopy:** Thin or thick blood smears stained with Wright's or Giemsa's stain.
- **Serology** is unreliable. Antibodies cross-react with Cardiolipin and give positive VDRL and Weil Felix test.
- **Animal inoculation** is tried only at high centres. White mice or young rats are inoculated with 1 mL of patient's blood intraperitoneally. Within 2 days, animal's blood shows the presence of organisms.

Treatment

Tetracycline, erythromycin and penicillin are effective.

Prevention

Avoid exposure to ticks and lice.

LYME DISEASE

It is named after the town of Lyme, Connecticut, where clusters of cases in children were identified. It is caused by *B. burgdorferi*. From some areas, *Borrelia garinii* and *Borrelia afzelii* are also isolated. It is transmitted to humans by the bite of *Ixodes dammini* and related tick.

Morphology

B. burgdorferi resembles other borreliae. Culture is difficult and is not usually recommended for diagnosis. **Barbour–Stoenner–Kelly medium (BSK II)**, a liquid media, is used to grow *B. burgdorferi*. Specimen from erythema migrans skin lesion gives best results on culture. It can also be grown from ticks. Addition of antibiotics like rifampin, fosfomycin (phosphonomycin) and amphotericin B helps in preventing contamination. Culture takes 2–3 weeks at 33°C.

The whole genome of *B. burgdorferi* has been sequenced. *B. burgdorferi* has linear chromosome and multiple plasmids. They code for several antigens like lipoproteins, including outer surface proteins OspA–F. OspA and OspB are primarily expressed in ticks and help in migration of *B. burgdorferi* from gut to salivary glands.

Pathology and Clinical Manifestation

Common clinical manifestation is flu-like illness with erythema migrans, fever, malaise, arthralgia and arthritis. Transmission occurs with tick bites. Borreliae adhere to proteoglycans on host cells through a glycosaminoglycan receptor. They cause local lesion on skin and then spread by lymphatics or blood to other organs.

Incubation period is 3 days to 4 weeks. Lyme disease occurs in three stages.

- **First stage:** Localised infection, erythema migrans, a flat red skin lesion near tick bite along with flu-like illness.
- **Second stage:** Disseminated disease, fever, myalgia, arthralgia, occurs 1 week after first stage.
- **Third stage:** Persistent infection occurs weeks to months later; arthritis, meningitis, facial nerve palsy, myopericarditis.

Laboratory Diagnosis

Smear examination usually is not useful. *B. burgdorferi* in tissue sections using immunodiagnostic staining can be demonstrated.

Specimen is blood for serologic tests. Culture is not done as it may be hazardous. Detection of *B. burgdorferi* DNA by molecular assays is recommended.

Treatment

For both early and late disease, doxycycline or amoxicillin for 20–30 days is the treatment of choice.

16

Chlamydia

Amita Jain

LEARNING OBJECTIVES

- General characteristics of *Chlamydia*
- Disease caused by:
  - *Chlamydia trachomatis*
- *Chlamydia pneumoniae*
- *Chlamydia psittaci*

INTRODUCTION

Chlamydiae are **obligate intracellular parasites**. There are three types of *Chlamydia* species causing human diseases: *Chlamydia trachomatis*, *Chlamydophila (Chlamydia) pneumoniae* and *Chlamydophila (Chlamydia) psittaci*. The major differences in these three types are shown in Table 16.1. The major characteristics are shown in Box 16.1.

REPLICATION OF CHLAMYDIAL AGENTS

Elementary body (EB) is the infectious particle. It measures around 0.3 μm in size. After entry into the host cell, the elementary body enlarges and changes in a large body called a reticulate body (RB). The RB grows inside a membrane-bound vacuole and divides repeatedly by binary fission to form elementary bodies, till entire vacuole is filled with elementary bodies. This vacuole forms a cytoplasmic inclusion. These elementary bodies are released from the host cell to infect new cells. The whole cycle takes ∼24–48 h.

> ### BOX 16.1 Major Characteristics of *Chlamydia*
>
> - Gram-negative bacteria lacking mechanisms for ATP production
> - Share a common group antigen
> - Multiply intracellularly
> - Cannot grow in cell-free media
> - Occur in two forms with in the host cell: elementary body (EB) and reticulate body (RB). Both these forms are seen as inclusion bodies within infected cells
> - Cell wall resembles that of Gram-negative bacteria
> - Elementary and reticulate bodies contain both DNA and RNA

CHLAMYDIAL ANTIGENS

Chlamydia is known to have genus-specific lipopolysaccharide antigens. Outer membrane proteins mainly contribute to species-specific or serovar-specific antigens. Specific antigens are usually not shared among *Chlamydia* species. C.

TABLE 16.1 Important Characteristics of Pathogenic Species of *Chlamydia*

	C. trachomatis	C. pneumoniae	C. psittaci
Inclusion body	Vacuolar	Dense	Large
Presence of glycogen in inclusion bodies	Yes	No	No
Susceptible to sulphonamides	Yes	No	No
Presence of plasmid	Yes	No	Yes
Serovars (number)	15 (A, B, Ba, C–K except I and J, and L1–L3)	1	4
Natural host	Humans	Humans	Birds
Mode of transmission	Person to person, mother to infant	Airborne, person to person	Airborne, bird excreta to humans
Major diseases	Trachoma, sexually transmitted diseases (STDs), infant pneumonia, lymphogranuloma venereum	Pneumonia, bronchitis, pharyngitis, sinusitis	Psittacosis, pneumonia, PUO (pyrexia of unknown origin)

trachomatis has 15 serovars: A, B, Ba, C–K except I and J, and L1–L3. *C. psittaci* has at least four serovars. *C. pneumoniae* has only one serovar.

IMMUNITY

Antibodies against Chlamydiae and immunisation have minimal protective effect against reinfection. Furthermore, prior infection or immunisation aggravates inflammation and scarring due to hypersensitisation seen in trachoma. Latent infection may persist in the presence of high antibody titers.

ANTIMICROBIAL AGENTS

Treatment with tetracyclines for long duration is the protocol. Tetracyclines, erythromycins and sulphonamides are effective in most clinical infections. Aminoglycosides, penicillins and cephalosporins have no effect on *Chlamydia*. Penicillin-binding proteins are intrinsically present in Chlamydiae; hence, cell wall inhibitors are not effective.

STAINING PROPERTIES

Giemsa Stain

It shows inclusions of *C. trachomatis*, that appear as dark purple compact dense masses near the nucleus. Elementary bodies stain purple, while reticulate bodies stain blue. Host cell cytoplasm stains blue.

Gram Stain

It is not used.

Lugol's Iodine

Inclusion of *C. trachomatis* has glycogen which stains brown (Fig. 16.1); *C. pneumoniae* or *C. psittaci* does not have glycogen, hence they do not stain.

CULTURE

Chlamydiae cannot grow in cell-free media but grows in McCoy cells treated with cycloheximide. Chlamydiae also grow in yolk sac of embryonated eggs.

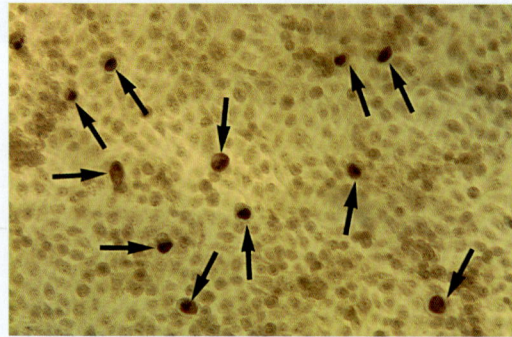

Fig. 16.1 Lugol's Iodine-Stained Cells Showing Inclusion Bodies. (*Source:* Centers for Disease Control and Prevention [CDC], Public Health Image Library [PHIL], Dr. E. Arum, Dr. N. Jacobs: Chlamydia Infections, ID# 6428, 1975.)

---

¹ Adapted from World Health Organization, Trachoma, Strategy, SAFE.

CHLAMYDIA TRACHOMATIS

C. trachomatis infection is seen in humans, and it grows in cell lines. The serovars associated with diseases are as follows:
- Endemic trachoma: A, B, Ba and C
- Sexually transmitted diseases (genital infections and inclusion conjunctivitis): D–K except I and J
- Lymphogranuloma venereum: L1, L2 and L3.

Trachoma

Trachoma is a chronic kerato-conjunctivitis, caused by serovars A, B, Ba and C. There are acute inflammatory changes in the conjunctiva and cornea in initial stage which progresses to scarring and blindness. Infection spreads via personal contact and from eyes or nasal discharge of infected persons by means of vector i.e. flies. Incubation period is 3–10 days. Mixed bacterial infection with *C. trachomatis* is common in trachoma. Lacrimation, mucopurulent discharge, conjunctival hyperemia and follicular hypertrophy are common manifestations. Common examination findings are keratitis, infiltrates and pannus with sequelae, that is, scarring of the conjunctiva, entropion and trichiasis.

Diagnosis

Diagnosis is clinical. However, laboratory confirmation can be made by smear examination/culture/serology/molecular methods. Typical inclusions are found that can be seen by microscopy in epithelial cells of conjunctival scrapings (Fig. 16.1). Both group and serovar-specific antibodies are detected by ELISA and immunofluorescence in serum and in eye secretions of patients. Molecular methods are available for diagnosis.

Treatment

Even single dose of azithromycin is effective for treatment. Erythromycin and doxycycline are not as often used as azithromycin. Topical therapy has no or limited use.

Epidemiology and Control

Trachoma is most prevalent in Africa, Asia and the Mediterranean basin where childhood infection and severe blinding disease are common. The WHO has initiated the SAFE programme to eliminate blinding trachoma and reduce clinically active disease.[1]

SAFE:	S: Surgery for deformed eyelids;
	A: Azithromycin therapy (Periodic)
	F: Face washing and hygiene
	E: Environmental improvement such as building latrines, access to water and decreasing the number of flies

Genital Infections and Inclusion Conjunctivitis

Genital chlamydial infection and inclusion conjunctivitis are sexually transmitted diseases. Genital infections spread by contact with infected sex partners. Inclusion conjunctivitis in adults occurs due to self-inoculation of conjunctiva with pathogen. The newborn acquires the infection during passage through an infected birth canal. Inclusion conjunctivitis starts 7–12 days after delivery as a mucopurulent conjunctivitis.

C. trachomatis serovars D–K except I and J cause sexually transmitted diseases, that is nongonococcal urethritis and epididymitis in sexually active men and urethritis, cervicitis and pelvic inflammatory disease in women. Proctitis and proctocolitis may occur in both sexes. Common manifestations are dysuria, nonpurulent discharge and frequency of urination.

Laboratory Diagnosis

Specimen include endocervical, vaginal, urethral, conjunctival swab, and biopsy tissue.

1. **Nucleic acid detection**: Commercial nucleic acid amplification tests are available. They are most commonly recommended tests and have good sensitivity and specificity.
2. **Direct fluorescent antibody (DFA)**: DFA detects species-specific antigen on the chlamydial major outer membrane protein (MOMP). Test is highly sensitive and specific.
3. **Enzyme-linked immunoassay (EIA)**: It detects the presence of genus-specific lipopolysaccharide antigens extracted from elementary bodies. Commercially available kits are used. Tests are highly sensitive and specific.
4. **Serology**: Anti-chlamydial antibody can be detected in genital secretions and serum. Serum antibodies are present in high titer. Due to high prevalence of chlamydial genital tract infections in population, antibody positivity in normal population is high. Hence, serologic tests to diagnose genital tract chlamydial infections are not considered useful.
5. **Culture**: Endocervical, vaginal, urethral or conjunctival swab is collected in Chlamydiae transport medium and is inoculated in cycloheximide-treated McCoy cells line at 35–37°C for 48–72 h. The monolayers are examined by direct immunofluorescence to visualise the cytoplasmic inclusions. Method is about 80% sensitive but 100% specific.

Treatment

Tetracyclines (e.g. doxycycline) given to both sex partners and infants is the most effective treatment. Azithromycin is effective. Topical tetracycline/erythromycin drops are used for inclusion conjunctivitis.

Epidemiology and Control

The ultimate control depends on safe sex practices, and early diagnosis and treatment of infected persons. Early diagnosis and treatment of the pregnant woman and her sex partner may protect newborn against chlamydial conjunctivitis. Chemoprotection does not work.

Pneumonia Due to *Chlamydia trachomatis*

About 10%–20% children born to infected mother may develop pneumonia 2–12 weeks after birth. Tachypnea, paroxysmal cough, absence of fever and eosinophilia are characteristic manifestations. A newborn with inclusion conjunctivitis with respiratory symptoms is highly suggestive of *C. trachomatis* infection. High IgM antibody titer to *C. trachomatis* is diagnostic. Erythromycin is effective and drug of choice.

Upper respiratory tract symptoms are common in adults with inclusion conjunctivitis. Lower respiratory symptoms are infrequent in adults.

Lymphogranuloma Venereum

Lymphogranuloma venereum is a sexually transmitted disease caused by any one of the three serovars (L1–L3) of *C. trachomatis* and is characterised by suppurative inguinal adenitis. After exposure, a small papule on external genitalia develops, which may ulcerate. Ulcer is painless and heals without notice. This follows enlargement of matted and painful inguinal lymph nodes. There is suppuration and pus discharge from multiple sinus tracts. Involvement of peri-rectal lymph nodes, lymphangitis and proctatitis is commoner in women. Symptoms like fever, headaches, meningismus, conjunctivitis, skin rashes, nausea and vomiting, and arthralgia are common. Management with effective antimicrobials in acute stage is necessary to prevent fibrosis which causes many sequelae.

Laboratory Diagnosis

Samples: Pus, buboes or biopsy material.
1. *Smear examination*: Usually not helpful.
2. *Culture*: On McCoy cell line can be useful.
3. *Serology*: Complement fixation and immunofluorescence tests can be done. Tests are nonspecific and are reactive with many chlamydial antigens.

Treatment

Sulphonamides, tetracyclines and erythromycin especially in early stage of disease are effective.

Epidemiology and Control

The infection occurs all over the world. High incidence has been reported from subtropical and tropical areas. The disease spreads by sexual contact. The measures used for the control of other sexually transmitted diseases also apply to the control of lymphogranuloma venereum.

CHLAMYDIA PNEUMONIAE

The TWAR strain of *C. pneumoniae* was first grown in laboratory in chick embryo yolk sac culture. It produces round, dense, glycogen-negative sulphonamide-resistant inclusions. Most infections are asymptomatic or associated with mild illness. Pharyngitis is common. Sinusitis, otitis media and atypical pneumonia occur at times.

Laboratory Diagnosis

1. Serology is the most sensitive method. A single IgM titer of 1:16, a single IgG titer of 1:512, and a fourfold rise in either the IgM or IgG titers in immunofluorescence assay is diagnostic.
2. *C. pneumoniae* can be grown in HL and HEp-2.
3. Direct detection of elementary bodies is insensitive.

Treatment

Macrolides, tetracyclines and fluoroquinolones are treatment of choice.

Epidemiology

Both endemic and epidemic infections are common. Worldwide, 30%–50% of adults have antibodies to *C. pneumoniae*. Transmission is airborne from person to person.

CHLAMYDIA PSITTACI

C. psittaci causes wide spectrum of respiratory and other manifestations, for example, mild illness, severe pneumonia and sepsis. *C. psittaci* grows in embryonated eggs, in mice and cell cultures. Species-specific antibodies neutralise the pathogen. *C. psittaci is transmitted* through the respiratory route, and involves lung. Bacteraemia occurs in first 2 weeks of the disease. Sputum sample gets positive after bacteraemia subsides and pathogen settles in lung. A patchy inflammation of the lungs like that seen in viral pneumonitis is usual. A person exposed to birds is at a risk of psittacosis. The incubation period is around 10 days. Manifestations are usually nonspecific like malaise, fever, anorexia, sore throat, photophobia and severe headache. Patients improve after few days of treatment. The mortality is high in untreated elderly cases.

Immunity in animals and humans is not complete. A carrier state in humans can persist for 10 years after recovery. During carrier state, the agent may continue to be excreted in the sputum.

Laboratory Diagnosis

Antigen detection by direct fluorescent antibody staining or by immunoassay can be done in respiratory specimen. Culture of *C. psittaci* can be dangerous, hence is not recommended. Molecular diagnosis by polymerase chain reaction is the preferred method. *C. psittaci* DNA is detected in respiratory tract specimens, vascular tissues, serum and mononuclear cells from peripheral blood.

Treatment

Azithromycin, clarithromycin, erythromycin and doxycycline in adults treat most *C. psittaci* infections.

Epidemiology

People in contact with birds are at risk. Birds may or may not develop diarrhoea. Birds are infected for their normal life span. Healthy birds' excreta may contain *Chlamydia*. Contaminated dry excreta may be the source of infection. Handling of infected tissues and inhalation of an infected aerosol can also cause infection in human. Birds like parrots kept as pets are also an important source of human infection.

17

Rickettsia

Amita Jain

LEARNING OBJECTIVES

- General characteristics of Rickettsia
- Differences in infections caused by different genera of Rickettsia
- Clinical features, epidemiology, laboratory diagnosis and management of rickettsial infections
- *Ehrlichia*

INTRODUCTION

Rickettsia is an obligate intracellular, Gram-negative bacterium. Genera *Rickettsia*, *Orientia*, *Coxiella* and *Ehrlichia* are causing diseases in human. They are transmitted to humans by arthropods (except Q fever). Arthropods are both vector and reservoir for the pathogen.

MORPHOLOGY

Rickettsiae are pleomorphic coccobacilli with Gram-negative cell wall. They can be stained with Giemsa and acridine orange, but do not stain well with Gram stain. Their cell wall contains surface proteins OmpA and OmpB. These proteins form the basis for their serotyping.

CULTURAL CHARACTERISTICS

Rickettsiae can be cultivated in yolk sacs of embryonated chick eggs and in cell culture. Routine laboratories do not perform culture due to biosafety issues. Growth addition of sulphonamides in growth media enhances their growth. Hence, sulphonamides are contraindicated in the treatment of rickettsial disease as they make symptoms more severe.

PHYSICAL PROPERTIES

They can be destroyed by heat, drying and common disinfectants. *Rickettsia prowazekii* and *Coxiella burnetii* are most resistant to drying and pasteurisation.

PATHOLOGY

Vasculitis in small vessels of skin and other organs occurs due to multiplication of organisms in endothelial cells of small blood vessels. Necrosis of endothelial cells and thrombosis of the vessel are followed by disseminated intravascular coagulation and vascular occlusion. Changes in brain and heart may also happen.

IMMUNITY AND ANTIBODY FORMATION

Infection is followed by partial immunity to reinfection, but relapses may occur. Antibodies take at least 2 weeks to appear; hence, serologic tests are useful after 2 weeks of infection.

CLASSIFICATION

Based on clinical features, rickettsiae are divided into the typhus group and spotted fever group (SFG). Q fever is caused by *C. burnetii*. Important details are provided in Table 17.1.

CLINICAL FINDINGS

Important clinical findings are listed in Table 17.1. Rickettsial infections are characterised by fever, skin rash, headache and enlargement of the spleen and liver. Rickettsial infections are difficult to diagnose; however, establishing the diagnosis and specific treatment is important because untreated cases may lead to multiple organ failures.

1. *Epidemic typhus* (*R. prowazekii*): Usually severe manifestations occur, fever last for about 2 weeks. Mortality in patients over 40 years of age is high in untreated patients. The rickettsiae can persist for many years in the lymph nodes of an infected host without any manifestations. **Brill–Zinsser disease** is a recrudescence of an old typhus infection.
2. *Endemic typhus* (*Rickettsia typhi*): It is a milder disease. Elderly patients may rarely have a severe manifestation.
3. *Scrub typhus*: It is the commonest rickettsial infection in India. Acute fever with breathlessness, cough, nausea, vomiting, myalgia and headache is the commonest manifestation. A painless ulcer (around 1 cm in size), with a

TABLE 17.1   Some Important Details of Rickettsial Pathogens

Agents	Disease	Geographic Distribution	Vector	Mammalian Reservoir	Eschar Present/Absent
Typhus group: fever, chills, myalgia, headache, rash					
Rickettsia prowazekii	Epidemic typhus (louse-borne typhus), Brill–Zinsser disease	Worldwide: South America, Africa, Asia,	Louse	Humans	No eschar, but severe illness
Rickettsia typhi	Murine typhus, endemic typhus, flea-borne typhus	Worldwide (small foci)	Flea	Rodents	Milder illness
Orientia tsutsugamushi	Scrub typhus	Asia, South Pacific, northern Australia	Mite	Rodents	**Eschar present**, lymphadenopathy
Spotted fever group: fever, headache, rash					
Rickettsia rickettsii	Rocky Mountain spotted fever	Western hemisphere (United States, South America)	Tick	Rodents, dogs	Systemic manifestations
Rickettsia akari	Rickettsial pox	United States, Korea, Russia, South Africa	Mite	Mice	Mild illness, **eschar present**
Rickettsia conorii	Mediterranean spotted fever, Israeli spotted fever, South African tick fever, African (Kenya) tick typhus, Indian tick typhus	Mediterranean countries, Africa, Middle East, India	Tick	Rodents, dogs	**Eschar present**
Q fever: headache, fever, fatigue, pneumonia					
Coxiella burnetii	Q fever	Worldwide	Airborne fomites, tick	Sheep, cattle, goats, others	No rash, complications frequent

black necrotic centre (resembling the mark of a cigarette burn), that is, eschar at the site of chigger bite, can be seen.

4. **Spotted fever group (SFG):** It resembles typhus except rashes appear on extremities first. Mortality varies from one causative agent to other. Mortality may be high in untreated cases.

5. **Rickettsial pox:** It is a mild disease with a rash resembling varicella and often has a black eschar.

6. **Q fever:** It is a febrile illness without rash and resembles influenza. Pneumonia, hepatitis and encephalopathy are common complications. Transmission occurs from inhalation of dust contaminated with rickettsiae or from aerosols in slaughterhouses. Fever is biphasic. Diagnosis is made by demonstrating rise in the titre of specific antibodies to *C. burnetii* phase 2 proteins. There is a high titre of antibodies to *C. burnetii* phase 1 proteins. Treatment with tetracycline is advocated for many months.

EPIDEMIOLOGY

The geographical distribution of Rickettsial pathogens is shown in Table 17.1. The seasonal distribution relates to their life cycle and habitat of vectors.

1. **R. prowazekii:** Life cycle runs in humans and the **human louse** (*Pediculus humanus corporis* and *Pediculus humanus capitis*). Infected louse transmits infection to a new host through contaminated faeces. Louse gets infected by biting an infected host. Transovarian transmission is not seen. Poor personal hygiene is known to be a risk factor.

2. **R. typhi:** In nature, infection is maintained between rat and rat flea. Humans are accidental hosts. Flea cannot transmit the rickettsiae transovarially.

3. **Orientia tsutsugamushi:** Infection is common in India. It is transmitted through the larval mites or 'chiggers' belonging to the family Trombiculidae. In nature, life cycle runs between mites and wild rats. Both act as reservoir as well. Scrub typhus is endemic in forest areas where bushes and small trees are frequently found. Incidence of scrub typhus is higher among rural population.

4. **Rickettsia rickettsii:** It is found in healthy wood ticks (*Dermacentor andersoni*) and is passed transovarially. Natural cycle runs between ticks and vertebrates, usually deer and rodents. Man is an accidental host.

5. **Rickettsia akari:** The vector is mite (species *Allodermanyssus sanguineus*). Natural cycle runs between mice and mite. Transovarial transmission occurs in the mite. Mite may act as a reservoir as well as a vector.

6. **C. burnetii:** Natural cycle runs in ticks with transmission to sheep, goats and cattle. Infection is an occupational hazard in individuals working with these animals. *C. burnetii* is transmitted by the respiratory route. It can be transmitted by ingestion of unpasteurised milk because cows which are infected may secrete pathogens in milk. The placentas of infected animal may be

infected, which may contaminate soil and can spread infection through aerosols.

PREVENTION AND CONTROL

There is no vaccine available. Control is possible by breaking the infection chain and treating patients with antibiotics.

LABORATORY DIAGNOSIS

As scrub typhus is common in India, we are focusing on diagnosis of scrub typhus. However, similar approach can be adapted for all the rickettsial infections.

1. **Serological assays**: Indirect immunoperoxidase assay (IPA) and immunofluorescence assay (IFA) are serological gold standards but are available only at reference laboratories, as they are expensive and technically challenging. ELISA-based assays for estimation of anti-*O. tsutsugamushi* immunoglobulin M (IgM) are now the first-line test. Antibodies are detectable after 1–2 weeks of illness.
2. **Weil–Felix test**: Rickettsiae share antigens with *Proteus* spp. This property makes the basis of this heterophile antibody test. Test demonstrates agglutinins to *Proteus vulgaris* strain OX19, OX2 and *Proteus mirabilis* OXK. The interpretation is shown in Table 17.2. Test lacks sensitivity and specificity; however, this is an inexpensive first-line test.
3. **Molecular assays**: Molecular diagnosis by PCR is available. It is a rapid and specific test, and is used to detect rickettsial DNA in whole blood/eschar samples.

TREATMENT

Tetracyclines are effective when given daily orally for long duration, that is at least 3–4 days after defervescence. They

TABLE 17.2 Interpretation of Weil–Felix Test

Disease	OX19	OX2	OXK
Scrub typhus	Negative	Negative	Positive
Epidemic/Endemic typhus	Positive	Negative	Negative
Rocky Mountain spotted fever	Positive	Positive	Negative
Rickettsial pox/Trench fever	Negative	Negative	Negative

are given even to children and pregnant women. Sulphonamides are contraindicated because they enhance the symptoms.

EHRLICHIA

The human pathogens in the group are listed in Table 17.3 and can cause disease in animals as well as humans. Tick is the vector, which transmits disease to human. These are small, Gram-negative, obligate intracellular bacteria that are taxonomically grouped with the rickettsiae. Ehrlichiae and chlamydiae are quite similar to each other because they both are found in intracellular vacuoles. However, unlike *Chlamydia*, they are able to synthesise ATP. They infect circulating leukocytes where they multiply within phagocytic vacuoles, forming inclusion like bodies. In human it produces nonspecific mild symptoms like fever, chills, headache, myalgia, nausea or vomiting, anorexia and weight loss. Laboratory diagnosis is made by demonstration of typical intracellular bodies/demonstration of *Ehrlichia* DNA/specific antibodies. Doxycycline is the treatment of choice. *Ehrlichia* is not reported from India.

TABLE 17.3 Common Species of *Ehrlichia* and Their Important Characteristics

Agents	Disease	Geographic Distribution	Vector	Mammalian Reservoir	Clinical Features
Ehrlichia chaffeensis	Human monocyte ehrlichiosis	South central, southeastern and western United States	Tick	Deer	Fever, headache, atypical white blood cells
Neorickettsia sennetsu	Human monocyte ehrlichiosis	Japan, Malaysia	Trematode, infected fish?	Mammals	Fever, headache, atypical white blood cells
Anaplasma phagocytophilium	Human granulocyte anaplasmosis	Upper midwestern, northwestern, and West Coast United States and Europe	Tick	Mice, other mammals	Fever, headache, myalgia
Ehrlichia ewingii	Human granulocyte ehrlichiosis	Midwestern United States	Tick	Dogs	Fever, headache, myalgia

Mycoplasma

Amita Jain

INTRODUCTION

Mycoplasmas are the smallest organisms. However, they are capable of independent free-living. They can replicate in artificial culture media. At least 15 of Mycoplasma species are isolated from human (Box 18.1).

In humans, mainly four species of Mycoplasma can cause infections:

1. *Mycoplasma pneumoniae* causes pneumonia and at times joint infections.
2. *Mycoplasma hominis* causes postpartum fever and uterine tube infections.
3. *Ureaplasma urealyticum* causes non-gonococcal urethritis in men and lung disease in premature infants.
4. *Mycoplasma genitalium* causes urethral infections.

BOX 18.1 Unique Characteristics of Mycoplasma

- Smallest free-living pathogens (125–250 nm in size), extremely small size and can pass through filters with 0.4 μm pore size.
- No well-formed cell wall present, instead a triple-layered membrane is present.
- Highly pleomorphic.
- Grow on complex but cell-free media.
- Addition of serum or cholesterol to the growth medium enhances growth.
- Resistance to penicillin because they lack the cell wall.
- Inhibited by tetracycline or erythromycin.
- Can reproduce in cell-free media (fried egg colonies).
- Have an affinity for mammalian cell membranes.
- Mycoplasmas use glucose as a source of energy, while ureaplasmas require urea for growth.
- Some species produce peroxides which haemolyse red blood cells.
- Many established animal and human cell culture lines carry mycoplasmas as contaminants.

MORPHOLOGY

Mycoplasmas are small organisms that cannot be visualised through naked eyes even in culture media. These cells vary in size, ranging from 15 to 250 nm in diameter. *M. genitalium* is the smallest of all Mycoplasma species; almost twice the size of certain large viruses. They lack a cell wall; hence shape is pleomorphic, indefinite and changes often. Pathogenic mycoplasma has flask-like or filamentous shapes and a polar tip, which helps in adhering to host cells.

CULTURE

They grow in both solid and liquid media to give rise to many different forms. Mycoplasmas grow in heart infusion peptone broth with either 30% human ascitic fluid or animal serum (horse, rabbit). Growth on solid media follows incubation at 37°C for 2–6 days. Colonies are round measuring 20–500 μm, have a granular surface and a dark centre which is depressed slightly and can be seen with a hand lens (fried egg appearance). A small square of agar with mycoplasma colony on it can be examined on a slide after staining with methylene blue.

ANTIGENIC STRUCTURE

Based on biochemical and serological characteristics, many antigenically distinct species of mycoplasmas are characterised of which at least 14 species have been isolated from human.

PATHOGENESIS

Pathogenic mycoplasmas have specialised polar tip structures made of proteins, adhesins and adherence-accessory proteins. The presence of proline in this apical structure influences the protein folding and adherence of mycoplasma

TABLE 18.1 Commonly Isolated Mycoplasma from Human Specimen and Their Disease Associations

Mycoplasma spp.	Disease Association	Recovered from Healthy Individuals (Yes/No)
M. pneumoniae	Atypical pneumonia	Not usually
M. hominis	Salpingitis and tubo-ovarian abscesses, postabortal or postpartum fever, occasionally arthritis	Yes, oropharynx
U. urealyticum	Non-gonococcal urethritis in men, lung infection in premature infants	Yes, genital tract of adults in reproductive age group
M. genitalium	Acute and chronic non-gonococcal urethritis, cervicitis, endometritis, salpingitis, infertility	Yes, genital tract of adults in reproductive age group
M. salivarium	Not known	Yes, oral cavity
M. orale	Not known	Yes, oral cavity

to the host cells. Those mycoplasmas, which lack the polar tip, often use alternative mechanisms to adhere to host cells. Following adherence, generation of hydrogen peroxide and superoxide radicals and other cytotoxic mechanisms cause cell damage.

DISEASES CAUSED IN HUMANS

Establishing an association of *Mycoplasma* with diseases in human is always not possible as mycoplasmas are also part of the normal flora of the mouth and other sites, for example genital, urinary and respiratory tracts (Table 18.1). Mycoplasmas' positivity of the genitourinary tract is directly related to the number of sex partners.

LABORATORY DIAGNOSIS

Specimens

Specimens include throat swabs, sputum, respiratory secretions, urethral secretions and genital secretions.

Microscopy

It is not rewarding as mycoplasmas are too small to be visualised by light microscope.

Culture

- It grows on special solid media (Eaton's Agar)/special liquid broth (pleuropneumonia-like organisms [PPLO] broth). Culture requires long incubation (3–10 days), 37°C temperature and 5% CO_2 (microaerophilic conditions). Media contains up to 20% horse serum.
- One or two subcultures may be needed.
- Colonies may have a "fried egg" appearance on agar, which can be visualised under microscope.

Serology

M. pneumoniae and *M. genitalium* are antigenically cross-reactive. Normal individuals demonstrate high false positivity; hence, rising antibody titer is to be demonstrated for diagnostic significance.

Antibodies can be demonstrated by one of the following methods:

- Complement fixation tests—using glycolipid antigens extracted with chloroform–methanol from cultured mycoplasmas.
- Haemagglutination inhibition tests—using tanned red cells with adsorbed mycoplasma antigens.
- Indirect immunofluorescence test—using mycoplasma grown in culture.

TREATMENT

Tetracyclines and erythromycins are drugs of choice in mycoplasmal pneumonia. Some ureaplasmas are resistant to tetracycline.

IMPORTANT MYCOPLASMA SPECIES AFFECTING HUMAN

M. pneumoniae

M. pneumoniae causes pneumonia, generally mild especially in children and adolescents. Clinical picture resembles pneumonia caused by other bacterial pathogens; hence, it is difficult to suspect clinically. It is usually treated with antibiotics effectively. Mortality is rare. Infection may be asymptomatic/mild/severe with neurologic and haematologic complications. Interstitial and peribronchial pneumonitis and necrotising bronchiolitis are commonly seen.

M. pneumoniae infections are endemic worldwide. Most of the cases are asymptomatic. Only one-third of infected individuals are symptomatic. Transmission occurs from person to person through respiratory route. The incubation period varies from 1 to 3 weeks. The onset is gradual. Fever, headache, sore throat and non-productive cough may be initial symptoms. Later cough may turn productive with blood-stained sputum. Initially, illness is mild but may turn severe if not treated. Recovery occurs slowly over 1–4 weeks. Mortality is rare and complications are uncommon.

Following infection antibodies are formed and cell-mediated immunity occurs. Antibodies are protective against fresh infection and therefore second attacks are not usual. The immune response-mediated damage adds to pathogenesis.

Other manifestations of *M. pneumoniae*—erythema multiforme, meningitis, meningoencephalitis, mono- and polyneuritis, myocarditis, pericarditis, arthritis and pancreatitis.

Laboratory Diagnosis

A clinical suspicion with slightly high leucocyte count is suggestive. Culture and microscopy have limited value. Serological tests may be of value. Usually molecular assays are diagnostic.

Treatment

Tetracyclines or erythromycins provide clinical improvement.

M. hominis

M. hominis is strongly associated with salpingitis and tubo-ovarian abscesses. Other than this, *M. hominis* is also associated with many diseases such as postabortal or postpartum fever, arthritis and upper urinary tract infections.

U. urealyticum

U. urealyticum requires 10% urea for growth. *U. urealyticum* is present in the female genital tract; however, its association with disease is weak. It is associated with non-gonococcal urethritis in men, and occasionally with lung disease in premature low birth weight infants.

M. genitalium

M. genitalium causes acute and chronic non-gonococcal urethritis and was first isolated from urethral cultures. In women, it causes a variety of infections, for example, cervicitis, endometritis, salpingitis and infertility. It is difficult to culture. Therefore, diagnosis is based on molecular and serological assays.

L-Phase Variants (L-forms)

L-phase variants (L-forms) are bacteria which are cell wall-defective. They are also referred to as spheroplast. They are not related to mycoplasmas. They are mutant bacterial forms, which lack cell wall; hence, they appear as non-rigid cells. They have a cell membrane, and can replicate and appear as colonies on solid media. Some L-phase bacteria are stable, while the other revert back to parental forms. They cannot be killed with cell wall acting antibacterial. They do synthesise some antigens of the cell wall of the parent bacteria (e.g. streptococcal L-forms produce M protein and capsular polysaccharide). Their pathogenic potential is not well known. However, they may help in maintenance of organisms in host.

19

Introduction to Viral Pathogens

Amita Jain

LEARNING OBJECTIVES

- Definitions of some common terminology
- Milestones in history of virology
- Properties, structure, classification and replication of viruses
- Diagnosis of viral infections

INTRODUCTION

Although viral diseases were recognised for centuries, virus was discovered only in the early 20th century. Understanding about viruses broadened due to advances in technology. Some important definitions regarding virology are summarized in Box 19.1. Viruses may have originated 4 billion years ago either by
1. regressive evolution—degenerate life forms, or
2. cellular origins—subcellular functional assemblies, or
3. independent entities—self-replicating molecules or primitive prebiotic 'RNA world'.

HISTORICAL MILESTONES

- 3700 BC: A hieroglyph from Memphis (Ancient Egypt) depicts a temple priest called RUMA, showing typical clinical signs of paralytic poliomyelitis.
- The Pharaoh Siptah ruled during 1200–1193 BC and died at the age of 20. His body was mummified, shows withered (wasted/thin) left leg with rigidly extended foot; classic signs of paralytic poliomyelitis.

- 1143 BC: Ramesses V. died of smallpox. Typical pustular lesions on skin and face could be seen.
- 1000 BC: Smallpox became endemic in China. The practice of variolation was developed in China. By AD 1520, smallpox reached America.
- 1796: Edward Jenner used the first vaccine against smallpox. He vaccinated an 8-year-old boy, James Philips, with material from a cowpox lesion (on the hand of a milkmaid named Sarah Nelmes). Six weeks later, Jenner challenged the boy by inoculating him with material from a real case of smallpox. The boy did not become infected.
- 1885: Louis Pasteur tested his pioneering rabies treatment on a human for the first time. He vaccinated a child, Joseph Meister, who had been bitten by a rabid dog and saved his life (post-exposure prophylaxis).
- First live attenuated vaccine developed was against yellow fever.
- 1880s: Robert Koch and Louis Pasteur jointly proposed the 'germ theory' of disease. Koch defined four postulates to establish aetiology of disease, known as Koch's postulates. Subsequently, Pasteur identified rabies caused by a 'virus', but in spite of this, he could not discriminate between bacterial and other agents of disease.
- In 1898: Iwanowski and Beijerinck, Russian botanist, showed that a filterable agent obtained from infected tobacco plant (tobacco mosaic disease) can transmit infection in new plants when exposed. These agents multiplied only in living cells. This was called contagium vivum fluidum (soluble living germ).
- Loeffler and Frosch demonstrated similar agents in foot-and-mouth disease in cattle.
- 1902: The famous Yellow Fever Commission headed by Walter Reed showed a filterable agent associated with human disease—the yellow fever.
- Yellow fever virus is the first recognised human virus.
- 1908: Karl Landsteiner and Erwin Popper established that a virus causes poliomyelitis.

BOX 19.1 Important Definitions

Virology is the branch of science that deals with viruses.

Virus means poisonous (Latin). This is the smallest and simplest of all forms of life. Virus has only one type of nucleic acid (either DNA or RNA) and proteins. It multiplies only when it is inside a live cell.

Viroids are small (200–400 nt) and circular RNA molecules. They have rod-like secondary structure. They have no capsid and envelope.

Virusoids are satellite molecules resembling viroid. They are larger than viroids (~1000 nt). They can multiply only in the presence of virus replication and get packaged into virus capsids as passengers.

Prions are infectious agents, which consist of a single type of protein molecule with no nucleic acid component.

- 1917: Frederick Twort and Felix d'Herelle discovered a virus infecting bacteria—bacteriophage. This greatly helped in understanding virus–host cell interaction.

PROPERTIES OF VIRUSES

1. Viruses are very small and infectious obligate intracellular organisms.
2. The virus genome is composed of either DNA or RNA. Viruses lack the genes for coding proteins essential for generation of energy and protein synthesis (ribosomes).
3. The virus genome directs the synthesis of viral proteins within an appropriate host cell.
4. Viruses do not have the biochemical or genetic potential to generate the energy required for biological processes, such as DNA replication. They are totally dependent on the host cell for these function.
5. Newly made viral components are assembled to form progeny virus particles. Other agents reproduce by division.
6. Newly formed virus particles spread infection to new cells.

THE VIRAL STRUCTURE (FIG. 19.1)

A virus particle consists of nucleic acid, surrounded by a capsid, which may be surrounded by an envelope.
1. **Viral genome (nucleic acid):** Viruses contain either DNA or RNA. Nucleic acid directs synthesis of viral proteins within host cells.
2. **Capsid (*nucleocapsid*):** Capsid is formed with small number of protein subunits, which are assembled into repeating, symmetrical structures known as *capsomere*. Their function is to protect nucleic acid. Large capsid is needed to enclose large genetic material.
   Capsid has three major classes of symmetry: (1) helical (rod-like), (2) icosahedral (sphere-like) and (3) complex.
3. **Envelope:** Envelopes are not present on all viruses. However, many animal viruses are surrounded by a lipid bilayer called envelope. Envelope is derived from the host cell membrane. During the process of virus budding and release from host cell, the part of cell membrane enveloping the viral particle gets detached from cell and becomes the outer coat of the virus. These viral envelopes later start expressing virally encoded proteins, which are often glycoproteins. Lipid envelopes are studded with surface projections (spikes or peplomers). Viruses, which contain envelopes, are usually less stable. Envelope plays a role in the process of virus attachment to host cell and entry/uptake. These are usually glysosylated by host systems prior to making them 'sticky'.

VIRUS NOMENCLATURE

One or more of the following properties/features make the basis of their nomenclature:
1. Diseases caused by them, for example, HIV and measles.

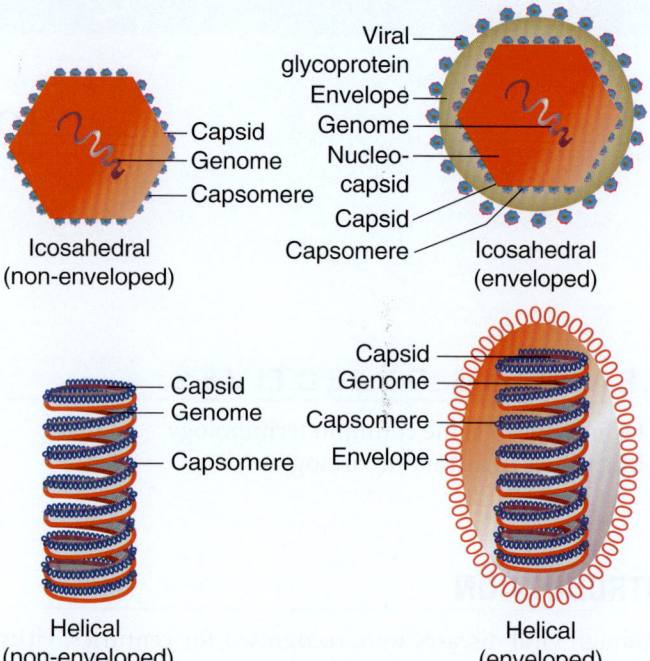

Fig. 19.1 Commonly Seen Structural Symmetry of Virus Particles.

2. Cellular changes produced during infection, for example, cytomegalovirus.
3. Anatomical site of infection or isolation, for example, adenovirus, enterovirus and rhinovirus.
4. Biochemical features, for example, retrovirus.
5. Geographical locations where they were discovered, for example, Japanese encephalitis virus.
6. People that discovered them, for example, Marburg virus.

CLASSIFICATION OF VIRUSES

One of the following viral features makes the basis of their classification:
- Type of nucleic acid found in the virion (RNA or DNA).
- Symmetry and shape of the capsid (helical/icosahedral/complex).
- The presence or absence of an envelope (enveloped/non-enveloped).
- Size of the virus particle.

International Committee on Viral Taxonomy

International Committee on Viral Taxonomy (ICTV) emphasised that uniformly viral genome should be the basis of viral taxonomy. It uses genomics for virus classification and is especially useful for animal viruses. Virus family names (usually Latinised) start with capital letters and end with the suffix—*viridae* (e.g. *Herpesviridae*). Often used interchangeably with the common names for viruses (e.g. *herpesviruses*) (Flowchart 19.1).

Rules for naming the viruses:

Order: ends in the suffix–virales (italics, first letter capitalized).

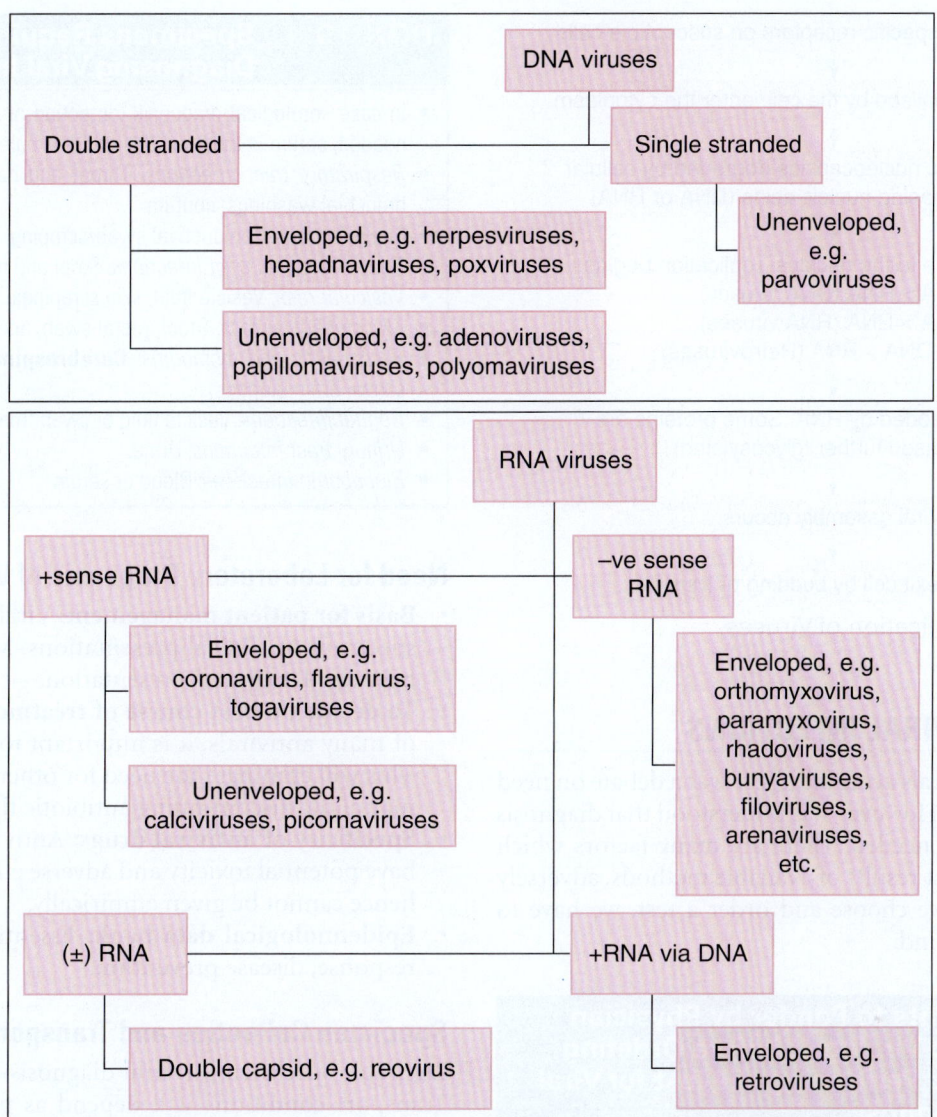

Flowchart 19.1 RNA and DNA Viruses Classified as per Genome Type.

Family: ends in the suffix–viridae (italics, first letter capitalized).

Subfamily: ends in the suffix–virinae (italics, first letter capitalized).

Genus: ends in the suffix–virus (italics, first letter capitalized).

TRANSMISSION AND REPLICATION OF VIRUSES

Modes of transmission of viruses are listed in Box 19.2. Viruses cannot replicate outside the host cell. In order to maintain themselves in nature, they stay within host cells. An infectious cycle includes virus attachment to host cell, entry/uptake, production of viral mRNA and proteins, genome replication, assembly and release of new particles (Flowchart 19.2).

Following viral replication, one of the following consequences of virus and host cell interactions occurs.

1. Viral proliferation and cell lysis, as seen in varicella zoster infection.

2. Latent infection (non-replicating viruses), as seen in infection with herpes group of viruses.

3. Persistent infection (viral replication with few or no symptoms), for example, hepatitis B infection.

4. Oncogenic infections (causes tumours), for example, human papillomavirus.

BOX 19.2 Mode of Transmission of Viruses

- **Direct personal contact,** e.g. herpesviruses and HIV.
- **Airborne spread,** e.g. chicken pox and influenza virus.
- **Parenteral spread,** e.g. HIV, hepatitis B and C and cytomegalovirus (CMV).
- **Through fomites,** e.g. enteroviruses and other sturdy drying-resistant viruses.
- **Through vectors,** e.g. West Nile virus (WNV), dengue virus (DV) and Japanese encephalitis virus (JEV).
- **Vertical transmission,** e.g. HIV, herpes simplex, cytomegalovirus and rubella (German measles).
- **Enteral spread** (foodborne), e.g. hepatitis A and gastroenteritis viruses.

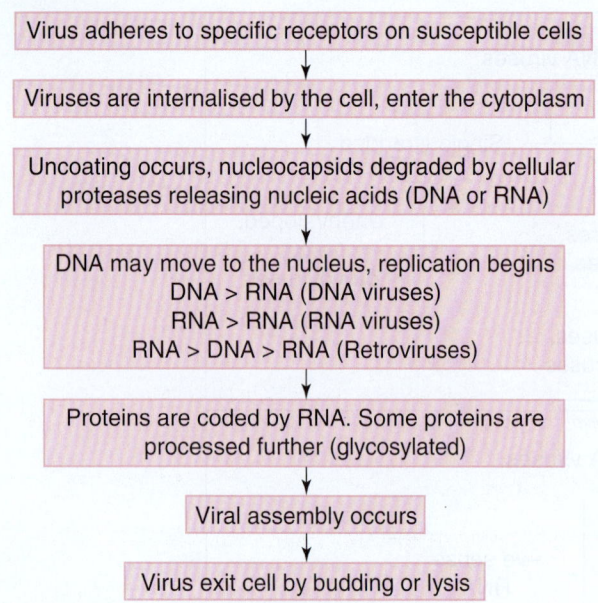

Flowchart 19.2 Replication of Viruses.

DIAGNOSIS OF VIRAL INFECTIONS

Till there were antivirals available, there was a debate on need for viral diagnostics. However, it is understood that diagnosis of viral infections is needed. There are many factors which may influence the test results of available methods, adversely (Box 19.3). Before we choose and order a test, we have to keep these facts in mind.

BOX 19.3	**Points to Ponder by Physician and Virologist before Ordering and Performing the Test for Viral Diagnosis**

- Selection of assays—appropriate, sensitive and specific
- Clinical data input should be available
- Clinical details which are important to know before applying a test
  - Date of onset of symptoms, and provisional diagnosis
  - Epidemiological details
  - Local epidemics if occurring at the time
  - History of contact with animals
  - Immunisation status
- Appropriateness of sample: Both in terms of time of collection and quality
- Transportation methods: As per the test required
- Knowledge on disease kinetics—will help decide the test
- Virus isolation can be influenced by
  - Host cell specificity
  - Selection of tissue/cell line
- Test result interpretation
- Lack of effective communication between clinician and virologist/laboratory
- Understanding that no diagnostic assay is perfect
  - Knowledge of false positive/negative results—a possibility with all currently available assays
- Quality control in diagnostic virology laboratory

BOX 19.4	**Specimens Required for Diagnosing Viral Infection**

- In case serological diagnosis (detection of antiviral antibodies) is needed, serum is the most desirable sample.
- *Respiratory tract infections*: Throat and nasal swabs, nasal and bronchial washings, sputum.
- *Eye infections:* Conjunctival swab/scraping, throat swab.
- *Gastrointestinal tract infections*: Stool and rectal swab.
- *Vesicular rash:* Vesicle fluid, skin scrapings.
- *Maculopapular rash*: Stool, rectal swab, and throat swab.
- *Encephalitis and meningitis*: **Cerebrospinal fluid,** stool, tissue, saliva, brain biopsy.
- *Genital infections*: Vesicle fluid or swab from floor of vesicle.
- *Urinary tract infections*: Urine.
- *Bloodborne infections*: Blood or serum.

Need for Laboratory Diagnosis of Viral Diseases

- **Basis for patient management:** Viral diseases have broad spectrum of clinical presentations. Many viral infections have similar clinical presentation.
- **To decide on best course of treatment:** With the advent of many antivirals, it is important to know the aetiology. This will eliminate the need for other diagnostic tests and may allow discontinuing antibiotic therapy.
- **Specificity of antiviral drugs:** Antivirals are specific and have potential toxicity and adverse effects. They are costly, hence cannot be given empirically.
- **Epidemiological data bank:** Essential for public health response, disease prevention.

Specimen Collection and Transport

Desirable specimen for viral diagnosis is listed in Box 19.4. Transport conditions will depend as per test requirement. Viral isolation has most stringent transport conditions, that is maintenance of cold chain and use of viral transport medium.

Diagnostic Techniques
Virus Isolation

1. **Virus isolation is usually not used for clinical diagnosis as this is time-consuming and technically challenging.** However, this is the most definitive method (the gold standard). This is the only way to characterize the virus. No single cell line can support all viruses; hence, variety of cell lines are required to be maintained in the laboratory. It is ordered in first few days of illness.
2. Viruses are highly sensitive to temperature; hence cold chain needs to be maintained during transportation. Viruses get inactivated at room temperature within 30 min.
3. Transport immediately in virus transport medium (VTM) under cold condition. VTM is an isotonic solution (Hank's balanced salt solution) containing gelatin, bovine serum albumin or foetal calf serum.
4. Body fluids can be directly transported.

Microscopy/Antigen Detection

1. Sample should contain sufficient cellular material.

Serology: Detection of Virus Specific Antibodies in Serum and Body Fluids

1. Paired serum sample is mandatory.
2. One sample early in the course of illness (within 1st week of onset of symptoms) (acute serum).
3. Second sample 10–14 days after first sample (convalescent serum).

Molecular Assays

Samples are collected and transported in the same manner as for virus isolation.

Common Tests Applied for Diagnosis of Viral Diseases

Microscopy

Light microscopy. Smears and slides prepared from cellular specimen like blister fluid/tracheobronchial aspirates or biopsy tissue can be examined for cytopathic effects (CPEs). Immunohistochemistry staining or in situ hybridisation methods can be applied to these smears.

Electron microscopy. Transmission electron microscopy and immune electron microscopy can be done at reference laboratories to study the structure of viral particles. Not recommended for routine diagnosis.

Antigen Detection Tests

Antigen detection methods are quick and reliable alternative to culture. Cell-bound antigens can be detected by the following:

- Immunofluorescence assay (IFA), for example, **respiratory viruses, HSV, VZV, CMV** etc.
- Immunoperoxidase staining, for example, **CMV**, **rabies**, etc.
- Enzyme-linked immunosorbent assay (ELISA): Soluble antigens (free antigen) are detected by ELISA, for example, **HBsAg, RSV, rotavirus and HIV-p24**.

Antibody Detection

Antibody detection methods are most widely used methods, although these tests ideally require paired sera. The advantages of these methods are listed in Box 19.5. Common methods of antibody detection are as follows:

- ELISA (IgG, IgM detection).
- Indirect immunofluorescence assay (IFA).
- Virus neutralisation assay (NT).
- Complement fixation test (CFT).
- Haemagglutination inhibition assay (HAI).

Nucleic Acid Detection (by Nucleic Acid Amplification Assay (NAAT)

This is a sensitive detection of specific viral nucleic acids. Advantages and disadvantages of these methods are listed in Box 19.6. Different methods are listed as follows:

> **BOX 19.5 Advantages of Antibody Detection Methods**
>
> - Rapid diagnosis
>   - Virus-specific IgM antibody—indicates current infection
>   - Rubella, measles, human parvovirus B19
>   - Virus specific antibody (any class)—indicates current or past infection
>   - HIV, HCV
> - Useful in defining virus specific immune status
>   - Infection with varicella zoster virus (VZV), cytomegalovirus (CMV), herpes simplex virus (HSV), measles virus (MV), rhinovirus (RV), B19V, hepatitis A virus, hepatitis B virus.
> - Antibody profile (against a panel of antigens)
>   - EBV infection

> **BOX 19.6 Advantages and Disadvantages of NAAT**
>
> | Advantages | Disadvantages |
> | --- | --- |
> | Can be used to detect viruses that are not cultivable | High sensitivity – false positivity |
> | Rapid identification | DNA/RNA extraction techniques are cumbersome |
> | Can be used to manage patients (e.g. HIV viral load estimation) | Only few assays are licensed for diagnostic use |

- Detection by DNA/RNA probes
- Dot blot hybridisation
- In situ hybridisation
- Amplification of nucleic acid
  - **Qualitative**
    - Polymerase chain reaction (PCR); most commonly used
    - Reverse transcriptase-PCR for RNA viruses (RT-PCR)
    - Multiplex PCR
    - In situ PCR
    - Nested PCR
  - **Quantitative:** Real-time PCR

Virus Isolation

There are three modalities for viral culture:

1. Animal inoculation
2. Chick embryo inoculation
3. Cell culture

Animal inoculation was the only available method initially. Mouse is the most widely used animal; however, no single animal is susceptible for all viruses.

Route of inoculation includes subcutaneous, intraperitoneal, intracerebral, intravenous, respiratory etc. (Box 19.7).

Chick embryo inoculation was introduced in 1930s. Developing chick embryo supports growth of several viruses.

BOX 19.7   Common Animal Models and Route of Inoculation Used for Virus Culture

- Rabies—young adult mice (intracerebral)
- Arbovirus—suckling mouse (intracerebral)
- Coxsackie viruses—suckling mouse (subcutaneous, intraperitoneal, intracerebral)
- Herpesviruses—suckling mouse (intracerebral)

TABLE 19.1   Routes of Chick Embryo Inoculation

Route	Virus
Chorioallantoic membrane (CAM)	Smallpox virus, herpes simplex virus
Allantoic cavity	Influenza
Amniotic cavity	Influenza
Yolk sac	Rabies

TABLE 19.2   Commonly Used Cell Lines for Virus Culture and Cytopathic Effects Observed

Virus	Cell Line	CPE
Enteroviruses (polio virus)	Vero	Shrinking of cells and piknosis of nucleus
Respiratory syncytial virus	Hep-2	Syncytia formation
Herpes simplex virus	Vero	Rounding and ballooning of cells

This is a good alternative to animals at least for certain viruses. Method is used for diagnosis and vaccine production. Routes of inoculation are listed in Table 19.1.

Tissue/Cell line inoculation. Enders and co-workers grew polio virus in primary monkey kidney cells cultured in a test tube in 1949. This was the beginning of a new era in virology. Previously explant cultures and organ cultures were used to grow viruses with limited success. Cells grow as a single layer on a supporting surface such as glass or plastic. Multiplication is limited by contact inhibition.

Different types of cell lines are as follows: primary, diploid and continuous cell lines.

Primary cell lines. Examples include primary monkey kidney cells (PMK), primary rabbit kidney cells and amniotic fibroblast culture.
- Cell line derived from culturing cells directly taken from an organ.
- Epithelial cells or fibroblasts.
- Do not have the ability to multiply indefinitely. Do not grow beyond 5–6 subcultures.

- Difficulty in preparing cell line as these are nutritionally demanding.
- High virus susceptibility.
  Diploid cell lines. Examples include MRC-5 (human embryonic fibroblast cells) and WI-38 (human embryonic lung fibroblast cells).
- Cell line capable of growing up to 50 subcultures and retain diploid number of chromosomes.
- Cells are fibroblasts in origin.
- Nutritionally demanding, difficult to maintain.
  Continuous cell lines. Examples include Vero cell line and HEp-2 cell line.
- Cell lines capable of indefinite multiplication.
- Original cells may be normal or malignant.
- Cells may be of human, animal or insect origin.
- Low virus susceptibility.
- Most widely used cell lines.
  Viral growth in cell lines is detected by the following:
  CPE: Visible results of viral infection.
  Immunofluorescence for viral antigens.
  Virus neutralisation assay.
  Haemagglutination assay.
  Haemadsorption test.
  PCR.

CPEs are examined using inverted light microscopes. Some examples are listed in Table 19.2. All viruses may not produce CPE. Common findings which are observed are: rounding/detachment from plastic flask, syncytia formation/fusion of cells, cell shrinkage, increased refractivity, aggregation of cells, loss of cell adherence, cell lysis/death, inclusion body formation etc.

20

Herpesviruses

Amita Jain

LEARNING OBJECTIVES

Structure, classification, replication and pathogenesis of Herpesvirus:
- *Herpes simplex viruses 1 and 2*
- *Varicella-zoster virus*

- *Cytomegalovirus*
- *Epstein–Barr virus*
- *Human herpesviruses 6, 7 and 8*

INTRODUCTION

Family *Herpesviridae* includes several human pathogens. Human herpesviruses are listed in Box 20.1. The characteristic biological properties of the family *Herpesviridae* are as follows:
1. Requires several enzymes for metabolism of nucleic acid (e.g. thymidine kinase), DNA synthesis (e.g. DNA helicase/primase) and protein processing (e.g. protein kinase)
2. Synthesis of viral genome and assembly of capsid occur in the host cell nucleus
3. Infection causes host cell destruction
4. Herpesviruses are latent viruses which get reactivated in immunocompromised patients' hosts
5. Some are oncogenic in nature

STRUCTURE

Herpesviruses are large viruses, around 150–200 nm in size. All herpesvirus virions appear similar by electron microscopy and have four structural elements (Fig. 20.1).
1. **Core**: It is the central part of virion, which holds a double-stranded DNA, and is surrounded by a protein coat.
2. **Capsid**: It surrounds the core. It has icosahedral symmetry. It is made up of 162 identical units called capsomeres.
3. **Envelope**: It surrounds the nucleocapsid. This is the outer most coat, that is derived from the nuclear membrane of the infected cell and contains viral glycoprotein spikes embedded in the envelope.

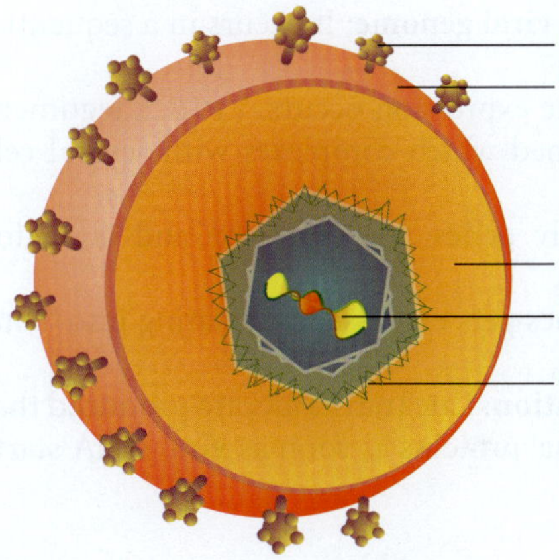

Fig. 20.1 Structure of Herpesvirus.

4. **Tegument**: The amorphous material between the capsid and envelope is called tegument. It consists of viral enzymes.

Herpesvirus genome encodes several, around 100 different proteins; some are structural and others are involved in nucleic acid metabolism, DNA synthesis, gene expression and protein regulation.

CLASSIFICATION

Members of the herpesvirus family are classified as shown in Table 20.1.

REPLICATION

It occurs in the following steps (Fig. 20.2):
1. **Entry in the host cell**: The glycoproteins on the virus envelope bind to the host cell.
2. **Fusion with the cell membrane:** Viral envelope fuses with cell membrane. Capsid enters in the cell cytoplasm from

---

BOX 20.1 Human Herpesviruses

- *Herpes simplex virus types 1 and 2*
- *Varicella-zoster virus*
- *Cytomegalovirus*
- *Epstein–Barr virus*
- *Herpesviruses 6 and 7*
- *Herpesvirus 8* (Kaposi's sarcoma-associated herpesvirus)

TABLE 20.1	Classification of Family *Herpesviridae*		
Subfamily	**Official Name of Members**	**Earlier Name of Members**	**Latency in Cells**
Alphaherpesvirinae • Fast-growing • Cytolytic • Establish latent infections in neurons	HHV-1 HHV-2 HHV-3	HSV-1 HSV-2 VZV	Neurons Neurons Neurons
Betaherpesvirinae • Slow-growing • May be cytomegalic	HHV-5 HHV-6 HHV-7	*Cytomegalovirus* HHV-6 HHV-7	Salivary glands, kidney Lymphocytes Lymphocytes
Gammaherpesvirinae • Infect lymphoid cells • Latent in lymphoid cells	HHV-4 HHV-8	EBV HHV-8	Lymphocytes Lymphocytes

EBV = Epstein–Barr virus; HHV = human herpesvirus; HSV = *herpes simplex virus*; VZV = *varicella-zoster virus*.

where it is transported to nucleus through a nuclear pore. Uncoating of viral DNA occurs in the nucleus and it gets associated with the nucleus.

3. **Expression of the viral genome**: It occurs in a sequential fashion.
   a. Initial viral gene expression occurs. VP16, a tegument protein, is formed which complexes with several cellular proteins.
   b. Immediate early genes are expressed and code for 'alpha' proteins.
   c. Early set of genes gets expressed translating into 'beta' proteins.
4. **Viral DNA replication**: Late transcripts are produced that give rise to 'gamma' proteins as soon as viral DNA starts replicating.

5. Herpesvirus-infected cells produce many viral proteins. Most of the alpha and beta proteins are non-structural enzymes, whereas gamma proteins are structural proteins.

HERPES SIMPLEX VIRUSES

INTRODUCTION

Herpes simplex viruses are of two types: type 1 and type 2 (HSV-1 and HSV-2). They have substantial genomic and antigenic homology. They grow rapidly and are cytolytic. They cause many types of clinical syndromes. HSV-1 often causes gingivostomatitis, keratoconjunctivitis, encephalitis, keratitis, cold sores, herpetic whitlow, eczema herpeticum and blisters. HSV-2 often causes genital herpes, neonatal infections and cutaneous herpes. Majority of herpetic infections above the waist are caused by HSV-1, while those below the waist are caused by HSV-2. The herpes simplex viruses remain latent in nerve cells and get reactivated often.

PATHOLOGY

HSV usually causes local disease. A brief period of viraemia disseminates virus in the body following which latency in the craniospinal ganglia is established. Recurrence following triggers like physical or psychological stress, infection, fever and irradiation can occur.

CLINICAL MANIFESTATIONS

The common clinical manifestations are shown in Table 20.2.

EPIDEMIOLOGY

- Worldwide in distribution
- Common in children between 6 months and 3 years of age
- Spread by contact

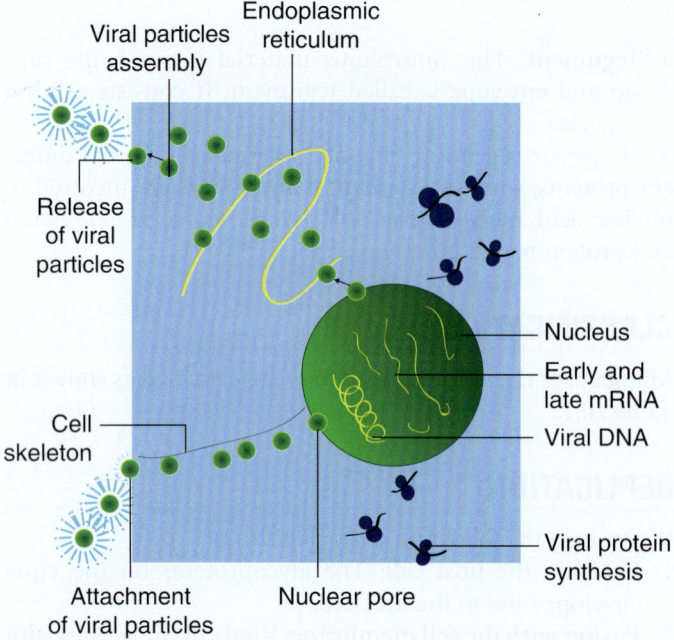

Fig. 20.2 Replication Cycle of Herpesvirus.

TABLE 20.2 Clinical Manifestations of HSV-1 and HSV-2

Clinical Conditions	Causative Virus (Mostly)	Primary Infection/ Reactivation	Clinical Features	Remarks
Acute gingivostomatitis	HSV-1	Primary infection	Painful ulcers with necrotic bases, fever and cervical lymphadenopathy are present	Heals within 2 weeks on its own
Herpes labialis (cold sore)	HSV-1	Recurrence of oral HSV	Tingling and itching at the site may occur followed by papules and then vesicles	
Ocular herpes	HSV-1	Primary	Keratitis, dendritic ulcers	
Ocular herpes	HSV-1	Recurrent	Keratitis, conjunctivitis, iridocyclitis, chorioretinitis, cataract	
Herpes genitalis	HSV-2	Primary/Recurrent (60%)	Genital painful ulcers. Bladder, sacral nerve routes, and meninges may be involved	Mortality rate is high without treatment
Meningitis and encephalitis	HSV-1	Primary	Temporal lobe is most commonly affected	
Neonatal encephalitis	HSV-2	Primary	Global involvement of the brain	Mortality rate of approximately 100% without treatment
Neonatal herpes	HSV-2	Primary	Mild disease localised to the skin to a fatal disseminated infection	Acquires infection perinatally during passage through the birth canal
Other manifestations	HSV-1 and HSV-2	Recurrent	Disseminated herpes simplex infections, e.g. eczema herpeticum, herpetic whitlow, zosteriform herpes simplex	

- Shed in saliva, tears, genital and other body secretions
- Majority of adults have antibodies against HSV-1
- Latency persists lifelong with occasional recurrences
- HSV-2 is acquired as a sexually transmitted disease in teenagers and adults
- Recurrent genital infections are frequent and may be symptomatic or asymptomatic
- No animal reservoirs or vectors are known to occur

LABORATORY DIAGNOSIS

- **Amplification of herpesvirus DNA (NAAT):** Routinely done for the diagnosis of herpes simplex encephalitis.
- **Tzanck smear:** Scrapings from the base of a vesicle are stained with Giemsa stain for demonstration of multinucleated giant cells, which are present in HSV-1, HSV-2 or varicella-zoster infections.
- **Immunofluorescence:** It can distinguish between HSV and varicella-zoster virus (VZV).
- **Virus isolation:** HSV-1 and HSV-2 take 1–5 days to grow on HeLa/Vero cell lines.
- **Serology:** Diagnostic value of serologic assays is limited due to persistence of antibodies for life, latency of virus and cross-reactivity among groups.
- **Electron microscopy** of vesicle fluid demonstrates virus. Morphologically HSV and VZV are indistinguishable.

MANAGEMENT

Acyclovir is the drug of choice. Early diagnosis and treatment can change the outcome, especially in cases of herpetic encephalitis. It is given in intravenous injections, oral formulations, cream and ophthalmic ointment (for herpetic keratoconjunctivitis) formulations. Oral preparations of famciclovir and valacyclovir are also tried.

VARICELLA-ZOSTER VIRUS

INTRODUCTION

VZV causes two types of manifestation. In children, it causes chickenpox (varicella), an acute disease in a virus-naive person. Varicella manifests as vesicles on skin and mucosa. In adults, it causes zoster (shingles). This is the reactivation of latent varicella virus present in sensory ganglia. It manifests as rash limited to the skin supplied by one sensory ganglion. These diseases manifestations are the result of differing host responses.

Structure of VZV is identical to HSV. Human are only the reservoirs. The virus easily grows in cell culture.

CHICKENPOX

The pathogenesis is shown in Flowchart 20.1. The rash in chickenpox is typical which is characterised by changes in

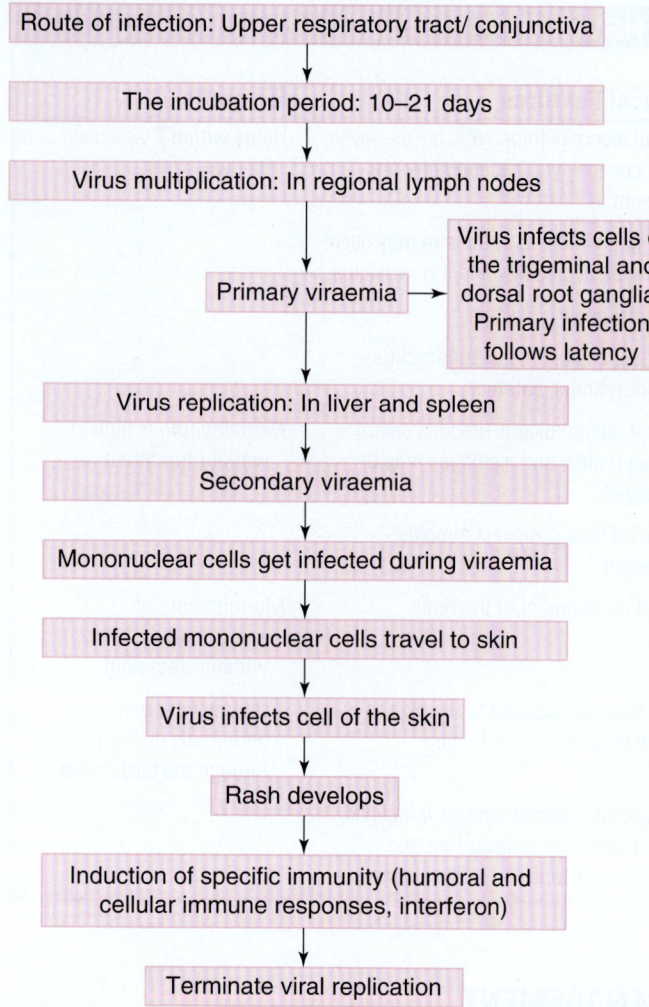

Flowchart 20.1 Primary Infection With Varicella-Zoster Virus (VZV).

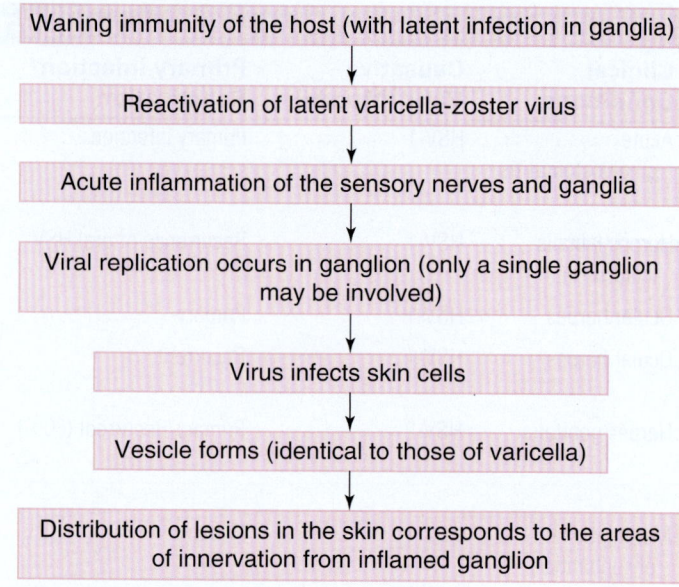

Flowchart 20.2 Reactivation of VZV.

epithelial cells, swelling and ballooning degeneration. There is surrounding oedema.

Clinical Picture of Chickenpox

- Malaise, fever, rash
- Rash first appears on the trunk followed by face, limbs and oral mucosa
- All stages of rash, that is macules, papules, vesicles and crusts may be seen in one patient at one time
- Rash lasts for about 5 days
- Encephalitis may occur in rare cases and can be life-threatening
- Neonates may contract infection from the mother just before or after birth
- Complications in normal individuals are rare and the mortality rate is very low
- Immunocompromised patients are at increased risk of complications. Varicella pneumonia is the most common complication. Disseminated intravascular coagulation

may occur. Severe, disseminated VZV disease is seen in children with leukaemia.
- Subclinical chickenpox is rare

ZOSTER

The pathogenesis of zoster is shown in Flowchart 20.2.

Clinical Picture of Zoster

- Zoster is often seen in immunocompromised persons
- Healthy young adults are rarely affected
- Severe pain in the area of skin or mucosa, followed by a crop of vesicles is common manifestations
- Nerves of trunk, head and neck are most commonly affected
- Most common complication is postherpetic neuralgia
- Infection with varicella provides lifelong immunity to varicella. Antibodies persist lifelong. Neutralising antibodies to varicella do not provide protection against zoster.

EPIDEMIOLOGY

Varicella and zoster occur worldwide. Chickenpox is highly communicable. Children are most commonly affected; however, adult cases do occur. Zoster demonstrates no seasonal prevalence. Spread is by airborne droplets and direct contact. Every case with rash is infectious. Contact infection is less common in zoster.

LABORATORY DIAGNOSIS

- Specimen: Vesicle fluid, skin scrapings, biopsy material and serum
- Tzanck smear: Scrapings or swabs of the base of vesicles show multinucleated giant cells

- Intracellular viral antigens demonstration: By immuno-fluorescence
- Virus isolation: Possible from vesicle fluid in first few days of illness
- Serology: A rise in specific antibody titre can be detected in the patient's serum by ELISA and other assays

VACCINE

A live attenuated, highly effective vaccine is available.

TREATMENT

No treatment required. Patients with severe infections need treatment. Varicella-zoster immunoglobulin may be preventive, however, it has no role once lesion have appeared. Several antiviral compounds, for example, acyclovir, valacyclovir, famciclovir and foscarnet can be used.

CYTOMEGALOVIRUS

INTRODUCTION

Cytomegaloviruses (CMV) are common infections, which are named after the classical inclusion bodies seen in virus-infected cells. CMV is highly specific to cell types and host species. Human cytomegalovirus is fastidious to culture in laboratory. In human fibroblasts, slow growth may occur.

Cytomegalovirus has large DNA genome (240 kbp), larger than that of HSV. Some of the important viral proteins found in CMV are:

- A cell surface glycoprotein
- Major immediate early promoter-enhancer

Cytomegalovirus produces a characteristic cytopathic effect. Perinuclear cytoplasmic inclusions and intranuclear inclusions in multinucleated giant cells are often seen (Fig. 20.3). Cytomegalic cells with inclusion bodies can be present in clinical samples.

PATHOGENESIS

Primary infection with CMV occurs through person-to-person contact. After incubation period of 4–8 weeks, systemic infection occurs involving lung, liver, oesophagus, salivary gland, colon, kidneys, monocytes and T and B lymphocytes. Subclinical infections and an infectious mononucleosis-like syndrome are common. Cell-mediated immunity is depressed in infected individuals leading to the persistence of viral infection. Lifelong latent infections are the rule. Cellular immunity may take long to recover after primary infection. Virus from latently and acutely infected individuals is shed intermittently in the urine and from pharynx for long.

From infected mother foetal and newborn infections occur and may be chronic and severe. The virus can be transmitted in utero, during delivery from exposure to virus in the mother's genital tract and from maternal breast milk.

In immunosuppressed hosts, for example, individuals receiving organ transplants, receiving chemotherapy and with AIDS, severe infections occur. CMV pneumonia is the most common complication. Reactivation is common in immunocompromised patients than in normal hosts.

CLINICAL PICTURE

- **Immunocompetent host:** Usually asymptomatic, occasionally spontaneous infectious mononucleosis syndrome occurs. These cases appear as heterophil-negative non-Epstein–Barr virus mononucleosis cases.
- **Immunocompromised hosts:** Pneumonia and interstitial pneumonitis are the commonest complications. Bronchiolitis in lung transplants cases, graft atherosclerosis after heart transplantation and *Cytomegalovirus*-related rejection of renal allografts occur.
- **Untreated AIDS patients:** Gastroenteritis and chorioretinitis.
- **Foetus:** Death in utero.
- **Newborns:** Cytomegalic inclusion disease of newborn, severe hearing loss, ocular abnormalities, mental retardation, CNS involvement, intrauterine growth retardation, jaundice, hepatosplenomegaly, thrombocytopaenia, microcephaly and retinitis.

CMV-specific antibodies of the IgM, IgA and IgG classes persist lifelong post infection, although they are not protective against reinfection. Maternal antibodies are not protective to foetus but may reduce chances of severe infection.

EPIDEMIOLOGY

Cytomegalovirus is endemic worldwide, without seasonal variation. Humans are the only known host for *Cytomegalovirus*. Antibody prevalence is high to moderate in adults in developed countries. Transmission requires close person-to-person contact. Oral and respiratory routes are the most common modes of transmission. Transmission does occur by blood transfusion, organ transplantation, sexual contact and transplacental routes.

LABORATORY DIAGNOSIS

Specimen: Blood and urine
- Polymerase chain reaction (PCR): It is a routine diagnostic test. Viral load estimations are important in predicting *Cytomegalovirus* disease.

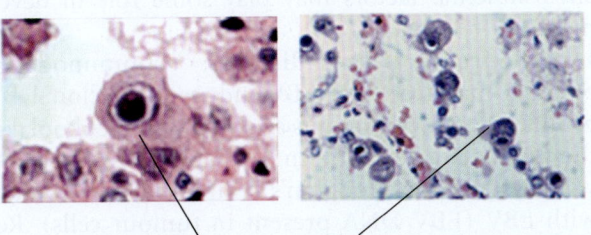

Intranuclear and cytoplasmic inclusion

Fig. 20.3 Cytopathogenic Changes in *Cytomegaloviruses* (CMV)-Infected Cells. (*Source:* Characterization of Cytomegalovirus Lung Infection in Non-HIV Infected Children in Viruses. PubMed, Fig. 6, 6(5):2038–51, May 2014.)

- Antigen detection assays: Monoclonal antibodies against viral antigens can be used to detect virus-positive leukocytes from patients.
- Serology: Anti-CMV IgG antibodies indicate past infection with CMV. Anti-CMV IgM antibodies suggest a current infection.
- Cell culture: It is too slow to be useful in clinical practice.

TREATMENT

Ganciclovir, a nucleoside structurally related to acyclovir, is the drug of choice. Foscarnet is recommended for treatment of retinitis. Acyclovir and valacyclovir may be tried in transplant patients.

CONTROL

No vaccine/specific control measures are available.

EPSTEIN–BARR VIRUS

INTRODUCTION

Epstein–Barr virus (EBV) is highly species-specific. EBV infects B lymphocyte. It enters B cells by binding to the receptor for the C3d component of complement (CR2 or CD21). EBV has potential to turn infected B cells, immortal. EBV-immortalised B lymphocytes express B cell activation products (e.g. CD23) and immunoglobulins. EBV needs not necessarily multiply before establishing latency, hence, infected B cells may not release virus. EBV multiplies in epithelial cells of the oropharynx, parotid gland and cervix.

Viral antigens are as follows:
- EBV nuclear antigens (EBNA1, 2, 3A-3C, LP)
- Latent membrane proteins (LMP1, 2)
- Small untranslated RNAs (EBER1, 2)

EBV is divided into two types: EBV-1 and EBV-2 on the basis of EBNAs, EBERs.

EBV antigens are divided into the following three classes:
1. Latent phase antigens (EBNAs and LMPs): Presence is suggestive of latently infected cells.
2. Early antigens (nonstructural proteins): presence is suggestive of viral replication.
3. Late antigens (viral capsid antigen and viral envelope glycoprotein): Suggestive of presence of virus and active virus replication

PATHOGENESIS

It is illustrated in Flowchart 20.3.

CLINICAL PICTURE

It causes acute infectious mononucleosis, and is associated with nasopharyngeal carcinoma, Burkitt's lymphoma, Hodgkin's disease, other lymphoproliferative disorders.
1. **Subclinical primary infections** in children are usual.

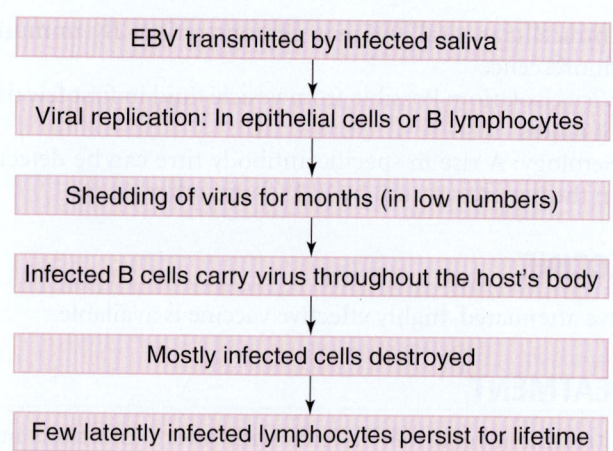

Flowchart 20.3 Pathogenesis of Epstein–Barr Virus (EBV) Infection.

2. **Acute infectious mononucleosis** (polyclonal stimulation of lymphocytes) is seen in young adults. Incubation period is 30–50 days. Symptoms last 2–4 weeks followed by recovery. Common symptoms include headache, fever, malaise, fatigue, sore throat, enlarged lymph nodes and spleen. Hepatitis is rare. Complications are rare in normal hosts.
   Lymphocytosis and presence of large, atypical T lymphocytes is often seen. EBV-infected B cells synthesise autoantibodies with heterophil antibody (reacts with antigens on sheep erythrocytes).
3. **Oral hairy leukoplakia**: It is an epithelial replicative infection of EBV. Appears as a wart-like growth on the tongue of immunocompromised individuals.
4. **Burkitt's lymphoma**: Tumour of the jaw in African children and young adults. More than 90% of African Burkitt's lymphomas and about 20% of Burkitt's lymphomas from rest of the world contain EBV DNA and express EBNA1 antigen. It is speculated that (1) EBV is involved in immortalising B cells, (2) malaria is a recognised cofactor, (3) characteristic chromosome translocations on immunoglobulin genes result in deregulation of expression of the c-*myc* proto-oncogene.
5. **Nasopharyngeal carcinoma** of epithelial cells is common in China. EBV DNA and EBNA1 and LMP1 antigens are found in nasopharyngeal carcinoma cells. Genetic and environmental factors may play some role in development.
6. **Lymphoproliferative diseases**: Immunodeficient patients manifest fatal EBV-induced polyclonal B cell proliferation. Lymphomas and oral hairy leukoplakia of the tongue are common in AIDS patients. Central nervous system non-Hodgkin's lymphomas are associated with EBV (EBV DNA present in tumour cells). Reed–Sternberg cells, a hallmark of Hodgkin's disease is seen in ~50% cases.
7. **Reactivations of EBV latent infections**: Virus is detected in saliva and blood cells. Usually reactivations are silent, sometimes with serious consequences.

TABLE 20.3	Interpretation of Serological Assays for Diagnosis of EBV Infections			
	Anti-capsid IgM	Anti-capsid IgG	Anti-Early Antigen Antibodies	Antibodies to EBNA Antigens
• Current infection	Positive	Negative	Positive	Negative
• Past infection	Negative	Positive	Negative	Positive (in some cases)
• Burkitt's lymphoma	—	—	Positive	—
• Nasopharyngeal carcinoma	—	—	Positive	—

LABORATORY DIAGNOSIS (TABLE 20.3)

Samples

Saliva, peripheral blood or lymphoid tissue.

Nucleic Acid Hybridisation

Nucleic acid hybridisation is the most sensitive test. EBER-RNAs are a useful diagnostic target.

Viral Antigens Detection

Viral antigens detection in lymphoid tissues and in nasopharyngeal carcinomas.

Viral Culture

Viral culture is laborious, technically challenging and time-consuming; hence, has no diagnostic application.

Serology

IgM and IgG antibodies to viral capsid antigen, antibodies to the early antigen, antibodies to EBNA and the membrane antigen and **nonspecific antibodies** (heterophil antibodies that agglutinate sheep cells) can be detected. Commercially available spot tests are available to detect these antibodies.

EPIDEMIOLOGY

EBV is common worldwide. Most of adults are seropositive. Transmission occurs by contact with oropharyngeal secretions. Infection occurs early in life, and is asymptomatic usually, resulting in permanent immunity to infectious mononucleosis. In adults, infection is manifested by infectious mononucleosis, mainly seen in developed world.

TREATMENT AND CONTROL

There is no EBV vaccine available. Acyclovir has controversial role in management.

HUMAN HERPESVIRUS 6

Human herpesvirus 6 (HHV-6) is T-lymphotropic virus and was first recognised in 1986. HHV-6 is widespread in population. The mode of transmission is via oral secretions. World over *human herpesvirus 6* is unrelated antigenically to the other human herpesviruses except with HHV-7. There are two antigenic groups of HHV-6, designated as A and B. 6B variant is the cause of roseola infantum, or 'sixth disease'. It is the mild common febrile childhood disease with rash. The virus grows well in B cells, glial cells, CD4 T lymphocytes, fibroblasts and megakaryocytes.

Human CD46 is the cellular receptor for the virus. Infections occur in early childhood. Infections persist for life. Virus can be detected in saliva. Reactivation is seen in transplant patients and during pregnancy.

HUMAN HERPESVIRUS 7

Human herpesvirus 7 (HHV-7), a T-lymphotropic virus was first isolated in 1990 from peripheral blood lymphocytes of a healthy individual. HHV-7 shares about 50% homology with HHV-6 at the DNA level. HHV-7 is found world over, with most infections occurring in childhood. Persistent infections are established in salivary glands. No established association with human disease.

HUMAN HERPESVIRUS 8

Human herpesvirus 8 (HHV-8), also known as Kaposi's sarcoma-associated herpesvirus (KSHV), is a lymphotropic virus, which was first detected in 1994 in specimens from Kaposi's sarcoma. HHV-8 contains many cellular regulatory genes involved in cell proliferation, apoptosis and host responses.

HHV-8 causes following diseases:
- Kaposi's sarcomas
- Vascular tumours of mixed cellular composition
- Lymphomas occurring in AIDS patients
- Multicentric Castleman's disease

KSHV is most probably sexually transmitted among men who have sex with men and is shed in saliva. Virus can be transmitted through organ transplants. Viral DNA can be detected in patient specimens. Serologic assays are available. Foscarnet, ganciclovir and cidofovir have activity against KSHV.

21

Poxviruses

Amita Jain

LEARNING OBJECTIVES

- Structure, composition, classification and replication of *poxviruses*
- Smallpox, Orf, cowpox and monkeypox
- Molluscum contagiosum
- Vaccinia virus and its uses

INTRODUCTION

Family Poxviridae includes a big group of viruses which are largest of all human viruses. Most poxvirus infections, for example, variola virus, the etiologic agent of smallpox, are characterised by a rash. In 1796, the first live vaccine was developed against smallpox. Eradication of smallpox could be possible by 1980. Smallpox is a category A agent as declared by the Centers for Disease Control and Prevention (CDC), Atlanta, USA, because of their great potential as bioterrorism-biowarfare agents.

STRUCTURE AND COMPOSITION

Poxviruses are large viruses and at times appear as featureless particles by light microscopy. Important structural features of poxviruses are as follows (Fig. 21.1):

- Large viruses
- Enveloped
- The extracellular virion possesses two envelopes (envelope and outer membrane), while the intracellular virus has only one envelope
- Contain dsDNA which is approximately 130–375 kb in size
- Brick-shaped virus particle

- Core is 'dumbbell' shaped which contains nucleic acid
- Outside the core there are two lateral bodies whose function is not known
- Contains more than 100 polypeptides, which are species specific and can illicit immune response
- Many enzymes are present in core which help in virus replication
- Replication occurs in cytoplasm

CLASSIFICATION

The poxviruses which infect vertebrate hosts fall into eight genera, with some antigenic relatedness. Most of the human poxviruses are contained in the genera *Orthopoxvirus* and *Parapoxvirus* (Table 21.1). All poxviruses share a common

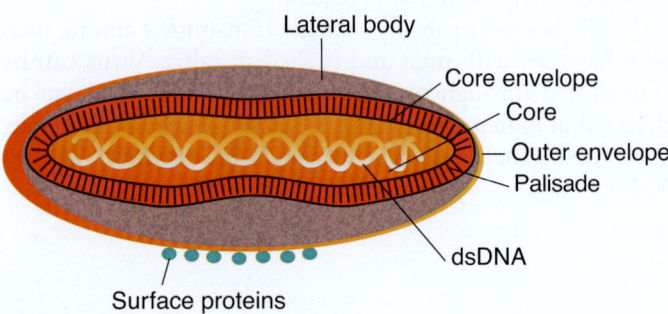

Fig. 21.1 Structure of Poxvirus.

Lateral body
Core envelope
Core
Outer envelope
Palisade
dsDNA
Surface proteins

TABLE 21.1 Human Pathogens of Family Poxviridae

Genus	Virus	Primary Host	Disease Produced
Orthopoxvirus	Variola	Humans	Smallpox
	Vaccinia	Humans	Localised lesion
	Buffalopox	Water buffalo	Human infections rare
	Monkeypox	Rodents, monkeys	Human infections rare
	Cowpox	Cows	Human infections rare
Parapoxvirus	Orf	Sheep	Human infections rare
	Pseudocowpox	Cows	
	Bovine papular stomatitis	Cows	
Molluscipox-virus	Molluscum contagiosum	Humans	Many benign skin nodules

nucleoprotein antigen, although immunisation with vaccinia virus does not protect against disease caused by other poxviruses.

REPLICATION

The replication of poxviruses is complex and takes place in the cytoplasm of infected cells. Replication occurs in following steps:

1. Uncoating: Virus particles fuse with the host cell membrane and cores of viral particles are released into the cytoplasm. Fusion requires virus-encoded protein, which is freshly synthesised.
2. With the help of viral RNA polymerase about half the viral genome is transcribed into **early mRNA**. This step occurs within the viral core. Synthesised mRNAs are released into the cytoplasm. Viral DNA gets released from the cores and codes for RNA and protein synthesis.
3. Viral DNA replication starts 2–6 h after infection in cytoplasm. Condensation of Viral DNA within the cytoplasm appears as inclusion bodies (**Guarnieri's inclusion bodies**).
   One inclusion body represents one virus particle.
4. With the onset of viral DNA replication following changes take place:
   a. The synthesis of early proteins is inhibited as they are expressed mainly within the core.
   b. Intermediate class of genes are expressed briefly followed by expression of the late class of genes.
   c. Late viral mRNA is mainly translated into large amounts of structural proteins and small amounts of other nonstructural viral proteins and enzymes.
5. Each infected cell produces around 10,000 virus particles. Some of the particles are released from the cell by budding. The majority of virus particles remain within the host cell.
6. Host cell metabolism is inhibited at this stage.

SMALLPOX

Introduction

Smallpox is an eradicated disease. Last case was seen in 1977, and it was declared in 1980 that world is free from smallpox. Currently, only two nations supposedly have smallpox viruses: the United States (CDC, Atlanta) and Russia (VECTOR, Moscow). The name smallpox was coined to discriminate it from syphilis, which was also called large pox. No animal reservoir exists.

Smallpox was caused by variola virus. Two species of variola were present: variola major and variola minor. Variola major was associated with high mortality, while variola minor caused mild illness.

History

Earliest evidence of smallpox is seen in Egyptian mummies. Smallpox killed nearly 1 billion people till date. The virus was used as a biological weapon against the Pontiac Indians by British.

In 1796, Edward Jenner developed a vaccine using the less virulent cowpox virus, which shares antigens with smallpox. Jenner made a very useful observation, which made a breakthrough in vaccine research. He observed that the milkmaids very rarely contracted smallpox. He noted that almost all of them had contracted cowpox infections as an occupational hazard. He guessed that cowpox is probably protecting against smallpox. Cowpox causes a mild infection in human; hence, he dared to inoculate his nephew with cowpox crusts. He named the process as variolation. This was an early approach to immunisation. Later variolation was first used as a protection against smallpox in the Far East followed by England. Variolation was associated with a very low fatality rate of <1%.

Today's smallpox vaccine is live attenuated vaccinia virus, which is similar to cowpox.

Epidemiology

Smallpox was a strict human pathogenic virus. Smallpox was transmitted through respiratory route and had an incubation period of 10–14 days. About 7 days postinfection, host was shedding virus, though the symptoms got relieved. Outbreaks occurred in clusters. Smallpox is highly contagious and was spread primarily by the respiratory route. It was also spread less efficiently through close contact with dried virus on clothes or other materials. Despite the severity of the disease and its tendency to spread, several factors contributed to its elimination (Box 21.1).

Pathogenesis

Smallpox infections were initiated by inhalation of nasal or oropharyngeal droplets. The incubation period was 10–14 days. After inhalation, smallpox virus replicated in the upper respiratory tract and spread to the local lymph nodes. Dissemination occurred via lymphatic and cell-associated viraemic spread. Viraemia occurs on days 3–4, with spread to the bone marrow and spleen, though remained asymptomatic. Secondary viraemia occurred around day 8. Dermal tissues got inoculated with simultaneous eruption of the characteristic 'pocks' during secondary viraemia.

BOX 21.1 Characteristics of Smallpox That Led to its Eradication

Viral characteristics
- Human were the only host (no animal reservoirs or vectors)
- Single serotype (immunisation protected against all infections)

Disease characteristics
- Consistent disease presentation with visible pustules (identification of sources of contagion allowed quarantine and vaccination of contacts)

Vaccine
- Immunisation with animal poxviruses protects against smallpox
- Stable, inexpensive and easy-to-administer vaccine
- Presence of scar indicating successful vaccination

Public health service
- Successful worldwide WHO programme combining vaccination and quarantine

Clinical Symptoms

Smallpox usually started with infection of the respiratory tract followed by the involvement of local lymph glands and viraemia. After a 5- to 17-day incubation period, high fever, fatigue, severe headache, malaise, vomiting, diarrhoea and excessive bleeding appeared followed by the vesicular rash in the mouth and on the body. Maculopapular lesions started appearing on extremities, face and buccal and pharyngeal mucosa which later involved trunk. Vesicles were the initial lesion which progressed to pustules following umbilication and scab formation. In contrast to chickenpox, smallpox rash was characterised by skin lesions which were in the same stage of evolution at a given time. The smallpox lesions healed with significant scarring.

ORF, COWPOX AND MONKEYPOX

Infection with the orf (poxvirus of sheep and goat) or cowpox (vaccinia) virus is often seen in animal handlers. Direct contact with the lesions on the animal is the source and mode of infection. A single nodular lesion usually on fingers or hand forms which may progress to haemorrhagic or granulomatous lesion. Lesion heals in 25–35 days. It is an occupational hazard. Monkeypox causes a milder version of smallpox and is not fatal to humans.

MOLLUSCUM CONTAGIOSUM VIRUS

Molluscum contagiosum is a human pathogenic virus. Infection is acquired through direct contact with lesions by sexual contact or fomites, for example, using the shared towels. Incubation period is 2–8 weeks. This virus replicates at the site of inoculation. The lesions produced are nodular or wartlike unlike pox lesions. Lesions begin as papules and then develop as pearl-like, umbilicated nodules with central white coloured plug. Lesions are commonly present on trunk, genitalia and proximal extremities. Usually, they occur in a cluster of 5–20 nodules.

VACCINIA VIRUS

Vaccinia virus is used as a vaccine vector. The virus is non-pathogenic but can form a scar at the inoculation site and can spread to regional lymph nodes which heal over 10–14 days. Lesions resolve by pustule formation followed by scabbing and healing. Immunity develops following healing of lesions and can persist for long may be up to 10 years.

22

Hepatitis Viruses

Amita Jain

LEARNING OBJECTIVES

- Aetiology of viral hepatitis and common causes
- Characteristic features of hepatitis viruses
- Clinical manifestation, epidemiology, diagnosis and management of HAV, HBV, HCV, HDV and HEV

INTRODUCTION

Hepatitis viruses produce acute inflammation of the liver resulting in fever, gastrointestinal symptoms such as nausea and vomiting, and jaundice. Most cases of acute viral hepatitis are caused by one of the aetiologies listed in Box 22.1.

BOX 22.1   Definition and Aetiology of Hepatitis

Hepatitis is an inflammation of the liver parenchyma due to either infectious (i.e. viral, bacterial, fungal and parasitic organisms) or non-infectious (e.g. alcohol, drugs, autoimmune diseases and metabolic diseases) causes. Viruses are the most common cause of hepatitis.
Infectious aetiology of hepatitis—includes viruses, bacteria, fungus and parasites.

Viruses
Common viral causes are as follows:
- Hepatitis viruses
  - Hepatitis A virus (HAV)
  - Hepatitis B virus (HBV)
  - Hepatitis C virus (HCV)
  - Hepatitis E virus (HEV)

Less common viral causes are as follows:
- Hepatitis D virus (HDV)
- Adenovirus
- Cytomegalovirus (CMV)
- Epstein–Barr virus (EBV)
- Herpes simplex virus (HSV)

Parasitic
- *Trypanosoma cruzi*
- *Leishmania* spp.
- *Plasmodium* spp.
- *Entamoeba histolytica*
- *Echinococcus granulosus*
- Liver flukes *Fasciola hepatica* and *Clonorchis sinensis*

Bacterial
- *Escherichia coli*
- *Klebsiella pneumoniae*
- *Neisseria meningitidis*
- *Neisseria gonorrhoeae*
- *Bartonella henselae*
- *Borrelia burgdorferi*
- *Salmonella* spp.
- *Brucella* spp.
- *Campylobacter* spp.
- *Treponema pallidum*
- *Coxiella burnetii*
- *Rickettsia* spp.

Fungal
- *Candida* sp. (mainly opportunistic)

Hepatitis A virus (HAV) causes infectious hepatitis

Hepatitis B virus (HBV) causes serum hepatitis

Hepatitis C virus (HCV) is the common cause of post-transfusion hepatitis

Hepatitis D virus (HDV) is a defective virus and causes acute hepatitis in HBV infected persons

Hepatitis E virus (HEV) causes enterically transmitted hepatitis

Differentiating features of these common hepatitis viruses are shown in Table 22.1. Histopathological lesions observed in the liver during acute disease are identical, irrespective of virus causing it.

HEPATITIS A VIRUS (HAV)

INTRODUCTION

HAV belongs to family *Picornaviridae* and genus *Hepatovirus*. It was provisionally classified as *Enterovirus* 72. But later, it was assigned to a new picornavirus genus, *Hepatovirus*.

TABLE 22.1   Important Characteristics of Hepatitis Viruses

Viruses	Hepatitis A	Hepatitis B	Hepatitis C	Hepatitis D	Hepatitis E
Family	*Picornaviridae*	*Hepadnaviridae*	*Flaviviridae*	Unclassified	Unclassified
Genus	*Hepatovirus*	*Orthohepadnavirus*	*Hepacivirus*	*Deltavirus*	*Hepevirus*
Virus particle, size and shape	27 nm, icosahedral	42 nm, spherical	60 nm, spherical	35 nm, spherical	30–32 nm, icosahedral
Envelope present or not	No	Yes (HBsAg)	Yes	Yes (HBsAg)	No
Stability in presence of physical and chemical substances	Heat- and acid-stable	Acid-sensitive	Ether-sensitive, acid-sensitive	Acid-sensitive	Heat-stable
Transmission	Faeco–oral	Parenteral, mother to foetus	Parenteral, mother to foetus	Parenteral, mother to foetus	Faeco–oral
Source of infection	Contaminated food/ water	Blood/Blood products	Blood/Blood products	Blood/Blood products	Contaminated food/ water
Prevalence in India	High	High	High	Low	High
Causes fatal disease	Can but rarely	Can but rarely	Can but rarely	Frequently	More commonly in pregnant women
Causes chronic disease	No	Yes, frequently	Yes, frequently	Yes, frequently	No
Carcinogenic potential	No	Yes	Yes	Not known	No
Prevention	Vaccine, safe drinking water, immunoglobulin	Vaccine, immunoglobulin	Blood donor screening	HBV vaccine	Safe drinking water

HBV = hepatitis B virus; HBsAg = hepatitis B surface antigen.

Only one serotype is known. Genomic sequence analysis divided HAV isolates into seven genotypes. It causes an acute, asymptomatic liver infection. No chronic infection is known to occur. Protective antibodies develop in response to infection, which confers lifelong immunity. There is no antigenic cross-reactivity with HBV or with the other hepatitis viruses.

VIRUS

It is a RNA virus, 27–32 nm in size, spherical with icosahedral symmetry and non-enveloped. Genome is a single-stranded RNA. It grows with difficulty in monkey kidney cell line and does not produce cytopathic effects. The virus is resistant to heat (60°C for 1 h), acid (pH 1.0 for 2 h), ether, lipid solvents and desiccation (1 month at 25°C). It is stable at −20°C for years but can be destroyed by autoclaving, boiling water for 5 min, ultraviolet light, formalin and sodium hypochlorite (1:100 dilutions).

TRANSMISSION

It occurs by feco–oral route, person-to-person contact, inadequately cooked oysters/clams and rarely by blood and blood products.

PATHOGENESIS

It is detailed in Flowchart 22.1.

Infection occurs through contaminated food or water, mostly

↓

Viraemia occurs soon after infection and persists till liver enzyme (alanine aminotransferase [ALT]) is elevated. Viraemia lasts for many weeks. RNA can be detected in blood and stool in acute phase of infection

↓

Virus multiplies in intestinal mucosal cells, enters blood flow and reaches liver

↓

Virus multiples in liver cells, causes impairment of liver function and causes hepatitis

↓

Virus is excreted in bile and shed in stool in large quantities for many weeks

Flowchart 22.1 Pathogenesis of Hepatitis A.

INCUBATION PERIOD

It is 15–50 days (average 30). An infected person is most infectious when virus concentration in stool is highest, which happens around 2 weeks before appearance of jaundice. Once jaundice appears, shedding in stool stops. There are no reports of chronic shedding of virus in stool, except in cases that present with relapse.

CLINICAL FEATURES

The disease caused is known as infectious hepatitis. Around 90% infections in children are subclinical. The proportion of clinical cases increases with age. Fulminant hepatic failure is rare. No chronicity is seen. The disease usually manifests as abrupt onset of symptoms which can include fever, malaise, anorexia, nausea, abdominal discomfort, dark urine and jaundice. Rare complications are fulminant hepatitis, cholestatic hepatitis and relapsing hepatitis. There are no chronic sequelae. Mortality rate is higher in adults as compared to children.

EPIDEMIOLOGY

Around 1.4 million cases per year occur worldwide. In developing countries, paediatric disease is common, while in developed nations disease is mainly seen in adults. Prevalence of hepatitis A is inversely related to socioeconomic status of the society. India has high prevalence of hepatitis A. Around 18% of all acute hepatitis cases are due to HAV.

LABORATORY DIAGNOSIS

The specimen required for laboratory confirmation is mostly blood. Commercial tests are available for the detection of total anti-HAV antibodies and anti-HAV IgM in serum.

Events occurring in a patient postinfection against timeline are shown in Fig. 22.1.

Anti-Hepatitis A Virus IgM Antibody

It is most commonly employed method, detected by enzyme immunoassays (ELISAs) and detectable by 2nd–3rd week of infection. Nearly 100% sensitivity is achieved by 3rd week. Antibodies persist for few weeks and then decline. The presence of IgM antibodies indicates active infection.

Anti-Hepatitis A virus IgG antibody appears after 3–4 weeks of infection and persists for years. The presence indicates immunity and is useful in knowing the vaccine efficiency.

Nucleic acid amplification assays are not routinely needed.

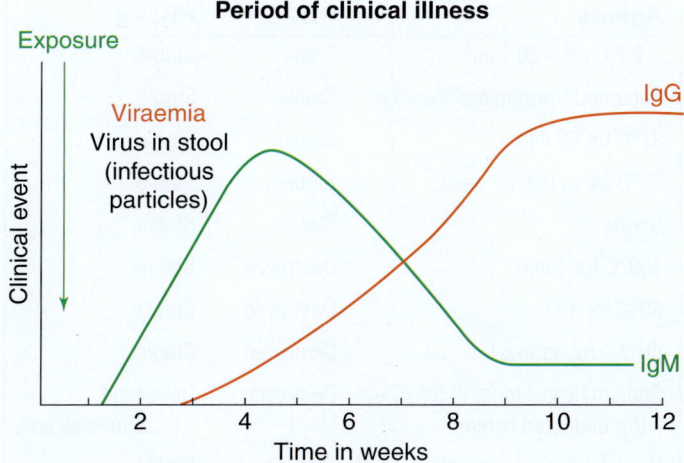

Fig. 22.1 Events in HAV Infection.

TREATMENT

It is mainly symptomatic. There are no effective antiviral drugs available.

PREVENTION

It can be easily achieved by
1. availability of safe drinking water;
2. maintaining public and personal hygiene;
3. providing knowledge to food handlers and
4. vaccination of high-risk individuals i.e.;
   a. children,
   b. food handlers,
   c. siblings of hepatitis A patients and
   d. young troops.

VACCINE

There is an inactivated vaccine available which is not yet included under Universal Immunization Program. The vaccine is highly immunogenic. About 97%–100% seroconversion with first dose and ~100% with second dose is seen. It is highly efficacious.

Immunoglobulins can be given pre-exposure to travellers to intermediate and high HAV-endemic regions and post-exposure (within 14 days).

HEPATITIS B VIRUS (HBV)

INTRODUCTION

Dr Baruch Blumberg received the Nobel Prize in Physiology/Medicine for discovering HBV in 1965. Originally, virus antigen, found in the blood of Australian aboriginal people, was named **Australia antigen**; later named hepatitis B surface antigen (**HBsAg**). This was an accidental discovery (electron microscopic finding) in serum of two intravenous drug users. Australia antigen also reacted with antibodies in the serum of an American haemophilia patient.

VIRUS

It is a DNA, spherical and enveloped (HBsAg) virus with icosahedral symmetry. The genome is double-stranded circular DNA. The virus grows in hepatocyte cell cultures with difficulty and does not produce CPE.
- Family: *Hepadnaviridae*
- Genus: *Orthohepadnavirus*

Dane Particle: D.S. Dane discovered virus particle in 1970 by electron microscopy, hence called Dane particles. It is a **complete virion and is infectious**. Size of complete virion is 42 nm. There are surface protein particles detectable in serum by electron microscopy which is 20 nm in size, spherical and tubular (Fig. 22.2).

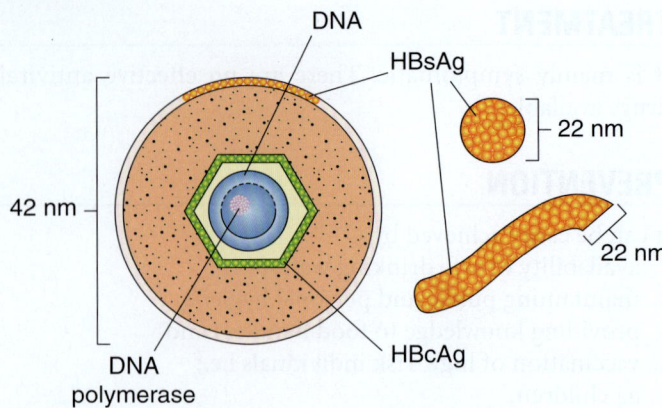

Fig. 22.2 Dane Particle.

Antigens of HBV

- **HBsAg** is the surface proteins of the virus particle clump together into
  - spherical particles—22 nm diameter
  - rods of variable length
  - Produced in larger quantities than required for assembly of virus particles, hence are found as free particle in serum of patients
  - Non-infectious particles
- **Hepatitis B core antigen (HBcAg)** is not detectable in the blood. This part of virus is covered by envelop and is detectable only in liver cells
- **Hepatitis B e-antigen (HBeAg): e (early)**—it is seen in blood only when virus multiplies in large numbers. This is an antigen which is not a part of virus structure

Genome of HBV

It is a circular DNA. This is unusual and partially double-stranded. There are two strands: one is long (3020–3320 nucleotide) and other is short (1700–2800 nucleotides) (Fig. 22.3). One end of larger strand is attached to viral DNA polymerase.

Whole genome of HBV was sequenced in 1980s. There are four open reading frames (ORFs) which encode four known genes: C, X, P and S.

- HBcAg is coded by ORF C without preC. HBcAg is not released in serum and is found in liver cell and within virus particle only.
- HBeAg is produced by proteolytic processing of the precore protein + C and is released in serum.
- DNA polymerase is encoded by gene P.
- HBsAg is encoded by one long ORF but contains three ORFs that divide the gene into three sections, pre-S1, pre-S2 and S. These three ORFs code for three different sized proteins: large, middle and small (pre-S1 + pre-S2 + S (large), pre-S2 + S (middle) or S (small)). Free Middle and small proteins molecules are released in serum, in excess. Large proteins are a part of virus particle only.
- The function of protein coded by gene X is not understood. This protein is supposedly associated with development of liver cancer.

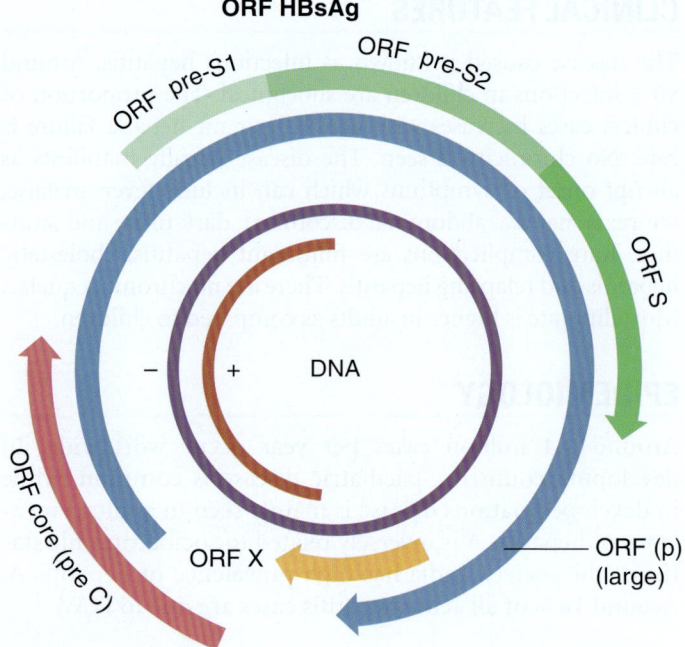

Fig. 22.3 Genome of HBV.

Stability and Sensitivity of HBV and HBsAg to Physical and Chemical Agents

The stability of HBsAg does not always coincide with that of the infectious agent because HBsAg may be present in serum of a patient without Dane's particle, which is the infectious agent. (Table 22.2).

Serotypes

HBV contains a group-specific antigen, *a*, and two subdeterminants, *d/y* and *w/r*. Based on which, four phenotypes of HBsAg are seen in nature: *adw, ayw, adr* and *ayr*.

TABLE 22.2 Stability and Sensitivity of HBV and HBsAg to Physical and Chemical Agents		
Agents	**HBV**	**HBsAg**
−20°C for >20 years	Stable	Stable
Repeated freezing and thawing	Stable	Stable
37°C for 60 min	Stable	Stable
25°C for at least 1 week	Stable	Stable
Drying	Stable	Stable
100°C for 1 min	Destroyed	Stable
60°C for 10 h	Destroyed	Stable
pH 2.4 for up to 6 h	Destroyed	Stable
Sodium hypochlorite, 0.5%/5% (for undiluted serum)	Destroyed	Destroyed (in diluted serum)
Ultraviolet irradiation of plasma or other blood products	Stable	Stable

Genotypes

There are eight genotypes (A–H) according to overall nucleotide sequence. Genotypes have a distinct geographical distribution which can be used in tracing evolution and transmission of virus.

- Eight genotypes—labelled A through H,
- Differ by at least 8% of the sequence—type F is most divergent, diverges by 14%
- Prevalence of each genotype varies from country to country. Commonly occurring genotypes are listed as follows:
  - Type A: Europe, Africa and Southeast Asia
  - Type B: Asia
  - Type C: Asia
  - Type D: Mediterranean area, Middle East and India
  - Type E: Sub-Saharan Africa
  - Type F: Central and South America
  - Type G has been found in France and Germany
  - Type H: Central and South America
  - Genotypes A, D and F: Brazil
  - All genotypes: United States

Subtypes of Genotypes

There are 24 known subtypes which differ by 4%–8% in genome.

- Type A: Aa (A1) and Ae (A2)
- Type B: Bj/B1 ('j' for Japan) and Ba/B2 ('a' for Asia). Type Ba has four clades (B2–B4)
- Type C: Cs (C1), Ce (C2)—six clades (C1–C6)
- Type D: Seven subtypes (D1–D7)
- Type F: Four subtypes (F1–F4). F1 has been further divided into 1a and 1b.

Significance of Genotypes

Genotypes are known to affect

- disease severity,
- course of disease,
- likelihood of complications,
- response to treatment and
- possibly response to vaccine.

REPLICATION

See Flowchart 22.2.

MODE OF TRANSMISSION: PARENTERAL

- Direct contact with blood and body fluids
- Sexual contact
- Mother to child—transplacental
- Blood and organ transplant

Incubation Period

60–90 days

PATHOGENESIS

One or more of the following clinical courses may be seen with HBV infection. Clinical outcome of HBV infection is shown in Flowcharts 22.3 and 22.4.

Enters the hepatocyte by endocytosis (binds to a receptor on cell surface)
Viral outer membrane fuses with host cell membrane
DNA and core proteins released in cell cytoplasm
Viral DNA transferred to the cell nucleus by host proteins (chaperones)
Partially double-stranded viral DNA converted to fully double-stranded DNA. HBV is a non-retroviral virus, which uses resverse transcription to convert in double stranded DNA
Double-stranded DNA transformed into covalently closed circular DNA (cccDNA)
cccDNA codes four viral mRNAs
Largest mRNA makes new copies of the genome, viral DNA polymerase and capsid core protein and viral
Form new virion particles
Virion release from the cell
Long mRNA is transported back to the cytoplasm, synthesises DNA via its reverse transcriptase activity and virion protein P

Flowchart 22.2 Replication of HBV.

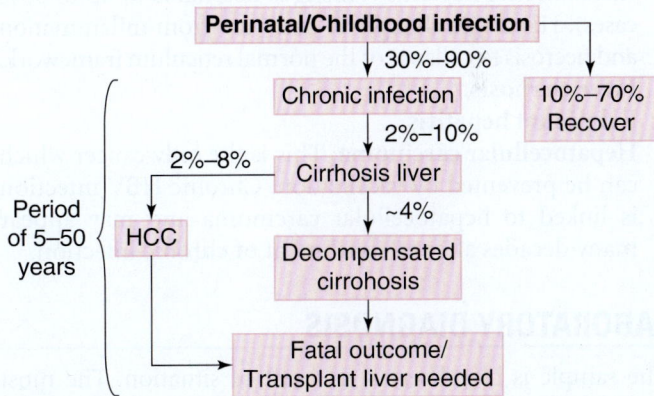

Flowchart 22.3 Clinical Outcome of HBV Infection in Children and Newborns.

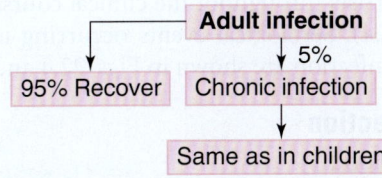

Flowchart 22.4 Clinical Outcome of HBV Infection in Adults.

- **Subclinical infection**
- **Asymptomatic carrier state:** Most of the HBV carriers have no demonstrable symptoms. The carrier state may be of short duration. Long-term carriers are also common. Virus may establish persistent infection in hepatocytes with or without an evidence of liver damage. About 8%–10% adult patients of acute hepatitis B infection develop persistent unresolved viral hepatitis. It is characterised by sporadically abnormal liver enzymes and hepatomegaly. Lobular architecture is maintained. Portal inflammation, swollen and pale hepatocytes (cobblestone arrangement) and slight fibrosis may be present. It usually does not progress to cirrhosis, and has a good prognosis.
- **Acute self-resolving hepatitis:** It is most common clinical manifestation of HBV infection and is usually self-limiting. Rarely, it can be fatal ($\sim$1% cases). Recovered person may become a long-term carrier of HBV. The virus enters blood through parenteral route, reaches liver, multiples in liver cells and causes impairment of liver function. This results in acute hepatitis. Acute hepatitis in HBV infection shows spotty parenchymal cell degeneration, necrosis of hepatocytes, a diffuse lobular inflammatory reaction and disruption of liver cell cords. There is reticuloendothelial (Kupffer) cell hyperplasia, periportal infiltration by mononuclear cells and cell degeneration. The damaged hepatic tissue is usually restored in 8–12 weeks. Viraemia lasts for many weeks.
- **Chronic hepatitis:** It is also known as persistent viral hepatitis seen in 8%–10% of adult patients following infection with HBV. Usually, it is a mild disease with sporadically abnormal liver enzymes values; however, some cases develop chronic active hepatitis (HBsAg is detectable in up to 50% cases). Histologic changes in liver vary from inflammation and necrosis to collapse of the normal reticulum framework.
- **Liver cirrhosis.**
- **Fulminant hepatitis.**
- **Hepatocellular carcinoma:** This is the only cancer which can be prevented by vaccination. Chronic HBV infection is linked to hepatocellular carcinoma and may appear many decades after establishment of chronic infection.

LABORATORY DIAGNOSIS

The sample is blood/serum in most of situation. The most useful detection methods are ELISAs for HBV antigens and antibodies and PCR for viral DNA. HBV DNA, HBsAg and HBeAg appear early in the incubation period. Communicability is highest at this time as HBV particles may be present in the blood in high numbers (up to 10^{10} particles/mL; Table 22.3). HBsAg is usually detectable 2–6 weeks before clinical features appear and persists throughout the clinical course of the disease (Table 22.4). Serological events occurring in acute and chronic HBV infection are shown in Figs. 22.4 and 22.5.

Antigen Detection

- HBsAg in blood
- HBeAg in blood
- HBcAg in liver cells

TABLE 22.3 Hepatitis B Laboratory Markers and Their Interpretations

Markers	Interpretations in Reference to HBV Infection
HBsAg	Positive in acute and chronic infections and carriers
IgM antibody to hepatitis B core antigen (IgM Anti-HBc, or IgM-HBcAb)	Positive in acute infections (even in 'window' phase of infection)
Antibody (IgG) against hepatitis B core antigen (Anti-HBc or HBcAb)	Positive in acute, resolved or chronic infection (absent in vaccine recipients)
Antibody against hepatitis B surface antibody (Anti-HBs or HBsAb)	Positive in cases with resolved infection and vaccine recipients
HBeAg	Positivity means viraemia indicator of infectivity
Antibody against hepatitis B e-antigen (Anti-HBe or HBeAb)	Resolving infection, low risk of transmission
HBV DNA	Quantitative marker of viral replication, assesses and monitors treatment of chronic HBV infection
Serum aminotransferase (ALT)	Elevated serum levels indicative of liver damage, used to monitor HBV disease progression

HBeAg = hepatitis B e-antigen; HBsAg = hepatitis B surface antigen; HBV DNA = hepatitis B virus DNA.

TABLE 22.4 Interpretation of Three Most Commonly Employed ELISA Tests in Reference to Staging of HBV Infection

HBsAg	Anti-HBsAg	Anti-HBcAg (IgM)	Interpretations
+	−	−	Early infection or HBsAg Carrier
−	−	+	Recent infection (window period), infection in remote past, 'low-level' HBV carrier, false-positive or non-specific reaction
+	±	+	HBV infection, either acute or chronic. HBeAg or HBV DNA estimation and anti-HBcAg (IgM) estimation will help
−	+	+(IgG)	Past infection/recovered, immunity to infection
−	+	−	Evidence of vaccination
−	−	−	Never infected with HBV

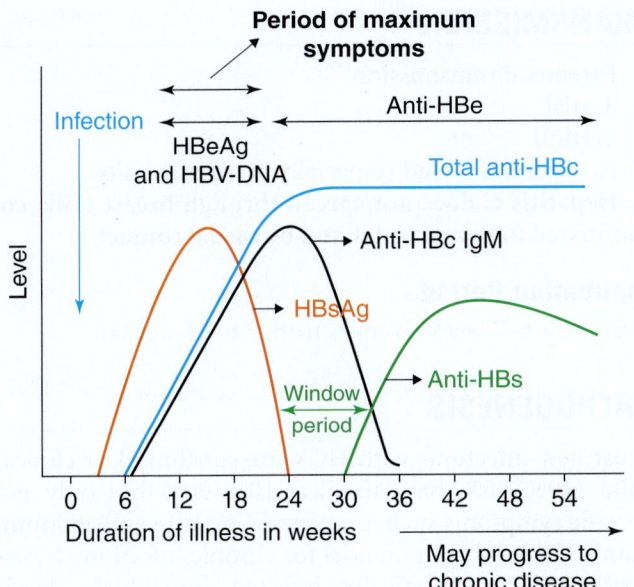

Fig. 22.4 Serological Sequence of Events in Acute HBV Infection.

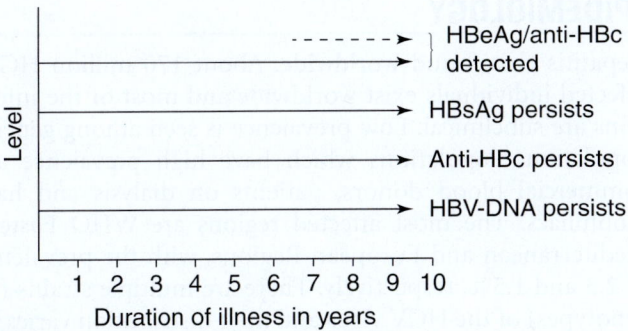

Fig. 22.5 Serological Sequence of Events in Chronic HBV Infection.

Antibody Detection

- Anti-HBsAg
- Anti-HBcAg: IgM antibody
- Anti-HBcAg: IgG antibody
- Anti-HbeAg

PCR–HBV DNA Detection

It is most sensitive method and is accepted for diagnosis

EPIDEMIOLOGY

HBV infection is worldwide in distribution, known as serum hepatitis. There are >250 million carriers of the virus worldwide, accounting for 1 million deaths annually. In India alone, there are ~34 million HBsAg carriers. Around 25% of carriers develop chronic active hepatitis.

HBV infection has no predilection for any age/sex/season. Subclinical infections are common, and these unrecognised infections represent the principal hazard to hospital personnel. Against each symptomatic person, there are four infected asymptomatic persons. There are definite high-risk groups such as parenteral drug abusers, jail inmates, health care personnel, multiply transfused patients, organ transplant patients, patient on haemodialysis, highly promiscuous persons and newborn infants born to mothers with hepatitis B. People have been infected by improperly sterilised syringes, needles or scalpels and even by tattooing or ear piercing. The number of cases of transfusion-associated hepatitis has been dramatically reduced, since screening of blood donors for HBsAg has become mandatory.

HBsAg can be detected in saliva, nasopharyngeal washings, semen, menstrual fluid and vaginal secretions as well as in blood. Transmission from persons with subclinical cases and carriers of HBsAg to homosexual and heterosexual long-term partners is known to occur.

TREATMENT

The *WHO guidelines for the prevention, care and treatment of persons living with chronic hepatitis B infection* lay out a simplified approach to the care of people living with chronic hepatitis B, particularly in settings with limited resources.

Use of two safe and highly effective medicines, tenofovir or entecavir, for the treatment of chronic hepatitis B is recommended with monitoring treatment response, using simple tests.

PREVENTION

Passive immunisation using specific hepatitis B immune globulin (HBIG) can be given soon after exposure. HBIG is not recommended for pre-exposure prophylaxis. There is a subunit recombinant vaccine available for pre-exposure prophylaxis, which is quite effective. Hepatitis B vaccination is recommended for:

- All infants
- All health care workers not previously vaccinated
- All pregnant women if not vaccinated
- All children and adolescents not previously vaccinated

Pre-exposure prophylaxis with a commercially available hepatitis B vaccine currently is recommended for all children as a part of national immunisation program of India.

Vaccine

A vaccine for hepatitis B has been available since 1982. The initial vaccine was prepared by purifying HBsAg from healthy HBsAg-positive carriers and treating the particles with virus-inactivating agents (formalin, urea, heat). They have been replaced by recombinant DNA-derived vaccines. Recombinant vaccine is made by combining viral DNA coding for HBsAg in yeast cells or in continuous cell lines. The vaccine formulated using this purified material is safe and has potency similar to that of vaccine made from plasma-derived antigen.

HEPATITIS C VIRUS (HCV)

INTRODUCTION

Studies had suggested that there were several non-A, non-B (NANB) hepatitis agents which were not related to HAV or HBV. The major agent was identified as HCV. Most cases of posttransfusion NANB hepatitis were caused by HCV.

VIRUS

HCV is a positive-stranded RNA virus, classified as family Flaviviridae, genus *Hepacivirus*. The genome is 9.4 kb in size and encodes a core protein, two envelope glycoproteins and several non-structural proteins. HCV has six major genotypes (clades) and more than 100 subtypes. Genotypes are at least >25% different from each other, while subtypes are >15% different from each other.

Genotypes of HCV

There are six genotypes (Table 22.5). The geographic variation is seen amongst various genotypes. Genotype 3 is more prevalent in intravenous drug users. Genotype 1b or 4 is associated with poor rate of interferon treatment response. The duration of therapy usually is ~1 year for genotype 1 and ~6 months for genotypes 2 and 3.

Methods of Genotyping

- Sequencing
- Real-time PCR
  - Multiplex PCR by Taqman chemistry
  - Melting curve analysis (SYBR green chemistry)
- Line probe assay by Innogenetics

TRANSMISSION

- Parenteral transmission
- Sexual
- Vertical
- Nosocomial spread (especially in dialysis units)

Hepatitis C does not spread through breast milk, contaminated food and water and by casual contact.

Incubation Period

Average is 6–7 weeks (ranges from 2 to 26 weeks).

PATHOGENESIS

Most new infections with HCV are subclinical or clinically mild (20%–30% has jaundice, 10%–20% has only non-specific symptoms such as anorexia, malaise and abdominal pain). There is high potential for chronic infection (>70%); 50%–70% of chronically infected individuals develop chronic liver disease. HVC is not a major cause of an acute or fulminant hepatitis but is an important cause of primary hepatocellular carcinoma. Serological sequence of events seen in HCV infection is shown in Fig. 22.6.

EPIDEMIOLOGY

Hepatitis C is found worldwide. About 170 million HCV-infected individuals exist worldwide and most of the infections are subclinical. Low prevalence is seen among general population. Populations which have high prevalence are commercial blood donors, patients on dialysis and haemophiliacs. The most affected regions are WHO Eastern Mediterranean and European Regions, with the prevalence of 2.3 and 1.5%, respectively. There are multiple strains (or genotypes) of the HCV virus and their distribution varies by region (Table 22.5).

Genotypes	Geographic Distribution	Important Population Affected/Response to Treatment
1a	Worldwide distribution	Poor response to IFN therapy, 48 weeks of therapy
1b	Europe, North America	Older persons; transfusion recipients; poor response to IFN therapy, 48 weeks of therapy
2	Mediterranean region, Asia	Good response to IFN therapy, 24 weeks of therapy
3	Europe	Injection drug users; good response to IFN therapy, 24 weeks of therapy
4	Middle East	Not known
5	South Africa	
6	Southeast Asia	

TABLE 22.5 Genotypes of HCV Infection and Their Distribution

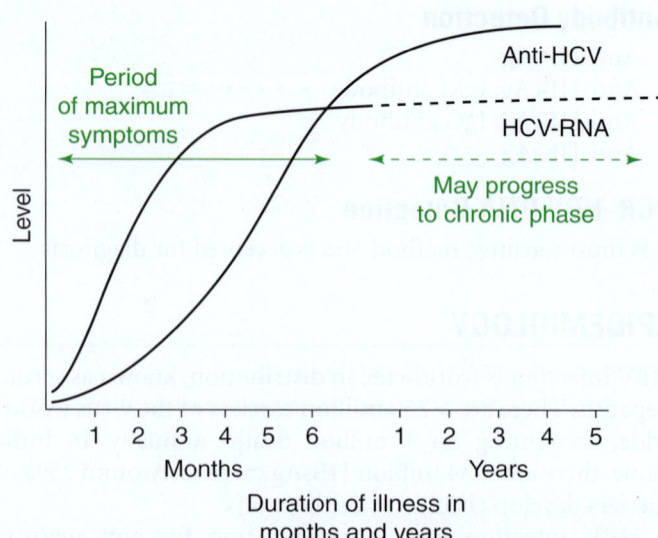

Fig. 22.6 Serological Sequence of Events Seen in HCV Infection.

LABORATORY DIAGNOSIS

Specimen: Serum

- Detection of anti-HCV antibodies: Serological assays are available for diagnosis of HCV infection. ELISAs detect antibodies to HCV. Anti-HCV antibodies can be detected in ~70% of patients at onset of symptoms.
- Reverse transcriptase polymerase chain reaction (RT-PCR): Nucleic acid-based assays (e.g. RT-PCR) detect the presence of circulating HCV RNA and are useful for monitoring patients on antiviral therapy. This is a confirmatory test used to confirm diagnosis in patients who test positive for anti-HCV antibodies.

THERAPY

The standard of care for hepatitis C management is changing rapidly. Sofosbuvir, daclatasvir and the sofosbuvir/ledipasvir provide positive response in almost 95% of patients. These are safe to use. Pegylated interferon and ribavirin are now seldom used.

PREVENTION

There is no vaccine available. Prevention is done by taking precautions against bloodborne infections and health education.

HEPATITIS D VIRUS (HDV)

INTRODUCTION

HDV is a defective RNA virus that requires HBV replication for its multiplication and acquires an HBsAg coat for transmission. Hepatitis D infection cannot occur in the absence of HBV.

VIRUS

HDV contains delta-Ag (HDAg) surrounded by an HBsAg envelope. It has a particle size of 35–37 nm and genome of single-stranded, circular, negative-sense RNA, 1.7 kb in size. It is the smallest of known human pathogens and resembles viroids. No homology exists with the HBV genome.

Antigens of HDV

An antigen–antibody system termed the delta antigen (delta-Ag) and antibody (anti-delta) is detected in some HBV infections. HDAg is the only protein coded by HDV RNA and is distinct from the antigenic determinants of HBV. It is often associated with severe forms of hepatitis in HBsAg-positive patients.

TRANSMISSION

The routes of HDV transmission are the same as for HBV.

SYMPTOMS

Chronic HBV carriers are at risk for infection with HDV. The coinfection or superinfection of HDV with HBV causes a more severe disease than HBV monoinfection.

Coinfection is a severe acute disease with low risk of chronic infection, while superinfection usually develop chronic HDV infection with high risk of severe chronic liver disease.

Acute hepatitis—simultaneous infection with HBV and HDV can lead to a mild-to-severe or even fulminant hepatitis, but recovery is usually complete.

Superinfection: HDV is a defective virus and causes superinfection in individuals infected with HBV chronically. It causes severe infection in 70%–90% of coinfected persons.

EPIDEMIOLOGY

HDV and HBV coinfection is seen in around 5% of HBV-positive individuals worldwide. Mediterranean, Middle East, Pakistan, Central and Northern Asia, Japan, Taiwan, Greenland and parts of Africa, the Amazon Basin and certain areas of the Pacific have high prevalence.

LABORATORY DIAGNOSIS

HDV infection is diagnosed by anti-HDV IgG and IgM and confirmed by detection of HDV RNA in serum.

TREATMENT

There is no specific treatment for acute or chronic HDV infection. Pegylated interferon alpha is the only drug effective against HDV; antiviral nucleotide analogues for HBV have no or limited effect on HDV replication. Treatment is given for >1 year. Sustained virological response is not common. Relapses are frequent.

PREVENTION

There is no specific vaccine available. The prevention of HBV transmission through hepatitis B immunisation is the only available method. Hepatitis B immunisation does not provide protection against HDV for those who are already HBV-infected.

HEPATITIS E VIRUS (HEV)

INTRODUCTION

Hepatitis E is a liver disease caused by the HEV. HEV is transmitted enterically and occurs in epidemic form in developing countries where water or food supplies are sometimes fecally contaminated. Poor sanitation is a known risk factor.

VIRUS

HEV is a small virus, not enveloped, 27–30 nm spherical/32–34 nm in diameter and very labile. The viral genome is a positive-sense, single-stranded RNA, 7.6 kb in size. The virus is classified in four genotypes: 1, 2, 3 and 4. Genotypes 1 and 2 are found only in humans.

TRANSMISSION

Humans are natural host, but antibodies to the HEV or closely related viruses are detected in primates and other animal species. Transmission is through Faecal–oral route. The virus is shed in the stools of infected persons and enters the human body through oro–faecal route. It is transmitted mainly through contaminated drinking water.

Other transmission routes include:
- Foodborne transmission from ingestion of products derived from infected animals
- Zoonotic transmission
- Transfusion of infected blood products
- Vertical transmission from a pregnant woman to her foetus
- Contaminated water or food supplies implicated in major outbreaks
- Ingestion of raw or uncooked shellfish can be the source of sporadic cases in endemic areas

Incubation Period

Average 40 days (range 15–60 days)

CLINICAL FEATURES

Mostly cases are asymptomatic or a very mild illness without jaundice that goes undiagnosed; is seen. Cases may also present as acute, sporadic or epidemic viral hepatitis, frequent in children. Symptoms are common in young adults aged 15–40 years. Jaundice, anorexia, hepatomegaly, abdominal pain and tenderness, nausea and vomiting with fever are common manifestations. Severity of illness increases with age. In rare cases, it can result in fulminant hepatitis and death. Overall case-fatality rate is 1%–3%, although in pregnant women it increases from 15% to 25%. Chronic sequelae are not identified. Serological sequence of events seen in HEV infection is shown in Fig. 22.7.

Common complications:
- Overall population mortality: 0.5%–4.0%
- Fulminant hepatitis frequent in pregnancy
- Greater risk of obstetrical complications
- 20% Mortality in third trimester
- Reactivation reported in immunocompromised people

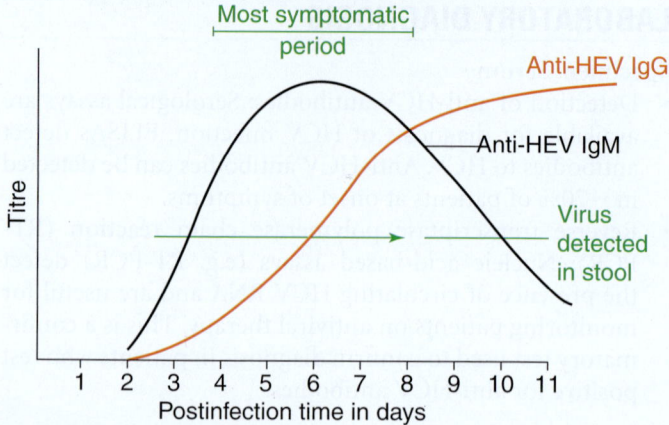

Fig. 22.7 Serological Sequence of Events Seen in HEV Infection.

EPIDEMIOLOGY

In 1955, about 29,000 cases of hepatitis occurred after sewage contamination of drinking water supply in Delhi, India. It was first time when HEV was described. Since then many outbreaks are reported in India. Usually, the infection is self-limiting and resolves within 2–6 weeks. Occasionally, a serious disease known as fulminant hepatitis (acute liver failure) develops. Mortality is not commonly seen except in pregnant women who may have a high (20%) mortality rate if fulminant hepatitis develops.

Most outbreaks are associated with faecal contamination of drinking water and person-to-person transmission.

LABORATORY DIAGNOSIS

Detection of anti-HEV IgM antibody detection by ELISA, in serum, is most commonly used method. RT-PCR to detect HEV RNA in serum and stool is tried.

TREATMENT

There is no specific treatment of hepatitis E infection. It is a self-limiting disease. Rarely hospitalisation is required for people with fulminant hepatitis. Pregnant women more often have a serious disease with fatal outcome. Ribavirin and interferon can be helpful in certain situations.

PREVENTION

It is the best approach. Methods used are same as applied for hepatitis A. HEV vaccine is under clinical trial.

23

Orthomyxoviruses and Paramyxoviruses

Amita Jain

ORTHOMYXOVIRUSES (INFLUENZA VIRUSES)

INTRODUCTION

Orthomyxoviruses are negative-sense, single-stranded RNA viruses whose genome is segmented. Antigenically three distinct types (A, B and C) and many subtypes within the type A viruses are recognized. There is no cross-immunity between different types. Genetic variation and genetic reassortment is commonly seen.

CLASSIFICATION

Family: Orthomyxoviridae
Genus: *Influenzavirus*
Species: Influenza virus A, influenza virus B, influenza virus C

Major differences in the major three types of viruses are seen in Table 23.1.

STRUCTURE

Fig. 23.1 shows the labelled line diagram of influenza virus. This is an enveloped virus with nine structural proteins.

Surface antigens: Envelope has two **surface antigenic** proteins: haemagglutinin (HA) and neuraminidase

Haemagglutinin (HA): HA is the major antigen. There are 16 antigenic types, which are responsible for evolution of the virus. HA is a protein, which helps in attachment of virus to host cell surface and initiates infection. Antibodies to HA neutralises the virus.

Neuraminidase (NA): There are nine antigenic types. This is a receptor-destroying sialidase enzyme, which removes sialic acid from glycoconjugates and facilitates release of virus from host cell. It does not allow viruses to self-aggregate, but

TABLE 23.1	Major Differences in Influenza Virus A, Influenza Virus B and Influenza Virus C		
	Influenza A	**Influenza B**	**Influenza C**
Severity of illness	Causes epidemics and pandemics, severe illness	Causes milder epidemics, usually not severe illness	Does not cause major disease, mild illness
Animal reservoir	Yes, infects both humans and other species	No, limited to humans	No, limited to humans
Segmented genome	Yes	Yes	Yes
Antigenic changes	Frequent antigenic variations, shift, drift	Infrequent antigenic variations, drift	Antigenically stable, drift
Human pandemics	Yes	No	No
Human epidemics	Yes	Yes	No
Amantadine, rimantadine, zanamivir	Sensitive	No effect	No effect
Surface glycoproteins	2	2	1

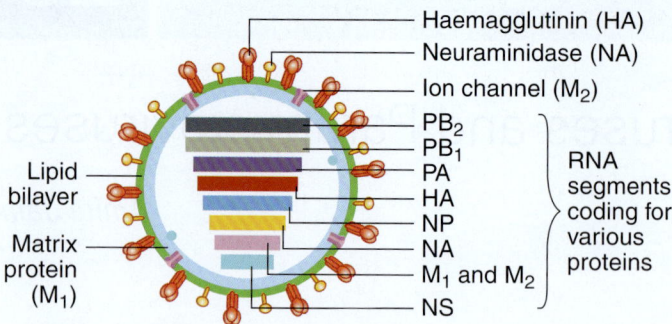

Fig. 23.1 Structure of Influenza Virus. (NP = nucleoprotein. M1 and M2 = matrix proteins; NS = non-structural protein; PB1 = endonuclease; RNA polymerase; PB2 = RNA polymerase (basic); PA = RNA polymerase (acidic).

helps the virus to move through mucin layer. Antibodies to NA modify the infection severity.

NOMENCLATURE

On 30 September 1971, WHO gave a system for nomenclature of influenza viruses. It consists of two parts: a strain designation and a description of H and N Ag.

1. Strain designation
   a. A description of the antigenic type of nucleoprotein (e.g. A, B or C)
   b. Host of origin (not mentioned for strains isolated from human)
   c. Geographic origin
   d. Strain number
   e. Year of isolation
2. Description of H and N Ag (e.g. H1N1)
3. Examples:
   a. A/Singapore/l/57(H2N2)
   b. A/Hong Kong/1/68(H3N2)
   c. A/turkey/Wisconsin/1/66(Hav5N2)
   d. A/duck/Ukraine/l/63(Hav7Neq2)

ANTIGENIC VARIATION AND REASSORTMENT OF INFLUENZA VIRUS GENOME

Of possible 144 (16 × 9) combinations, a minimum of 16 HA subtypes and 9 NA subtypes are known to exist in nature. Until 2015, only three combinations affected humans.

Antigenic variation in the viral genome occurs by either antigenic drift or antigenic shift.

Antigenic Shift

It is the sudden, complete or major change in genome. This phenomenon usually occurs due to genetic recombination of human and animal/avian influenza viruses. Result of this recombination may be a novel subtype different from both parent viruses (e.g. pandemic H1N1, emergence in 2009). If 'novel subtype' is readily transmissible from person to person, it may cause pandemics. Evidence suggests that human

influenza viruses causing last three pandemics, including 2009 H1N1 pandemic, contained gene segments related to avian influenza viruses.

Antigenic Drift

It causes gradual antigenic change over a period. It occurs due to 'point mutations' in genes. The selection pressure caused by immunity in host population is the main reason. The phenomenon is responsible for emergence of different strains of same type, causing frequent influenza epidemics, necessitating reformulations of seasonal influenza vaccines.

DISEASE CAUSED

They cause epidemic acute respiratory disease known as **Influenza**, characterised by fever, cough and systemic symptoms.

Case definition of influenza-like illness (ILI) (given by World Health Organization (WHO)): An acute respiratory infection with measured fever of $\geq 38°C$ and cough/rhinorrhoea with onset within the last 10 days.

Complications seen in these cases are listed in Box 23.1.

EPIDEMIOLOGY

It is world-wide in distribution. Attack rates during epidemics are 10%–20% in general population and >50% in closed populations. Attack rates are lower among adults. It affects all ages and both sexes. High case fatality ratio (CFR) is seen during epidemic in old people, children and persons with diabetes, chronic heart disease, renal and respiratory diseases etc.

- Sporadic cases occur throughout the year
- Outbreaks (primarily influenza A): occur during season, may be every year
- Major epidemics may be seen every 2–3 years
- Pandemics are rare and may occur at an interval of 10–15 years or more. Major human pandemics are listed in Table 23.2

> ### BOX 23.1   Complications of Influenza Infection
>
> Complications often seen in infants, senior citizens and people with chronic disorders are as follows:
> - Pneumonia
> - Croup (in young children)
> - Primary influenza virus pneumonia
> - Secondary bacterial infection, e.g. *Streptococcus pneumoniae, Staphylococcus aureus, Haemophilus influenzae*
> - Myositis
> - Cardiac complications
> - Encephalopathy
> - Eye syndrome
> - Guillain–Barré syndrome (GBS) etc.

TABLE 23.2	Major Human Pandemics Seen in Past			
Pandemic Year	Common Name	Virus Subtype	Age Group Commonly Affected	Estimated Excess Mortality
1918	Spanish flu	H1N1	Young adults	20–50 million
1957–1958	Asian flu	H2N2 (avian)	All age group	1–4 million
1968–1969	Hong Kong flu	H3N2 (avian)	All age group	1–4 million
2009–2010	Influenza A (H1N1 2009)	H1N1 (swine)	All age group	100,000–400,000

Influenza pandemic 2009 was a global outbreak of a new strain of influenza A virus subtype H1N1. The pandemic started in April 2009 and was officially named the **'novel H1N1/pdm H1N1'**, commonly called **'swine flu'**. **Infection is seen in pigs, but is** rarely fatal in pigs. Infection cannot spread by consuming pork products.

Reservoir of Infection

Humans are primary reservoir of infections. Other reservoirs are animals and birds (pigs, horses, dogs, cats, domestic poultry, water birds, wild birds etc.).

Source of Infection

A case or subclinical case

Incubation Period

18–72 h (shortest for an infectious disease)

Communicability

A case is usually infectious up to 3–5 days from clinical onset in adults and up to 7 days in young children. Patient is most infectious on first day of illness, as maximum viral shedding occurs on that day. Overcrowding enhances transmission which is mainly air-borne (droplet infection and droplet nuclei may spread the virus through direct contact of contaminated objects).

Seasonality

In temperate zones, influenza epidemic occurs in winter; in tropics, it occurs in rainy season. Sporadic cases may be seen during any month of the year.

IMMUNITY

Antibodies appear in about 7 days after an attack and reach maximum level in 2 weeks; drop to pre-infection level in 8–12 months. Antibodies to 'HA' neutralise the virus and antibodies to 'NA' modify the infection. Interferon- and cell-mediated immunities play an important role in recovery. Tissue repair may take some time. After 5 days of infection, most of the symptoms usually disappear, but cough and weakness may continue. Symptoms usually disappear within 2 weeks.

LAB DIAGNOSIS

Method of choice is real time PCR for detection of matrix gene, which is influenza virus type specific. Use of >1 target gene assays is carried out for correct identification.

Important gene targets for diagnosis are as follows:
- Type A influenza matrix gene
- Haemagglutinin gene specific for influenza A (H1N1)swl virus
- Haemagglutinin gene specific for seasonal influenza A H1/H3 and other subtypes

The following protocols are currently available:
- Influenza A type-specific real time PCR
- Real time reverse transcriptase polymerase chain reaction (rRT-PCR) protocol for the detection and characterisation of influenza A (H1N1)

Respiratory specimens which can be tested include the following:
- **Throat and nasal swab (preferred)**
- Bronchoalveolar lavage
- Tracheal aspirates
- Nasopharyngeal or oropharyngeal aspirates as washes
- Nasopharyngeal or oropharyngeal swabs

Swab sticks preferred are those with synthetic tip (such as polyester or Dacron) and aluminium or plastic shaft. Swabs with cotton and wooden shafts should not be accepted. Specimens collected with calcium alginate swabs are acceptable. Samples should be collected as soon as possible, soon after symptoms begin and before antiviral medications are administered.

Other method for diagnosis is detection of viral antigens in respiratory epithelial cell by immunofluorescence assay, though it is less sensitive. Virus can be easily grown on chick eggs and cell lines like Madin-Darby Canine Kidney (MDCK) cells.

TREATMENT

Center for Disease Control, USA (CDC) recommends that diagnosed cases of H1N1 pandemic influenza should be treated with oseltamivir (Tamiflu) and zanamivir (Relenza). Three influenza antiviral medications approved by the US Food and Drug Administration (FDA) recommended for use include: oral oseltamivir (Tamiflu), inhaled zanamivir (Relenza) and intravenous peramivir (Rapivab) (3 November 2016). Everyone who is infected does not require treatment.

BOX 23.2   Influenza Patients Who Need Treatment

- Children younger than 2 years
- Adults aged 65 years and older
- Persons with chronic pulmonary, cardiovascular, renal, hepatic, haematological, neurologic and metabolic disorders (including diabetes mellitus)
- Persons with neurodevelopment conditions (e.g. disorders of the brain, spinal cord, peripheral nerve and muscle, such as cerebral palsy, epilepsy, stroke, intellectual disability, moderate-to-severe developmental delay, muscular dystrophy or spinal cord injury)
- Persons with immunosuppression (e.g. caused by medications or by HIV infection)
- Women who are pregnant or postpartum (within 2 weeks after delivery)
- Persons aged younger than 19 years who are receiving long-term aspirin therapy
- Persons who are morbidly obese (i.e. body mass index is equal to or greater than 40)
- Residents of nursing homes and other chronic care facilities

BOX 23.3   Population Who Need/Does Not Need Influenza Vaccine

Persons who need vaccination
- All children 6 months to 18 years old
- Anyone ≥50 years old
- Women who are planning to get pregnant during the flu season
- Anyone who lives or works with infants under 6 months old
- Residents of nursing homes or day care facilities
- Anyone with chronic medical conditions, such as diabetes and asthma
- Health-care personnel who have direct contact with patients, such as doctors and nurses
- Contacts of anyone in any of the high-risk groups mentioned previously

Persons who should not be vaccinated
- Infants under 6 months old
- Anyone who is severely allergic to eggs and egg products
- Anyone who has ever had a severe reaction to a flu vaccination
- Anyone with GBS
- Anyone with a fever

Persons who need treatment are listed in Box 23.2. Drugs should be judiciously used as emergence of drug resistance is frequent (Box 23.3).

Antibiotics have no effect on the virus but help prevent bacterial pneumonia and other secondary infections in influenza cases.

PREVENTION

Antiviral chemoprophylaxis is recommended up to 48 h after exposure to an infectious person. After that, it has no role. Widespread or routine use of antiviral medications for chemoprophylaxis is not recommended so as to limit the possibilities of antiviral-resistant emergence. Indiscriminate use of chemoprophylaxis might promote resistance to antiviral medications, reducing antiviral medication availability for treatment of persons at higher risk for influenza complications or those who are severely ill.

Types of vaccine: Three types of influenza vaccines are available:

Killed virus: Injectable vaccine (egg-based)

Live virus: Live attenuated influenza virus strains, nasal spray

Virus subunit: HA extracted from recombinant virus

Trivalent vaccines are recommended for use every year by WHO, specifying the virus to be used.

For example in the 2016– 2017 influenza season (northern hemisphere winter), recommendation was as follows (**25 February 2016: WHO**):

an A/California/7/2009 (H1N1)pdm09-like virus

an A/Hong Kong/4801/2014 (H3N2)-like virus

a B/Brisbane/60/2008-like virus

Quadrivalent vaccines contain an additional strain of Influenza B virus (B/Phuket/3073/2013-like virus) in trivalent vaccine.

PARAMYXOVIRUSES

INTRODUCTION

Paramyxoviridae is a family of RNA virus containing many members (genus) which can be distinguished by the activities of the viral attachment protein. Structurally these are quite similar to each other. These are large viruses consisting of a negative-sense RNA genome in a helical nucleocapsid. These are enveloped viruses. Envelope contains a viral attachment protein (haemagglutinin–neuraminidase) and a fusion glycoprotein (F).

STRUCTURE

These are single-stranded, negative-sense RNA viruses with RNA-directed RNA polymerase. These are enveloped particles, 150–300 nm in diameter, pleomorphic, tube-like and helically symmetrical (Fig. 23.2). Although paramyxovirus

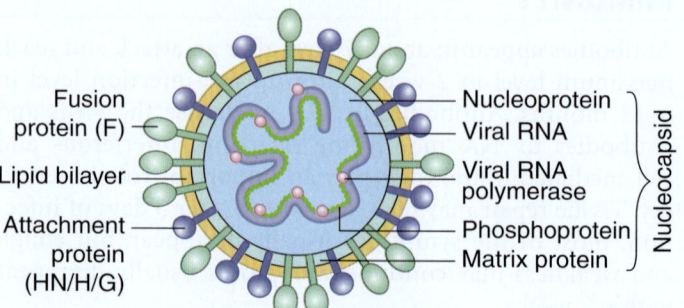

Fig. 23.2 Structure of Paramyxovirus.

genomes are structurally similar, the order of the protein-coding regions differs for each genus. The nucleocapsid consists of the negative-sense, single-stranded $5–8 \times 10^6$ bp RNA associated with the following:

1. Nucleoprotein (NP): The NP protein helps maintain genomic structure
2. Polymerase phosphoprotein (P): The P protein facilitates RNA synthesis
3. Large (L) protein: The L protein is the RNA polymerase
At the base of a double-layered lipid envelope matrix protein (M) is present. The envelope contains two glycoproteins as follows:

- **Fusion (F) protein:** It promotes fusion of the viral and host cell membranes and helps virus entry into the host cell
- **Viral attachment protein:** It is either haemagglutinin–neuraminidase [HN] or haemagglutinin [H] or G protein

CLASSIFICATION

The Paramyxoviridae include two subfamilies: *Paramyxovirinae* and *Pneumovirinae* which have many genera and viruses as shown in Table 23.3.

Replication of paramyxoviruses occurs as shown in Flowchart 23.1.

PARAINFLUENZA VIRUSES

INTRODUCTION

Parainfluenza viruses were discovered in the late 1950s. They usually cause mild respiratory symptoms. They are especially associated with laryngotracheobronchitis (croup). They have five serotypes: 1, 2, 3, 4a and 4b. There is no common group

HN, H or G proteins present on virus envelope bind with sialic acid (receptor) on the cell surface of host cells (respiratory epithelium)
↓
Nucleocapsid is transferred in cell cytoplasm along with enzymes of RNA replication
↓
Transcription of mRNAs occurs in host cell cytoplasm
↓
Synthesis of L, N and NP proteins to form nucleocapsids also occurs in host cell's cytoplasm
↓
The glycoproteins are synthesised and processed like cellular glycoproteins in cytoplasm
↓
Nucleocapsid is associated with the M proteins on viral glycoprotein
↓
Mature virions bud from the host cell plasma membrane and exit the cell

Flowchart 23.1 Replication of Paramyxoviruses.

antigen among these five subtypes. They are closely related to mumps virus and possess haemagglutination and neuraminidase (HN) surface proteins.

PATHOGENESIS

Parainfluenza viruses infect epithelial cells of the upper respiratory tract with occasional lower respiratory tract infection. Parainfluenza viruses rarely cause viraemia. The virus can cause giant cell formation and cell lysis. Diseases include

TABLE 23.3	Classification and Important Characteristics of Members of *Paramyxoviridae* Family					
Subfamilies	Paramyxovirinae				Pneumovirinae	
Genera	*Respirovirus*	*Rubulavirus*	*Morbillivirus*	*Henipavirus*	*Pneumovirus*	*Metapneumovirus*
Human viruses	Parainfluenza 1 and 3 viruses	Parainfluenza 2, 4a, and 4b viruses; mumps virus	Measles virus	Hendra, Nipah (zoonotic)	Respiratory syncytial virus	Human metapneumovirus
Nucleocapsid (nm)	18	18	18		13	13
Large glycoprotein	HN	HN	H	G	G	G
Haemagglutinin	+	+	+	−	−	−
Neuraminidase	+	+	−	−	−	−
Fusion protein	+	+	+	+	+	+
Haemolysis	+	+	+	−	−	−
Inclusion bodies	Cytoplasmic	Cytoplasmic	Cytoplasmic/ nuclear	NO	Cytoplasmic	−

cold-like symptoms, bronchitis (inflammation of bronchial tubes) and croup (laryngotracheobronchitis).

Infection induces protective immunity of short duration. The cell-mediated immune response provides protection. IgA responses are protective but short-lived. Reinfection is often seen, which is attributable to circulation of multiple serotypes and short-lived immunity.

EPIDEMIOLOGY

Parainfluenza viruses are ubiquitous. The virus spreads from person to person through respiratory route. Host range is limited to humans. Incubation period is 2–6 (up to 10) days. Infants and children younger than 5 years usually get primary infection. Re-infections may occur throughout life. Older children and adults generally experience milder infections than those seen in young children. Most children recover within 48 h. A parainfluenza virus infection in infants may be more severe than infections in adults. Viruses can cause outbreaks in nurseries and paediatric wards, as it spreads easily. Virus is easily inactivated by dryness, alcohol and acid (enveloped).

CLINICAL MANIFESTATIONS

Clinical manifestations of parainfluenza virus infection are shown in Box 23.4.

LABORATORY DIAGNOSIS

- **Real time-PCR** techniques are becoming the method of choice to detect and identify parainfluenza viruses from respiratory secretions.
- **Detection of antigen:** Presence of virus-infected cells in aspirates or in cell culture is indicated by the finding of syncytia and is identified with immunofluorescence.
- **Virus isolation:** Virus may be readily isolated from nasopharyngeal aspirates and throat swabs on MDCK/LLCMK2 cells. The haemagglutinin of the parainfluenza viruses promotes haemadsorption and haemagglutination. The serotype of the virus can be determined through the use of specific antibody to block haemadsorption or haemagglutination (haemagglutination inhibition).
- **Serology:** A retrospective diagnosis may be made by serology.

BOX 23.4 Clinical Manifestations of Parainfluenza Virus Infection

Respiratory tract syndromes
- Coryza
- Pharyngitis
- Mild bronchitis
- Wheezing
- Bronchiolitis
- Pneumonia
- Most notable: Croup (laryngotracheobronchitis)

MANAGEMENT

Treatment of croup caused by parainfluenza viruses is symptomatic. No specific antiviral agents are available for treatment. No live attenuated vaccine is available.

MUMPS VIRUS

INTRODUCTION

Mumps virus is the cause of acute viral parotitis and orchitis, described by Hippocrates in 5th century. Their viral aetiology was described by Johnson and Goodpasture in 1934. Mumps outbreaks were frequent among military personnel in pre-vaccine era. The virus is most closely related to parainfluenza virus 2, but there is no cross-immunity with the parainfluenza viruses. There is only one antigenic type (serotype). Virus is rapidly inactivated by chemical agents, heat and ultraviolet light.

PATHOGENESIS

Virus is transmitted through respiratory route, infects epithelial cells of respiratory tract and spreads systemically by viraemia. Replication of virus occurs in nasopharynx and regional lymph nodes. Viraemia occurs 12–25 days after exposure. Infection of parotid gland, testes, and central nervous system commonly occurs. Principal symptom is swelling of parotid glands. The parotid gland gets infected either via parotid duct or by viraemia. Interferon production, specific cellular and humoral immunity develop after infection. Interferon limits virus spread and multiplication. Interferon production ceases as virus levels decrease and antibodies and cell-mediated immunity (CMI) appear. IgM class-specific antibodies develop within the first 3 days after onset of symptoms and persist for approximately 2–3 months. IgG ab appear a few days later and persist for life. Circulating antibodies provide lifelong protection. IgG class is transplacentally transferred to newborn and persists in declining titres during the first 6 months.

EPIDEMIOLOGY

Mumps is a communicable disease with only one serotype and it infects only humans. Direct person to person contact and respiratory droplets are the modes of transmission. Unvaccinated people are at risk; infection occurs in 90% of people by the age of 15, if not vaccinated. Virus may cause asymptomatic shedding. Host range is limited to humans. The virus is released in respiratory secretions from patients who are asymptomatic for the 7-day period before clinical illness. Living or working in close contact with infected person promotes the spread of the virus. Local outbreaks are seen in institutions, boarding schools and military camps. Virus has large enveloped virion that is easily inactivated by dryness and acid.

CLINICAL PICTURE

Mumps infections are often asymptomatic. Commonest clinical manifestation is bilateral parotitis with fever. Complications may occur with or without parotitis. Mumps orchitis may cause sterility.

Complications

CNS involvement: 10%–15% cases
Orchitis: 20%–50% in adult and adolescent males
Pancreatitis: 2%–5% cases
Deafness: 1/20,000 cases
Death: 1–3/10,000 cases

LABORATORY DIAGNOSIS

RT-PCR remains the test of choice these days. Virus can be recovered from saliva, urine, pharynx and cerebrospinal fluid. Virus can be detected from saliva up to a week and urine up to 2 weeks after the onset of symptoms.

Mumps virus grows well in cell culture (monkey kidney cells). Presence of multinucleated giant cells is commonly seen in infected cells. The haemadsorption of guinea pig erythrocytes also occurs on virus infected cells due to the viral haemagglutinin.

Confirmation by serologic testing can be done. A four-fold increase in the virus-specific IgG antibody level or the detection of mumps-specific IgM antibody indicates active infection.

TREATMENT, PREVENTION AND CONTROL

Antiviral agents are not available. Mumps vaccine is usually given as MMR vaccine and is also available as monovalent mumps vaccine. Protective efficacy is ~95% (detailed in the end of chapter).

RESPIRATORY SYNCYTIAL VIRUS

INTRODUCTION

RSV was first isolated from a chimpanzee in 1956. It is one of the most common causes of acute respiratory tract infection in infants and young children, causing death. Infection by the age of 2 years is the rule.

PATHOGENESIS

RSV initiates a localised infection in respiratory tract. Infection generally is confined to the epithelium of upper respiratory tract, but may involve lower respiratory tract. Mainly ciliated mucosal epithelial cells of nose, eyes and mouth are infected. Virus does not cause viraemia or systemic spread, but induces syncytia formation. The damage to respiratory epithelium is caused by direct cell injury followed by immunological cell damage. Smaller airways get blocked due to necrosis of the lining cells and mucus plugs. Airways of young infants get easily obstructed by plugs. Pathogenesis of bronchiolitis may be immunologic or directly due to viral cytopathology. Bronchiolitis during the first year of life may be a risk factor for later development of asthma. Pneumonia is caused due to spread of virus from cell to cell. Bronchiolitis is inflammatory, and pneumonia is interstitial.

Virus is shed in respiratory secretions. It starts with appearance of clinical symptoms and continues till the neutralising antibodies appear in bronchial secretions.

IMMUNITY

Bronchiolitis is most likely mediated by host's immune response. Degree of illness varies with age and immune status of host. Protective immunity is mainly elicited by F and G proteins. Virus spreads both extra-cellularly and by fusion of cells to form syncytia, thus, humoral antibodies that do not penetrate intracellularly cannot restrict infection. Natural immunity does not prevent reinfection, and vaccination with killed vaccine appears to enhance the severity of subsequent disease. Maternal antibody does not protect infant from infection. Non-specific defence, such as virus-inhibitory substances in secretions probably contribute to resistance and recovery. Resistance to re-infection and repeat illness depend mainly on the presence of neutralising antibody on mucosal surfaces.

EPIDEMIOLOGY

RSV is distributed worldwide and prevalent in young children. Disease is endemic, reaching epidemic proportions occasionally. Almost all children have been infected by 2 years of age. About 1% of cases need hospitalisation. Source of infection is respiratory tract of humans. Incubation period is about 4–5 days. Host range is limited to humans. Contagious period precedes symptoms and may occur in absence of symptoms. Virus is secreted in respiratory secretions for many days, especially by infants. Infection is transmitted through person to person contact and air-borne route. RSV infections almost always occur in the winter, every year. The introduction of the virus into a nursery or paediatric ward infects almost every incumbent. Infants usually have a more severe disease than in older children. RSV infects infants during the first few months of life despite the presence of maternal serum antibodies. By the age of 4 years, almost all the children have exposure to RSV. Outbreaks are occasionally seen in nursing homes, housing elderly people.

CLINICAL SYNDROMES

Symptoms vary from a common cold to pneumonia. This is most common cause of severe lower respiratory infection in infants. Common manifestations in infants are:

- Bronchiolitis which is usually self-limiting, but it can be severe in an infant. RSV accounts for 50%–90% cases
- Pneumonia/bronchopneumonia (5%–40%)
- Upper respiratory tract infection

- Croup
- Apnoea
- Otitis media

In older children/adults common manifestations are:

- Upper respiratory tract infection
- Bronchitis
- Pneumonia (elderly(

Infants at risk of severe infection:

1. Infants with congenital heart disease
2. Infants with underlying pulmonary disease
3. Immunocompromised infants
4. Premature infants

LABORATORY DIAGNOSIS

- The presence of the viral genome in infected cells and nasal washings by RT-PCR
- Detection of the viral antigen by immunofluorescence and ELISA
- Seroconversion or a four-fold or greater increase in the antiviral IgG antibody titre
- Culture for diagnostic purposes is not needed

TREATMENT, PREVENTION AND CONTROL

In otherwise healthy infants, treatment is supportive. Ribavirin, a guanosine analogue, is used in cases with severe infection. Infected children must be isolated. Control measures include hand washing and wearing gowns, goggles and masks. These measures may be needed for hospital staff attending children. No vaccine is currently available for RSV prophylaxis. RSV immunoglobulin can be given to infants at risk of severe infection.

MEASLES

INTRODUCTION

Measles is a highly contagious viral illness, first described in 7th century. Disease affected nearly all children universally in pre-vaccination era. Frequent and often fatal in the developing parts of the world, it kills about 1.0 million children in a year.

PATHOGENESIS AND IMMUNITY

Measles is highly contagious and is transmitted through respiratory droplets of infected person. Measles causes cell fusion, leading to the formation of giant cells. This helps the virus pass directly from cell to cell, thus avoiding antibody-mediated neutralisation. Virus infects epithelial cells of respiratory tract. It also infects conjunctiva, respiratory tract, urinary tract, lymphatic system, blood vessels and central nervous system. Infection usually leads to cell lysis, but persistent infections without lysis can occur in certain cells, for example, the brain cells. During the incubation period, measles causes a decrease in eosinophils and B and T cells. CMI is essential to control infection. Antibody is not sufficient to control infection because virus may spread cell to cell.

- Maculopapular rash is the most common feature, which happens due to a T-cell response against virus-infected epithelial cells of capillaries.
- Encephalitis can be caused due to either direct infection of neurons, or immune-mediated postinfectious encephalitis.
- Subacute sclerosing panencephalitis (SSPE) is caused by a defective variant of measles virus generated during the acute disease. The SSPE virus acts as a slow virus and causes cytopathologic effect in neurons. Symptoms appear many years after acute disease and outcome is nearly always fatal. Cell-mediated immunity plays the most important role in causing symptoms as well as controlling disease.

EPIDEMIOLOGY

Measles is still common in unvaccinated population. Outcome may be fatal in immunocompromised and malnourished children with measles. Only known reservoir is human. Transmission is air-borne and communicability starts 4–5 days before to 4–5 days after rash onset. The measles virus has only one serotype and infection usually manifests as symptoms. Epidemics tend to occur in 1- to 3-year cycles in unvaccinated communities. Preschool non-vaccinated children are most susceptible.

CLINICAL SYNDROMES

Following presentations may occur: Characteristic maculopapular rash, cough, conjunctivitis, coryza, photophobia, Koplik's spots (red spot on palate and oral mucosa)

Rash characteristics:

- Appears 2–4 days after prodrome and about 14 days after exposure
- Maculopapular rash becomes confluent
- Begins on face and head
- Persists for 5–6 days
- Fades in order of appearance

Complications

Otitis media, croup, bronchopneumonia, diarrhoea and encephalitis are the complications.

Atypical measles: Very intense rash mostly in distal areas, such as limbs; possible vesicles, petechiae, purpura or urticaria

SSPE: It is a rare complication of measles infection, slowly progressing, fatal neurodegenerative disease of CNS affecting children and young adults, which happen due to the persistence of defective measles virus in brain cells. It is manifested usually after 7–15 years following primary measles and affects all parts of the brain, hence, called as panencephalitis. Incidence is rare: 1:300,000 to 1:1000, 000 cases of measles. Exact pathogenic mechanism of SSPE remains unclear. SSPE patients will have high levels of measles specific-antibody in blood and CSF. Measles infection occurring before 1 year of age is a significant risk factor. Invariably fatal disease with central nervous system manifestations (e.g. personality, behaviour and memory changes; myoclonic jerks; spasticity and blindness) is the norm.

Efficacy of MMR vaccine	95% (range: 90%–97%)
Duration of immunity	Lifelong
Schedule	1 Dose

LABORATORY DIAGNOSIS

- Clinical diagnosis can be easily made.
- Measles RNA can be easily detected by RT-PCR in pharyngeal cells or urinary sediment. Respiratory tract secretions, urine, blood and brain tissues are the recommended specimens for lab confirmation. It is best to collect the specimen in very early phase of disease.
- Measles antigen can be detected with immunofluorescence in aforesaid specimen.
- The measles virus can be grown in primary human or monkey cell cultures.
- Classical giant cells are demonstrated in Giemsa-stained smears prepared from the upper respiratory tract and urinary sediment.
- Antiviral IgM can be detected after appearance of rash. Seroconversion or a four-fold increase in the titre of measles-specific IgG antibodies between sera obtained during the acute stage and the convalescent stage is diagnostic.

TREATMENT, PREVENTION AND CONTROL

A live attenuated measles vaccine is in use since 1963. Exposed susceptible people who are immunocompromised should be given immune globulin to lessen the risk and severity of clinical illness. Antivirals are not required.

Measles, Mumps and Rubella Vaccine (Live)

Freeze-dried live attenuated strains of the following:
- Edmonston–Zagreb measles virus propagated on human diploid cell culture
- L-Zagreb mumps virus propagated on chick embryo fibroblast cells
- Wistar RA 27/3 rubella virus propagated on human diploid cell culture

The reconstituted vaccine contains, in single dose of 0.5 mL (diluted in sterile water for injection), not less than
- 1000 $CCID_{50}$ of measles virus,
- 5000 $CCID_{50}$ of mumps virus,
- 1000 $CCID_{50}$ of rubella virus.

HUMAN METAPNEUMOVIRUS

Avian pneumovirus (aMPV, Turkey rhinotracheitis virus) and human metapneumovirus (hMPV) are pathogens of birds and humans, respectively, associated with respiratory tract infections; aMPV and hMPV have been classified into a new genus referred to as *Metapneumovirus*. In 2001, researchers in the Netherlands reported unidentified pathogen associated with ARTI in Europe, America, Asia, Australia and South Africa in individuals of all ages. Its identity was unknown until recently because it is difficult to grow in cell culture. Use of RT-PCR method was helpful in detection and distinction of the pneumoviruses from other respiratory disease viruses. The virus is ubiquitous and almost all 5-year-old children have experienced a virus infection and are seropositive. Seronegative children, elderly persons and immunocompromised people are at risk of acquiring the disease. Incidence varies from 1.5% to 25%. It plays an important role in paediatric URTI/LRTI with epidemiological and clinical features similar to RSV and influenza. It is usually a mild disease presenting with cough, sore throat, runny nose and high fever. Approximately 10% of patients with metapneumovirus have severe manifestations of wheezing, dyspnoea, pneumonia, bronchitis or bronchiolitis. Infection may be associated with significant morbidity and mortality in preterm infants and children with underlying clinical conditions. Laboratory identification of the virus can be performed by RT-PCR. Supportive care is the only therapy as no specific treatment is available for these infections.

NIPAH AND HENDRA VIRUSES

Hendra virus was discovered in 1994 in Australia. Nipah virus was first time isolated from patients after an outbreak of severe encephalitis in Malaysia and Singapore in 1998. Both viruses resemble in many ways. Both have broad host ranges, including pigs, humans, dogs, horses, cats and other mammals. The human is an accidental host for these viruses, but the outcome of human infection is severe with high mortality rate. Disease signs include flu-like symptoms, seizures and coma. For Nipah virus, the reservoir is a fruit bat (flying fox). The virus can be detected in fruits bitten by infected bats.

24

Arboviruses

Amita Jain

LEARNING OBJECTIVES

- Classification, common features, structure, transmission, epidemiology, pathogenesis, clinical features, laboratory diagnosis and management of :
  - Dengue virus (DENV)
  - Yellow fever virus
  - Zika virus
- Japanese encephalitis virus (JEV)
- Chikungunya virus
- West Nile virus (WNV)
- Kyasanur forest disease (KFD) virus
- Crimean–Congo haemorrhagic virus
- Hantavirus

INTRODUCTION

The word 'arboviruses' stands for 'arthropod-borne viruses'. These viruses infect vertebrate host. Arthropods are vectors and transmit the infection. Once infected arthropods remain infected lifelong and can transmit infection transovarian as well. **Most of the arboviruses are named after the disease/ geographical area of their first isolation.**

CLASSIFICATION

The classification of Arboviruses is presented in Table 24.1. More than 450 viruses exist and about 100 of them are human pathogens. Arboviruses of human importance are listed in Box 24.1.

VECTORS

Vectors are usually mosquitoes/ticks (*Ixodes* spp.)/sandfly.

Mosquitoes are vectors for Japanese encephalitis (JE), dengue, yellow fever, St Louis encephalitis, eastern equine encephalitis (EEE), western equine encephalitis (WEE) Venezuelan equine encephalitis (VEE) etc.

Ticks are vectors for Crimean–Congo haemorrhagic fever, various tick-borne encephalitis, etc.

Sandflies are vectors for Sicilian sandfly fever and Rift Valley fever.

ANIMAL RESERVOIR

In many cases, the actual reservoir is not known. The following animals are implicated as reservoirs:

Birds: JE, St Louis encephalitis, EEE and WEE
Pigs: JE
Monkeys: Yellow fever
Rodents: VEE and Russian spring summer encephalitis

TRANSMISSION CYCLE

Two types of transmission cycles are seen in arboviruses (Fig. 24.1). Both cycles may be seen with some arboviruses, such as yellow fever; while in some, only one type of cycle is seen.

TABLE 24.1	Classification of Arboviruses				
Single-Stranded Positive Sense RNA Viruses		**Single-Stranded Negative Sense RNA Viruses**		**Double-Stranded RNA Viruses**	
Family	Genus	Family	Genus	Family	Genus
Togaviridae	***Alphavirus***	***Bunyaviridae***	• ***Orthobunyavirus*** • ***Nairovirus*** • ***Phlebovirus*** • ***Tospovirus***	Reoviridae	• *Orbivirus* • *Coltivirus*
Flaviviridae	***Flavivirus***	Rhabdoviridae	Vesiculovirus		

Note: Viruses of human importance are presented in bold.

BOX 24.1 Arboviruses of Human Importance

- **Togaviridae**
- ***Genus Alphaviruses***
  - *Chikungunya virus*
  - *Eastern equine encephalitis (EEE) virus*
  - *Western Equine Encephalitis (WEE) virus*
  - *Venezuelan Equine Encephalitis (VEE) virus*
  - *Sindbis virus*
- **Flaviviridae**
- ***Genus Flaviviruses***
  - *Yellow fever virus*
  - *Dengue virus*
  - *Japanese encephalitis (JE) virus*
  - *West Nile encephalitis virus*
  - *Kyasanur forest disease virus*
  - *Zika virus*
  - *Murray Valley encephalitis virus*
  - *Russian spring summer encephalitis virus*
  - *St Louis encephalitis virus*
- **Bunyaviridae**
- Orthobunyavirus
  - *California encephalitis virus*
  - *Bunyamwera virus*
- Phlebovirus
  - *Sandfly fever virus*
  - *Rift Valley fever virus*
- Nairovirus
  - *Crimean–Congo haemorrhagic fever virus*
- Hantavirus
  - *Hantaan virus (Korean haemorrhagic fever virus)*

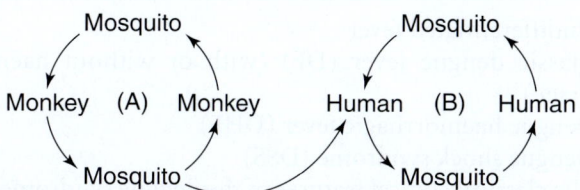

Fig. 24.1 (A) Sylvatic and (B) Urban Dengue Virus Transmission Cycles.

1. **Man to arthropod vector to man (urban cycle),** which is seen in dengue, urban yellow fever, etc. Either man or arthropod vector can serve as reservoir. In vector, transovarian transmission may take place. Sylvatic cycle runs between animals and arthropod vectors in forests. Once the virus comes to areas inhabited by humans, cycle changes to urban.

2. **Animal to arthropod vector to man,** which is seen in JE, EEE, WEE, jungle yellow fever, etc. In these cases, reservoir is in an animal and virus is maintained in nature in a transmission cycle involving arthropod vector and animal. Man becomes infected incidentally.

DISEASES CAUSED

Following types of manifestations are usually noted:
- **Fever and rash:** usually a non-specific illness resembling a number of other viral illnesses, such as influenza, rubella and enterovirus infections
- **Encephalitis:** for example, EEE, WEE, St Louis encephalitis and JE
- **Haemorrhagic fever:** for example, yellow fever, dengue and Crimean–Congo haemorrhagic fever

In this chapter, we will discuss arboviruses of importance for Asian region.

DENGUE VIRUS (DENV)

Dengue virus belongs to the family Flaviviridae and has positive sense, single stranded, encapsulated RNA genome containing three structural and seven non-structural protein genes (Fig. 24.2).

SEROTYPES AND GENOTYPES

Dengue virus (DENV) is very diverse. It has four known serotypes: DEN-1, DEN-2, DEN-3 and DEN-4. There is antigenic similarity amongst serotypes, but cross-immunity lasts only a few months. Each serotype provides specific lifetime immunity, but poor cross-immunity. All serotypes can cause severe and fatal disease.

Within each serotype, there are distinct three to five genetic groups (genotypes) and lineages. Some genetic variants may be more virulent or have greater epidemic potential than others.
- DEN-1: Genotype I, II and III
- DEN-2: Genotype I, II, III, IV, V and sylvatic
- DEN-3: Genotype I, II and III
- DEN-4: Genotype I, II, III, IV and V

VECTORS

Aedes aegypti is the commonest vector. *Aedes albopictus* can also be a possible vector. It has white bands or scale patterns

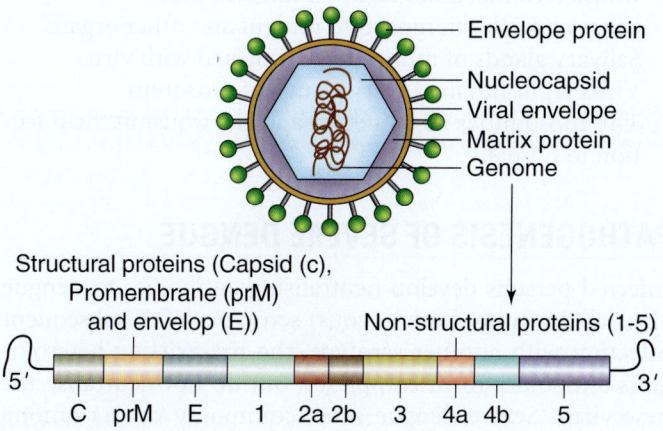

Fig. 24.2 Genome and Structure of Dengue Virus.

on legs and thorax and lives in proximity to human habitations in urban areas. It breeds mostly in man-made containers, needs only few drops of water to lay eggs and is a daytime feeder. Peak biting periods are morning and dusk. It bites multiple people during each feeding period. High temperature, precipitation and humidity favour vector breeding. Larva survives in dry conditions as well.

EPIDEMIOLOGY

Dengue is common in poor urban areas, suburbs and the rural areas; however, has now expanded to more affluent urban and rural areas as well. Dengue haemorrhagic fever (DHF) was first recognised in the 1950s during dengue epidemics in the Philippines and Thailand, although currently it is endemic in whole of Asia. Till 1970, severe dengue epidemics were reported from only nine Asian countries.

Dengue has spread to more than 100 countries in last 50 years, affecting more than 50% of world's population. Common risk factors are rainy season, high temperature, relative humidity, degree of urbanisation and quality of vector control services in urban areas.

There are several reasons which helped in dengue expansion:
- Extensive vector infestation
- Declining vector control
- Unreliable water supply systems
- Increasing use of non-biodegradable containers
- Poor solid waste disposal
- Increased air travel
- Increasing population density in urban areas

REPLICATION

Following steps lead to dengue replication and transmission in sequential order:
- Virus transmitted to human through mosquito saliva with mosquito bite
- Virus infects white blood cells and lymphatic tissues
- Virus replicates in target organs, such as lymphatic tissue
- Viraemia occurs
- Uninfected mosquito feeds on infected blood
- Virus replicates in mosquito midgut and other organs
- Salivary glands of mosquito get infected with virus
- Virus replicates in salivary glands of mosquito
- This mosquito is now infected and can transmit the infection to human

PATHOGENESIS OF SEVERE DENGUE

Infected persons develop neutralising antibodies to dengue virus of the same (homologous) serotype. With subsequent infection with another serotype, the pre-existing heterologous antibodies form complexes, but do not neutralise the new virus. Severe dengue most commonly occurs among patients with secondary DENV infections and infants. **Antibody-dependent enhancement (ADE)** is the hypothesised mechanism for severity of infection in secondary dengue, although the exact mechanisms are not clear. ADE is the process in which DENV coated with non-neutralising antibodies is engulfed by macrophages and other cells of same lineage. Infected cells release vasoactive substances which cause *increased vascular permeability and haemorrhagic manifestations*. This phenomenon is described as **cytokine tsunami**.

Antibodies typically protect humans from viruses in three ways:
1. Neutralisation (antibody blocks virus interaction with host cell)
2. **Opsonisation** (antibody coats virus and typically targets it for uptake by **macrophages** and neutrophils)
3. Antibody-dependent cellular cytotoxicity (ADCC)

In dengue, non-neutralising heterotypic IgG anti-DENV antibodies produced during a person's first DENV infection (or sub-neutralising level of antibodies in the case of infants who acquired IgG passively in utero) can form antibody–DENV complexes in the secondary infection that can allow uptake of DENV by macrophages. DENV then replicates in these macrophages thereby increasing viral production.

Viral Risk Factors for DHF Pathogenesis

- Virus strain (genotype): Some genotypes have more strong epidemic potential. The level of viraemia and infectivity of some genotypes is more than others.
- Virus serotype: DHF risk is greatest for DEN-2, followed by DEN-3, DEN-4 and DEN-1.

CLASSICAL SYNDROMES OF DENGUE VIRUS INFECTION

- Undifferentiated fever
- Classic dengue fever (DF) (with or without haemorrhage)
- Dengue haemorrhagic fever (DHF)
- Dengue shock syndrome (DSS)

The classical clinical features of the dengue syndromes are listed in Box 24.2.

In year 2009 World Health Organization (WHO) has revised its classification of dengue illness for better management of patients (Fig. 24.3).

LABORATORY DIAGNOSIS

1. Mostly diagnosis is done by the following:
   a. Detection of anti-DENV IgM using enzyme-linked immunosorbent assay (ELISA) test in patient's serum.
   b. Detection of NS-1 antigen using ELISA test in patient's serum.
   Commercial kits are available for both the tests.
2. Virus isolation from patient's sera can be easily done either in mosquito cell line C6/36 or by mosquito inoculation.
3. Detection of viral RNA in patient's sera is a sensitive test.

BOX 24.2 Clinical Characteristics of Dengue

Classical Dengue Fever
- Fever
- Headache (orbital)
- Muscle and joint pain
- Nausea/vomiting
- Rash
- Haemorrhagic manifestations

Haemorrhagic Manifestations of Dengue
- Skin haemorrhages, petechiae, purpura and ecchymoses
- Gingival bleeding
- Nasal bleeding
- Gastrointestinal bleeding, haematemesis and melena
- Haematuria
- Increased menstrual flow

Dengue Haemorrhagic Fever
Four necessary criteria of DHF are as follows:
- Fever or recent history of acute fever
- Haemorrhagic manifestations

- Low platelet count (100,000/mm³ or less)
- Objective evidence of 'leaky capillaries':
  - *Elevated haematocrit (20% or more over baseline)*
  - *Low albumin*
  - Pleural or other effusions

Dengue Shock Syndrome (DSS)
Evidence of circulatory failure is manifested indirectly by all of the following criteria of DSS:
- Rapid and weak pulse
- Narrow pulse pressure or hypotension for age
- Cold, clammy skin and altered mental status
- Frank shock is direct evidence of circulatory failure

Danger Signs in Dengue Haemorrhagic Fever
- Abdominal pain: intense and sustained
- Persistent vomiting
- Abrupt change from fever to hypothermia with sweating and prostration
- Restlessness

(*Source*: Adapted from Dengue haemorrhagic fever: Chapter 2: Clinical diagnosis.)

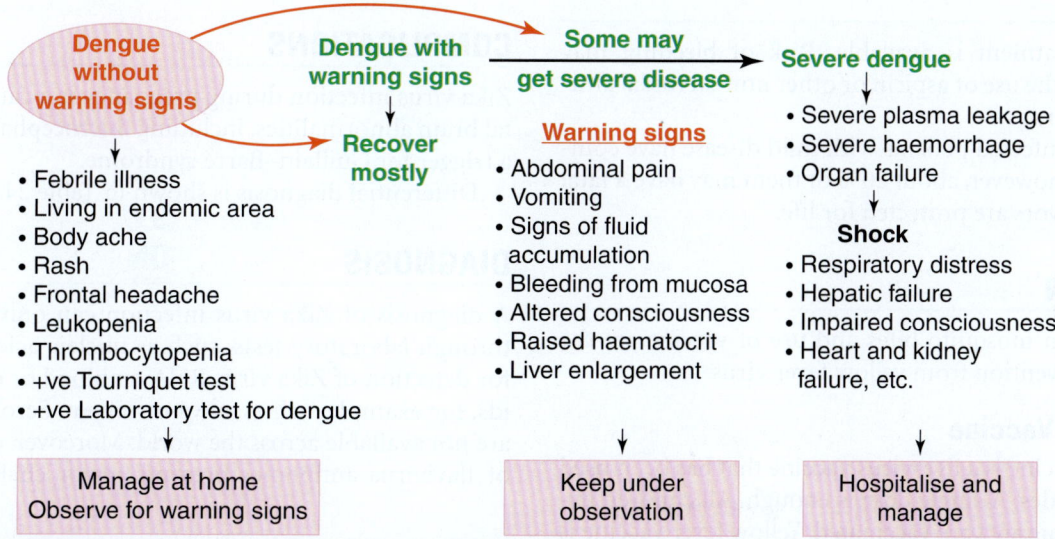

Fig. 24.3 World Health Organization Classification of Dengue as per 2009 Recommendation. (*Source*: WHO, Dengue: Guidelines for diagnosis, treatment, prevention and control, Figure 1.4, 2009.)

TREATMENT

No specific antiviral therapy is available. Fluid management, rest and paracetamol (avoid aspirin and non-steroidal anti-inflammatory drugs) is the line of management. Monitoring blood pressure, haematocrit, platelet count and level of consciousness is essential.

DENGUE VACCINE

Presently no vaccine is available. Field testing of an attenuated tetravalent vaccine, which may provide protection against all four serotypes, is underway. Effective, safe and affordable vaccine may be available in the near future.

YELLOW FEVER VIRUS

INTRODUCTION

Yellow fever virus belongs to genus *Flavivirus*. This virus is quite similar to the dengue virus. It is transmitted by mosquito bite. Illness spectrum ranges from a self-limited febrile illness to severe haemorrhagic fever with hepatitis. Yellow fever virus is related to other flaviviruses, for example, WNVs and JEVs. Yellow fever virus is not found in India. It is commonly found in tropical and subtropical areas in South America and Africa.

TRANSMISSION

It is transmitted to people primarily through the bite of infected *Aedes* mosquitoes. Mosquitos bite either an infected primate or an infected human host to acquire infection, which can be transmitted to a new host. Period of viraemia starts 3–4 days before and lasts up to 4–6 days after symptoms appear.

Incubation period: 3–6 days.

CLINICAL FEATURES

Most infected individuals are either asymptomatic or have a mild illness. In persons who develop symptoms, initial symptoms include sudden onset of fever, chills, general body aches, nausea and vomiting and malaise. Most persons improve after the initial presentation. The severe form is characterised by high fever, jaundice, bleeding and eventually shock and multiple organ failure.

TREATMENT

No specific treatment is available. Risk of bleeding may increase due to the use of aspirin or other non-steroidal anti-inflammatory drugs.

Majority of infected persons with mild disease have complete recovery; however, about 20% of them may have a fatal outcome. Survivors are protected for life.

PREVENTION

Prevention from mosquito bites and use of vaccine are the methods of prevention from yellow fever virus.

Yellow Fever Vaccine

Yellow fever vaccine is a live-virus vaccine that has been used for several decades. A single dose is enough. All children living in an endemic area are vaccinated. Yellow fever vaccine is required for travellers to certain countries.

ZIKA VIRUS

INTRODUCTION

Zika virus is a mosquito-borne flavivirus. It is transmitted through the bite of an infected *Aedes* species mosquito (*A. aegypti* and *A. albopictus*), similar to the dengue virus.

HISTORY

Zika was first identified in Uganda in 1947 in monkeys and later identified in humans in 1952 in Uganda and the United

Republic of Tanzania. Africa, Americas, Asia and the Pacific have reported some outbreaks. Mostly patients present with mild non-specific manifestations. The first large outbreak of Zika infection was seen in Federated States of Micronesia in 2007. Brazil reported an association between Zika virus infection and Guillain–Barré syndrome and microcephaly in 2015. In 2018, outbreak of infection was reported from India, mainly from Rajasthan, Gujarat and Madhya Pradesh.

TRANSMISSION

Zika virus is primarily transmitted to people through the bite of an infected *Aedes* mosquito. Sexual transmission of Zika virus is also possible. Maternal to foetal transmission through *intrauterine and perinatal modes is possible*. Probable transmission through blood transfusion, organ and tissue transplant, fertility treatment, laboratory exposure and breast feeding is being investigated.

The incubation period is only few days.

CLINICAL FEATURES

The symptoms are mild, lasting for 2–7 days and similar to dengue, including fever, skin rashes, conjunctivitis, muscle and joint pain, malaise and headache.

COMPLICATIONS

Zika virus infection during pregnancy is a cause of congenital brain abnormalities, including microcephaly. Zika virus is a trigger for Guillain–Barré syndrome.

Differential diagnosis is shown in Table 24.2.

DIAGNOSIS

A diagnosis of Zika virus infection can only be confirmed through laboratory tests, such as nucleic acid amplification for detection of Zika virus RNA on blood or other body fluids, for example, urine, saliva or semen. Serological test kits are not available across the world. Moreover, cross-reactivity of flavivirus antibodies presents major challenge in result

TABLE 24.2	Differential Diagnosis of Zika Virus Disease	
Bacterial	**Viral**	**Parasitic/ Non-infectious**
Scarlet fever	Dengue	Malaria
Meningococcaemia	Chikungunya	Rheumatological diseases
Typhoid	Rubella	Postinfectious arthritis
Bacterial septicaemia	Influenza	
Leptospirosis	West Nile disease	
Rickettsial diseases		

interpretation (plaque reduction neutralisation test helps in verifying MAC-ELISA results).

TREATMENT

No specific treatment is available. Plenty of rest, enough fluids and pain and fever medicines are enough.

PREVENTION

Protection against mosquito is the key. No vaccine is available.

JAPANESE ENCEPHALITIS VIRUS

INTRODUCTION

JEV is also a mosquito-borne flavivirus. It is transmitted by *Culex* mosquito.

EPIDEMIOLOGY

The first case of JE was documented in 1871 in Japan; hence, the name Japanese encephalitis virus. JEV is one of the common causes of encephalitis in Indian subcontinent. It is geographically restricted and is endemic in 24 countries in South-East Asia and Western Pacific regions. The annual incidence of clinical disease varies both across and within endemic countries, ranging from <1 to >10 per 100,000. Mostly children are affected. In endemic areas, adults attain immunity by natural subclinical exposures.

Most of the infections are asymptomatic. Those with encephalitis may show mortality rate as high as 30% and permanent neurologic or psychiatric sequelae in 30%–50% of cases.

JE was first recognised in India in 1955 in North Arcot district of Tamil Nadu. Since 1972, JE has spread to more areas. West Bengal, Uttar Pradesh, Assam, Manipur, Bihar, Andhra Pradesh, Pondicherry, Karnataka, Goa, Kerala and Maharashtra are afflicted districts. The spread of JEV in new areas has been correlated with agricultural development and intensive rice cultivation supported by irrigation programmes.

In Asia, JEV transmission season varies from region to region. Most often infection occurs during the rainy season and preharvest period in rice-cultivating regions. The disease is predominantly found in rural and peri-urban settings, where humans live in closer proximity to vertebrate hosts.

Major outbreaks of JE occur every 2–15 years. JE transmission intensifies during the rainy season when the vector population increases.

VECTORS

Vectors for JE are mosquitoes (*Culex tritaeniorhynchus* and *Culex vishnui*). They breed in rice fields and good rainfall encourages breeding. Highest prevalence is seen during

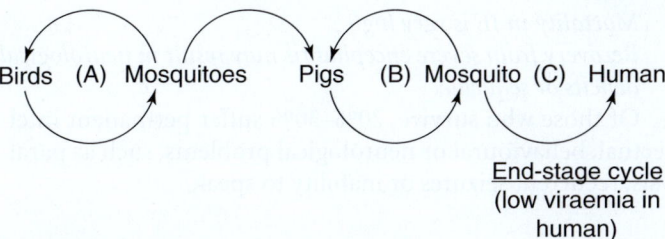

Fig. 24.4 Transmission Cycle of JEV: (A) Maintenance Cycle, (B) Multiplying Cycle and (C) End Stage.

monsoon season. Highest biting activity is from dusk to midnight. They prefer to feed on pigs.

Reservoir Hosts

Reservoir hosts are wading birds. They carry the virus; viraemia is present but no clinical disease is seen.

Amplifier Host

Amplifier host is pig (swine). Virus amplifies in pigs. Transmission of disease is facilitated by susceptible pig population, crowded conditions, paddy field where wading birds are often present and proximity to mosquito populations.

TRANSMISSION

The transmission cycle exists between mosquitoes, pigs and/or water birds (enzootic cycle). *Culex* mosquito bites infected animal and transmits the same infection to a new host. Human can be an accidental host. As the viraemia in human is usually mild, they cannot infect a feeding mosquito (Fig. 24.4).

PATHOGENESIS

Primary multiplication of virus occurs probably in fibroblasts. Virus then reaches regional lymph nodes. A brief period of viraemia lasts 2–3 days. In some individuals, virus invades the central nervous system; usually the grey matter of the brain is involved. Lesions are seen in thalamus, substantia nigra, cerebral cortex, cerebellum, Ammon's horn and anterior horn of spinal cord.

CLINICAL FEATURES

Most JEV infections are asymptomatic or very mild. Only 1 in 250 infections result in severe illness.

Prodromal stage: 2–3 days
- *High-grade fever, headache and malaise*

Acute encephalitis stage: Lasts for usually 3–4 days or longer
- *Focal asymmetric neurological deficits*
- *Fever, headache, vomiting and meningeal irritation*
- *Seizures and/or other abnormal movements*
- *Asymmetrical spontaneous eye movements or doll's eye movements*
- *Cranial nerve palsies*

- *Mortality in JE is very high*
- *Recovery from severe encephalitis may result in neurological deficits or sequelae*

Of those who survive, 20%–30% suffer permanent intellectual, behavioural or neurological problems, such as paralysis, recurrent seizures or inability to speak.

LABORATORY DIAGNOSIS

Specimen is blood/cerebrospinal fluid (CSF).

Serological tests (sensitive after 5–7 days of illness):
- *IgM capture ELISA*
- *Haemagglutination inhibition (HI) test*
- *Neutralisation test*

Virus isolations and propagation:
- *Tissue culture: **mosquito cell line C6/36***
- *Infant mouse inoculation*
- *Mosquito inoculations*

Antigen detections:
- *Immunofluorescence test*
- *Antigen capture ELISA*

PCR is a relatively insensitive test for laboratory confirmation of JE, hence it is not recommended for diagnosis.

TREATMENT

There is no antiviral treatment for patients with JE. Treatment is supportive to relieve symptoms and stabilise the patient.

PREVENTION AND CONTROL

Mosquito control is the key. Pigs should be kept away from inhabited areas.

Vaccine

The live attenuated SA14-14-2 vaccine manufactured in China is recommended for use in endemic countries by World Health Organization, since 2013. There is little evidence to support a reduction in JE disease burden from interventions other than the vaccination of humans.

Presentation: Five-dose or single-dose vial, (lyophilised powder requiring reconstitution with supplied diluent), single 0.5 mL dose given by subcutaneous injection

Boosters: Boosters are unlikely to be required as with most live attenuated vaccines, it is thought one dose will provide lifelong protection.

Studies have already documented the ongoing protection from a single dose for a minimum of 5 years in a JE-endemic area.

CHIKUNGUNYA VIRUS

INTRODUCTION

Chikungunya (chik-en-gun-ye) was first described during an outbreak in southern Tanzania in 1952. This is a mosquito-borne febrile illness. The name 'chikungunya', in the Kimakonde language, means 'to become contorted', and describes the stooped posture of patient due to severe arthralgia. The disease can be clinically confused with dengue and Zika virus infection.

EPIDEMIOLOGY

Chikungunya has been identified in over 60 countries in Asia, Africa, Europe and the Americas. Disease was first recognised as a human pathogen during the 1950s in Africa. It has become endemic in South Asia, Africa and India. First outbreak occurred in 1952 on the Makonde Plateau, border between Tanganyika and Mozambique. First published report is from Africa in 1955 by Marion Robinson and W.H.R. Lumsden. One large epidemic occurred in Malaysia in 1999.

Significant urban outbreaks of chikungunya fever in India occurred in late 1970–80s. Epidemic resurgence of CHIKV occurred in India during 2005–06 after a gap of 32 years. Since then disease has become endemic in most of India.

VECTORS

Vectors are *A. aegypti* and *A. albopictus* mosquitoes; same vectors as for dengue and yellow fever.

Reservoir

Non-human primates in Africa are the reservoir; no animal reservoir is found in India.

TRANSMISSION

Transmission is maintained in nature by man–mosquito–man cycle. There is no known mode, other than the mosquito bite.

Incubation period: 2–12 days, usually 3–7 days; viraemia lasts for 5 days (infective period).

CLINICAL FEATURES

Very similar to those of dengue but no haemorrhagic/shock syndrome is seen in Chikungunya as seen in dengue infections.

- **Arthralgia or arthritis** is marked and most important feature (lasting several weeks):
  Joint pain with or without swelling
  The small joints of the lower and upper limbs are commonly affected
  Migratory polyarthralgia usually without much effusions
  Larger joints may also be affected (knee, ankle)
  Pain may be worse in the morning and better by evening
  Joints may be swollen and painful to touch
  Some patients have incapacitating joint pains
  Arthritis may last for weeks or months
- Sudden onset of **fever and chills**; high-grade fever (40°C or 104°F)

- Headache, nausea, vomiting and abdominal pain
- Low back pain and rash
- Silent CHIKV: inapparent infections in children
- Conjunctival suffusion and mild photophobia
Persons predisposed to serious manifestations:
- Pregnant women
- Elderly people
- Newborns
- Women in general
- Diabetics
- Immunocompromised patients
- Patients with severe chronic illnesses

LABORATORY DIAGNOSIS

Several methods can be used for diagnosis as follows:
1. **IgM capture ELISA:** IgM antibody levels are the highest 3–5 weeks after the onset of illness and persist for about 2 months.
2. **Nucleic acid amplification** by PCR and reverse transcription polymerase chain reaction (RT-PCR): Samples collected during the first week after the onset of symptoms should be tested by both serological and virological methods (RT-PCR).
3. **Four-fold or more rise in HI antibody.**
4. **Virus isolation:** The virus may be isolated from the blood during the first few days of infection by inoculation in:
   a. Infant Swiss albino mice
   b. *Vero and BHK-21 cell lines*

TREATMENT

There is no specific antiviral drug treatment for chikungunya. Treatment is symptomatic in relieving the joint pain using antipyretics, optimal analgesics and fluids.

PREVENTION AND CONTROL

Protection from mosquito bites is the only prevention available. There is no chikungunya vaccine available.

WEST NILE VIRUS

INTRODUCTION

WNV is a member of the *Flavivirus* genus and closely resembles JEV, antigenically. It is a neurotropic virus, which causes encephalitis. This virus infects humans, birds, mosquitoes, horses and other mammals.

EPIDEMIOLOGY

WNV is commonly found in Africa, Europe, the Middle East, North America and West Asia. Recently cases are being reported from India.

VECTORS

Mosquitoes of the genus *Culex* are the principal vectors of WNV. WNV is maintained in mosquito populations through vertical transmission (adults to eggs).

Reservoir Hosts

Birds are the reservoir hosts. Members of the crow family (Corvidae) are particularly susceptible.

Other Hosts

Horses, just like humans, are 'dead-end' hosts, meaning that while they become infected, they do not spread the infection. It can cause severe disease and death in horses. Vaccines are available for use in horses but not for human use.

TRANSMISSION

Natural transmission cycle is maintained between birds and mosquitoes. Mosquitoes get infection by feeding on infected birds. Virus circulates in mosquito's blood, which eventually gets into the mosquito's salivary glands. Human gets infection following bites from infected mosquitoes. Rarely, organ transplant, blood transfusions and breast milk from infected host can transmit the infection.

The incubation period is between 3 and 14 days.

CLINICAL FEATURES

Infection with WNV is asymptomatic in around 80% of infected people. Rest can get either febrile illness or encephalitis. Serious illness is common in people over the age of 50 and some immunocompromised persons.

LABORATORY DIAGNOSIS

- Detection of IgG antibody by ELISA: Seroconversion in two serial specimens collected at least 1 week apart is diagnostic
- Detection of IgM antibody capture ELISA in CSF and serum sample is diagnostic. Serum IgM antibody may persist for more than a year
- Neutralisation assays
- Viral detection by RT-PCR assay
- Virus isolation by cell culture

TREATMENT AND VACCINE

Treatment is supportive.

No vaccine is available. Only control is through the prevention of mosquito bite.

KYASANUR FOREST DISEASE VIRUS

INTRODUCTION

Heavy mortality in langur and bonnet monkeys was noted in March 1955 in the forests of Shimoga district, Karnataka state, India. The mortality in monkeys was followed by high incidence of acute prostrating febrile illness among the villagers residing in neighbouring area. Disease had 10%–15% fatality rate. Finally a virus was isolated from monkeys, man and ticks and was named Kyasanur forest disease virus. Since then yearly 40–1000 cases are reported in that area.

VIRUS

A member of family Flaviviridae, this is a neurotropic virus, similar to Russian spring summer encephalitis virus.

VECTORS

Ticks, mainly *Haemaphysalis spinigera*, are the vectors. However, the virus has been isolated from 16 species of ticks. These are highly anthropophilic ticks which bite the people who trespass forest.

VERTEBRATE HOST

Humans are accidental and dead-end hosts. Other hosts are as follows:
- Langur monkey (*Presbytis entellus*)
- Bonnet monkey (*Macaca radiata*)
- Small rodents
- Neutralising antibodies are found in a number of other animals and birds, but they are not the major reservoir

TRANSMISSION CYCLE

It is shown in Fig. 24.5.
Incubation period: 2–7 days.

CLINICAL FEATURES

Sudden onset of chills, frontal headache, high fever (40°C) for 12 days or longer with severe myalgia, cough, diarrhoea,

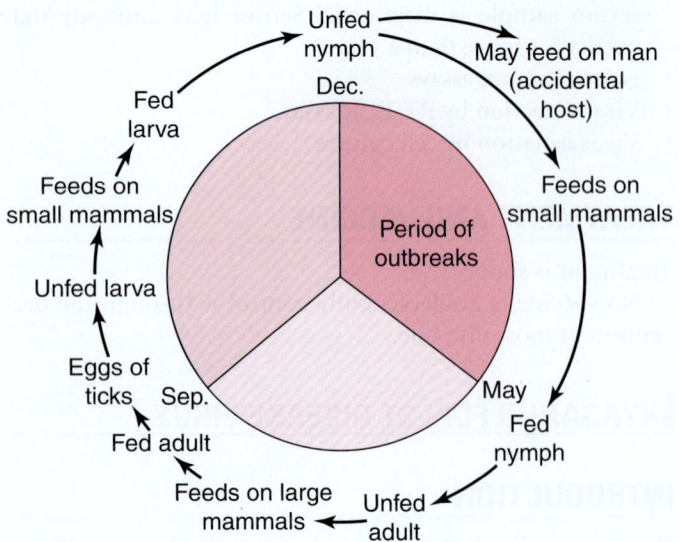

Fig. 24.5 Transmission Cycle of KFD Virus.

vomiting and photophobia are commonly seen. Disease has prolonged convalescent period. Relapse after 1–2 weeks of afebrile period can be seen. The second phase lasts for 2–12 days with neck stiffness, mental disturbance and giddiness.

VACCINE

Formalin-inactivated chick embryo tissue culture vaccine, which is developed by National Institute of Virology, Pune, is available to restricted population and about 70% seroconversion is seen.

BIOSAFETY CONCERNS

During laboratory investigations, over 100 laboratory persons got infected. KFD virus is ranked as Risk Group 4 pathogen. Protection of laboratory workers is very important.

CRIMEAN–CONGO HAEMORRHAGIC FEVER VIRUS

INTRODUCTION

Clinical entity was described in 1944–45 in Crimea during World War II. A CCHF outbreak was reported in Gujarat, India, in 2011, characterised by a zoonotic origin and a person-to-person spread in hospital setting. During December 2010, National Institute of Virology, Pune, had detected Crimean–Congo haemorrhagic fever virus-specific IgG antibodies in livestock serum samples from Gujarat and Rajasthan states in India. During January 2011, Crimean–Congo haemorrhagic fever virus was confirmed in a nosocomial outbreak, in Ahmedabad, Gujarat, India. Retrospective investigation of suspected human samples confirmed that the virus was present in Gujarat state, earlier to this outbreak. Antibodies were observed during and after the outbreak in human beings, ticks and domestic animals (buffalo, cattle, goat and sheep) from Gujarat. Case fatality rate was from 5% to 80%. Further spread of the disease was curtailed by the following:
- *High index of clinical suspicion*
- *Early laboratory diagnosis*
- *Containment measures*

Presence of haemagglutination inhibition antibodies have been detected in animal sera from Jammu and Kashmir, the western border districts, southern regions and Maharashtra state in India. In 2012, this virus was again reported in human beings and animals. Phylogenetic analysis showed that all the four isolates of 2011, as well as the S segment from specimen of 2010 and 2012 were highly conserved and clustered together in the Asian/Middle East genotype IV.

VIRUS

CCHF virus circulates in an enzootic tick–vertebrate–tick cycle.

VECTORS

Hyalomma spp. *Ticks* are the vectors.

BIOSAFETY

Minimum biosafety level 3 laboratories are required for handling clinical samples for laboratory diagnosis.

HANTAVIRUS

INTRODUCTION

Haemorrhagic fever with renal syndrome (HFRS, later renamed hantavirus disease) first came to the attention during the Korean war when over 3000 UN troops were afflicted. Disease was described by the Chinese about 1000 years earlier. In 1974, the causative virus was isolated from the Korean striped field mice and was called Hantaan virus. In 1995, a new disease entity called hantavirus pulmonary syndrome (HPS) was described.

VIRUS

Enveloped ssRNA virus forms a separate genus in the family Bunyaviridae. Virion is ~98 nm in diameter with a characteristic square grid-like structure. Genome consists of three RNA segments: L, M and S.

RESERVOIR

Hantavirus is the most widely distributed zoonotic rodent-borne virus. Unlike other members in orthoviruses, its transmission does not involve an arthropod vector.

CLINICAL SYNDROMES

It causes two clinical syndromes:
1. HFRS in Asia
2. HPS in Americas

LABORATORY DIAGNOSIS

Serological diagnosis: IF, HAI, SRH and ELISAs have been developed.

Direct detection of antigen: It is more sensitive than serology tests in the early diagnosis of the disease. The virus antigen can be demonstrated in the blood or urine.

RT-PCR: It is found to be of great use in diagnosing HPS.

Virus isolation: Isolation of virus is done from urine early in hantavirus disease. Isolation of virus from blood is less consistent.

Immunohistochemistry: It is useful in diagnosing HPS.

Enteroviruses and Rhinoviruses

Amita Jain

PICORNAVIRUSES

The term 'Pico' means very small in Greek. Picornaviruses are most diverse (more than 200 serotypes) and are among the 'oldest' known **RNA viruses**. Foot-and-mouth disease virus was one of the first viruses to be recognised by Loeffler and Frosch in 1898. Most important member of picornaviruses is poliovirus. Poliomyelitis as a viral disease was first recognised by Landsteiner and Popper in 1909 (though not isolated until the 1930s). Poliomyelitis is an acute infectious disease, which affects the central nervous system, mainly motor neurons in the spinal cord, resulting in flaccid paralysis. Poliovirus has served as a model picornavirus in many laboratory studies of the molecular biology of picornavirus replication.

Picornaviridae is a family of viruses listed in Table 25.1. This division is based on physical properties (particle density and pH sensitivity), nucleotide sequence and serological relatedness.

Enteroviruses

Enteroviruses are classified in Table 25.2.

TABLE 25.1	Members of Family Picornaviridae
Aphthovirus	7 Serotypes (F&MD virus)
Cardiovirus	2 Serotypes (cardiovirus)
Enterovirus	111 Serotypes (poliovirus)
Hepatovirus	2 Serotypes (hepatitis A virus)
Rhinovirus	105 Serotypes (rhinovirus)
Parechovirus	3 Serotypes (parechovirus)
Total:	~230 Viruses

TABLE 25.2	Classification of Enteroviruses
Polioviruses	3 Serotypes (1–3)
Coxsackieviruses group A	23 Serotypes (1–24, except 23)
Coxsackieviruses group B	6 Serotypes (1–6)
Echoviruses	32 Serotypes (1–34, except 10 and 28)
Enteroviruses (unnumbered)	4 Serotypes (68–71)

POLIOVIRUSES

INTRODUCTION

Polio is an ancient disease. Hieroglyphics and Egyptian mummies which date to 3000 BC have evidence of polio. Several epidemics of polio have occurred across the world in 19th and 20th centuries. One of the famous victims was Franklin Roosevelt (President of USA). Vaccine for polio prevention was developed in 1955 and World Health Organization's global eradication initiative was started in 1983. By year 2014, 80% world was declared polio free.

STRUCTURE OF POLIOVIRUS

Polioviruses are small, 28–30 nm in size, non-enveloped, positive-sense RNA viruses with icosahedral symmetry. Capsid has 60 capsomers (Fig. 25.1). There are four important viral proteins: VP1, VP2, VP3 and VP4. VP1 and VP3 are surface proteins, which help in attachment to host cell and are major targets for neutralising antibody binding.

Virus is rapidly inactivated by heat, formaldehyde, chlorine and ultraviolet light. It is stable at acid pH, is ether resistant and gets inactivated at 55°C for 30 min; 1 mol/L Mg^{++} can prevent heat inactivation of poliovirus.

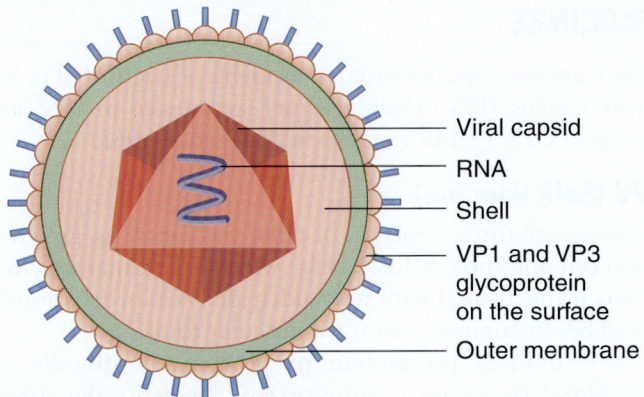

Fig. 25.1 Structure of Poliovirus.

Antigenic types: There are three poliovirus serotypes: poliovirus 1, 2 and 3, which are antigenically distinct. There is no cross-protective immunity among serotypes. They have identical physical properties but only 36%–52% nucleotide homology.

NATURAL PROGRESSION OF POLIO

Humans are infected through oral cavity with contaminated food or water. Features related to poliovirus infection are seen in Box 25.1. Virus infects the epithelial cells of the pharynx and small intestine from where it reaches local lymph nodes, that is tonsils, lymph nodes in neck and Peyer's patches, followed by viraemia. Virus multiplies mainly in gut and seasonal infections are common in humans, which remain undiagnosed. It may infect cells of the central nervous system and spread along axons reaching spinal cord. Virus multiplies in cytoplasm of anterior horn cells of spinal cord causing destruction of nerve cells and classical lower motor neuron-type acute flaccid paralysis (AFP). Poliovirus does not multiply in muscle in vivo. Some of the neurons may recover completely.

CLINICAL MANIFESTATIONS

Subclinical/Asymptomatic Infection (90%–95%)

Seen in up to 95% of all poliovirus infections. Ratio of subclinical to paralytic illness varies from 50:1 to 1000:1 (usually 200:1). Infected persons shed virus in the stool, which is also the potential source of infection.

BOX 25.1	**Important Features of Poliovirus Infection**

Reservoir: Human
Mode of transmission: Person-to-person via the faecal–oral route
Portal of entry: Oral (contaminated water)
Incubation period: 6–20 days (3–35 days)
Communicability: Highly infectious
Highly infectious: 7–10 days before and after the onset of symptoms
Poliovirus may be present in the stool from 3 to 6 weeks

Abortive (4%–8%) Poliomyelitis

Seen in approximately 4%–8% of poliovirus infections. It presents as minor, non-specific illness without any evidence of CNS invasion. Patient usually recovers within a week. Commonly influenza-like illness, sore throat and fever, gastrointestinal disturbances, that is nausea, vomiting, abdominal pain, constipation and diarrhoea are seen.

Non-Paralytic Infection (Aseptic Meningitis)

(In 1%–2% cases): Presents with signs and symptoms of abortive poliomyelitis with stiffness and pain of the back of neck. It occurs in 1%–2% of polio infections and typically lasts from 2 to 10 days. Patient gets complete recovery.

Paralytic (1%) Polio Infection

Is less than 1% of all polio infections, which presents as (1) spinal, (2) bulbar or (3) bulbospinal manifestations.
- **Spinal polio:** Approximately 80% cases, asymmetric paralysis (legs)
- **Bulbar polio:** 1%–2% cases, weakness of muscles innervated by cranial nerves
- **Bulbospinal polio:** Approximately 20% cases, combination of bulbar and spinal paralysis

Paralytic symptoms begin 1–10 days after prodrome and progress for 2–3 days. Once temperature returns to normal, no further paralysis is seen. The limb becomes floppy and lifeless, which is known as acute flaccid paralysis. Paralysis is usually asymmetrical with no sensory losses or changes in cognition. The muscles of the legs are affected more often than the arm muscles. Extensive paralysis, involving trunk, thorax and abdomen, can result in quadriplegia. Brain stem involvement results in reducing breathing capacity and difficulty in swallowing and speaking. Without ventilator support, it can result in death. Patients may recover with some degree of muscle function. Maximal recovery usually occurs within 6 months. Remaining weakness or paralysis is usually permanent.

Mortality

Mortality is 2%–5% in children and 15%–30% in adults. Highest mortality (25%–75%) is seen in patients with bulbar involvement.

Post-Polio Syndrome

Occurs after an interval of 30–40 years. About 25%–40% of people who had paralytic poliomyelitis, experience new muscle pain, exacerbation of existing weakness, development of new weakness or paralysis. These manifestations are due to physiological and ageing changes and not due to persistence of poliovirus.

EPIDEMIOLOGY

Poliomyelitis used to occur worldwide. The most important strategies adopted for its eradication are listed in Box 25.2. All age groups were susceptible; children were more susceptible than adults. Human is the only natural host. The virus spreads by faecal–oral route. Favourable conditions for spread are poor hygiene, overcrowding, etc. Most children are infected by 5 years of age. For eradication of polio from India and world, WHO/ CDC has made certain special efforts (Box 25.3).

Methods for Diagnosis

- **Molecular diagnosis:** RT-PCR assay is offered to detect poliovirus RNA. Its sensitivity is superior to other methods. This method can distinguish wild virus and vaccine virus.
- **Virus isolation:** Can be done on primary monkey kidney (PMK) cells, Vero cells and L-20B cells (transgenic mouse cell line). Poliovirus produces typical enteroviral CPE characterised by rounding and shrinking of cells with pyknosis of the nuclei.
- **Serology:** Virus neutralisation assay is done to demonstrate four-fold rise in titre of serotype-specific anti-polio antibodies. It is very cumbersome and thus not offered by most diagnostic laboratories; rarely used for diagnosis. Occasionally it can be used for immune status screening for immunocompromised individuals.

TREATMENT

Symptomatic management and use of ventilator is the treatment given. Earlier iron lungs were used to manage patients with brain stem involvement. No antiviral is available.

VACCINES

There are two types of vaccines available – inactivated poliovirus vaccine (IPV) (Salk vaccine) introduced in 1955 and trivalent OPV (Sabin vaccine) introduced in 1963.

IPV (Salk Vaccine)

Contains all three serotypes of polio vaccine virus, grown in Vero cell line (now MRC-5 cells) which is a human cell line. Virus is inactivated with formaldehyde. Vaccine is administered by subcutaneous or intramuscular route.

IPV induces production of protective antibodies in the blood (humoral immunity) and prevents the spread of poliovirus to CNS. Low level of gut immunity is only induced. Advantages and disadvantages of IPV are listed in Box 25.4.

OPV (Sabin Vaccine)

Contains live attenuated strains of P1, P2 and P3 in 10:1:3 ratio. Vaccine virus replicates in the intestinal mucosa and lymphoid cells and is excreted in the stool up to 6 weeks. Vaccine viruses may spread from recipient to contacts. It induces humoral immunity and produces a local mucosal immunity (IgA type) in gastrointestinal tract. This also limits multiplication of 'wild' poliovirus in gut. Advantages and disadvantages of OPV are listed in Box 25.5.

Immunisation schedule:
- In the first year of life, 4 doses of OPV coupled with DPT vaccine
- Additional doses of OPV during all national immunisation days (Pulse Polio Programme) till 5 years of age

OTHER ENTEROVIRUSES

- Coxsackieviruses group A: Type 1–24 (except 23)
- Coxsackieviruses group B: Type 1–6
- Echoviruses: Type 1–33 (except 10 and 28)
- Enteroviruses: Type 68–71

COXSACKIEVIRUSES

Coxsackieviruses are a large subgroup of the enteroviruses. Coxsackieviruses were discovered by inoculation of specimen from two diseased children into newborn mice, in year 1948 and were named coxsackie after the name of the town in New York State, where they were discovered. There are two groups: A and B. Types A and B were identified on the basis of the changes seen in infected mice and changes in cell culture. Each type has a type-specific antigen. All strains from group B and one from group A (A9) share a common group antigen. Cross-reactivity is demonstrated between several group A viruses but no common group antigen is found.

Virus has been recovered from the blood, throat and stool (up to 5–6 weeks) of naturally infected hosts. The incubation period of coxsackievirus infection ranges from 2 to 9 days.

They produce a variety of illnesses in humans. Clinical manifestations of Coxsackie A and B virus infections are listed in Box 25.6.

Laboratory Diagnosis

- **Virus culture:** Specimens (throat swab, serum, CSF and stool) are inoculated into tissue cultures and also into suckling mice. In tissue culture, a cytopathic effect appears within 5–14 days. The virus is identified by the pathologic lesions and by immunologic means. Due to the difficulty of the technique, virus isolation in suckling mice is rarely attempted.
- **Nucleic acid detection:** Reverse transcription-PCR tests are sensitive and specific.
- **Serology:** Neutralising antibodies appear early during the course of infection, tend to be specific for the infecting virus and persist for years. Serologic tests are difficult to evaluate.

Immunity

In humans, neutralising antibodies are transferred passively from mother to foetus. Multiple infections with these viruses are common. Antibodies to various coxsackieviruses are found in serum collected from persons all over the world, increasingly so with age; hence antibodies are common in adults.

Coxsackieviruses have been isolated from human faeces, pharyngeal swabs, sewage and flies, all around the world. The

---

BOX 25.6 **Clinical Manifestations of Coxsackie A and B Viruses**

- CNS manifestations:
  - Aseptic meningitis
  - Paralysis (A7, A9)
  - Encephalitis
- Skin- and mucosa-related:
  - Herpangina (severe febrile pharyngitis, vesicles on palate, pharynx, tonsils or tongue)
  - Hand, foot and mouth disease (A16)
  - Exanthematous fever
- Acute haemorrhagic conjunctivitis (A24)
- Acute lower respiratory tract infections
- Hepatitis

Clinical manifestations of Coxsackie B virus include the following:
- CNS manifestations:
  - Aseptic meningitis
  - Paralysis
  - Encephalitis
- Respiratory manifestations:
  - URTI
  - LRTI
- Cardiac manifestations:
  - Pleurodynia
  - Myocarditis
  - Pericarditis
- Hepatitis
- Diabetes mellitus

---

most frequent types of coxsackieviruses are types A9 and B2–B5. There are no vaccines or antiviral drugs currently available for prevention or treatment of diseases caused by coxsackieviruses.

ECHOVIRUSES (ENTERO CYTOTOXIC HUMAN ORPHAN)

Echoviruses produce cytopathic changes in cell culture, are non-pathogenic for newborn mice and subhuman primates and were accidentally discovered in human faeces, unassociated with human disease during epidemiological studies of polioviruses. Total 32 types of echoviruses were detected; types 1–34 except echovirus 10 and 28, which were later reclassified as other viruses. Hence these two numbers are not used. Clinical syndromes which may be caused by echoviruses are listed in Box 25.7.

NEWLY IDENTIFIED PICORNAVIRUSES

Numbered enteroviruses (68–72) are not classified as poliovirus/coxsackievirus/echovirus. Box 25.8 lists the diseases caused by newly identified Picornaviruses.

- **Enterovirus 70:** is neurovirulent. It caused epidemic acute haemorrhagic conjunctivitis in India during 1969–74.
- **Enterovirus 71:** causes epidemic aseptic meningitis; encephalitis; hand, foot and mouth disease, etc.

BOX 25.7 Clinical Manifestations of ECHO Viruses Infections

- CNS manifestations:
  - Aseptic meningitis
  - Paralysis
  - Encephalitis
- Exanthems
- Respiratory manifestations:
  - URTI
- Cardiac manifestations:
  - Pleurodynia
  - Myocarditis
  - Pericarditis
- Hepatitis

BOX 25.8 Clinical Manifestations of Enteroviruses Type 68–71 Infection

- CNS manifestations:
  - Aseptic meningitis
  - Paralysis (71)
  - Encephalitis (71)
- Skin- and mucosa-related:
  - Herpangina (severe febrile pharyngitis, vesicles on palate, pharynx, tonsils or tongue) (71)
  - Hand, foot and mouth disease (71)
- Acute haemorrhagic conjunctivitis (70)
- URTI

- **Enterovirus 72:** had now been assigned to hepatoviruses, known as hepatitis A virus.

LABORATORY DIAGNOSIS

Specimens

Stool, throat washing (swab), vesicular fluid in VTM, **CSF, pleural fluid, paired blood samples**

Types of Tests Used for Diagnosis of Newer Picornaviruses

- **Animal inoculation:** Suckling mice
- **Tissue culture:** Typical enteroviral CPE rounding and shrinking of cells with pyknosis of the nuclei
  Susceptible cell lines are: PMK cells, Vero cells and RD cells
- **Serology:** Virus neutralisation assay
- **Neutralising antibody assay:** Four-fold rise in titre of serotype-specific antibodies
- **Molecular diagnosis:** RT-PCR assays to detect viral RNA

PARECHOVIRUS

Parechovirus contains three species; of which, types 1 and 2 were originally classified as echoviruses 22 and 23. Parecho-

viruses are highly divergent from enteroviruses. Protein homology with enteroviruses is around 30% only. The capsid contains three proteins, while enteroviruses have four proteins. Parechoviruses replicate in the respiratory and gastrointestinal tract. They cause diseases similar to other enteroviruses, such as aseptic meningitis, encephalitis, respiratory diseases and neonatal diseases.

RHINOVIRUSES

Rhinoviruses are the common cold viruses. More than 100 species are known. Human rhinoviruses are numbered sequentially. Human rhinovirus 87 is the same serotype as human enterovirus 68. They use either intercellular adhesion molecule-1 (ICAM-1) as receptor (major group) or low-density lipoprotein receptor (LDLR) family as receptors (minor group). Rhinoviruses are similar to enteroviruses but are acid-labile and are more thermostable than enteroviruses. Nucleotide sequence identity among all rhinoviruses is more than 50%. Humans and primates are the only hosts. They can be grown in a number of human cell lines, for example, MRC-5 cell lines at 33°C.

Rhinoviruses rarely cause lower respiratory tract disease. Common manifestation is the common cold. The virus enters via the upper respiratory tract. Replication is limited to the surface epithelium of the nasal mucosa. Nasal secretion increases in quantity and in protein concentration. High titres of virus in nasal secretions can be found 2–4 days after exposure. Illness may persist after that, but virus becomes undetectable. It has been seen that if viral load is high, disease is severe.

The incubation period is from 2 to 4 days and the illness lasts for about 7 days. Non-productive cough may persist for 2–3 weeks. Adults usually have ~2 attacks each year. Usual symptoms include sneezing, nasal obstruction, nasal discharge and sore throat with or without fever, accompanied with headache, mild cough, malaise and a chilly sensation. Antibody develops 7–21 days after infection. Neutralising antibody is detected in serum and nasal secretions of most persons, post-rhinovirus infection. Recovery is not dependent on antibody. Serum antibody persists for years but decreases in titre.

The disease occurs throughout the world. Prevalence rates are lowest in summer. The virus is transmitted through close contact, by respiratory secretions. Transmission to susceptible persons occurs by contact with contaminated hands or objects. Rhinoviruses can survive for hours on contaminated environmental surfaces. Infection rates are highest among infants and children (preschool-aged and school-aged children) and decrease with increasing age. Secondary attack rates in the family vary from 30% to 70%. In a single community, multiple rhinovirus serotypes cause outbreaks of disease in a single season. There are usually a limited number of serotypes causing disease at any given time.

No specific prevention method or treatment is available.

26

Human Immunodeficiency Virus

Amita Jain

LEARNING OBJECTIVES

- Classification, structure, genome, types and subtypes, transmission, epidemiology, pathogenesis, diagnosis, prevention and treatment of Human Immunodeficiency Virus (HIV)

INTRODUCTION

For the first time on 5 June 1981, a new clinical condition 'Acquired Immunodeficiency Syndrome' was reported by Centre for Disease Control (CDC), Atlanta. Condition was reported in homosexual men and intravenous drug abusers in New York and California. These individuals developed opportunistic infections (OIs), *Pneumocystis (carinii) jiroveci* pneumonia and Kaposi's sarcoma. In 1983, Luc Montagnier from Pasteur Institute, France and in 1984 Robert Gallo from National Institute of Health, USA identified the causative agent and named them lymphadenopathy associated virus (LAV) and T-cell lymphotropic virus III, respectively. Later in 1986, the International Committee on Taxonomy of Viruses named it as Human Immunodeficiency Virus (HIV). Serological tests for diagnosis became available in 1985.

CLASSIFICATION

The family Retroviridae is divided into two subfamilies: *Orthoretrovirinae and Spumaretrovirinae*. HIV is a *Lentivirus*. Infection with HIV takes several years to manifest (word 'lenti' means 'slow'), hence the name. Genus *Lentivirus* belongs to subfamily *Orthoretrovirinae* and includes species **Human immunodeficiency virus**. Other species included in genus *Lentivirus* are *Simian* and *Feline* immunodeficiency viruses. All of them cause immunodeficiency.

STRUCTURE

HIV is an enveloped RNA virus (Fig. 26.1). The bilayered envelope is derived from the host cell membrane and contains glycoprotein 160 (gp160). Gp160 has two subunits: gp120 and gp41. Outer gp120 binds with co-receptors on the surface of human CD4 T cells. Gp41 is membrane bound, and helps in exposing co-receptors on human cells, at the time of binding of viral particle to cell surface. Envelope surrounds the matrix, made up of matrix protein (p18). Capsid (core) is made up of protein p24. Capsid contains two strands of single-stranded positive sense ribonucleic acid (RNA) and enzymes required for viral replication (reverse transcriptase [RT], integrase and protease).

GENOME

It comprises two segments of RNA, approximately 9.1 kB in size, and contains nine genes and long terminal repeats (LTR; Box 26.1). Viral proteins coded by some of the genes (*env*, *gag* and *pol*) and antibodies against them make the basis of the most of diagnostic tests. The details of genes and the proteins they code are shown in Box 26.1. Fig. 26.2 shows the location of various genes on the genome of HIV.

HIV TYPES, SUBTYPES AND SUB-SUBTYPES

Based on antigenic and molecular diversity, HIV is divided into two types: HIV-1 and HIV-2. HIV is a rapidly mutating and rapidly multiplying virus. Strains isolated from different geographic area may vary from each other. The reverse transcriptase of HIV lacks proofreading property, hence mutations are common.

HIV-1 is divided into types, groups, subtypes and sub-subtypes, and its classification is seen in Flowchart 26.1.

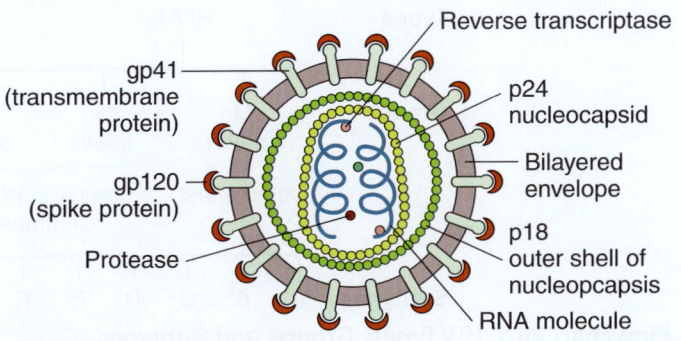

Fig. 26.1 Structure of HIV.

BOX 26.1   Genes and Antigens of HIV-1

Structural genes: *gag*, *pol* and *env*

- *gag* gene: Codes for core and shell antigens of virus.
  - *Core antigens:*
    - p24 (p27); major protein
    - p15
    - p55 (p56) precursor protein, which splits in p15, p18 and p24
  - *Shell antigen*
    - p18 (p16); nucleocapsid protein
- *env* gene: Codes for envelop antigens
  - *Envelop antigens*
    - gp120 (gp106): Major glycoprotein, spike antigen
    - gp41 (gp36): Transmembrane protein
- *pol* gene: Codes for viral enzymes
  - p31 (p31/34), p51 (p53), p66 (p68): Enzymes required for viral replication, e.g. reverse transcriptase, protease and endonuclease. Protein is expressed as precursor protein which gets cleaved.

Non-structural and regulatory genes:

- *tat* (trans-activating gene): Upregulates expression of viral genes
- *nef* (negative factor gene): Downregulates viral replication
- *rev* (regulator of virus gene): Upregulates expression of structural genes
- *vif* (viral infectivity factor): Accessory genes
- *vpu* (viral protein unique): Accessory genes
- *vpr* (viral protein R): Accessory genes (accessory genes are important in virus replication, regulation and host–cell interactions)
- LTR (long terminal repeats): Switches to control the production of new virus particles. Triggered by either viral or host proteins.

Note: HIV-2 proteins that are different are mentioned in parentheses.

HIV-1 is divided into three main groups: M, N and O. M group is the major group which is further divided into subtypes A–K. Circulating recombinant forms (CRFs) are the result of recombination, which may occur if more than one subtype infects the same cell. These recombinant forms may circulate in nature. These subtypes may be further divided into sub-subtypes, for example, subtypes A and F are classified into sub-subtypes A1, A2 and F1, F2, respectively. HIV-1 subtypes show geographical preferences. HIV-1 subtype C is the most predominant subtype in India.

HIV-2 is less diverse, but subtypes A–H are proposed.

VIRAL REPLICATION

HIV binds to receptors on immunocompetent cells and multiplies intracellularly in the following steps:

- **Receptor on host cell:** CD4 proteins on the surface of T lymphocytes, other cells, for example, dendritic cells, macrophages and microglial cells.
- **Second or co-receptors:** Receptors for chemokines (CCR5 or CXCR4).
- **Entry into host cells:** Gp120 of HIV binds to the CD4 receptor and co-receptor CXCR4 and CCR5.
- **Binding and fusion:** Attachment of gp120 to a CD4 molecule leads to conformational changes. Co-receptor (CCR5/CXCR4) gets exposed and virus envelope fuses with host cell membranes. Gp41 helps in fusion of viral envelope to host cell membrane. Viral RNA is released in the host cell cytoplasm.
- **Reverse transcription:** Viral reverse transcriptase enzyme transcribes the single-stranded viral RNA into a double-stranded c-DNA. c-DNA moves to the nucleus.

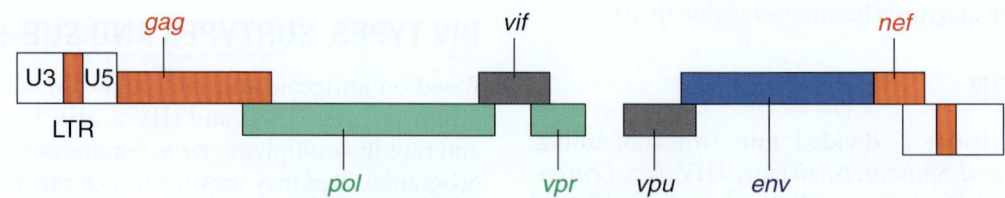

Fig. 26.2 Structural Location of Genes on HIV Genome.

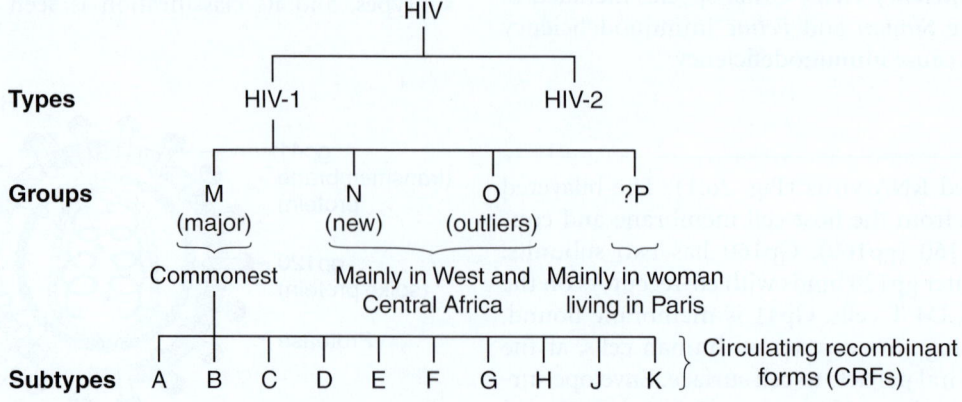

Flowchart 26.1 HIV Types, Groups and Subtypes.

- **Integration:** c-DNA gets integrated (pro-virus) with host genome with the help of enzyme integrase. The provirus may remain latent for several years.
- **Transcription:** In an activated host cell, host's RNA polymerase makes new copies of viral mRNA, which moves from host cell nucleus to protoplasm.
- **Protein synthesis and assembly:** In the host cell cytoplasm, viral mRNAs code for viral proteins. Viral protease slices polyproteins to individual proteins. These proteins get assembled with viral RNA to form new viral particles.
- **Budding:** New viral particles bud out from host cell, and while budding they cover themselves with host cell membrane, which forms envelope.

PATHOGENESIS

HIV is excreted in blood and body fluids (semen, vaginal secretions, breast milk, cerebrospinal fluid, amniotic fluid and synovial fluid) of an infected individual. When these fluids are injected in a new host, HIV infection occurs. Both intracellular and extracellular viral particles can be infectious. Gp120 present on viral envelope binds to the CD4 receptors and co-receptor X4 (CXCR4) and R5 (CCR5) present on host cells. HIV can infect helper T cells, monocytes, microglial cells, oligodendrocytes, and astrocytes of the brain, retinal cells in the human eye and other cells which are positive for receptors. Doubling time of HIV replication is 0.3 days. Replication is exponential in first 3 weeks after infection. HIV resides within immune cells. Immunity to virus develops during this time. Anti-HIV IgM antibodies appear 3 weeks after infection. However, in the initial phase of infection when the viral load is low and antibody titer is also low, both remain undetectable. This period is known as 'Window/Acute phase'. This period may be as short as 3 weeks or as long as 3 months.

After entry in the circulating lymphocytes and lymph nodes, the T lymphocytes get infected. Viral replication starts within 72 h of entry into the cells. HIV replicates better in activated lymphocytes, which get activated post infection. Infected cells form syntitium, and thus spread the infection to new cells. The spread of virus to newer cells and lymph nodes causes viraemia and shedding of the virus in genital secretions. Infected cells get destroyed ultimately, causing lowering of host's immune response.

HIV-2 replicates slower than HIV-1, hence immunosuppression in HIV-2 infection takes longer to manifest.

Immunopathogenesis

Decreased immunity of the host forms the basis of different symptoms and disease. Chief mechanisms which cause immunodeficiency are listed as follows:

1. HIV-specific cytotoxic T lymphocytes (CTL), and complement fixing and neutralising HIV-specific antibodies help to slow the HIV replication rate in acute stage.
2. Later CD4-positive cells and memory T cells get depleted rapidly due to cytotoxicity (apoptosis and immune-mediated killing), causing reversal of T4:T8 ratio.
3. Infected CD4-positive cells do not release required amount of cytokines, for example, interleukin-2, gamma interferon etc.
4. Due to lack of cytokines and antigen-specific antibodies, functions of monocyte and macrophages like chemotaxis and phagocytosis get decreased.
5. Cell-mediated immunity (CMI) is also suppressed due to lack of T4 and cytokines.
6. Hypergammaglobulinaemia occurs due to polyclonal activation of plasma cells.
7. Host immune response to any foreign antigen is lowered due to lack of CMI and humoral immune response.

NATURAL PROGRESSION

Clinically, HIV infection is staged in four stages:
- Group I—acute phase
- Group II—chronic asymptomatic phase
- Group III—generalised persistent lymphadenopathy
- Group IV—AIDS

Group I: Acute Infection Phase/Primary Infection or Window Period

It usually lasts for 3–6 weeks. Adaptive immune response develops during this period. Following events are seen in this phase in sequential manner:
- Virus dissemination to lymphoid organs starts soon after infection.
- Destruction of CD4 cells follows.
- Viraemia reaches its peak.
- CD4 T-cell populations are depleted (in the second to third week of infection).
- Hyperactivation of immune system followed by decrease in viraemia and viral titre in genital secretion (viraemia is lowest by 10th week of infection).
- Appearance of anti-HIV cytotoxic CD8 T lymphocytes, more on mucosal surface.
- Clinical symptoms are self-limiting.

Group II: Chronic Asymptomatic Phase

HIV causes latent infection. This phase may last up to 10 years. Once the immune response against HIV mounts, viraemia decreases. In spite of the fact that virus multiplication is restricted during this phase, virus is never eliminated from the body. Anti-HIV antibodies may provide selection pressure and help in emergence of new mutants. The genetic constitution, age, immune status of host, occurrence of OI, plasma virus load and so on may determine the time taken for progression of disease. After about 6–12 months of infection, latency phase is attained. Plasma viral load is set at a level, and virus is maintained in lymph nodes. The CD4 cell destruction occurs at a constant rate during this period. Once immune system is exhausted, the viral load in plasma increases and disease progresses to AIDS.

BOX 26.2   Manifestations in AIDS and Other Associated Illnesses

Respiratory symptoms: Cough, dyspnoea and fever present. *Mycobacterium tuberculosis*, viral and fungal causes common.

Gastrointestinal symptoms: Oral thrush, stomatitis, gingivitis, esophageal candidiasis, hairy leukoplakia, cryptosporidiosis and other parasitic, bacterial, fungal and viral intestinal infections causing diarrhoea.

Central nervous system symptoms: Toxoplasmosis, cryptococcosis, viral bacterial and fungal infection of brain causing meningoencephalitis.

Malignancies: Kaposi's sarcoma, lymphoma.

Dementia: Encephalopathy causing progressive dementia.

Skin lesions: Herpetic lesions, candidiasis, xeroderma, molluscum contagiosum, impetigo etc.

Group III: Generalised Persistent Lymphadenopathy

There is generalised lymphadenopathy present without any ascribable cause for the same.

Group IV: Acquired Immune Deficiency Syndrome (AIDS)

CD4 cell count gets less than 200/cmm and/or AIDS defining illnesses start appearing. It may last up to 3 years and without treatment patient usually dies. It is characterised by high plasma viral load, immune suppression, frequent OIs, malignancies, both the quantitative and qualitative dysfunction of the T lymphocytes and other immune system cells, for example dendritic cells (DCs), natural killer (NK) cells and macrophages. Common manifestations of AIDS are shown in Box 26.2.

Beyond the above four groups of HIV illness, CDC, USA, has also divided AIDS in some subgroups, which are listed in Box 26.3.

Paediatric AIDS

Babies born to HIV-positive mothers can get infected pre-, peri- and postnatal. Chances of perinatal transmission are highest. In comparison to adults, babies develop early and more serious secondary infections and malignancies. Most of them do not survive longer than a year.

BOX 26.3   Subgroups of HIV Disease

- **Subgroup A:** AIDS related complex (ARC)
- **Subgroup B:** Neurological diseases
- **Subgroup C:** Secondary infectious diseases
  - C1: Specified infections listed, e.g., *P. carinii pneumonia*, cryptosporidiosis, herpes etc.
  - C2: Other specified secondary diseases, e.g., hairy leukoplakia, salmonellosis.
- **Subgroup D:** Secondary cancers.
- **Subgroup E:** Others.

Patterns of Disease Progression

The disease progression from HIV asymptomatic infection to AIDS varies from individual to individual. Following patterns are seen.

Typical Progressors

In total, 80%–90% persons are typical progressors, and show the typical progression from stage I to IV of HIV disease. These individuals may survive up to 10 years without intervention.

Rapid Progressors

A total of 5%–10% persons living with HIV/AIDS (PLHA) develop AIDS symptoms within 1–3 years after primary HIV-1 infection.

Long-term Non-progressors (LTNP)

Rarely (<5%) some cases do not progress to AIDS even without treatment. The viral load in LTNP is set at around 5000 copies/mL. Their immune responses seem to keep the virus in check. They have some mutations in co-receptors also. A subset of LTNP is natural controllers, known as **elite controllers**. Viral load in elite controllers remains less than 50 copies/mL of blood.

Host, viral and environmental factors which influence the progression of HIV infection from the asymptomatic stage to AIDS are listed in Box 26.4.

OPPORTUNISTIC INFECTIONS

Tuberculosis is the most common OI reported in India. Other commonly reported OIs are as follows:

- Candidiasis
- Cryptosporidiosis
- Toxoplasmosis
- *Pneumocystis jirovecii* pneumonia

BOX 26.4   Factors Influencing the Progression of HIV Infection

Viral factors
- Viral fitness
- Co-receptor usage
- Generation of escape mutants
- Latency

Host factors
- Genetic factors like HLA and chemokine receptor genes polymorphism; certain types are associated with rapid disease progression
- HLA types A24, B35, B8 and C4 are known to be associated with rapid disease progression
- B27 and B57 associated with slow disease progression
- Mutations in the co-receptor genes CCR5, CCR2, SDF and CX3CR1

Environmental factors
- Nutrition, e.g., vitamin A deficiency
- Co-infections: Tuberculosis and hepatitis B (HBV) accelerate the disease progression

- CMV infections
- Herpes infections
- Salmonella infection
- Campylobacter infections
- Nocardia and actinomycosis infections
- Isosporiasis

TRANSMISSION

Transmission of HIV occurs through the following routes:
- **Sexual route:** Penetrative sexual acts (vaginal, anal or oral), both heterosexual and homosexual.
- **Parenteral route:** By contaminated blood transfusion, IV drug use, organ transplantation, by sharing of needles and syringes.
- **Vertical transmission:** From mother to child during pregnancy, child birth or by breast feeding.

High-risk Group

High-risk group is the population most vulnerable to the HIV infection:
- Sex workers, both male and female
- Men having sex with men
- Transgenders
- Drug abusers

EPIDEMIOLOGY

As per a World Health Organization report, 35 million people were infected with HIV worldwide, in 2013. First, HIV infection in India was reported in 1987–88 in Chennai and the first AIDS case was reported in Mumbai. Since then the HIV epidemic spread to all high-risk groups and to general population as well.

HIV-1 is closely related to Simian Immunodeficiency Virus (SIV) and is the commonest type seen worldwide. HIV-2 is geographically restricted, and relatively uncommon.

DIAGNOSIS

Demonstrating the Presence of Virus or Viral Products in the Host

Advocated for diagnosis either in acute stage of illness, or if there are some doubts on serological testing.
- **Nucleic acid amplification test (NAAT)** is the most sensitive test and can also be used to detect HIV nucleic acids copy numbers. This test is also required to monitor the response to treatment. This is best test for diagnosis of HIV/AIDS in neonates.
- **P24 antigen detection** is advocated either in acute stage or in late stage.
- **CD4 T-cell enumeration** is also used for monitoring response to therapy.

Detecting Host Response to the Virus

These tests demonstrate HIV-specific antibodies.

Serological tests. Commonly used platforms are as follows:
- Enzyme-linked immunosorbent assays (ELISA)
- Rapid tests
- Western blots (WBs)
  Platforms less often used are as follows:
- Chemiluminescence immunoassays (CIA)
- Immunofluorescence assays
- Line immunoassays

Serological tests, either rapid platforms or ELISA, are most commonly used. National AIDS Control Programme (NACO) of India recommends testing in algorithmic fashion. Depending upon the use algorithm varies (Flowchart 26.2).

ANTIRETROVIRAL TREATMENT

Highly active anti-retroviral treatment (HAART) recommended for these patients is helpful in
- reducing viral replication;
- reducing frequency of OI;
- prolonging latent period;
- prolonging longevity and
- improving quality of life.

Antiretroviral Drugs

Three classes of drugs are used in combination as advocated by NACO from time to time. These classes of drugs are as follows:
- **Nucleoside/Nucleotide reverse transcriptase inhibitor** (NRTI) causes DNA chain termination.
- **Non-nucleoside reverse transcriptase inhibitor** (NNRTI) inhibits the HIV reverse transcriptase enzyme.
- **Protease inhibitor** (PI) binds to protease and prevents viral maturation.

Newer Classes of Anti-HIV Drugs

- **Fusion inhibitors:** Stop the integration of virus DNA with host DNA.

PREVENTION AND CONTROL

Prevention and control is a priority for the national programme. NACO emphasises on risk reduction.
- **Among general population:** Through information, education and communication (IEC).
- **Among high-risk groups:** Through targeted intervention (TI) programmes.

Components of TI Programmes

- Promotion of condom use
- Peer educators/outreach workers working with IEC
- Treatment of STIs
- Pre-exposure prophylaxis
- Male circumcision
- Vaginal microbicides

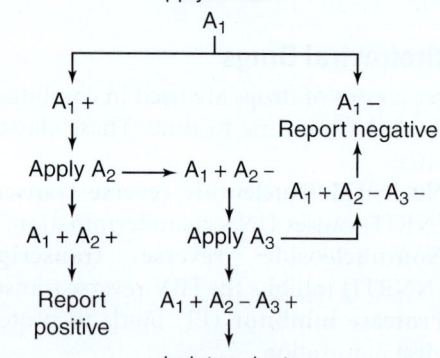

Strategy 1
(for transfusion/transplant safety)
Apply single test only (A_1)

A_1

$A_1 +$ → Consider positive

$A_1 -$ → Consider negative

Strategy 2A
(for surveillance only)
Apply 2 tests

A_1

$A_1 +$ → Apply A_2

$A_1 -$ → Consider negative

$A_1 + A_2 -$ → $A_1 -$

$A_1 + A_2 +$ → Report positive

Strategy 2B
(diagnosis of a case suggestive of AIDS/HIV)
Apply 3 tests

A_1

$A_1 +$ → Apply A_2

$A_1 -$ → Report negative

$A_1 + A_2 +$ → Report positive

$A_1 + A_2 -$ → Apply A_3

$A_1 + A_2 - A_3 -$ → $A_1 -$

$A_1 + A_2 - A_3 +$ → Indeterminate
(follow-up testing required after 4–6 weeks)

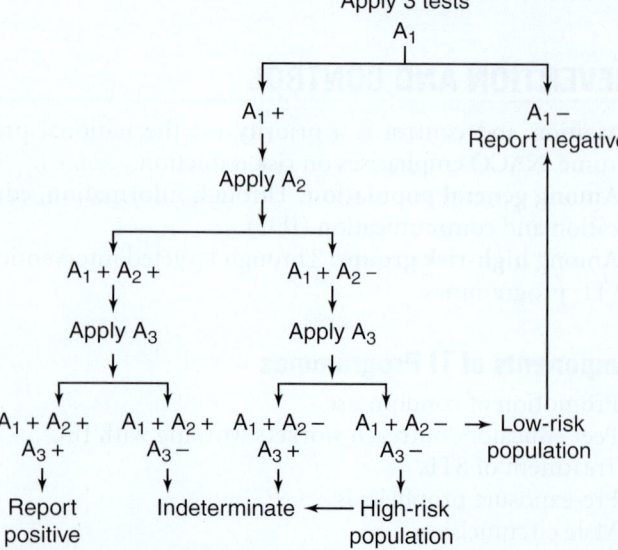

Strategy 3
(for screening of asymptomatic individuals)
Apply 3 tests

A_1

$A_1 +$ → Apply A_2

$A_1 -$ → Report negative

$A_1 + A_2 +$ → Apply A_3

$A_1 + A_2 -$ → Apply A_3

$A_1 + A_2 + A_3 +$ → Report positive

$A_1 + A_2 + A_3 -$ → Indeterminate ← High-risk population

$A_1 + A_2 - A_3 +$ → Indeterminate ← High-risk population

$A_1 + A_2 - A_3 -$ → Low-risk population

Flowchart 26.2 Strategies for Confirmation of Test Results, as Recommended by NACO.

Occupational Exposure and Postexposure Prophylaxis (PEP)

Persons working in health care setting may get exposed to biological material which may be a source of HIV infection. Risk of exposure varies with the type of exposure. Hollow needles, device visibly contaminated with patient's blood, deep injury, large volume of blood involved in the exposure and known high viral load in blood at the time of exposure are associated with high risk.

If timely (<2 h and up to 72 h) PEP is given, risk is reduced. It is important to note that at present there are no vaccines available.

Practices That Reduce Risk of Exposure Are as Follows:

- Use of personal protective equipment (PPE), that is gloves, gowns/aprons, masks and goggles while handling all potentially infectious material.
- Handling sharp objects (needles, lancets, scalpels, etc.) with care.
- Use disposable needles only.
- Never recap needles.
- Never break/bend needles by hand.
- Do not leave used needles/sharps unattended.
- Dispose sharps in a puncture-resistant container containing 1% sodium hypochlorite solution.
- Thoroughly wash hands with water and soap after removing gloves.
- Thoroughly wash hands with water and soap immediately after any contamination of skin surfaces.
- Decontaminate work surfaces with 0.1% sodium hypochlorite solution.
- No mouth pipetting, eating, drinking or smoking in the work area.
- Immunisation against HBV.

DOS AND DON'TS IN CASE OF EXPOSURE

Don'ts

- Do not do the following:
  - Panic
  - Place the pricked finger into the mouth reflexively
  - Squeeze blood from wound
  - Use bleach, alcohol, iodine, antiseptic, detergent etc. on the site of prick

Dos

- Stay calm
- Remove gloves
- Wash exposed site thoroughly with running water and mild soap
- If eyes or mouth got splashed, irrigate with water
- Consult PEP officer, he/she will take care of it

27

Rabies Virus

Amita Jain

Amita Jain

LEARNING OBJECTIVES

- Structure and properties of rhabdovirus
- Pathogenesis, clinical features, epidemiology, laboratory diagnosis and prophylaxis of rabies

INTRODUCTION

Rabies virus is a medically important member of the family *Rhabdoviridae*. Rest of the members are widely distributed in nature, but are of no medical importance. Rabies is a zoonotic disease, widely rampant in wild and unvaccinated animals. It is transmitted through the bite of an infected animal and causes infection of the central nervous system (CNS). The disease is almost always fatal. Postexposure prophylaxis (PEP) is effective and available.

CLASSIFICATION

Family: Rhabdoviridae
    Genus *Lyssavirus*: Rabies viruses
    Genus *Vesiculovirus*: Vesicular stomatitis viruses

STRUCTURE

Rabies virus is an enveloped bullet-shaped virus measuring around 75 × 180 nm. Envelope is a bilayered membrane derived from host cell. Outer envelope has peplomers (also known as spikes), which are composed of trimers of the viral glycoprotein (G). Inner layer is attached with matrix protein, which encompasses nucleocapsid. Nucleocapsid is helical, containing single-stranded, negative-sense RNA and RNA-dependent RNA polymerase (Fig. 27.1).

ANTIGENIC PROPERTIES

Single serotype exists in nature; however, strains from different geographic area may vary to some extent. Minor variations occur in genome which translates in variation in major antigens, that is glycoprotein and nucleoprotein. Glycoproteins are surface proteins, hence can be useful in immunodiagnostics and contribute to pathogenesis.

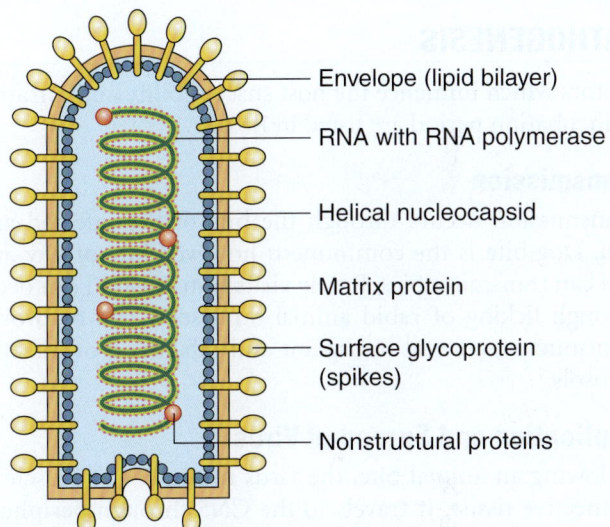

Fig. 27.1 Structure of Rabies Virus.

Labels: Envelope (lipid bilayer); RNA with RNA polymerase; Helical nucleocapsid; Matrix protein; Surface glycoprotein (spikes); Nonstructural proteins

SUSCEPTIBILITY TO PHYSICAL AND CHEMICAL AGENTS

Rabies virus survives at 4°C for weeks, but gets inactivated in dry ice as CO_2 is toxic for this virus. Ultraviolet radiation, sunlight, heat (1 h at 50°C), lipid solvents, trypsin, detergents and acidic and alkaline solutions are toxic for rabies virus.

SUSCEPTIBILITY TO ANIMALS

Rabies virus can infect all warm-blooded animals, including humans. Foxes, coyotes and wolves have high susceptibility, while some animals, such as dogs, sheep and cats have moderate susceptibility. The virus is present in nervous system, saliva, urine, lymph, milk and blood of infected animals. Highest concentration is found in saliva because virus adapts best in salivary glands. Some animals like dogs show the symptoms of the disease, while some other virus carriers like vampire bats do not show any symptoms.

STREET VIRUS

Fresh strains are isolated in the laboratory from the sample. These strains multiply slowly, infect both neural and extra-neural tissues (salivary gland) and form inclusion bodies. These strains are pathogenic.

FIXED VIRUS

Serial brain-to-brain passage of street virus in rabbits yields a mutant virus known as fixed virus. Virus multiplies rapidly and loses its potential to grow in extra-neural tissue and form inclusion bodies. These strains are used for vaccine production and are non-pathogenic.

PATHOGENESIS

Factors which influence the host susceptibility and duration of incubation period are listed in Box 27.1.

Transmission

Transmission occurs through the bite of the infected animal. Dog bite is the commonest; however, bite of any animal can transmit rabies. Rarely virus transmission can occur through licking of rabid animal on a scratched or broken skin/mucosa, corneal transplant and inhalation of infective aerosols.

Replication and Spread of Virus

Following an animal bite, the virus multiplies in muscle or connective tissue. It travels to the CNS through peripheral nerves present at neuromuscular junctions. Acetylcholine receptors at neuromuscular junction serve as receptor for the virus. It multiplies in the CNS. Neuroinvasiveness of rabies virus is because of G-glycoprotein.

Spread of rabies virus occurs via peripheral nerves (**Flowchart 27.1**). It can spread to any tissue, but the highest concentration is seen in salivary glands. Rabies virus has not been isolated from the blood of infected persons.

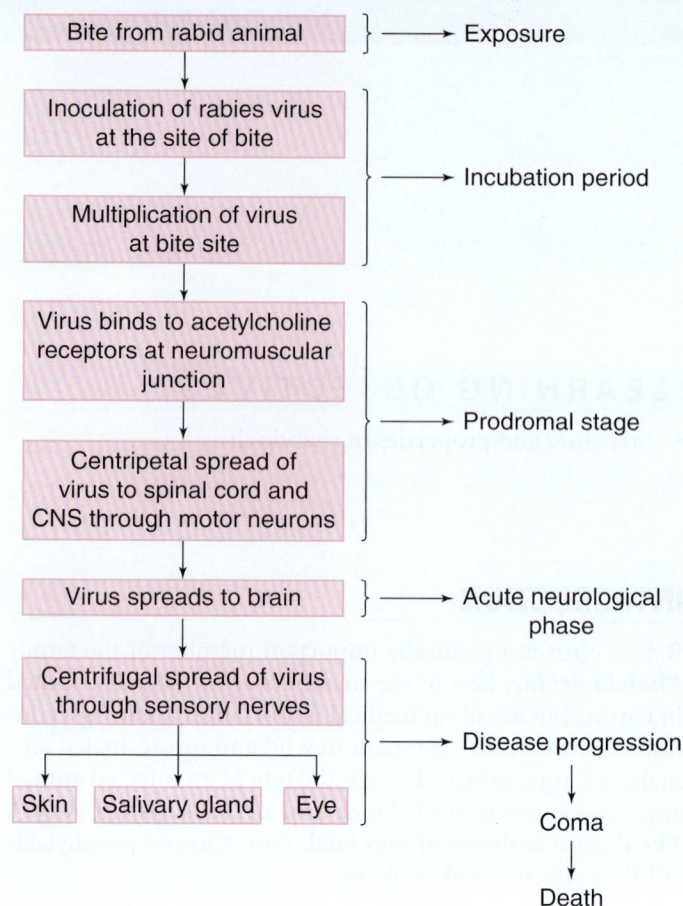

Flowchart 27.1 Pathogenesis of Rabies.

BOX 27.1	Factors Influencing Susceptibility to Infection and Incubation Period of Rabies

- Age of the patient
- Immune status of the patient
- Genetic constitution of the patient
- Virulence of viral strain
- Amount of inoculum
- Severity of injury
- Distance of bite site from CNS (bites on face or head have highest attack rate; bites on leg have lowest attack rate)

Rabies virus attaches to cells via its glycoprotein spikes to acetylcholine receptor. After entering the cell, the single-stranded viral RNA codes for the five virion proteins: nucleocapsid (N), polymerase proteins (L, P), matrix (M) and glycoprotein (G). The RNA molecule gets replicated and forms a particle with newly synthesised proteins. Envelope is acquired by budding through the plasma membrane. The viral matrix protein makes a layer on the inner side of the envelope; whereas, the viral glycoprotein presents as spikes on the outer side of the envelope.

Incubation Period

Incubation period is variable. It can be as short as 7 days to as long as many years; on an average, generally it is about 4 weeks.

Negri Bodies

Viral nucleocapsids get collected in an infected nerve cell's cytoplasm, which stains as eosinophilic inclusion bodies known as Negri body. These are seen only in some of the infected cells, but if detected are diagnostic of rabies. They are demarcated, spherical and 2–10 μm in diameter. Seller's technique (using basic fuchsin and methylene blue in ethanol) is used for staining.

CLINICAL FINDINGS

Ultimate presentation is acute, fulminant and fatal encephalitis. Three phases of disease is seen in a human case as follows:

Prodromal phase: Short phase, lasting for 2–10 days. Patient complains of sensation around bite wound, some non-specific symptoms, that is flu-like symptoms and at time neurological signs and come.

Acute neurologic phase: Lasts about 2–7 days. The disease course is slower, but recovery and survival are not seen. Common symptoms are listed in Box 27.2.

Convulsive seizures or coma and death: Mostly due to respiratory paralysis.

In dogs: Disease can manifest as furious or dumb rabies. Both forms are equally infectious and fatal.

EPIDEMIOLOGY

Rabies is a preventable zoonotic disease and affects all warm-blooded animals. Rabies is prevalent throughout the world except in islands. Except Australia and Antarctica where it is not found, it is endemic in many countries, although many countries are free from rabies. World Health Organization (WHO) opines that a country that has no indigenously acquired case of human or animal rabies within 2 years period is rabies-free. In Asia, many countries suffer from fatal rabies. In India, rabies is not a notifiable disease, in spite of the fact that an estimated 36% of all deaths from rabies occur in India. As per 2015 World Health Organization's report, India reports about 18,000–20,000 cases of rabies per year. Rabies incidence in India is not declining since a decade. Insufficient dog vaccination, uncontrolled stray animals and poor postexposure prophylaxis (PEP) have mainly contributed to high incidence of rabies in India. Children between 5 and 15 years of age, from poor socioeconomic strata, are mainly affected. People have poor knowledge about preventive measures and vaccination.

LABORATORY DIAGNOSIS

Specimen: A biopsy specimen is usually taken from the skin of the neck at the hairline. Postmortem, brain biopsy or corneal biopsy is done. Saliva is a good sample for demonstrating viral nucleic acid.

Demonstration of antigens by immunofluorescence or immunoperoxidase staining using antirabies monoclonal antibodies is tried most often. For detection of Negri bodies in the brain or the spinal cord tissue, postmortem is diagnostic. Absence of Negri bodies does not exclude rabies.

Demonstration of nucleic acids by reverse transcription-polymerase chain reaction is developed to be used as diagnostic test.

Viral isolation is done by intracerebral inoculation in suckling mice, only in specialised laboratories.

Demonstration of antibodies to rabies can be detected by immunofluorescence or neutralisation tests. Antibodies are found in serum of vaccinated individuals as well; however, presence of antibodies in CSF is seen only in infected cases.

Following animal bite, all animals should be held for observation for at least 10 days. If they show any signs of rabies, they should be killed and lab confirmation should be done. If they appear normal after 10 days, they may be considered non-rabid.

PROPHYLAXIS

As rabies is almost always fatal and no specific treatment is available, prophylaxis is one of the main steps of managing an animal bite.

Domestic Animal Bite

- Healthy vaccinated animal and stays healthy for 10 days: No PEP needed
- Suspected rabid/rabid: Vaccinate immediately
- Unknown status (escaped): Vaccinate immediately
- Wild animal bite: It is considered rabid; vaccinate immediately.

Postexposure Prophylaxis

Following three steps are mandatory:

1. **Management of animal bite wound:** All bites and wounds should immediately be thoroughly washed with soap and water. This is a very essential step and even if patient reports late, wound toilet should be performed.
2. **Passive immunisation with rabies immunoglobulin (RIG):** Administration of specific antibody at the site of inoculation will inhibit virus multiplication. It neutralises inoculated virus to some extent and host gets some time to potentiate immune response against vaccine. Both equine and human antibodies are available. Dose of equine antibodies is 40 µg/kg body weight and of human antibodies is 20 µg/kg body weight. Most of the volume should be injected around wound and if any amount is left, it can be injected intramuscularly in deltoid muscle.

   First two steps are very helpful in immunocompromised individuals.
3. **Active immunisation** with antirabies vaccine.

Types of Available Vaccines

In the past, neural vaccines were used, which were prepared from the brain of infected individuals. However, due to serious neurological complications these vaccines are not available and are no longer used.

There are two vaccines available for human use in India and both are equally effective and safe.

1. **Cell culture vaccines (CCVs):** Fixed rabies virus strain was adapted to grow in the diploid human fibroblast cell line, for example WI38 or MRC-5. Cultured virus is inactivated and concentrated before use. No serious anaphylactic or encephalitic reactions have been reported.

2. **Purified duck embryo vaccine (PDEV):** The CCV virus is grown in embryonated duck eggs. It produces more viruses and is cheaper to manufacture.

A recombinant viral vaccine consisting of vaccinia virus with rabies surface glycoprotein gene is successfully used in animals.

Postexposure Prophylaxis Schedule

Most frequently used schedule is as follows:

Total doses: 5–6

Days of injection: 0, 3, 7, 14 and 30 days post bite; one booster can be given later

Duration of protection: 5 years (for any fresh incident within 5 years, give a booster and after 5 years, give full course)

Route of injection: Intramuscular or subcutaneous on deltoid region

(Note: New syringe and needle to be used, not the same which was used for injecting RIG. Do not inject on the same deltoid on which RIG was injected.)

Dose: At least 2.5 IU/dose (volume may vary from 0.5 to 1.0 mL)

Intradermal injection: Intradermal injection is an alternate plan. Only 0.1 mL of vaccine is inoculated on both the deltoid muscles on day 0, 3, 7 and 28. Day 14 is missed. This is equally effective and economical; however, injection should be given intradermally. If this gets in subcutaneous region, the doses are too low to be effective.

Pre-Exposure Prophylaxis

It is recommended only for persons at risk, such as animal handlers and veterinary doctors. Antibody titres of vaccinated individuals should be monitored periodically and boosters should be given when required. Three doses either on 0, 7 and 21 days or on 0, 28 and 56 days are given. Each dose should be at least 2.5 IU.

TREATMENT

Once rabies develops, there is no treatment. Rarely, complete recovery from rabies with supportive management is demonstrated.

28

Miscellaneous DNA Viruses

Amita Jain

LEARNING OBJECTIVES

- Structure, classification, replication, pathogenesis, clinical manifestations, laboratory diagnosis and epidemiology of adenovirus:
  - Adenoviruses
- Human parvovirus B19
- Polyomaviruses

ADENOVIRUSES

INTRODUCTION

Adenovirus was first isolated in 1953 from cell lines derived from adenoid tissue, hence the name. Adenoviruses are DNA viruses, which have been widely used as models for cancer induction and molecular studies in human. Till date more than 57 serotypes are identified, of which at least one third cause human infections. Mostly infections are subclinical.

STRUCTURE

These are icosahedral, non-enveloped, 70–90 nm in diameter with 252 capsomeres in the form of three major types: fibre, penton and hexon-based proteins. DNA is linear and double-stranded. Capsomeres form projection fibres which projects from each of 12 vertical surfaces (Fig. 28.1). These fibres help in the attachment of virus to host cells.

CLASSIFICATION

Adenoviruses are classified into the following two genera:
- *Aviadenovirus:* Infects birds
- *Mastadenovirus:* Infects mammals. All human strains are found in *Mastadenovirus.*

On the basis of their physical, chemical and biologic properties, and tumorigenic potential human adenoviruses are divided into seven groups (A–F; Table 28.1).

REPLICATION OF VIRUS

Adenoviruses replicate only in cells of epithelial origin.
- They attach to host cells with viral fibre and CAR (coxsackie-adenovirus receptor, a member of the immunoglobulin gene superfamily structures) interaction, following internalisation of viral particles.
- Uncoating of viral DNA happens in the host cell cytoplasm.
- DNA is released at the nuclear membrane.

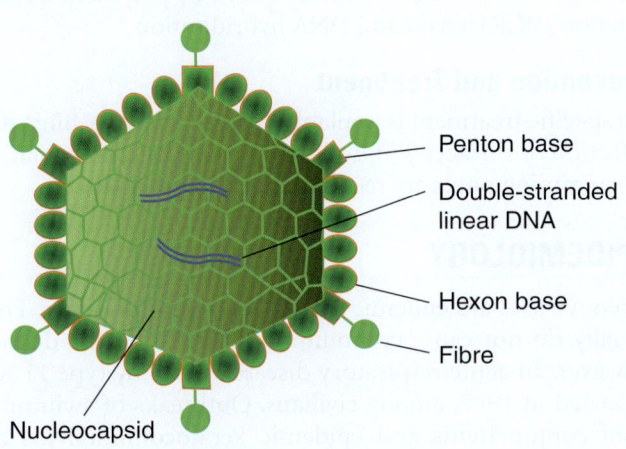

Fig. 28.1 Structure of Adenovirus.

Labels: Penton base; Double-stranded linear DNA; Hexon base; Fibre; Nucleocapsid

TABLE 28.1	Classification of Adenovirus	
Group	Serotypes	Tumourigenic Potential
A	12, 18, 31	High
B	3, 7, 11, 14, 16, 21, 34, 35, 50, 55	Moderate
C	1, 2, 5, 6, 57	Low or none
D	8–10, 13, 15, 17, 19, 20, 22–30, 32, 33, 36–39, 42–49, 51, 53, 54, 57	Low or none
E	4	None
F	40, 41	None
G	52	Not known

- The replicative cycle is divided into two events: early and late.
  - Early events: Occur before the onset of viral DNA synthesis. Host cell enter the S phase and become ready for viral replication. More than 20 early proteins, mostly non-structural are synthesised in early events.
  - Late events: Viral DNA replication occurs in host cell nucleus. Late genes (L) code for more than 18 segments of m RNA, which are transported to cytoplasm for protein synthesis.
- Viral assembly: Occurs in the nucleus, followed by the release of viral particles. Infected cell shows cytopathogenic changes.

PATHOGENESIS

1. Infect epithelial cells of the respiratory tract, eye, gastrointestinal tract, urinary bladder and liver.
2. Inhibits host immune response to escape the destruction of infected cells.
3. Causes cytopathic changes in human cell cultures.
4. Cytopathic changes: Rounding, enlargement and aggregation of affected cells. Rounded intranuclear inclusions may be seen even in biopsy specimen at times.
5. Group C viruses may persist as latent infections in lymphoid tissue.

DISEASES CAUSED

The association of disease types with virus types is shown in Table 28.2. Some important facts about the disease causing adenoviruses are as follows:
- Adenoviruses 1–7 are most common prevalent types worldwide.
- Only about one-third of serotypes are associated with human diseases.
- A single serotype may cause more than one types of disease.
- One type of clinical manifestation may be caused by more than one serotype.

Respiratory Infections

Respiratory infections are common in infants and children. Adenoviruses types 3, 7 and 21 are common cause. Severe

TABLE 28.2 Common Viral Types and Associated Disease	
Diseases Caused	**Virus Types**
Respiratory types	1, 2, 3, 5, 7
Gastroenteritis types	40, 41
Early childhood infections	1, 2, 5, 6
School going children	3, 7
Adult infections	4, 8, 19
Respiratory infections in army barracks	3, 4, 7

pneumonia is seen at times, but usually these are considered as mild infections.

Eye Infections
- Mild ocular involvement may be part of the respiratory adenovirus infection.
- Swimming pool conjunctivitis with types 3 and 7 may occur.
- Epidemic keratoconjunctivitis caused by types 8, 19 and 37 may occur in adults and children as well. It is highly contagious and may spread by the contaminated objects. Disease is self-limiting.

Gastrointestinal Infections

Types 40 and 41 are known to cause infantile gastroenteritis. These can be easily detected in stool by PCR, though are difficult to grow on cell culture.

Acute Haemorrhagic Cystitis

Types 11 and 21 may cause acute haemorrhagic cystitis in children (especially boys).

Infections in Immunocompromised Patients
- Severe pneumonia
- Adenovirus hepatitis
- Myocardial adenovirus infections
- Gastrointestinal infections

LABORATORY DIAGNOSIS

Specimen

Specimen can be collected from the site of infection, for example, urine, throat swab, conjunctival swab, stool, rectal swab etc.

Virus Isolation

Virus isolation can be done on HEp-2, HeLa and other human epithelial cell lines. Virus can be identified by cytopathogenic changes, immunofluorescence tests, HI and Nt tests.

Detection of Viral DNA

The detection of viral DNA can be done by polymerase chain reaction (PCR) assays and DNA hybridisation.

Prevention and Treatment

No specific treatment is available. Careful hand washing, disinfection of surfaces with sodium hypochlorite, chlorination of swimming pools are recommended to prevent it.

EPIDEMIOLOGY

Adenoviruses are endemic worldwide, round the year. They usually do not cause community outbreaks of the disease; however, an acute respiratory disease caused by type 11 was reported in 1997, among civilians. Outbreaks of swimming pool conjunctivitis and epidemic keratoconjunctivitis are also reported.

The virus can spread through direct contact, hand-to-eye transfer, by the faecal-oral route, by respiratory droplets, or by contaminated fomites. Most of the adenovirus infections are subclinical.

ADENOVIRUSES IN BIOTECHNOLOGY

Adenoviruses are the common choice of agents for gene delivery, cancer therapy, gene therapy and other studies.

HUMAN PARVOVIRUSES

Parvoviridae family (*Parvum* 'small'; Latin) contains many pathogenic animal viruses and Parvovirus B19 (B19V) is one of its members known to be pathogenic in humans. B19V is a member of the *Erythroparvovirus* genus. B19V was discovered in 1975 during evaluations of assays for hepatitis B surface antigen using panels of serum samples. Sample 19 in panel B (hence B19) gave a 'false positive' result due to the presence of 23-nm particles resembling parvoviruses. Parvoviruses have small icosahedral capsids of about 25 nm. They have a small genome of single-stranded DNA of approximately 5600 nucleotides.

An association of B19V with significant clinical disease was not made until 1981, but it is now known that B19V infection has a wide variety of disease manifestations dependent on the immunologic and haematologic status of the host (Table 28.3). Half of the infected patients have only non-specific flu-like symptoms, one-fourth are asymptomatic and remaining one-fourth have symptoms. Rash and arthropathy associated with parvovirus B19 infection generally coincide with the measurable serum antibody production and are thus presumed to be at least partially immune mediated.

The choice of diagnostic test depends on the clinical presentation. Table 28.4 lists the available tests and shows the positivity status of various tests in different clinical conditions associated with B19V infection. There is no specific treatment available. Mostly, patients do not require even the symptomatic treatment.

TABLE 28.4 Diagnosis of B19V Infection in Various Clinical Syndromes

Disease	IgM	IgG	B19 DOT BLOT	B19 PCR
Fifth disease	+++	++	−	+
Polyarthropathy syndrome	++	+	−	+
Transient aplastic crisis	±	±	++	++
Persistent anaemia	±	±	++	++
Hydrops/congenital infection	±	+	±	++
Previous infection	−	++	−	±

POLYOMAVIRUSES

INTRODUCTION

Polyomaviruses are icosahedral, non-enveloped, DNA viruses with double-stranded DNA. They have oncogenic potential and may cause tumours in human and animals. They have been associated with the neurological and renal diseases in humans. They replicate in nucleus and stimulate DNA synthesis. Members of human importance are SV40, BK virus and JC virus. The host range for polyomaviruses is highly restricted.

BK AND JC

BK and JC viruses are prevalent in humans worldwide. Humans acquire infection during the childhood and antibodies persist lifelong. In latent stage, healthy individuals carry these viruses in their kidney and lymphoid tissue, and may be shed in urine. During immunocompromised stage virus gets reactivated, for example, in renal transplant cases. Diseases caused by BK and JC viruses are listed in Table 28.5.

TABLE 28.3 Common Manifestations of B19V Infection

Disease	Acute or Chronic	Host
Fifth disease	Acute	Normal children
Polyarthropathy syndrome	Acute/chronic	Normal adults
Transient aplastic crisis	Acute	Patients with increased erythropoiesis
Hydrops fetalis or congenital anaemia	Acute or chronic	Foetus (<20 weeks)
Persistent anaemia	Chronic	Immunodeficient or immunocompromised patients

TABLE 28.5 Clinical Associations of BK and JC Viruses

Disease	Causative Virus	Population Affected Commonly
Haemorrhagic cystitis	BK virus	Bone marrow transplant recipients
Polyomavirus-associated nephropathy	BK virus	Renal transplant recipients
Progressive multifocal leukoencephalopathy (PML)	JC virus	Immunocompromised persons, especially with depressed cell-mediated immunity
Human brain tumours (probable association)	JC virus	Newborn hamsters

JC virus was first time isolated in 1971 from the brain of a patient with progressive multifocal leukoencephalopathy (PML). Once exceedingly rare, the disease is now seen in about 5% of patients with AIDS.

SV40

SV40 is highly tumourigenic in experimentally inoculated hamsters and in transgenic mice. The prevalence of SV40

infections in humans is not known. SV40 may cause a PML-like disease in rhesus monkeys and replicates in both monkey and human cells. It can transform many types of cells in culture. SV40 was first noticed as contaminant of live and killed poliovirus vaccines that had been grown in the monkey cells. Millions of people worldwide received SV40-contaminated vaccines during 1955 and 1963. SV40 DNA has been detected in human tumours, including brain tumours, mesotheliomas, bone tumours and lymphomas.

29

Miscellaneous RNA Viruses

Amita Jain

LEARNING OBJECTIVES

- Classification, structure, nomenclature, pathogenesis, epidemiology and diagnosis of rotaviruses and other viruses causing diarrhoea:
  - Caliciviruses: *Norovirus* and *Sapovirus*
  - Astrovirus
  - Adenovirus
- Structure, classification, human infections, laboratory diagnosis and management of the following:

- Human coronaviruses
- SARS and MERS
- Ebola virus
- Marburg virus
- Lassa virus
- South American haemorrhagic fever viruses
- Lymphocytic choriomeningitis (LCM) viruses

ROTAVIRUS AND OTHER VIRUSES CAUSING DIARRHOEA

Diarrhoea can be caused by bacteria, parasites, viruses and fungi. Viruses could be established as the cause of diarrhoea, following demonstration of viruses by electron microscopy in the stool of patients. It is well acknowledged that viruses are found in the faeces of even normal people. Some of these diarrhoeogenic viruses can be grown in cell cultures, while some are not cultivable. Viruses, such as adenoviruses and rotaviruses are present in stool in large numbers and can be easily identified by electron microscopy. Table 29.1 lists the important viruses causing diarrhoea and their important properties. Virus infection produces local antibodies in gut along with systemic antibodies, including in breast milk. However, their role in protection is not certain, except for the fact that babies who are breast-fed have low incidence of diarrhoea.

ROTAVIRUS

Rotavirus belongs to the family ***Reoviridae***. The classification of Reoviridae is shown in Flowchart 29.1. *Rotavirus* is predominantly pathogenic genera causing human infection. Rotaviruses are fastidious and are difficult to grow in cell culture.

Structure

The virion is icosahedral, 60–80 nm in diameter, non-enveloped but triple-layered. It has two concentric capsid layers, which hold double-stranded RNA in 11 segments. These segments code for different viral and non-structural proteins (Figs. 29.1 and 29.2). Most segments encode a single polypeptide. Virus expresses six structural viral proteins (VPs) and five non-structural proteins (NSPs). Based on antigenic character of VP6, rotaviruses are classified into seven species, that is, A–G. Group A is most often pathogenic

TABLE 29.1	Important Properties of Viruses Causing Diarrhoea			
Properties	**Rotavirus**	**Calicivirus**	**Adenovirus**	**Astrovirus**
Cultivable	No	No	Yes	No
Identifiable on electron microscopy	Yes	No	Yes	Yes, star-shaped present in large number
Genome	Double-stranded RNA, 11 segmented	Single-stranded RNA, non-segmented	DNA	Single-stranded RNA, non-segmented
Virion size	Icosahedral, 72 nm	Icosahedral, 27–40 nm	Icosahedral, 27–40 nm	Icosahedral, 28–30 nm
Strains causing diarrhoea	A–C antigenic group, A most common	Norovirus, Sapovirus	40 and 41 serotypes	5 serotypes
Association with outbreak	Endemic diarrhoea	Yes	Endemic diarrhoea	Endemic diarrhoea
Vaccine available	Yes	No	No	No

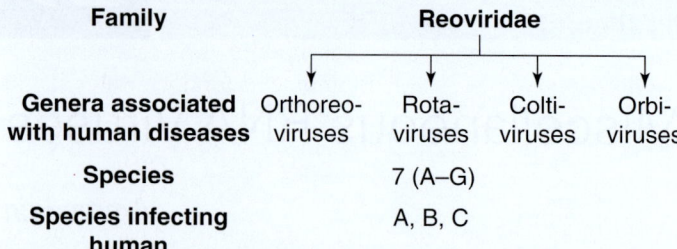

Flowchart 29.1 Members of Family Reoviridae Which Are of Human Importance.

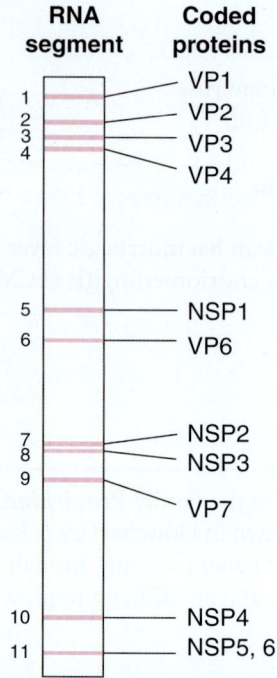

Fig. 29.1 RNA Segments of Rotavirus As Per Their Migration Pattern on Polyacrylamide Gels Strip Post-Electrophoresis and Proteins That Each One of Them Codes.

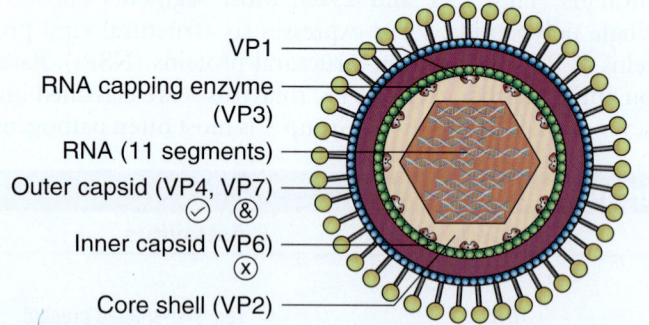

Fig. 29.2 Rotavirus Depicting Location of Structural Viral Proteins *(VP)* and Their Role in Classification of Rotaviruses. ⊗ VP6 = basis for subgroups (A–G); ⊘ VP4 = basis for P serotypes; & VP7 = basis for G serotypes.

to human. These groups can be typed by using common methods, such as immunofluorescence, ELISA and immune electron microscopy. VP4 and VP7 make the outer capsid proteins, hence are the vaccine target. Antibodies against them are used to neutralise the virus particle.

Nomenclature

Recently it was proposed to name rotaviruses as the following: RV group/species of origin/country of identification/common name/year of identification/G- and P-type. G serotype specificity is based on VP7, while P serotype specificity is based on VP4. Total 14 G types are known to exist, of which 10 are of importance to human infections. Four G types (G1–G4) are important cause of human infections.

Resistance to Physical and Chemical Methods

Rotaviruses are resistant to heat, acid pH and alcoholic disinfectants. They can be inactivated by 95% ethanol, phenol and chlorine.

Pathogenesis

Rotaviruses can infect both humans and animals. Subclinical infections are common. Rotaviruses infect enterocytes in the villi of the small intestine, multiply there and damage fluid and electrolyte transport mechanisms. NSP4 works as a viral enterotoxin. Impaired sodium and glucose absorption cause diarrhoea. Diarrhoea, fever and vomiting are common symptoms. The virus particles are present in large numbers in stool and have also been found in the vomitus. Seroconversion occurs following infection.

Epidemiology

Rotaviruses are mainly endemic. They are the commonest cause of diarrhoea in children under 5 years, especially those needing hospitalisation. Incubation period is 1–3 days. Subclinical infections are common. Utmost cleanliness can protect against the infection.

Diagnosis

Commercial immunoassays are available. Reverse transcriptase PCR is also an accepted method of diagnosis. Demonstration of segmented genome on polyacrylamide gel electrophoresis was used for a long time as a gold standard.

Prevention and Treatment

Two oral rotavirus vaccines are approved for use in children: (1) Rotarix (GlaxoSmithKline), given in a two-dose series to infants and (2) RotaTeq (CSL Limited/Merck and Co., Inc.) given as three doses.

There is no specific treatment. Patients are managed symptomatically.

CALICIVIRUSES

Caliciviridae group of viruses are important agents of viral gastroenteritis in humans. The most important members for human infections are *Norovirus* and *Sapovirus*. These are small viruses with a single-stranded RNA genome (Flowchart 29.2).

Noroviruses

Norwalk virus was first time seen by Kapikan in 1972 by immune electron microscopy in stool sample from a case

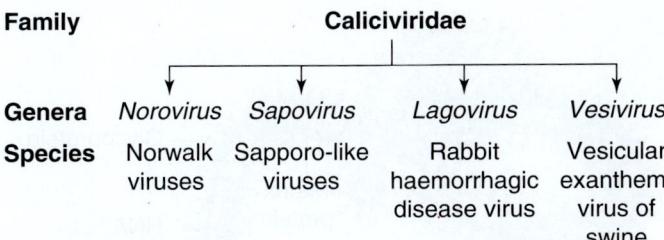

Family		Caliciviridae		
Genera	*Norovirus*	*Sapovirus*	*Lagovirus*	*Vesivirus*
Species	Norwalk viruses	Sapporo-like viruses	Rabbit haemorrhagic disease virus	Vesicular exanthem virus of swine

Flowchart 29.2 Members of Family Caliciviridae.

of gastroenteritis. These are small RNA viruses, which are non-cultivable. Noroviruses are considered as the most common cause of viral gastroenteritis in adults. These cases are usually mild, recover spontaneously and rapidly, but are highly communicable. As few as 10 virus particles can be infectious. Transmission can occur through various modes, mainly by faeco–oral route. Virus can survive in the presence of chlorine and at 50°C. Disease has an incubation period of 24–48 h. Chief symptoms are diarrhoea, nausea, vomiting, fever and malaise. Virus is most frequently associated with outbreaks of water-borne, food-borne and shellfish-associated gastroenteritis.

Reverse transcriptase-polymerase chain reaction in clinical specimens is the method of choice for diagnosis. Electron microscopy and immunoassays can be helpful.

Treatment is symptomatic. No specific antiviral is available.

Sapoviruses

For the first time in 1976, Sapoviruses were found in human faeces. They are morphologically similar to *Noroviruses*. There are four human serotypes, of which three are associated with outbreaks of diarrhoea and vomiting. Vomiting is usually more prominent than diarrhoea. The incubation period is 24–72 h. Virus is excreted in faeces and vomitus. Virus can be demonstrated by molecular assays in clinical samples. Other tests are less sensitive.

ADENOVIRUSES

Adenoviruses 40 and 41 are diarrheogenic viruses. The adenoviruses can be divided into five subgenera (A–E). Two serotypes 40 and 41 belong to a new subgenus F. These two subtypes are difficult to grow in cell culture. In children, excretion of adenoviruses has been found to be prolonged and asymptomatic. However, at times they have also been associated with outbreaks. Adenoviruses 40 and 41 are not recovered from the respiratory tract, while rest of the types are isolated from respiratory tract.

ASTROVIRUSES

Astroviruses cause diarrhoeal illness, have star-like shape and is about 28–30 nm in diameter. They have been named so because of their star-shaped configuration. Genome is single-stranded, positive-sense RNA, 6.4–7.4 kb in size. There are eight serotypes of human astroviruses. Astroviruses are found in large numbers in the stool of babies with diarrhoea. They may also be found in the stools of normal children and are transmitted by the faecal–oral route. Contaminated food or water is the main source of infection. They are pathogenic for infants and children. Immunocompromised adults may get the infection.

CORONAVIRUSES

INTRODUCTION

The name coronaviruses (CoV) is given to this group of viruses because crown-like spikes are present on their surface, resembling solar corona. Coronaviruses are limited to epithelial cells of the respiratory or gastrointestinal tract. These are species-specific. Most of these viruses cause respiratory illness and are acid-labile. Some of them which are acid resistant may cause gastrointestinal manifestations. These viruses are difficult to grow on cell lines and are cultured either in animals or on organ culture of human embryonic trachea, where it was grown for the first time in 1960s.

STRUCTURE

It is a large (120–160 nm in size), enveloped, single-stranded positive-sense RNA (27–32 kb) virus. Nucleocapsid is helical. There are widely spaced, 20-nm-long club- or petal-shaped glycoprotein projections (spike/peplomer) on the envelope (Fig. 29.3).

CLASSIFICATION

Coronaviruses are divided into four subgroups: alpha, beta, gamma and delta. Following coronaviruses can commonly infect humans:

1. **Common Coronaviruses:**
   a. 229E (alpha coronavirus)
   b. NL63 (alpha coronavirus)
   c. OC43 (beta coronavirus)
   d. HKU1 (beta coronavirus)

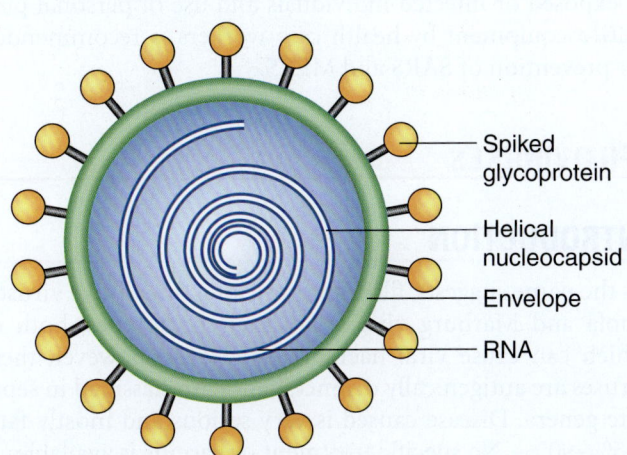

Fig. 29.3 Structure of Coronavirus.

2. **Newer Coronaviruses:** New coronaviruses which cause severe illness are as follows:
   a. MERS-CoV (beta coronavirus)
   b. SARS-CoV (beta coronavirus)

DISEASES CAUSED

Cold-like illness: Similar to those produced by rhinoviruses, but usually afebrile. The incubation period is about 1 week.

Gastroenteritis: Similar to that of rotavirus infections, it is usually demonstrated in animals.

Severe acute respiratory syndrome (SARS): SARS was first reported in 2003 in China. Chinese horseshoe bats are natural reservoirs. Incubation period is short, usually within a week. Clinically it is characterised by serious respiratory illness, leading to respiratory distress. Most of the cases need ventilator support. Death occurs due to respiratory failure in about 10% cases.

Middle East respiratory syndrome (MERS): MERS was first reported in 2012 in Saudi Arabia. This virus is supposed to have come to human from camels. Patients develop severe respiratory distress often with renal failure. Most of them need ventilator support.

LABORATORY DIAGNOSIS

PCR: Preferred test to detect CoV RNA in respiratory secretions, occasionally in serum and plasma.

Antigen detection: CoV antigen can be demonstrated in cells of respiratory epithelium using ELISA.

Serology: It is not used commonly for lab diagnosis.

EPIDEMIOLOGY

They are worldwide in distribution. They are air-borne infections and are common cause of cold. Infections are self-limiting, except SARS and MERS. Mortality with SARS and MERS is high and death occurs due to respiratory failure. Most of adults have seropositivity for antibodies against CoV.

TREATMENT, PREVENTION AND CONTROL

No specific treatment is available. Isolation and quarantine of exposed or infected individuals and use of personal protective equipment by health care workers is recommended for prevention of SARS and MERS.

FILOVIRUSES

INTRODUCTION

As the name suggests, filoviruses are long thread-like viruses. Ebola and Marburg viruses are their prototypes, both of which can cause viral haemorrhagic fever; however, these viruses are antigenically distinct, hence are classified in separate genera. Disease caused is very serious and mostly fatal (25%–90%). No specific treatment or vaccine is available.

Filoviruses are enveloped RNA viruses which belong to the family *Filoviridae*. These are pleomorphic long filamentous

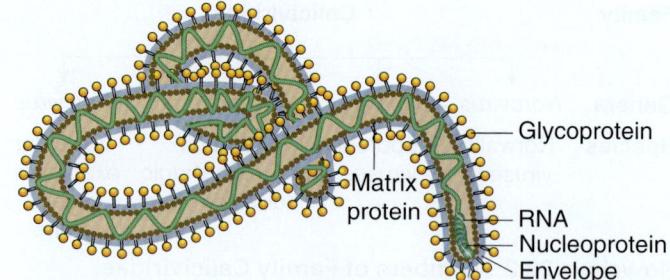

Fig. 29.4 Structure of Filoviruses.

thread-like viruses (Fig. 29.4). Filoviruses need biosecurity Level 4 for handling as they are highly virulent. Heat, phenolic disinfectant, bleach, alcohols and gamma irradiation can destroy them.

Reservoir of infection is not clearly known, but rodent and bat have transmitted infections to human in the past. Outbreaks were reported in African monkeys. Transmission to health care workers is one of the most important sources of infection, which can be controlled by strict adherence to infection control practices and training of the hospital staff.

Laboratory diagnosis can be made by demonstration of antigens or viral-specific antibodies in patient's sera by ELISA. Reverse transcriptase PCR can detect viral RNA in clinical samples. Virus can be grown on Vero and MA-104 monkey cell lines.

EBOLA VIRUS

It was named after Ebola River, on the sides of which, the first case was noted in 1976. After this case, epidemics were seen in Sudan and Zaire. More than 500 cases occurred with ~80% mortality. Latest outbreaks were seen during 2013–16 in West Africa, resulting in more than 11,000 deaths. These viruses were antigenically distinct and are divided into four subgroups: Zaire, Sudan, Reston and Ivory Coast. Zaire and Sudan subtypes are highly virulent.

MARBURG VIRUS

It was first detected in Germany and Yugoslavia, among laboratory persons working with African green monkeys imported from Africa. Patients suffered from viral haemorrhagic fever with high mortality. Person-to-person transmission occurred; outbreaks from time to time are reported from African subcontinent.

ARENAVIRUSES

INTRODUCTION

Arenaviruses are named so because of their sand granule-like morphology on electron microscopy (*Arena* [*Latin*] = Sand). These are enveloped RNA viruses containing two segments of RNA. They cause infections in rodents, except few which cause human viral haemorrhagic fever. Lassa, Junin, Machupo, Guanarito, Sabia, Whitewater Arroyo and LCM viruses are known to cause disease in human. Human can get infection on coming in contact with infected rodent. Human-to-human transmission occurs through aerosols and contaminated body fluids.

LASSA FEVER

First case of Lassa fever occurred in 1969 in Lassa, a Nigerian village. Lassa fever is a high-mortality haemorrhagic fever with multi-organ involvement. Mortality in pregnant women is very high. It is endemic in western African countries. Diagnosis is made by detection of antiviral IgM and IgG antibodies by ELISA. Viral RNA can be detected using reverse transcriptase-polymerase chain reaction assays. House rat is the reservoir. Human-to-human transmission occurs by contact. Contacts of patient in hospital may get infected. Ribavirin is of some help in the treatment of Lassa fever.

SOUTH AMERICAN HAEMORRHAGIC FEVERS

Junin haemorrhagic fever (Argentine haemorrhagic fever) is a common problem in Argentina. Rodents are the reservoir.

Person working in maize field can get exposed. Cases occur each year; mortality is frequent.

Machupo haemorrhagic fever (Bolivian haemorrhagic fever) was first detected in Bolivia in 1962. Case fatality is high.

Guanarito virus was first reported in 1990. This causes **Venezuelan haemorrhagic fever**, a high mortality disease.

Sabia virus was also reported in 1990. Fatal haemorrhagic fever due to this virus was reported in Brazil.

LYMPHOCYTIC CHORIOMENINGITIS

Lymphocytic choriomeningitis (LCM) virus is transmitted to human from mice, its natural vector. Human-to-human spread is not seen. First case was seen in 1933. Haemorrhagic fever, aseptic meningitis or influenza-like illness can be seen. Severe infection and mortality is rare. ELISA for antiviral IgM and IgG antibodies is the diagnostic test. Reverse transcriptase-polymerase chain reaction demonstrates viral RNA. Virus grows in Vero cells.

Other Miscellaneous Viruses

Amita Jain

LEARNING OBJECTIVES

- **Oncogenic viruses**
  - DNA and RNA tumour viruses
  - Mechanism of oncogenesis of DNA and RNA tumour viruses
  - Retroviruses
  - Human T-cell lymphotropic virus (HTLV)
- Polyomaviruses
- Papillomaviruses
- **Slow virus infections and prion diseases**
  - Subacute sclerosing panencephalitis
  - Progressive multifocal leukoencephalopathy
  - Visna

ONCOGENIC VIRUSES

INTRODUCTION

Oncogenic viruses (tumour viruses) induce tumour development in susceptible hosts. Both DNA and RNA viruses can induce tumour (Table 30.1). Proving a causal relationship between a viral agent and a cancer type may be difficult, as establishing Koch's postulates is not possible. Tumourigenic property of a virus can be strongly suspected by a relationship of high viral infection and tumour of a certain type in a population, property of inducing tumour by a type of virus in an animal model in laboratory and transformation of cells in which virus grows.

MECHANISM OF VIRAL ONCOGENESIS

Oncogenesis is a multistep process taking a long time. 'Immortalisation', 'hyperplasia' and 'preneoplastic' stages happen before 'neoplasia' is established. Oncogenic viruses are capable of transforming a normal cell so that it multiplies uncontrollably and changes in a cancer cell. Viruses alone are hardly ever capable of making such changes. Host and environmental factors play equally important role. Genome of some viruses directly act as transforming gene (direct acting), while some alter the expression of host gene to transform the host cell (indirect acting). Table 30.2

TABLE 30.1 Viruses Known to Have Association with Tumours

Virus	RNA/DNA Virus	Associated Human Cancer
Human papillomaviruses	DNA	• Genital tumours • Squamous cell carcinoma • Oropharyngeal carcinoma
EB virus	DNA	• Nasopharyngeal carcinoma • Burkitt's lymphoma • Hodgkin's disease • B-cell lymphoma
Human herpesvirus 8	DNA	Kaposi's sarcoma
Hepatitis B virus	DNA	Hepatocellular carcinoma
HTLV	RNA	Adult T-cell leukaemia
Human immunodeficiency virus	RNA	AIDS-related malignancies
Hepatitis C virus	RNA	Hepatocellular carcinoma

TABLE 30.2 Mechanisms of Pathogenesis of Tumour Viruses

DNA Viruses	RNA/Retroviruses
• DNA of the viral genome may become integrated into the host cell DNA and encode viral oncoproteins • Non-permissive host cells do not support viral growth and may get transformed by DNA viruses	• Retroviruses carry enzyme reverse transcriptase • RNA genome of the virus is reverted back to DNA copy (proviral DNA) • Proviral DNA gets integrated in the host cell DNA • Integrated DNA copy directs synthesis of proteins required for transforming host cell • Hepatitis C virus does not form provirus

describes important features of oncogenesis by DNA and RNA viruses.

Important **interactions of tumour viruses and host cells** required for oncogenesis are as follows:

- Cell susceptibility to viral infections
- Persistent infections in infected host cells
- Retention of tumour virus DNA/RNA in host cell
- Introduction of a new 'transforming gene' into the cell (direct acting), or the alteration of expression of a pre-existing cellular gene or genes (indirect acting) by infecting tumour virus
- Avoid detection and recognition by host immune system; different viral evasion strategies, immunosuppression in host and avoiding immune surveillance mechanisms to eliminate the neoplastic cells are chief mechanisms.

In this chapter we discussed about HTLV (retroviruses), polyomavirus and human papillomaviruses (HPVs). Rest of them will be discussed in respective chapters.

HUMAN T-LYMPHOTROPIC VIRUS (HTLV-1)

HTLV-1 infects mature T cells, and is expressed at very low levels in infected individuals. Important and unique features of retroviruses are listed in Box 30.1. The virus replicates with proliferation of T cells, thus maintaining them in host cells. They do not have an oncogene, instead they have *tax*, a transregulating gene which alters the viral genes, thus altering their functions. Altered genes contribute to oncogenesis.

There are several subtypes of HTLV-1. The major ones are subtypes A, B and C. The virus is distributed worldwide. HTLV-associated diseases are mainly seen in southern Japan, Melanesia, the Caribbean, Central and South America, and parts of Africa.

BOX 30.1   Important Features of Retroviruses

Structure: Spherical, 80–110 nm in diameter, helical nucleoprotein, icosahedral capsid
Genome: Single-stranded, linear, positive-sense RNA, 7–11 kb in size
Proteins: Reverse transcriptase enzyme
Envelope: Present; glycoprotein and lipid
Antigens: Glycoproteins in the viral envelope, and virion core
Maturation: Virions bud from plasma membrane

BOX 30.2   Important Features of Polyomavirus

Structure: Small viruses (diameter 45 nm), non-enveloped; icosahedral symmetry (45 nm in diameter)
Genome: Double-stranded circular DNA, 5 kbp
Proteins: Three structural proteins
Envelope: None
Replication: Occurs in nucleus
Oncogenesis: Viral oncoproteins interact with cellular tumour suppressor proteins. Virus stimulate cell DNA synthesis.
Clinical importance: Can cause human neurologic and renal disease and human cancer

---

HTLV-2 has been isolated from human, but no disease association has been proved.

POLYOMAVIRUSES

Polyomaviruses are widespread in nature with 77 recognised species. Their important features are listed in Box 30.2. *Alpha, Beta, Delta* and *Gamma*, genus of polyomaviruses, are commonly infecting human. Most of these infections are asymptomatic. They often persist as latent infections in a host without causing disease. Some members of the family are oncoviruses, while some are associated with other human diseases (Table 30.3).

PAPILLOMAVIRUS

In the past, papillomaviruses were parts of family Papovaviridae (Papilloma, Polyoma, Vacuolating viruses). Currently, the papillomaviruses are included in Papillomaviridae family. Family Papillomaviridae is divided into 16 genera, of which *Alpha, Beta, Gamma, Mupa* and *Nupa* papillomaviruses infect human.

Papillomaviruses are diverse with more than 100 HPV types. They are non-cultivable. Characterisation is based on molecular methods with at least 10% diversity in L1 genes sequence. Properties of the papillomaviruses are shown in Box 30.3.

Pathogenesis

Papillomaviruses infect epithelial cells of the skin and mucous membranes. Transmission of viral infections occurs by close

---

TABLE 30.3   Important Polyomaviruses with Known Clinical Correlation

Species	Proposed Genus	Virus Name	Clinical Correlation (If Any)
Human polyomavirus 5	Alpha	Merkel cell polyomavirus	Merkel cell cancer
Human polyomavirus 8	Alpha	Trichodysplasia spinulosa polyomavirus	Trichodysplasia spinulosa
Human polyomavirus 1	Beta	BK polyomavirus	Polyomavirus-associated nephropathy; haemorrhagic cystitis
Human polyomavirus 2	Beta	JC polyomavirus	Progressive multifocal leukoencephalopathy
Human polyomavirus 6	Delta	Human polyomavirus 6	HPyV6-associated pruritic and dyskeratotic dermatosis (H6PD)
Human polyomavirus 7	Delta	Human polyomavirus 7	HPyV7-related epithelial hyperplasia

BOX 30.3   Important Properties of Papillomaviruses

Virion: Icosahedral, 55 nm in diameter
Genome: Circular double-stranded DNA of 8 kbp, several open reading frames (Table 30.4)
Proteins: Two structural proteins
Envelope: Not enveloped
Replication: Occurs in nucleus
Characteristics: Significant cause of human cancers, available vaccine with good prevention, infects epithelial cells only

TABLE 30.5   Papillomavirus Genes and Their Functions

Gene	Function
L1	Major capsid protein (structural)
L2	Minor capsid protein (structural)
E1	Transcription factor: Helicase and episomal DNA replication (non-structural)
E2	Transcription factor: Regulates viral copy number (non-structural)
E4	Viral particle release (non-structural)
E5	Stimulates cell proliferation, prevents cell differentiation, downregulates MHC 1 expression
E6	p53 inactivation, induces malignant transformation
E7	Rb inactivation, induces malignant transformation

contact because viral particles are released from the surface of lesions. Genital infections are sexually transmitted. Papillomaviruses are associated with several lesions (Table 30.4). HPV types 16 and 18 are considered to be high cancer risk types. Many HPV types are benign. Cervical cancer and its association with HPV 16 and 18 is most well established.

Pathogenesis of Cervical Cancer

Cervical cancer develops slowly, over years to decades. HPV infects epithelial cells of cervix. Persistent infection with a high-risk HPV (e.g. types 16, 18) is a necessary component. Viral DNA gets integrated with host cell genome and persists. This is known as episomal form. Host immunologic factors also play role in pathogenesis. How viral proteins influence oncogenesis is shown in Table 30.5. Two viral early proteins, E6 and E7 (Table 30.5), are oncogenic and alter the control of host cell cycle. E5, E6 and E7 also inhibit the host cell immune response (Fig. 30.1). The E1 and E2 help viral replication.

Epidemiology

Squamous cell cancers of cervix, anal canal, penis and oropharynx are known to have association with HPV infections. Infection occurs in adolescents and young adults. Children may get infection during birth from contaminated secretions. Over 70% of cervical cancers are associated with HPV-16 or HPV-18 infection. HPV-6 and HPV-11 are associated

with Laryngeal papillomas in children and benign genital condylomas. Immunosuppressed patients have high incidence of genital warts and cancers.

Prevention

A quadrivalent recombinant HPV vaccine is available. Vaccine is a most cost-effective way to reduce anogenital HPV infections. The vaccine contains particles derived from HPV types 6, 11, 16 and 18. Vaccine is to be given to adolescent girls and boys. It is not effective once infection has already occurred.

SLOW VIRUS INFECTIONS AND PRION DISEASES

INTRODUCTION

Occasionally, a slow and progressive infection with some viruses causes chronic degenerative diseases of brain.

TABLE 30.4   Association of HPV Types and Clinical Lesions

HPV Type	Clinical Lesion	Oncogenic Potential
16, 18[a]	Cervical cancer, high-grade dysplasia, carcinomas of mucosa, genital, laryngeal and oesophageal	High risk, common types
30, 31, 33, 35, 39, 45, 51–53, 56, 58, 59, 66, 68, 73, 82	Cervical cancer, high-grade dysplasia, carcinomas of mucosa, genital, laryngeal and oesophageal	High risk, rare types
1, 2, 4, 27, 57	Plantar warts, common skin warts	Benign
3, 10, 28, 49, 60, 76, 78	Skin lesions	Low
5, 8, 9, 12, 17, 20, 36, 47	Epidermodysplasia verruciformis	Mostly benign
6, 11, 40, 42–44, 54, 61, 70, 72, 81	Anogenital condylomas; laryngeal papillomas; mucosal dysplasias and intraepithelial neoplasias	Low
7	Hand warts of butchers	Low

[a]Vaccine available.

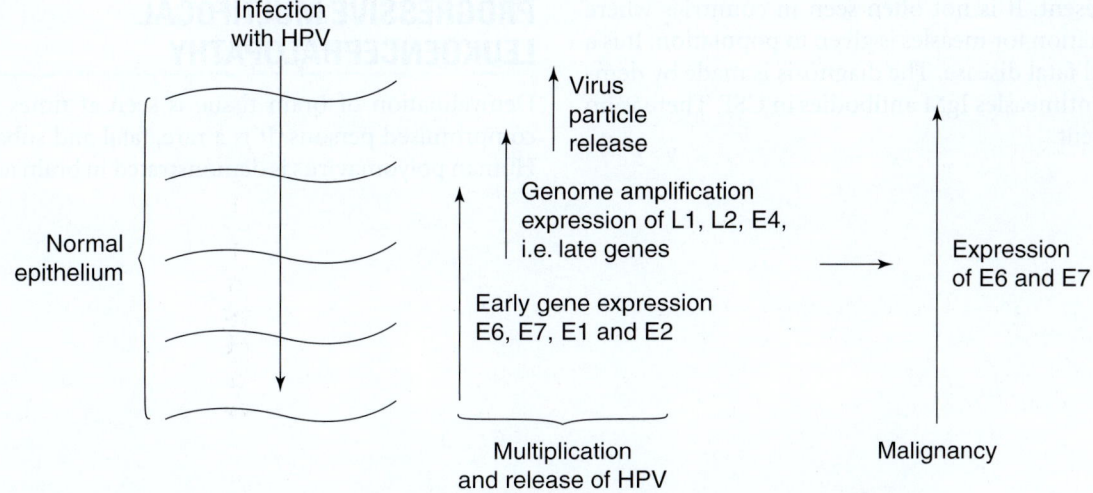

Fig. 30.1 Gene Expression and HPV Expression as Seen in HPV-Induced Carcinogenesis.

Slow viruses are divided into three types: A (Lentiviruses), B (Prion) and C (others; Table 30.6). These diseases are chronic, with incubation period in years and usually fatal. Most of the time there is limited immune response against these illnesses, which may not be protective. Some of the examples are visna, Creutzfeldt–Jakob disease (CJD), kuru, subacute sclerosing panencephalitis (SSPE) and progressive multifocal leukoencephalopathy (PML).

SUBACUTE SCLEROSING PANENCEPHALITIS

SSPE is a rare disease caused by chronic persistent infection of neurons and glial cells of brain by measles virus. These measles viruses are defective viruses and cannot be grown in culture alone. However, once infected brain cells are cultured, it grows in them. Serological evidence of past measles virus

TABLE 30.6   Slow Viruses and Prion Diseases: Common Properties

Group	Common Diseases	Host	Causative Organism	Incubation Period	Clinical Features	Comments
A	Visna	Sheep	Lentivirus	2 years	Demyelination of brain, fatal	Fatal
	Medi	Sheep	Lentivirus	2 years	Haemorrhagic pneumonia, fatal	Fatal
B: Prion diseases	CJD	Human	Prion*	1 year approximately, after transplant of infected cornea, etc.	Subacute presenile encephalopathy	Fatal in end
	Kuru	Human, seen in New Guinea	Prion	5–10 years, introduced in human through cannibalism	Encephalopathy, tremor, progressive cerebral ataxia	Kuru means tremor, fatal in months
	Scrapie	Sheep	Prion	In years, transmission vertical	Encephalopathy	Prion disease prototype, fatal
	Mad cow disease	Cattle	Prion	In years, feed contaminated with scrapie-infected meat	Encephalopathy	Fatal
C	SSPE	Human with latent measles infection	Measles virus	In years	Subacute sclerosing panencephalitis (SSPE)	Fatal after 2–3 years
	PML	Immunosuppressed persons	Human polyomavirus	In years	Demyelination, progressive degeneration of motor function	Fatal in few months

*Prion is not a virus. This lacks nucleic acid. It contains proteins only (Prion). It can be transmitted from one individuals to other.

infection is present. It is not often seen in countries where effective vaccination for measles is given to population. It is a progressive and fatal disease. The diagnosis is made by demonstration of antimeasles IgM antibodies in CSF. There is no specific treatment.

PROGRESSIVE MULTIFOCAL LEUKOENCEPHALOPATHY

Demyelination of brain tissue is seen at times in immunocompromised persons. It is a rare, fatal and subacute disease. Human polyomavirus is demonstrated in brain autopsy tissue.

Introduction to Parasites

Parul Jain

LEARNING OBJECTIVES

- Glossary of terms
- Protozoa and helminths

- Introduction, sources of infection, routes of entry into body, pathogenicity, classification and laboratory diagnosis

GLOSSARY OF TERMS

Glossary of terms related to parasitology (study of parasites) is given in Box 31.1.

INTRODUCTION

Parasites are classified into **protozoa** (unicellular organisms) and **helminths** (multicellular organisms).

Protozoa

The word *protozoa* is derived from Greek words '*protos*' meaning first and '*zoia*' meaning animals. The phylum protozoa consist of unicellular (single-celled) eukaryotes. This single cell acts like an animal. They have a complex internal structure and carry out complex metabolic activities. Some protozoa have a surface layer called pellicle that maintains the distinctive shape (*Trypanosomes* and *Giardia*). In some protozoa (amoeba), the cytoplasm is differentiated

---

BOX 31.1 Glossary of Terms

1. **Host:** An organism which harbours the parasite.
   a. **Definitive host:** An organism that harbours the adult stage of the parasite.
   b. **Intermediate host:** An organism that harbours the larval stage of the parasite.
   c. **Paratenic host:** Host in which the parasite remains viable without undergoing any further development.
   d. **Reservoir host:** An organism that harbours the parasites and acts as a source of infection for others.
2. **Parasite:** A living organism which receives nourishment and shelter from another organism where it lives.
   a. **Ectoparasite:** Parasites that live on the body surface of the host and can either cause disease or act as vectors for other parasites.
   b. **Endoparasite:** Parasites that live within the body of the host.
3. **Host parasite relationship:**
   a. **Symbiosis:** Both host and parasite are benefitted or live harmoniously so that none of the partner suffers any harm from the association.
   b. **Commensalism:** Parasite is benefitted without causing injury/benefit to its host. A commensal is capable of leading an independent life.
   c. **Parasitism:** Only the parasite derives benefit at the cost of any injury to the host.
4. **Protozoology:** Biological study of protozoans.
5. **Stages in life cycle of protozoa:**
   a. **Trophozoite:** Actively feeding and multiplying stage.
   b. **Cyst:** Stage with a protective membrane or thickened wall.
6. **Helminthology:** Study of parasitic worms and their effect on hosts.
7. **Stages in life cycle of helminths:**
   a. **Adults:** Capable of sexual reproduction.
   b. **Larva:** Developmental/sexually immature stage of life cycle.
   c. **Eggs:** Containing zygote and/or embryo.
   d. **Cysts:** Larval stage usually encapsulated in tissues of intermediate hosts.
8. **Helminths on the basis of their reproductivity:**
   a. **Oviparous:** Production of eggs, discharged from uterus of female.
   b. **Ovoviviparous:** Production of eggs, which hatch prior to discharge from the uterus of a female.
   c. **Viviparous:** Lay embryos.
   d. **Parthenogenesis:** Ability to produce offspring without fertilisation of eggs.
9. **Helminths on the basis of sexual types:**
   a. **Monoecious/hermaphrodite:** Both the sexes are present in the same worm.
   b. **Dioecious:** Separate sexes.

TABLE 31.1	Differences Between Protozoa and Helminths	
Characteristics	**Protozoa**	**Helminths**
Cellularity	Unicellular, single-cell performs all the functions	Multicellular; have differentiated organ systems
Life cycle	Have simple life cycles. They often multiply within the human host and therefore, usually a single brief exposure (e.g. bite of a mosquito) results in intense parasite loads	Have complex life cycles. They often have a developmental stage outside the human body (either as free living or in another mammalian host). The worm burden[a] increases only after repeated exogenous exposures
Stimulation of host immune response	Usually do not elicit eosinophilia in infected humans	Elicit eosinophilia in human tissues and blood

[a]Worm burden: Number of adult worms within a human host.

into ectoplasm (outer layer) and an endoplasm (inner layer containing organelles). Some protozoa have a cytostome (cell mouth) for ingesting food particles and water.

Some protozoans have a worldwide distribution and are the leading causes of mortality and morbidity (*Plasmodium* spp., *Leishmania* spp., *Entamoeba histolytica*) while others are geographically restricted (*Giardia lamblia*, *Cryptosporidium* spp.).They cause a spectrum of disease varying from asymptomatic infections to life-threatening disease. The manifestations depend upon species and strain of the parasite and host resistance. The infections may cause mild symptoms in persons with intact immune systems but may be life threatening in immunocompromised people, such as those with AIDS.

Helminths

The word *helminth* is derived from the Greek word '*helmins*' meaning parasitic worm. Helminths are multicellular parasites that may exist as free living organisms or as parasites of plants, animals or human hosts. Human helminthic parasites belong to two phyla: Nemathelminthes/Nematodes/roundworms and Platyhelminthes/flatworms that include cestodes/tapeworms and trematodes/flukes.

There are several differences between protozoa and helminths (Table 31.1).

SOURCES OF INFECTION

The parasites may be transmitted to human beings by different agencies, such as (1) by ingesting the infectious stage (eggs/cysts) along with contaminated food or water (e.g. eggs of *Ascaris lumbricoides*, *Taenia solium*) or through fingers contaminated with soil; (2) through raw or undercooked contaminated food (e.g. *T. solium*); (3) through vectors (e.g. *Plasmodium* spp., *Leishmania donovani*); (4) through contact with infected animals (e.g. *Toxoplasma gondii*) and (5) through autoinfection, without the agency of external source (e.g. *Strongyloides stercoralis*, *Capillaria philippinensis*) (Flowchart 31.1).

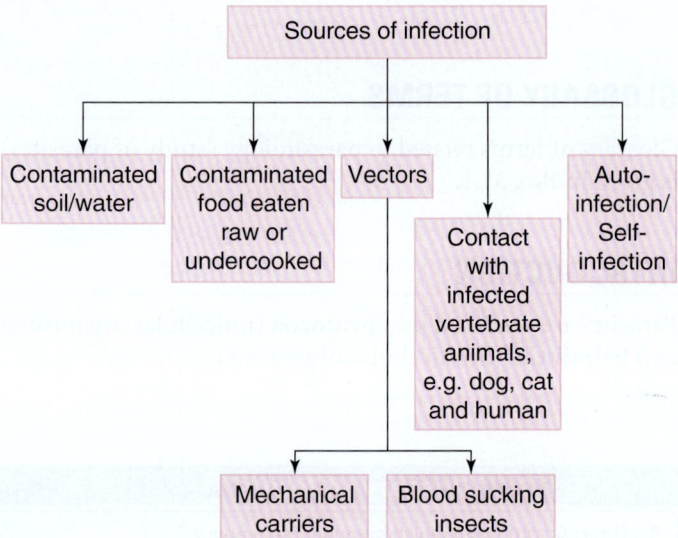

Flowchart 31.1 Sources of Parasitic Infections.

ROUTES OF ENTRY INTO BODY

The parasites may enter the human body through the following routes:
- Oral (e.g. *A. lumbricoides*, *Trichinella spiralis*)
- Skin penetration (e.g. hookworms)
- Sexual contact (e.g. *Trichomonas vaginalis*)
- Congenital (*T. gondii*)
- Iatrogenic (*T. gondii*)
- Inhalation (*Cryptosporidium* spp. and *Giardia intestinalis*)

PATHOGENICITY

Parasites can harm human body by the following mechanisms:
- Mechanical damage/obstruction (*A. lumbricoides*, hookworms)
- Traumatic damage/perforation (*A. lumbricoides*, *Dracunculus medinensis*, *E. histolytica*)
- Lytic necrosis (*Trypanosoma cruzi*, *E. histolytica*)

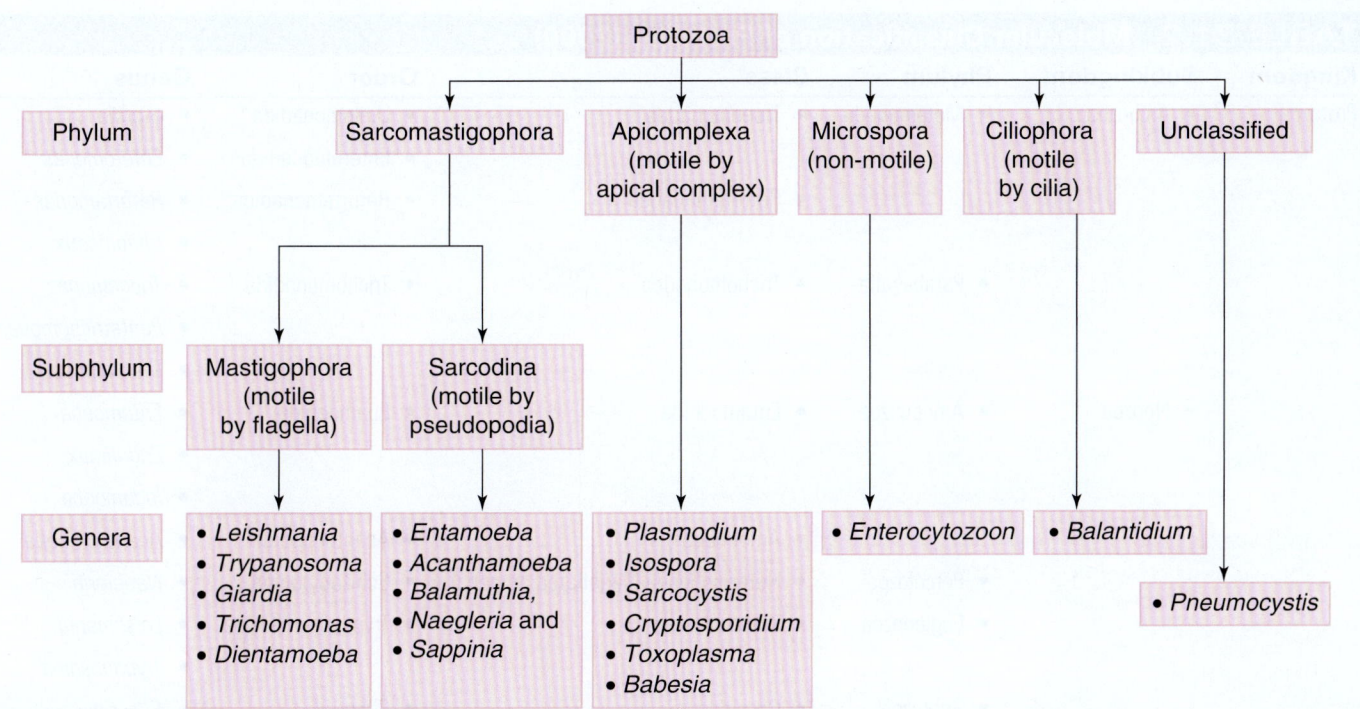

Flowchart 31.2 Traditional Classification of Protozoa (1980s). (*Source*: Adapted from Levine ND, Corliss JO. A newly revised classification of the Protozoa. J Protozoo 1980;27:37–58.)

- Inflammatory damage/allergic manifestations (*Toxocara* spp.)
- Competition for specific nutrients (*Toxocara* spp., hookworms, *Taenia* spp.)
- Neoplastic changes (e.g. liver flukes, *Schistosoma haematobium*)
- Secondary infection (e.g. *D. medinensis*)

CLASSIFICATION

Protozoa and helminths can be classified by several criteria. The protozoa have been classified on the basis of (1) organs of locomotion (traditional classification), (2) nucleotide sequences in RNA, DNA or both (molecular classification) and (3) location of adult worm in the human body. Likewise, helminths have been classified on the basis of (1) morphology, (2) nucleotide sequences in RNA, DNA or both (systemic classification) and (3) anatomic location of helminths in human body.

Classification of Protozoa

1. **Traditional classification**: The traditional classification (1980s) was based on organs of locomotion (Flowchart 31.2).
2. **Molecular classification**: This classification was based on ribonucleic acid (RNA) and protein sequences of the organisms using advanced molecular techniques. Genus *Microsporum* and *Blastocystis* were earlier considered as *Protozoa* but on the basis of molecular studies, *Microsporum* and *Blastocystis* have now been classified under the Kingdom *Fungi* and *Chromista*, respectively (Table 31.2).

3. **Classification on the basis of anatomic location of protozoa in human body**: This classification is given in Flowchart 31.3.

Classification of Helminths

1. **Traditional classification**: The medically important worms are traditionally classified according to their external shape (Flowchart 31.4). Helminths are classified into Platyhelminthes and Nemathelminthes/Nematodes. Platyhelminthes are flat worms (Greek, platy: flat, helminths: worms) with no specialised circulatory or respiratory system and with no body cavity (acoelomates). A definite head end is present. The phylum Platyhelminthes is divided into two classes: Cestodes (segmented tapeworms) and Trematodes (unsegmented leaf-like flukes). Nematodes are evolutionarily more developed than flatworms. They have a body cavity, a complete alimentary canal starting in mouth and ending in anus; are dioecious and have a rudimentary nervous and excretory system.
2. **Systemic classification (based on Anderson et al., 1974)**: Systemic classification of helminths is given in Table 31.3.
3. **Classification on the basis of anatomic location of helminths in human body**: This classification is given in Flowchart 31.5.

LABORATORY DIAGNOSIS OF PARASITES

The different modalities for laboratory diagnosis and their advantages and disadvantages are listed in Table 31.4.

TABLE 31.2 Molecular Classification of Protozoa (2000)

Kingdom	Subkingdom	Phylum	Class	Order	Genus
Protozoa	• Archezoa	• Metamonada	• Trepomonadea	• Diplomonadida	• *Giardia*
				• Enteromonadida	• *Enteromonas*
			• Retortamonadea	• Retortamonadida	• *Retortamonas*
					• *Chilomastix*
		• Parabasalia	• Trichomonadea	• Trichomonadida	• *Trichomonas*
					• *Pentatrichomonas*
					• *Dientamoeba*
	• Neozoa	• Amoebozoa	• Entamoebida	• Euamoebida	• *Entamoeba*
					• *Endolimax*
					• *Iodamoeba*
			• Amoebaea	• Acanthopodida	• *Acanthamoeba*
		• Percolozoa	• Heterolobosea (flagellated amoeba)	• Schizopyrenida	• *Naegleria*
		• Euglenozoa	• Kinetoplastea	• Trypanosomatidae	• *Leishmania*
					• *Trypanosoma*
		• Sporozoa	• Coccidea	• Eimeriida	• *Eimeria*
					• *Toxoplasma*
					• *Cryptosporidium*
					• *Cyclospora*
					• *Cystoisospora*
					• *Sarcocystis*
				• Haemosporida	• *Plasmodium*
				• Piroplasmida	• *Babesia*
		• Ciliophora	• Litostomatea	• Vestibuliferida	• *Balantidium*
Fungi		• Microspora	• Microsporea	• Microsporida	• *Enterocytozoon*
					• *Encephalitozoon*
					• *Pleistophora*
					• *Trachipleistophora*
					• *Brachiola*
					• *Microsporum*
Chromista	• Chromobiota	• Bigyra	• Blastocystea		• *Blastocystis*

(*Source*: Adapted from Cox FEG, Wakelin D, Gillespie SH, et al., editors. Topley and Wilson's microbiology and microbial infections: parasitology, 10th ed. Wiley; 2007.)

General Considerations for Diagnosing Intestinal Parasites

Relevant Information

A detailed history of signs and symptoms should always accompany the specimen. The information should also contain any recent travel history or immunocompromised status.

Sample Collection and Transportation

Freshly passed stool collected in a clean dry container is the sample of choice for intestinal parasites. Sample should not be collected from bedpan or diaper and should not be contaminated with urine or water. Label the container with patient's name, unique identification number and date and time of passage of stool.

Transport the samples immediately to the laboratory since any delay (>1–2 h of passage of stool) will cause morphological distortion of the parasites, for example, disintegration of amoebic or flagellate trophozoites or will make identification difficult (e.g. eggs of hookworm may hatch to release larvae which cause confusion with *Strongyloides* larvae). In the laboratory, the samples must be processed as

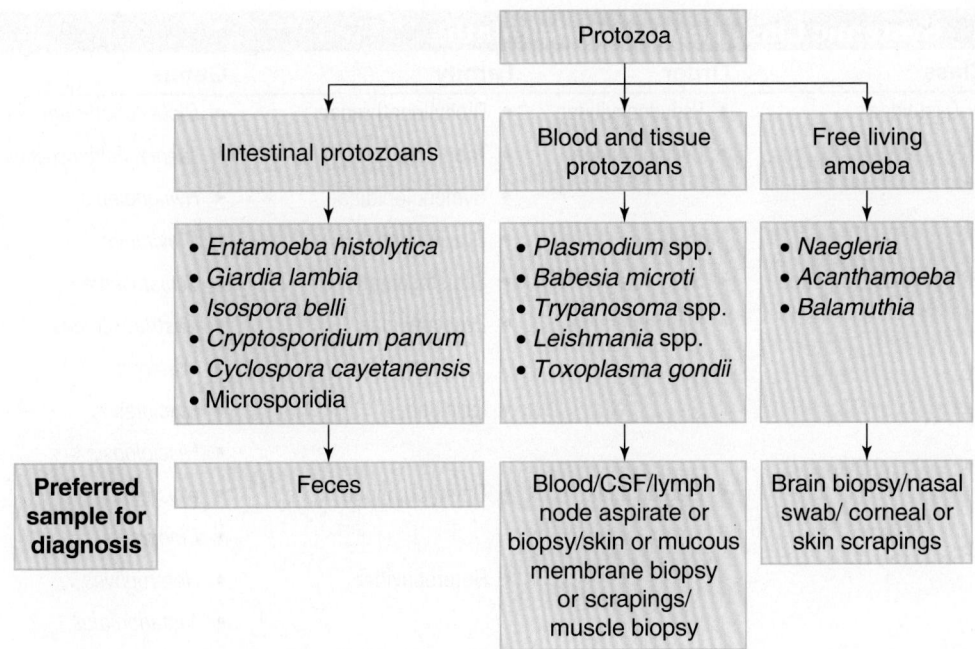

Flowchart 31.3 Classification of Protozoa on the Basis of Location in Host Body.

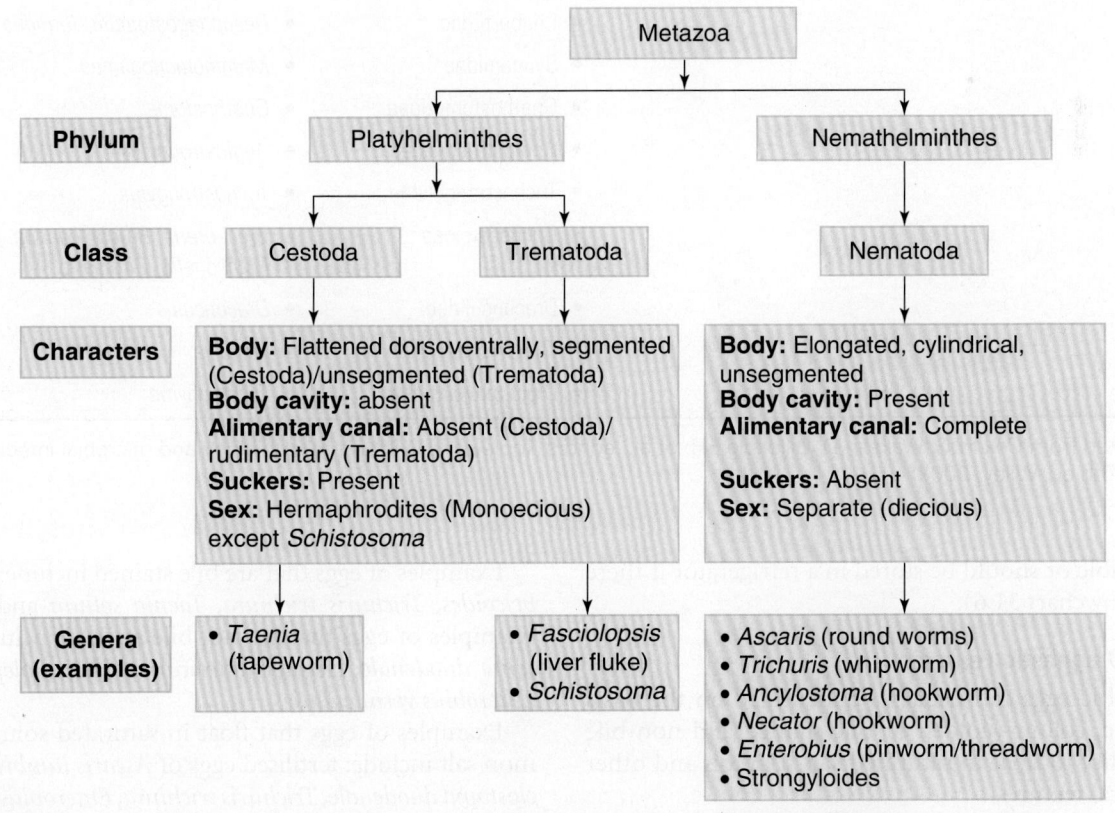

Flowchart 31.4 Classification of Metazoa/Helminths According to Their External Shape.

TABLE 31.3 Systemic Classification of Helminths

Phylum	Class	Order	Family	Genus
Platyhelminthes	• Cestoidea	• Pseudophyllidea	• Diphyllobothriidae	• *Diphyllobothrium, Spirometra*
		• Cyclophyllidea	• Taeniidae	• *Taenia, Echinococcus*
			• Hymenolepididae	• *Hymenolepis*
			• Dipylidiidae	• *Dipylidium*
	• Trematoda/Digenea	• Strigeida	• Schistosomatidae	• *Schistosoma*
		• Echinostomida	• Zygocotylidae	• *Gastrodiscoides*
				• *Watsonius*
			• Fasciolidae	• *Fasciola*
				• *Fasciolopsis*
		• Plagiorchiida	• Opisthorchiidae	• *Opisthorchis*
				• *Clonorchis*
			• Heterophyidae	• *Heterophyes*
				• *Metagonimus*
			• Paragonimidae	• *Paragonimus*
Nematoda	• Adenophorea		• Trichinellidae	• *Trichinella*
			• Trichuridae	• *Trichuris, Capillaria*
	• Secernentea		• Oxyuridae	• *Enterobius*
			• Ascarididae	• *Ascaris, Toxocara*
			• Anisakidae	• *Anisakis*
			• Ancyclostomatidae	• *Ancyclostoma, Necator*
			• Strongyloididae	• *Strongyloides*
			• Chabertiidae	• *Oesophagostomum, Ternidus*
			• Syngamidae	• *Mammomonogamus*
			• Gnathostomatidae	• *Gnathostoma*
			• Angiostrongylidae	• *Angiostrongylus*
			• Trichostrongylidae	• *Trichostrongylus*
			• Onchocercidae	• *Wuchereria, Brugia, Loa loa, Onchocerca, Mansonella, Dirofilaria*
			• Dracunculidae	• *Dracunculus*
			• Thelaziidae	• *Thelazia*
			• Dioctophymatidae	• *Dioctophyme*

(*Source:* Adapted from Cox FEG, Wakelin D, Gillespie SH, et al., editors. Topley and Wilson's microbiology and microbial infections: parasitology, 10th ed. Wiley; 2007.)

soon as possible or should be stored in a refrigerator if there is a delay (Flowchart 31.6).

Suggestive Diagnostic Features

The helminthic eggs can usually be identified on the basis of size, shape, colour (golden or bile stained and non-bile stained), ability to float in saturated salt solution and other morphological features.

Examples of eggs that are bile stained include: *Ascaris lumbricoides, Trichuris trichiura, Taenia solium* and *T. saginata*. Examples of eggs that are not bile stained include: *Ancyclostoma duodenale, Necator americanus, Hymenolepis nana* and *Enterobius vermicularis*.

Examples of eggs that float in saturated solution of common salt include: fertilised eggs of *Ascaris lumbricoides, Ancyclostoma duodenale, Trichuris trichiura, Enterobius vermicularis,*

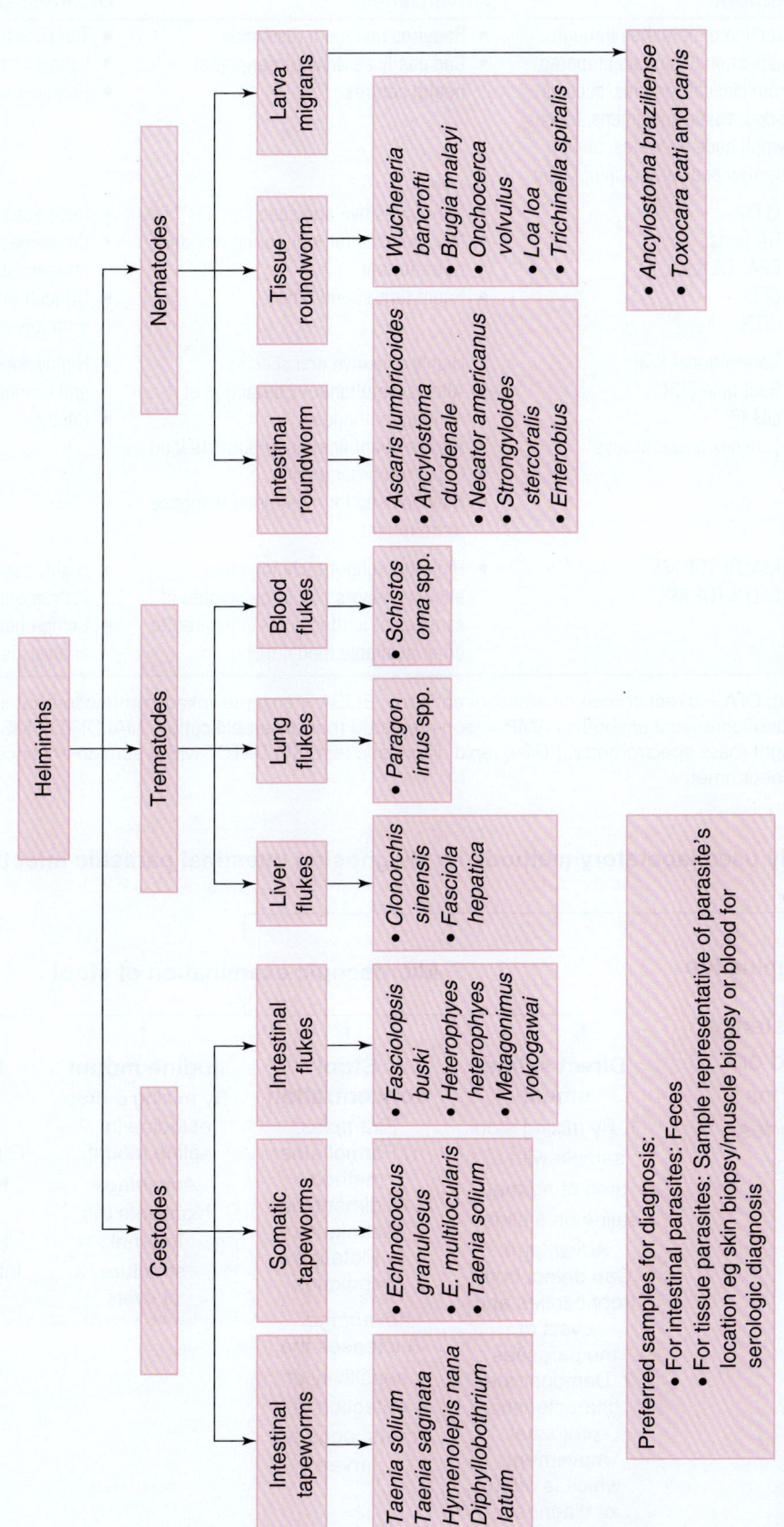

Flowchart 31.5 Classification of Helminths on the Basis of Their Anatomic Location in Human Body.

TABLE 31.4   Modalities Available for Laboratory Diagnosis of Parasites

Technique	Methods	Advantages	Disadvantages
Microscopy	Detection of parasites through inspection of smears prepared from clinical samples, such as blood, tissue specimens, faeces, lymph node aspirates, bone marrow and cerebrospinal fluid	• Requires minimum resources • Can easily be done at peripheral health centres	• Time consuming • Labour intensive • Requires trained personnel
Serology-based assays (detecting antigen/antibody)	• ELISA • HA tests • DFA/IFA tests • CFTs • RDTs	• More sensitive and specific • May be used for monitoring response to treatment • Rapid turnaround time	• Resources are required • Cross-reactivity common between antigenically related organisms • Difficult interpretation in cases with low signal-to-noise ratio
Molecular-based approach (detecting nucleic acid of a parasite)	• Conventional PCR • Real time PCR • LAMP • Luminex-based assays	• Highly sensitive and specific • Allows simultaneous detection of multiple pathogens • Allows quantification of parasite load • Rapid turnaround time • May be used for monitoring response to treatment	• Highly sophisticated instruments and trained personnel are required • Costly
Proteomic-based approach (analyses expression of proteins)	• MALDI-TOF MS • SELDI-TOF MS	• High throughput technique that allows analysis of a large number of samples at limited costs compared to other available modalities	• Highly sophisticated and costly instruments are required • Comprehensive database for analysis is required

CFT = complement fixation test; DFA = direct immunofluorescent antibody; ELISA = enzyme-linked immunosorbent assay; HA = haemagglutination; IFA = indirect immunofluorescent antibody; LAMP = loop-mediated thermal amplification; MALDI-TOF MS = matrix-assisted laser desorption ionization time-of-flight mass spectrometry; RDT = rapid diagnostic test; SELDI-TOF MS = surface-enhanced laser desorption ionisation time-of-flight mass spectrometry.

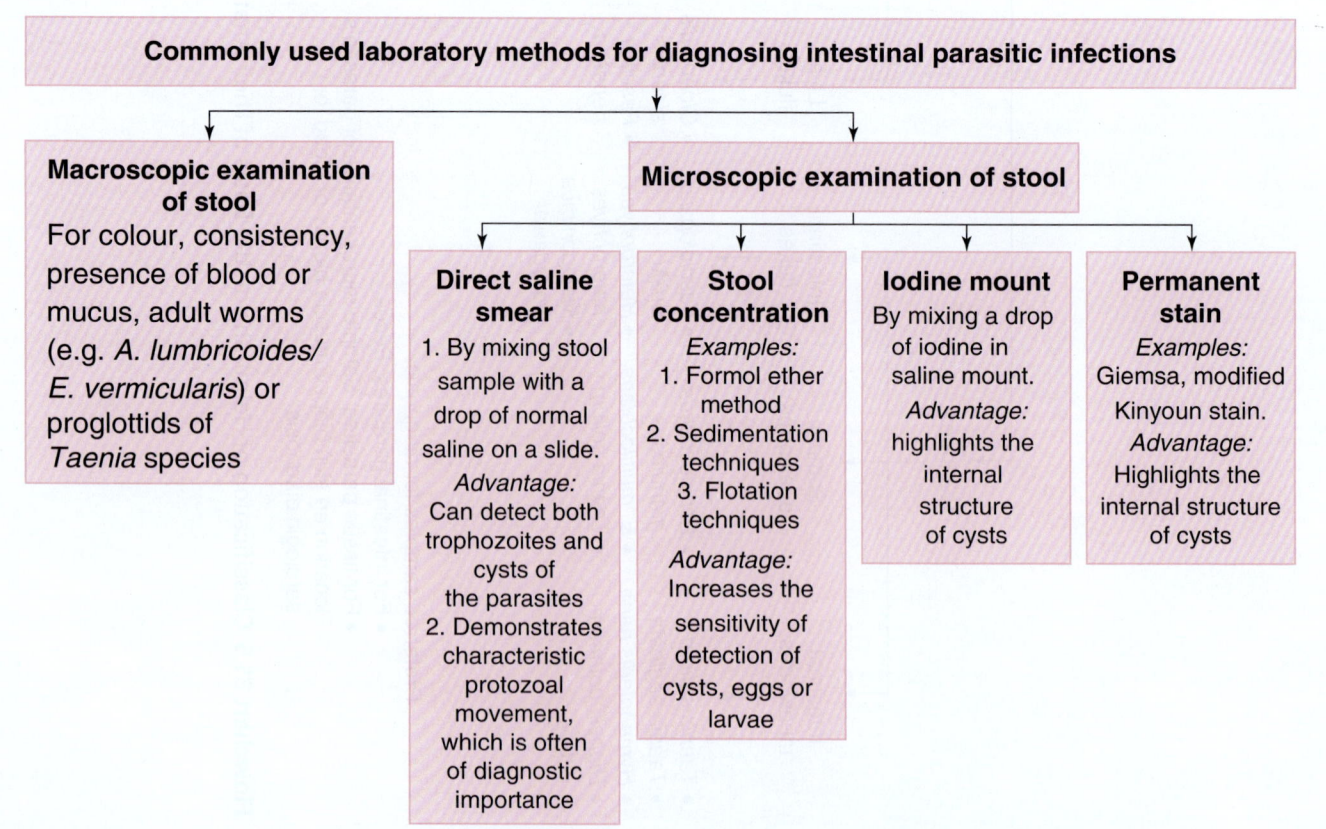

Flowchart 31.6 Commonly Used Laboratory Procedures for Diagnosis of Intestinal Parasitic Infections.

and *Hymenolepis nana*. Examples of eggs that do not float in saturated solution of common salt include: unfertilised eggs of *Ascaris lumbricoides*, *Taenia solium* and *Taenia saginata*.

Diagnosis of Blood Parasites

Thick and Thin Blood Smears

Thick and thin blood smears are used for diagnosing blood parasites. Thick smears are more sensitive for detecting the parasite and thin smears are required for species identification. The stains used are Romanowsky stains, such as Giemsa stain, Jaswant Singh Bhattacharya (JSB) stain, Field's stain, Leishman's stain and Wright's stain. These stains are neutral stains composed of a mixture of oxidized methylene blue (azure) dyes and Eosin Y. The azure dyes are basic dyes and bind to the acidic nuclei staining them blue to purple colour. The red dye eosin is basic and is attracted to the alkaline cytoplasm, producing red colouration.

Quantitative Buffy Coat (QBC)

This test is used to detect malaria parasites and microfilaria. Blood is collected in a capillary tube coated internally with acridine orange stain. The tube is centrifuged and then examined under UV rays.

Blood Concentration

It is useful for detection of microfilaria in blood. Different concentration methods are sedimentation method, cytocentrifugation, Knott's centrifugation etc.

Protozoa

Parul Jain

Life cycle, epidemiology, clinical manifestations, laboratory diagnosis, treatment and prevention of common protozoa affecting humans include the following:

- Amoebae
  - *Entamoeba* spp.
  - Free living amoebae
- Flagellates
  - *Leishmania* spp.
  - *Trypanosoma* spp.
  - *Giardia* spp.
  - *Trichomonas* spp.
  - Other flagellates
- Sporozoa
  - *Plasmodium* spp.
  - *Babesia* spp.
  - *Toxoplasma* spp.
  - *Cryptosporidia* spp.
  - *Cyclospora* spp.
  - *Cystoisospora* spp. (formerly called Isospora)
- Sarcocystis
  - Ciliophora and Chromista
  - Balantidiasis
  - Chromista

AMOEBAE

INTRODUCTION

Amoeba derives its name from a Greek word 'amoibe' meaning change. Amoebae constantly change their shape due to pseudopodia (the organ of locomotion). On the basis of habitat, the amoebae are classified into intestinal amoebae (*Entamoeba histolytica, Entamoeba dispar, Entamoeba moshkovskii, Entamoeba hartmanni, Entamoeba gingivalis, Endolimax nana, Iodamoeba butschlii*) and free-living amoebae (*Acanthamoeba, Balamuthia, Naegleria* and *Sappinia*). Taxonomically Amoebae belong to the kingdom Protozoa, subkingdom Neozoa, phyla Amoebozoa and Percolozoa (refer Chapter 31: Introduction to Parasites; Molecular Classification, 2000).

ENTAMOEBA SPECIES

INTRODUCTION

Several species of *Entamoeba* can infect humans but *E. histolytica* (tissue lysing amoeba) is the only pathogenic species that can cause amoebiasis (intestinal and extraintestinal infection). *E. histolytica* is morphologically similar to nonpathogenic species like *E. dispar and E. moshkovskii*. These species can be differentiated from other *Entamoeba* species on the basis of size, number and morphology of nuclei of the cysts. Other nonpathogenic species that can reside in human intestines include *Entamoeba coli, E. hartmanni, Entamoeba polecki, E. nana* and *I. butschlii*.

LIFE CYCLE

The life cycle of *E. histolytica* is given in Flowchart 32.1.

Host: *E. histolytica* needs a single host (human) to complete its life cycle.

Stages: The two stages are: the **cyst form (quadrinucleate)** and the **trophozoite form** (20–50 μm motile forms). The cyst stage is hardy and survives the harsh environmental conditions and is therefore the infectious form. The trophozoite form is easily killed on exposure to air or stomach acid and is therefore noninfectious. However, trophozoites can be infectious during oral anal sexual contact.

Depending upon several host factors, such as age, sex, nutritional status, intestinal motility, and overall immunity, the trophozoites may show different clinical courses: 1, 2 or 3 (Flowchart 32.1).

EPIDEMIOLOGY

- *E. histolytica* has been overestimated to infect 10% of the total world's population. Most of these prevalence studies were based on microscopy of stool and therefore could not distinguish cysts and trophozoites of *E. histolytica* from those of nonpathogenic species, *E. dispar* or *E. moshkovskii*. The prevalence of asymptomatic *E. histolytica* infection is region dependent and the incidence is higher in developing countries.
- Persons predisposed to amoebic colitis (complication of amoebiasis) include very young patients, malnourished patients, pregnant women, immunocompromised persons, male homosexuals and institutionalised persons.

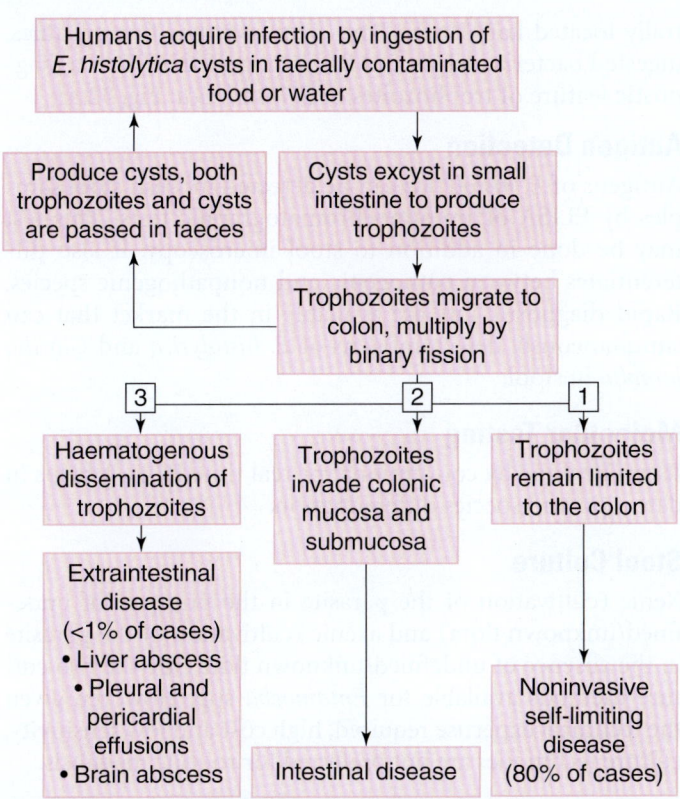

Flowchart 32.1 Life Cycle of *E. histolytica*.

- Amoebiasis, particularly amoebic liver abscess, is more common in men than in women.

CLINICAL MANIFESTATIONS AND PATHOGENESIS

The clinical manifestations depend on location of trophozoites, that is whether they remain restricted to the intestine (intestinal amoebiasis) or spread to liver (amoebic liver abscess) or metastasise to distant parts of the body (metastatic amoebiasis).

Intestinal Amoebiasis

Men and women are equally affected.
- **Asymptomatic cyst passers**: All infections with *E. moshkovskii* and *E. dispar* are asymptomatic and intestinal infection with *E. histolytica* is asymptomatic in 80% cases.
- **Amoebic diarrhoea without dysentery**: It is the most common disease manifestation.
- **Amoebic dysentery**: Amoebic dysentery occurs when trophozoites invade the colonic mucosa by secreting proteolytic enzymes and produce characteristic **flask-shaped ulcers** (broad base with a narrow neck; Fig. 32.1). Large numbers of trophozoites are secreted in stool along with blood and mucus.
- **Complications of intestinal amoebiasis**: Some patients may also develop acute necrotising colitis, paralytic ileus, amoeboma (amoebic granuloma), colonic mucosal sloughing or intestinal perforation (Fig. 32.1). Fever is usually absent and leucocytosis is rare. Fulminant colitis rarely occurs but is associated with mortality in 40% cases.

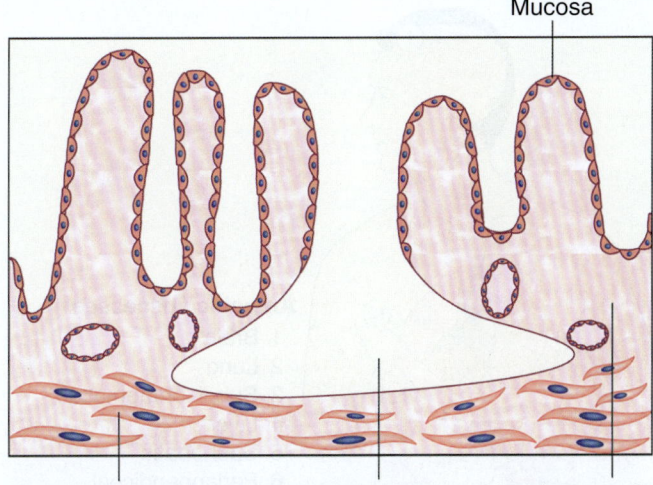

Fig. 32.1 A Flask-Shaped Amoebic Ulcer. (*Source:* Pathology for Dental Students, Geetika Khanna Bhattacharya, Fig. 10.31, Page 127, 2nd ed., RELX India Pvt. Ltd., 2017.)

Amoebic Liver Abscess

It is more commonly seen in men than women. The disease occurs when trophozoites penetrate through colon and reach the liver through portal circulation. The abscess is usually single and generally involves the right lobe of liver. The abscess wall, is composed of:
- inner central zone of necrotic hepatocytes without amoebae;
- middle zone of degenerative hepatocytes, leucocytes, erythrocytes and few amoebic trophozoites and
- outer zone of healthy hepatocytes invaded by trophozoites.
  The abscess is filled with **anchovy sauce pus**, thick chocolate brown-coloured pus comprising necrotic hepatocytes, no pus cells and occasional trophozoites.

Clinical Presentation of Hepatic Amoebiasis

The characteristic presentation is acute onset of symptoms (<10 days) including fever, right upper quadrant pain and hepatic tenderness. The disease may also present subacutely with prominent weight loss, fever or abdominal pain. The WBC count, ESR and alkaline phosphatase levels are usually raised. Most individuals with amoebic liver abscess do not have concurrent intestinal signs and symptoms. Therefore, careful history of dysentery within the previous year should be determined. It is often also not associated with the presence of trophozoites in stool. Its differential diagnosis includes pyogenic abscess, echinococcal cyst and hepatoma.

Metastatic Amoebiasis

It occurs as a result of direct extension of liver abscess. If the abscess ruptures through the diaphragm, liver abscess may lead to the following:
- **Pleuropulmonary amoebiasis** presenting as cough, pleuritic chest pain and shortness of breath.
- **Empyema, bronchohepatic fistulas** presenting as coughing up of contents of liver abscess containing copius amounts of brown sputum and trophozoites.
- **Peritoneal signs and shock**

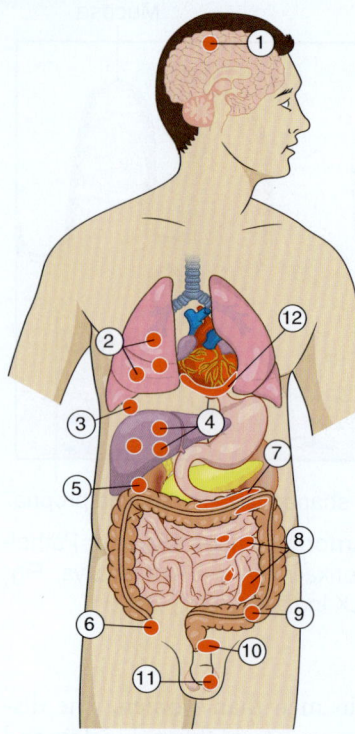

Amoebic abscesses
1. Brain
2. Lung
3. Subdiaphragmatic
4. Liver
5. Subhepatic
6. Periappendiceal

7. Amoebic ulcers
8. Peritonitis
9. Amoeboma
10. Anorectal fistula
11. Cutaneous amoebiasis
12. Pericardial effusion, amoebic pericarditis

Fig. 32.2 Complications of Amoebiasis. (*Source:* Infectious Disease Online, Pathology of Amebiasis (Entamoeba Histolytica Infection), Extraintestinal Amebiasis.)

- **Pericarditis** presenting as chest pain, pericardial rub, tachypnoea, dyspnoea or cardiac tamponade.
- Rare complications include cerebral abscess, cutaneous amoebiasis, genital or urinary disease (Fig. 32.2).

LABORATORY DIAGNOSIS

Sample: Freshly passed stool is the ideal sample. Other samples that may be collected are rectal exudates, colonoscopically collected rectal ulcer tissue or pus aspirate in extraintestinal amoebiasis.

Microscopic Examination

It can be done by wet mount, iodine mount or trichrome staining of stool samples from patients. The trophozoites or cysts of *E. histolytica* can be seen. But this method has a limited value since *E. histolytica* cysts cannot be differentiated from those of *E. moshkovskii* and *E. dispar*. The method is also only 25%–60% sensitive. Sensitivity of the procedure can be increased by: (1) examining at least three stool samples and (2) by formalin ether concentration of the stool samples.

Morphology of Cysts

Cysts are spherical, usually measuring 12–15 μm and have four nuclei (are quadrinucleate) with centrally placed endosomes (karyosomes) and fine chromatin that is peripherally uniformly distributed. Chromatoid bodies (elongated bodies with blunt ends) may sometimes be found. Immature cysts may contain one to three nuclei (Fig. 32.3A–C; Box 32.1).

Morphology of Trophozoites

Trophozoites usually measure 15–20 μm, have a single nucleus with peripherally arranged chromatin and a cen-

trally located karyosome. The cytoplasm has fine granules, ingested bacteria, debris and RBCs (ingested RBCs is a diagnostic feature of trophozoites of *E. histolytica*; Fig. 32.4).

Antigen Detection

Antigens of *E. histolytica* can be detected in fresh stool samples by ELISA or immunochromatographic tests. This test may be done in addition to stool microscopy. It also differentiates between pathogenic and nonpathogenic species. Rapid diagnostic kits are available in the market that can simultaneously detect antigens of *E. histolytica* and *Giardia lamblia* in stool.

Molecular Testing

It can be done by conventional or real time PCR. It helps in detection and species differentiation.

Stool Culture

Xenic (cultivation of the parasite in the presence of undefined/unknown flora) and axenic (cultivation of the parasite in the absence of undefined/unknown flora) are the two culture methods available for *Entamoeba* spp. However, given the technical expertise required, high cost and low sensitivity, culture techniques are not preferred for routine diagnosis.

Serology

In extraintestinal disease (e.g. amoebic liver abscess), specific antibodies against *E. histolytica* can be detected in patient's blood.

TREATMENT

- Asymptomatic patients with infection with *E. dispar* or other nonpathogenic species should not be treated. But since *E. dispar* is a marker of faeco–oral contamination, appropriate preventive measures should be taken.
- For asymptomatic *E. histolytica* infections, treatment of choice is a **luminal agent** (paromomycin or iodoquinol). For symptomatic intestinal infections or extraintestinal disease, treatment of choice is nitroimidazoles (metronidazole or tinidazole) followed by treatment with a luminal agent (paromomycin or iodoquinol).
- Amoebic liver abscess less than 10 cm can be treated by therapy with metronidazole followed by treatment with a luminal agent. For left lobe abscess more than 10 cm or impending rupture or non-responding to medical treatment within 3–5 days, drainage of liver abscess should be performed. Surgical intervention may be required in some cases.

PREVENTION

- Amoebiasis can be prevented by improved sanitation to prevent faecal contamination of food and water, health education, early treatment of carriers and water treatment.
- Cysts in water are killed by boiling for more than 1 min and those in vegetables are killed by washing them with detergent and soaking in acetic acid or vinegar for at least 10–15 min.

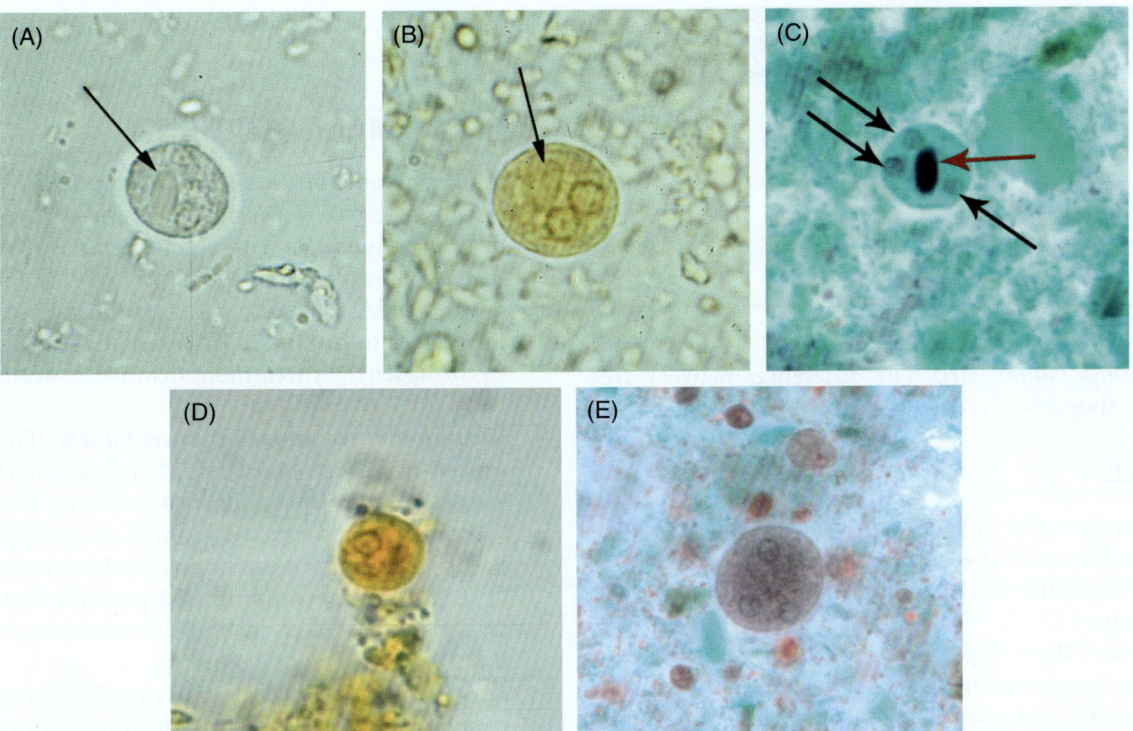

Fig. 32.3 (A) Cyst of *E. histolytica/E. dispar* in an unstained concentrated wet mount of stool. Notice the chromatoid bodies with blunt, rounded ends *(arrow)*. (B) Cyst of *E. histolytica/E. dispar* in a concentrated wet mount stained with iodine. Notice the chromatoid body with blunt, rounded ends *(arrow)*. (C) Cyst of *E. histolytica/E. dispar* stained with trichrome. Three nuclei are visible in the focal plane *(black arrows)*, and the cyst contains a chromatoid body with typically blunted ends *(red arrow)*. The chromatoid body in this image is particularly well demonstrated. (D) Cyst of an *E. hartmanni* in a wet mount, stained with iodine. (E) Mature cyst of *E. coli*, stained with trichrome. Eight nuclei can be seen between the two focal planes. Also, above the cyst, a trophozoite of *E. nana* can be seen. (*Source*: (A to C) Centers for Disease Control and Prevention (CDC), DPDx - Laboratory Identification of Parasites of Public Health Concern, Parasite Biology, Amebiasis. (D and E) Centers for Disease Control and Prevention (CDC), DPDx - Laboratory Identification of Parasites of Public Health Concern, Parasite Biology, Intestinal Amebae.)

BOX 32.1	**Differences Between Cysts of *E. histolytica* and Those of Other Species**

1. Cysts of *E. histolytica* are spherical, measure 12–15 μm, and have four nuclei.
2. *E. hartmanni* also has quadrinucleate cyst but is smaller (3–12 μm) (Fig. 32.3D).
3. *E. coli* cysts can have up to eight nuclei (Fig. 32.3E).
4. *E. gingivalis* does not form cysts.

- Vaccine against *E. histolytica* is under development, targeting Gal-lectin, serine-rich protein and the 29-kDa reductase antigen.

FREE-LIVING AMOEBAE

INTRODUCTION

Free-living amoebae of genera *Acanthamoeba, Balamuthia, Naegleria* and *Sappinia* live in lakes, swimming pools and air conditioning and heating ducts and can cause accidental or opportunistic infections in humans. Important species include *N. fowleri* (the only pathogenic species of the genera),

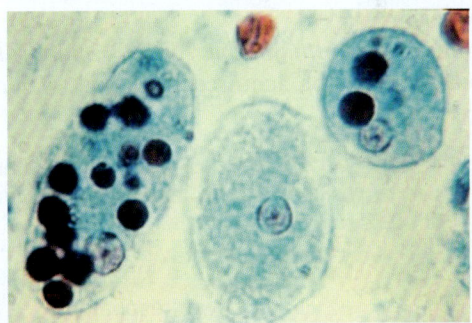

Fig. 32.4 Trophozoites of *E. histolytica* With Ingested Erythrocytes Stained With Trichrome. The ingested erythrocytes appear as dark inclusions. The parasites above show nuclei that have the typical small, centrally located karyosome, and thin, uniform peripheral chromatin. (*Source*: Centers for Disease Control and Prevention (CDC), DPDx - Laboratory Identification of Parasites of Public Health Concern, Parasite Biology, Amebiasis.)

Acanthamoeba polyphaga, Acanthamoeba lugdunensis, Acanthamoeba lenticulata, Acanthamoeba castellanii, Acanthamoeba culbertsoni, Acanthamoeba astronyxis, Acanthamoeba hatchetti, Acanthamoeba rhysodes, Acanthamoeba divionensis

	Free-Living Protozoa	Pathogenic Protozoa
Insect vectors required	None	Yes
Existence of human carrier state	None	Yes
Relation between poor sanitation and their transmission	None	Yes

and Balamuthia mandrillaris. The differences between free living and pathogenic protozoa are shown in Box 32.2.

LIFE CYCLE

Stages: *Acanthamoeba* spp. and *B. mandrillaris* have two stages in the life cycle: **cysts and trophozoites** (multiply by mitosis in which the nuclear membrane does not remain intact; Flowchart 32.2).

N. fowleri has three stages in its life cycle: cysts, trophozoites (multiply by promitosis, in which the nuclear membrane remains intact) and flagellated forms. The free-living trophozoites infect humans during water-related activities, such as swimming or underwater diving. Amoebae penetrate the olfactory mucosa, invade the cribriform plate and ascend via olfactory nerves to the brain causing **primary amoebic meningoencephalitis (PAM)** in healthy individuals.

EPIDEMIOLOGY

- *N. fowleri* occurs in all continents except Antarctica, in warm freshwater bodies such as ponds, swimming pools, lakes and rivers but not in seawater. Most humans acquire infection during recreational activities in warm freshwater.
- Infections by *Acanthamoeba* spp., *B. mandrillaris* and *Sappinia* spp. can be acquired from soil, water and air.

These infections, though infrequent, have a worldwide distribution.

CLINICAL MANIFESTATIONS

- ***Primary amoebic meningoencephalitis* (PAM):** PAM is caused by *N. fowleri* (the brain-eating amoeba) in healthy individuals with recent history of swimming in fresh hot water. The prodrome lasts for few days with fever, headache, nausea and vomiting, which rapidly progresses to fulminant disease with seizures and coma. In the absence of treatment death may occur within 1–2 weeks.
- ***Granulomatous amoebic encephalitis* (GAE):** GAE is caused by *Acanthamoeba* spp. and *B. mandrillaris.* It is a disease entity that follows a subacute course and has a prodrome of weeks to months. It presents as a space occupying lesion of the brain. *Acanthamoeba* spp. causes GAE in immunocompromised patients, and *B. mandrillaris* can cause GAE in both immunocompromised and immunocompetent persons.
- ***Keratitis:*** *Acanthamoeba* spp. can also cause keratitis in contact lens wearers. The infection is often associated with corneal trauma, breaches in contact lens hygiene or use of contaminated lens solution.

LABORATORY DIAGNOSIS

N. fowleri

- Diagnosis can be made by microscopic examination of the wet mounts and Giemsa stained smears of CSF. Motile trophozoites with characteristic morphology (10–35 µm in size with granular and vacuolated cytoplasm) may be detected in wet mount.
- Trophozoites may also be detected in tissue or biopsy specimens (Fig. 32.5). Cysts are not found in humans.
- Diagnosis can also be made by PCR or by antigen testing from CSF, tissue or biopsy specimens.

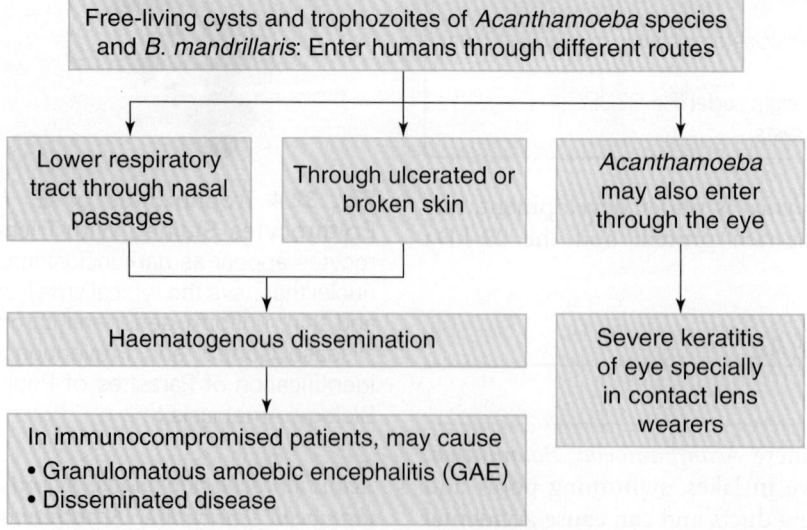

Flowchart 32.2 Life Cycle of *Acanthamoeba* spp. and *B. mandrillaris.*

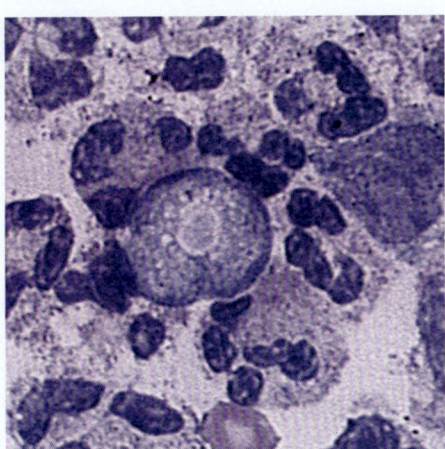

Fig. 32.5 Trophozoite of *N. fowleri* in CSF, Stained With Haematoxylin and Eosin (H&E). (*Source*: Centers for Disease Control and Prevention (CDC), DPDx - Laboratory Identification of Parasites of Public Health Concern, Parasite Biology, Free Living Amebic Infections.)

- Amoebae can also be grown on culture plate covered with bacteria that serve as a source of food for *N. fowleri*.

Acanthamoeba spp.

- Diagnosis can be made by demonstration of trophozoites (Fig. 32.6) or cysts (Fig. 32.7) in CSF, brain, eyes, lungs and other organs. The biopsy material may be stained with haematoxylin and eosin (H&E) or trichrome stain. The cysts measure 10–25 µm in size and have two walls: exocyst (wrinkled fibrous outer wall) and endocyst (hexagonal/polygonal/star-shaped/spherical inner wall). The cysts have a single nucleus with a large karyosome (Fig. 32.7).
- Indirect fluorescent antibody technique (IFAT) using specific antisera may be used for speciation.
- In vitro cultivation of *Acanthamoeba* spp. onto 1.5% non-nutrient agar (NNA) plates containing a layer of *E. coli* can also be used for diagnosis.

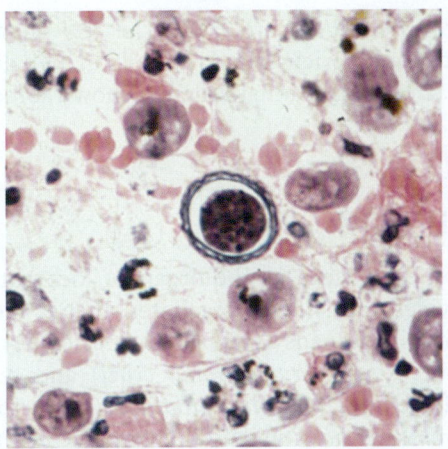

Fig. 32.7 Cyst of *Acanthamoeba* spp. From Brain Tissue, Stained With H&E. (*Source*: Centers for Disease Control and Prevention (CDC), DPDx - Laboratory Identification of Parasites of Public Health Concern, Parasite Biology, Free Living Amebic Infections.)

- Real time and conventional PCRs may also diagnose the infections.

B. mandrillaris

- Diagnosis is made by microscopically demonstrating trophozoites or cysts in human tissues. These are similar to those of *Acanthamoeba* spp. (Fig. 32.8).

TREATMENT

- GAE caused by *Acanthamoeba* spp. is fatal in most cases, but that caused by *B. mandrillaris* may be treated with a combination of flucytosine, pentamidine, fluconazole, sulphadiazine and either azithromycin or clarithromycin.

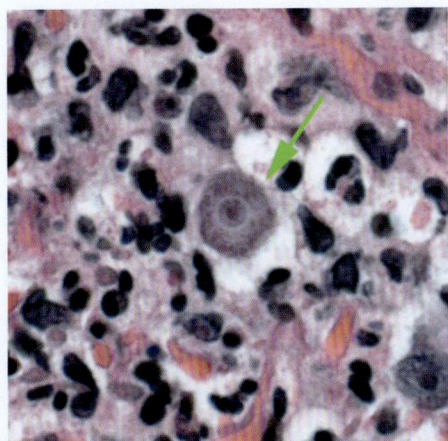

Fig. 32.6 Trophozoite of *Acanthamoeba* spp. in Tissue, Stained With H&E. (*Source*: Centers for Disease Control and Prevention (CDC), DPDx - Laboratory Identification of Parasites of Public Health Concern, Parasite Biology, Free Living Amebic Infections.)

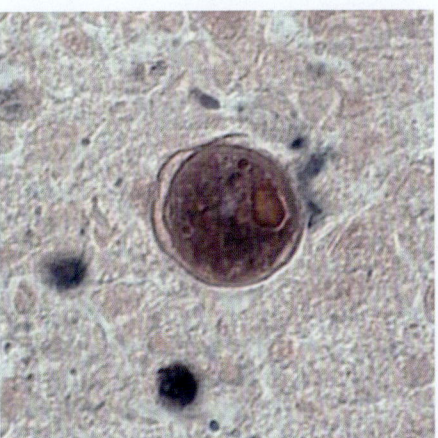

Fig. 32.8 Cyst of *B. mandrillaris* in Brain Tissue, Stained With H&E. (*Source*: Centers for Disease Control and Prevention (CDC), DPDx - Laboratory Identification of Parasites of Public Health Concern, Parasite Biology, Free Living Amebic Infections.)

- Most cases of PAM are fatal. The few survivors had received amphotericin B and rifampin in high doses. Miltefosine, an investigational drug has shown some promise.

FLAGELLATES

INTRODUCTION

Flagellates bear flagella as the organ of locomotion. They can be divided into **haemoflagellates**, that is those which live in blood in humans (*Leishmania* species, *Trypanosoma brucei* complex and *Trypanosoma cruzi*) and the **lumen-dwelling flagellates** (*Giardia intestinalis*, *Trichomonas* species, *Enteromonas hominis* and *Chilomastix mesnili*) based on their habitat. Taxonomically, flagellates belong to phyla Metamonada and Parabasalia in subkingdom Archezoa and to phylum Euglenozoa in subkingdom Neozoa (refer Chapter 31: Introduction to Parasites).

LEISHMANIA SPECIES

INTRODUCTION

Leishmania spp. are unicellular, eukaryotic, diploid and obligatory intracellular protozoa that primarily affect the

---

BOX 32.3   Spectrum of Leishmaniasis

- Visceral leishmaniasis: It affects internal organs, such as spleen, liver and bone marrow
- Cutaneous leishmaniasis causing skin sores
- Mucocutaneous leishmaniasis

---

reticuloendothelial system of the host. The parasite is vector borne, transmitted by the bite of female phlebotomine sandflies. These protozoa cause leishmaniasis, a diverse spectrum of syndromes that is broadly classified into three categories (Box 32.3).

Leishmania has two subgenera *L. (Leishmania)* and *L. (Viannia)*. More than 20 species can cause human infection; the most important ones are *Leishmania donovani* complex, *Leishmania mexicana* complex, *Leishmania tropica*, *Leishmania major*, *Leishmania aethiopica* and the subgenus *Viannia*. The classification of Leishmania is shown in Table 32.1.

LIFE CYCLE

The life cycle of *Leishmania* species is given in Flowchart 32.3.

TABLE 32.1   Some Important Facts of *Leishmania*

Species	Vector	Reservoir	Clinical Syndrome	Endemic Region
L. donovani[a]	Sand fly (*Phlebotomus argentipes*)	Humans[c]	VL, PKDL	India and some other parts of South Asia
	Sand fly (*P. orientalis, P. martini*)	Humans,[c] rodents, canines	VL, PKDL	Parts of Africa
	Sand fly (*P. perniciosus*)	Dogs, foxes, jackals	VL	Middle East, Africa
L. infantum[a]	Sand fly (*P. perniciosus*)	Dogs, foxes, jackals	VL, CL	Mediterranean, Middle East, Central Asia, China
	Sand fly (*P. turanicus*)	Humans,[c] dogs, foxes	VL	Central and North Asia
L. tropica[a]	Sand fly (*P. sergenti*)	Humans[c]	CL, LR	Western India, North and East Africa
L. major[a]	Sand fly (*P. papatasi*)	Rodents	CL	India, Middle East, Central and Western Asia
L. aethiopica[a]	Sand fly (*P. longipes*)	Hyraxes	CL, DCL	Parts of Africa
L. Viannia braziliensis complex[b]	Sand fly (*Lutzomyia* spp.)	Forest rodents	CL, MCL	South and Central America
L. Mexicana complex[b]	Sand fly (*Lutzomyia* spp.)	Forest rodents	CL, MCL, DCL	Central America and North of South America
L. chagasi (new world variant of *L. infantum*)[b]	Sand fly (*Lutzomyia* spp.)	Dogs, foxes	VL, CL	Central and South America

CL, Cutaneous leishmaniasis; *DCL*, diffuse cutaneous leishmaniasis; *LR*, leishmaniasis recidivans; *MCL*, mucocutaneous leishmaniasis; *PKDL*, post-kala azar dermal leishmaniasis; *VL*, visceral leishmaniasis.
[a]Cause old world Leishmaniasis.
[b]Cause new world Leishmaniasis.
[c]Can be transmitted from human to human through a vector (otherwise it is a zoonotic disease).

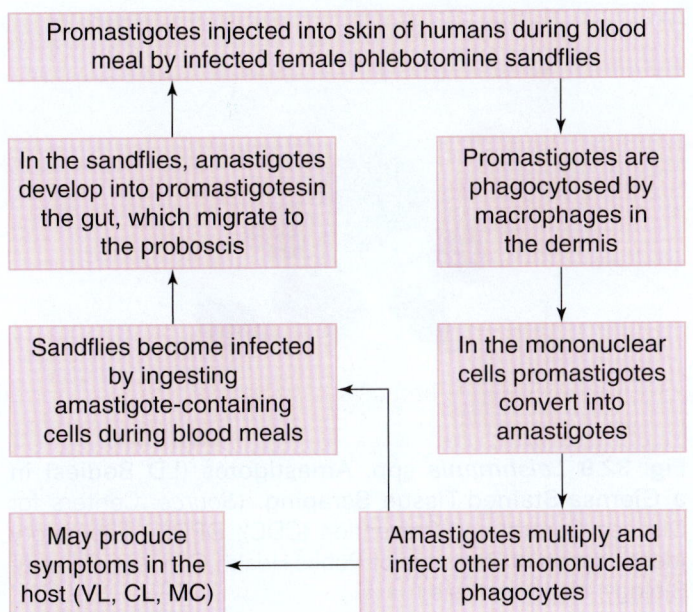

Flowchart 32.3 Life Cycle of *Leishmania* spp. (VL = visceral leishmaniasis, CL = cutaneous leishmaniasis, ML = mucocutaneous leishmaniasis).

Host: *Leishmania* species is digenetic, hence it requires two hosts to complete its life cycle – vertebrate host and insect vector.

Stages: *Leishmania* species has a dimorphic life cycle, that is it exists in two forms:

1. **Promastigote form** (elongate motile form found in sandflies and culture) and
2. **Amastigote form** (oval or round intracellular forms found in humans and is the nonmotile, multiplying stage).

Viannia subgenus: develops in the hindgut of sandflies.

Leishmania subgenus: develops in the midgut of sandflies.

EPIDEMIOLOGY

The leishmaniases are widely distributed across tropical, subtropical and temperate regions of the world and are found in all the continents except Australia and Antarctica. More than 30 species of sandflies transmit the protozoa in different parts of the world. These sandflies are most active from dusk to dawn. Transmission may be anthroponotic (human to sandfly to human transmission, i.e. humans are required to maintain the transmission cycle in nature) or zoonotic (animal reservoir to sandfly to humans, i.e. humans are not required to maintain the transmission cycle in nature). According to WHO, the disease is more common in areas with poverty, famine, malnutrition, illiteracy, lack of sanitation and hygiene and large scale migrations. It is more common in villages and outskirts of smaller towns than in big cities.

CLINICAL MANIFESTATIONS

Several clinical syndromes are caused by *Leishmania* spp. depending on the host, agent and environment factors. They are: cutaneous (infection of macrophages in dermis), muco-cutaneous (infection of macrophages in naso-oropharyngeal mucosa), visceral forms (infection throughout the reticulo-endothelial system), post-kala-azar dermal leishmaniasis (PKDL) and leishmaniasis recidivans.

Cutaneous Leishmaniasis

It is also known as 'the oriental sore' or 'Delhi boil' and is the most common form of leishmaniasis. Different species cause the old world (Asia, Middle East, the Mediterranean, Africa) and the new world (America) disease (Table 32.1). It is most frequently caused by *L. major*, *L. tropica* or *L. aethiopica* but *L. (Viannia)* and *Leishmania chagasi* can also cause cutaneous leishmaniasis.

The disease causes localised, single or multiple cutaneous ulcers, persisting for months to years. The lesions usually begin as papules that later develop into nodular plaques and ulcerative lesions. The painless ulcers are usually found in exposed parts of the skin and are surrounded by a raised or rolled border. The ulcers may heal leaving permanent scars, which cause social stigma.

Leishmaniasis Recidivans

It occurs occasionally several years after the primary sore has healed. It is a granulomatous response to *L. tropica* infection and is characterised by new scaly, erythematous papules developing on a previously healed sore.

Visceral Leishmaniasis

It is also known as 'kala-azar' and is caused by *L. donovani* complex (including *L. donovani*, the only species found in India and *L. infantum/L. chagasi*). India is the country worst affected by this parasite and together with Nepal, Bangladesh, Sudan and Brazil contributes to 90% of world's cases of VL. The parasite chiefly affects the reticuloendothelial system (including spleen, liver and bone marrow) and causes a disease with acute, subacute or chronic onset. The clinical manifestations of the disease include recurrent fever, weight loss with progressive emaciation, nontender soft hepatosplenomegaly (usually spleen is massively enlarged), pancytopenia, dry skin, loss of hair and greyish discolouration of skin of face, feet, hands and abdomen (hence called kala-azar, a Hindi term meaning 'black fever'). Lymphadenopathy may sometimes be seen. If untreated, severe cases are often fatal (may be 100% in 2 years). Leishmaniasis is an important opportunistic infection in people with HIV infection and may present with atypical manifestations, such as involvement of the gastrointestinal tract.

Post-kala-azar Dermal Leishmaniasis

PKDL is a condition in which *L. donovani* invades the skin cells and causes skin lesions typically on the face, 1–2 years after recovery of kala-azar. PKDL occurs in 5%–10% cases of VL within 2–4 years of treatment in South Asia (including India) and usually requires additional treatment. Patients with chronic PKDL are an important reservoir of infection.

The early sign of PKDL is an erythematous butterfly rash on mouth and cheek, aggravated by exposure to sunlight. It is characterised by a macular, maculopapular or nodular rash that starts from face and then spreads to arms and trunk and later resemble those of lepromatous leprosy. After a period of months to years, these progress to papules, nodules and verrucous forms.

Mucocutaneous Leishmaniasis

It is also known as 'espundia' and is caused by *L. (Viannia) braziliensis* complex or other New World *Leishmania* species belonging to *Viannia* subgenus (*Leishmania panamensis, Leishmania guyanensis*) or by *Leishmania amazonenesis*. In a small proportion of patients with primary cutaneous leishmaniasis, the parasites disseminate from the skin to mucosa of the nose, pharynx, oral cavity or larynx months to years after the skin lesions have healed. The soft tissues and cartilage of soft palate, nose and lips may be eroded. Simultaneous presence of skin and mucosal lesions has been observed in some cases. The disease has been linked to inadequate or no treatment of the primary lesions.

HOST IMMUNE RESPONSE

The immunology of leishmaniasis is complex and bipolar, similar to leprosy and has two extremes depending upon the type of T helper subset response (Table 32.2).

DIAGNOSIS

Clinical History

In an endemic area, fever of more than 2 weeks duration that does not respond to antibiotics or antimalarials, and is associated with anaemia, progressive leucopoenia or thrombocytopenia and hypergammaglobulinaemia is suggestive of leishmaniasis.

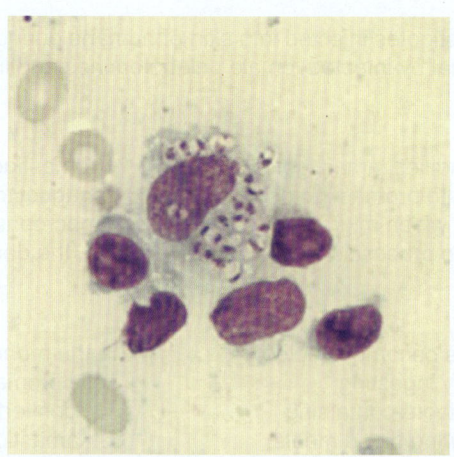

Fig. 32.9 *Leishmania* spp. Amastigotes (LD Bodies) in a Giemsa-Stained Tissue Scraping. (*Source*: Centers for Disease Control and Prevention (CDC), DPDx - Laboratory Identification of Parasites of Public Health Concern, Parasite Biology, Leishmaniasis.)

Laboratory Diagnosis

Sample: Tissue samples from representative sites, such as from skin lesions in CL or from spleen, bone marrow or lymph nodes in VL are the samples of choice.

Microscopy

The gold standard for diagnosis of leishmaniasis is demonstration of amastigotes inside macrophages (**Leishman Donovan (LD) bodies**) in smears of tissues. The amastigotes are oval to round in shape, measuring 1–5 µm by 1–2 µm in size. When stained with Giemsa or Leishman stain, the large nucleus stains red and cytoplasm appears pale blue. A characteristic deep red or violet rod-shaped **kinetoplast** (containing multiple copies of mitochondrial DNA) is visible at right angle to the nucleus. The amastigotes are usually found within the monocytes and macrophages, but few may also be found extracellularly (Fig. 32.9).

Culture

It is not usually required for diagnosis but may be done for research purposes on monophasic medium (Schneider's

TABLE 32.2	Host Immune Response	
	T Helper 1 Response	**T Helper 2 Response**
Cytokines produced	Interleukin-2 (IL-2), interferon γ (IFN γ)	IL-10 and IL-4
Effect on macrophages	Macrophages activated which kill amastigotes	Macrophages inhibited thereby enhancing survival and growth of the parasite
Response observed	Individuals able to control the infection due to successful immune response, CL, leishmaniasis recidivans, and in patients after treatment for VL	Individuals with active VL and diffuse CL
Leishmanin skin test (delayed-type hypersensitivity response to leishmanial antigens)	Positive	Negative

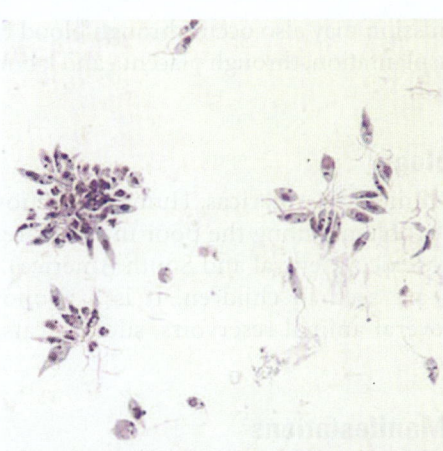

Fig. 32.10 *Leishmania* spp. Promastigotes From Culture. (*Source*: Centers for Disease Control and Prevention (CDC), DPDx - Laboratory Identification of Parasites of Public Health Concern, Parasite Biology, Leishmaniasis.)

insect medium, M199 medium) or biphasic medium (NNN medium, Novy-MacNeal-Nicolle medium and Tobies medium). The promastigotes are visible in the culture medium and are characterized by a flagellum and a kineto-plast anterior to the nucleus (Fig. 32.10).

Serology

Several serological tests that detect specific IgG antibodies are available for diagnosis of kala-azar, including direct agglutination test (DAT), rk39 dipstick test and ELISA. The commonly used 'Napier's aldehyde test' and 'Chopra's antimony test' are non-specific tests to detect hypergammaglobulinaemia. Most of these tests do not differentiate between active and latent infections. These tests also have a limited value in CL.

Molecular Methods

These may be used for diagnosis as well as for species differentiation.

Leishmanin Test (Montenegro Test)

It is a delayed hypersensitivity skin test to a suspension of killed promastigotes of *L. donovani* injected intradermally. An induration of more than 5 mm after 72 h is taken as positive, which indicates prior exposure to Leishmania antigen. This test can be used for epidemiological surveys.

TREATMENT

The drugs that can be used for treatment of leishmaniasis are pentavalent antimonials, such as sodium stibogluconate, pentamidine isethionate, amphotericin B, liposomal amphotericin B and miltefosine.

PREVENTION

- No vaccines or drugs are available for prevention. The only way of prevention is to protect oneself from bites of sandflies by using bednets, insect repellents and insecticides.
- Kala-azar elimination programme has been implemented in the WHO South-East Region and has shown a good response. For example Bangladesh reported >9000 new cases in 2006, which decreased to 600 in 2014–15.
- In India, kala-azar elimination programme was initiated by the Government of India in endemic areas in 1990–91 based on the following strategies: vector control; early diagnosis and treatment; information education and communication; and capacity building.

TRYPANOSOMA SPECIES

INTRODUCTION

The genus *Trypanosoma* contains several species. But human disease is known to be caused by only three species: *T. cruzi*, *T. brucei gambiense* and *T. brucei rhodesiense*. Human diseases caused by *Trypanosoma* spp. are: Chagas disease (American trypanosomiasis) caused by *T. cruzi* and sleeping sickness (African trypanosomiasis) caused by *T. b. gambiense* and *T. b. rhodesiense*.

The trypanosomes are flagellated protozoans belonging to subkingdom Neozoa, phylum Euglenozoa, class Kinetoplastea and order Trypanosomatida. The trypanosomes are divided into two major groups, based on their course of development in the vector, that is Stercoraria and Salivaria (Table 32.3).

CHAGAS DISEASE

Also known as American trypanosomiasis, it is a zoonotic disease caused by the protozoa *T. cruzi* and is transmitted to humans by blood sucking triatomine bugs/reduviid bugs/kissing bugs. Depending on stage of the disease, presentation may be acute (occurs after initial infection), or chronic, which in turn has two phases: (1) the indeterminate phase with no symptoms but with subpatent parasitaemia and detectable antibodies and (2) the symptomatic phase with cardiac or gastrointestinal symptoms that may be fatal.

Life Cycle (Flowchart 32.4)

Host: *T. cruzi* is digenetic, that is it needs two hosts to complete its life cycle: a mammalian host and an insect vector (reduviid bugs/kissing bugs/triatomine bugs). Humans are

TABLE 32.3	Groups of Trypanosomes			
Group	**Development in Vector**	**Transmission to Mammals**	**Stage Found in Humans**	**Example**
Stercoraria	In the hindgut of vector	By contamination	Amastigote stage	*T. cruzi*
Salivaria	In the salivary glands of vector	By inoculation	Trypomastigote stage	*T. brucei* complex

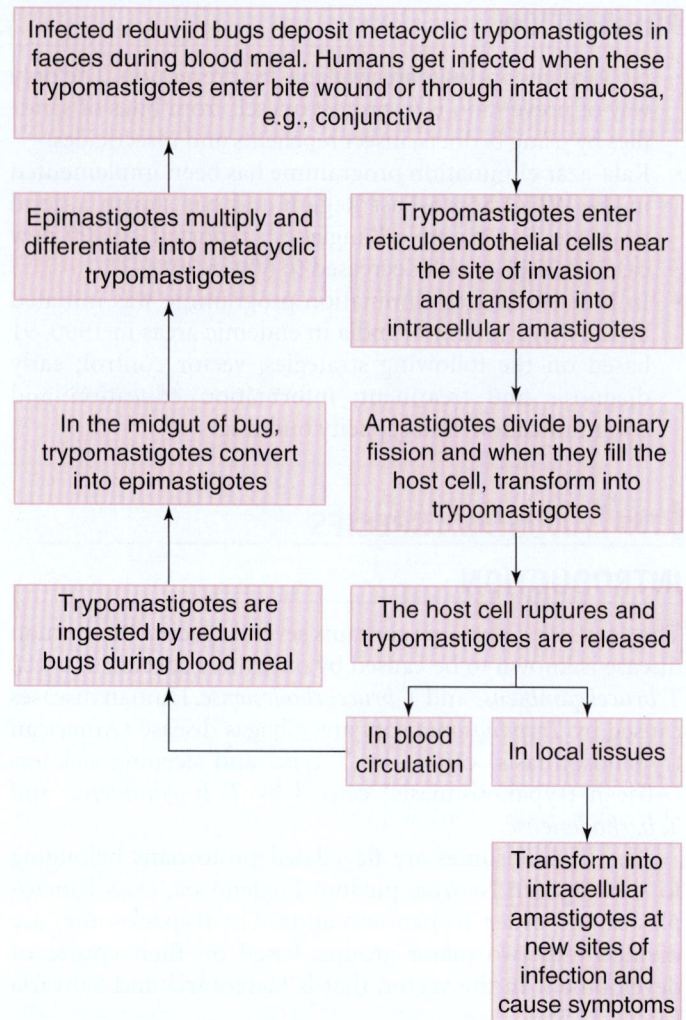

Flowchart 32.4 Life Cycle of *T. cruzi*.

the accidental hosts and are not necessary for natural transmission of its life cycle. The parasite may be maintained in a non-human mammalian host.

Developmental Stages (Box 32.4)

Within these vertebrate and invertebrate hosts, *T. cruzi* passes through three morphologic forms: amastigote, epimastigote and metacyclic trypomastigote.

Transmission

Metacyclic trypomastigote is the infective stage. The parasite is transmitted mainly through faeces of infected bugs

---

BOX 32.4   Developmental Stages of *T. cruzi*

- **Amastigotes**: It is found in cells of mammalian hosts (e.g. skeletal and cardiac muscles, nervous tissue and reticuloendothelial cells), is **nonmotile** and the **multiplying stage** of the parasite.
- **Metacyclic trypomastigotes**: It is an **infective** stage and is found in bloodstream, in the lumen of rectum of the vector insect, and in culture.
- **Epimastigotes**: It is found in mammalian hosts intracellularly, in digestive tract of the vector in multiplying forms and in culture.

---

but transmission may also occur through blood transfusion, organ transplantation, through placenta and laboratory accidents.

Epidemiology

T. cruzi is limited to Americas. Human infection is a public health problem among the poor in rural areas of Latin America (Mexico, Central and South America). Most new infections are seen in children. It is a zoonotic disease and has several animal reservoirs, such as cats, dogs and rodents.

Clinical Manifestations

- *Acute phase of the disease*: It is usually asymptomatic. '**Chagoma**' may appear about 1 week after invasion by the parasites. It consists of a nodular lesion or furuncle at the site of inoculation. '**Romana's sign**', unilateral oedema of palpebrae and periocular tissues may occur in patients in whom the parasite enters through the conjunctiva. These acute signs may be followed by fever, anorexia, oedema of face and lower extremities, generalised lymphadenopathy and hepatosplenomegaly. Heart failure, though rare may cause death in acute Chagas disease. Meningoencephalitis may occur rarely in children. In most patients, symptoms resolve spontaneously over few weeks to months to enter into asymptomatic/indeterminate phase of chronic Chagas disease.
- *Symptomatic chronic Chagas disease* occurs years to decades after initial infection and may cause cardiomyopathy (the main cause of death in patients), megaoesophagus, megacolon, weight loss. These complications may be fatal.
- *Congenital trypanosomiasis*: It can occur rarely, manifested as still birth, low birth weight, myocarditis or neurologic manifestations.

Laboratory Diagnosis

For Acute Chagas Disease

Sample: blood/CSF/tissue biopsy

Microscopic examination:

- **Direct examination**: the motile organisms may be detected in anticoagulated blood or buffy coat by examination under a cover slip or in microhaematocrit tubes.
- **Giemsa-stained smears**: (1) trypomastigotes can be detected in thick- and thin-blood smears. Trypomastigotes measure 12–30 μm in length, have large terminal or subterminal kinetoplast, a centrally placed nucleus, an undulating membrane and a flagellum running along the undulating membrane and leaving the body at its anterior end (Fig. 32.11). (2) Amastigotes are found in tissues and have a single nucleus, a kinetoplast and are morphologically indistinguishable from those of *Leishmania* species (Fig. 32.12).

For Chronic Chagas Disease

Microscopic examination of blood is not useful because trpomastigotes are not found in blood. Diagnosis is made

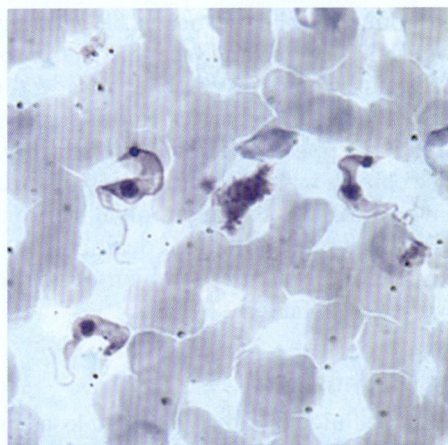

Fig. 32.11 Three *T. cruzi* Trypomastigotes in a Thin Blood Smear Stained With Giemsa. (*Source*: Centers for Disease Control and Prevention (CDC), DPDx - Laboratory Identification of Parasites of Public Health Concern, Parasite Biology, American Trypanosomiasis.)

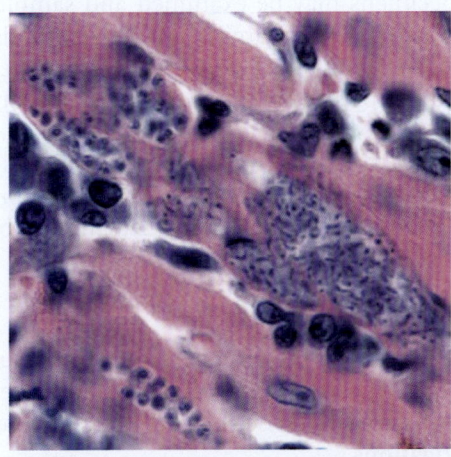

Fig. 32.12 *T. cruzi* Amastigotes in Heart Tissue. The section is stained with H&E. (*Source*: Centers for Disease Control and Prevention (CDC), DPDx - Laboratory Identification of Parasites of Public Health Concern, Parasite Biology, American Trypanosomiasis.)

by serologic tests: the indirect fluorescent antibody (IFA) test and enzyme immunoassay (EIA).

For suspected transfusion or transplant transmitted infections, congenital transmissions and laboratory acquired infections, exposures can be monitored using PCR in blood or heart tissue.

For screening of blood in endemic areas, serologic tests are used for screening blood in blood banks.

Treatment

The treatment of Chagas disease is unsatisfactory. Only two drugs, nifurtimox and benznidazole are available for treatment.

Prevention

The disease can be prevented by vector control using methods such as insecticide spray, improvements in housing and

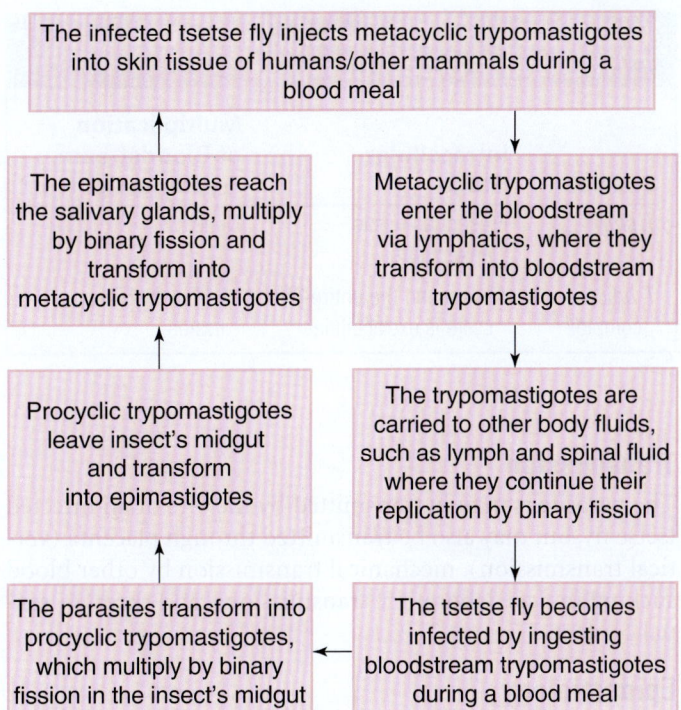

Flowchart 32.5 Life Cycle of *T. brucei* Complex.

health education. Besides this, serologic screening of donated blood is already in place in blood banks of endemic areas.

HUMAN AFRICAN TRYPANOSOMIASIS

It is also called 'sleeping sickness' and is a vector-borne parasitic disease caused by haemoflagellate protozoan parasites belonging to *T. brucei* complex. It is transmitted to humans by tsetse fly (genus *Glossina*). The trypanosomes initially cause a febrile disease, which if left untreated may lead to progressive neurologic impairment and death, months or years later.

There are two subspecies of *T. brucei* complex that are morphologically indistinct. They cause two epidemiologically and clinically distinct disease patterns:

1. *T. brucei gambiense*: It causes West African sleeping sickness (*gambiense*) (98% of the reported disease)
2. *T. brucei rhodesiense*: it causes East African sleeping sickness (*rhodesiense*).
3. *T. brucei brucei* is the third member of the complex but does not cause human disease.

Life Cycle

The life cycle of *T. brucei* complex is given in Flowchart 32.5. Differences in life cycle of *T. cruzi* and *T. brucei* are shown in Table 32.4.

Host: *T. brucei* complex needs two hosts to complete its life cycle (digenetic) – man or other vertebrate host and tsetse fly. The main reservoir for *T. b. gambiense* is human and that for *T. b. rhodesiense* is wild game (hence history of contact becomes very important), though other animals may also be infected.

TABLE 32.4   Differences in Life Cycles of *T. cruzi* and *T. brucei* Complex

Species	Intracellular Stage	Multiplication of Bloodstream Trypomastigotes
T. cruzi	Present as tissue amastigotes	None
T. brucei complex	Absent and the entire life cycle is extracellular	Multiply by binary fission

Transmission

The parasite is mainly transmitted by the bite of an infected tsetse fly, but may also be transmitted through placenta (vertical transmission), mechanical transmission by other blood sucking insects, laboratory transmissions, and through sexual contact.

Epidemiology

The disease is found in 36 countries of sub-Saharan Africa where the vector tsetse fly is found. The people living in rural areas in endemic regions and depending on agriculture, fishing, animal husbandry or hunting are the most exposed to the disease. According to the World Health Organization, *T. b. gambiense* is found in 24 countries in west and central Africa and accounts for 97% of the reported cases of sleeping sickness. *T. b. rhodesiense* is found in 13 countries in eastern and southern Africa and accounts for 3% cases. Both forms of the disease are present only in Uganda, but in different zones.

Clinical Manifestations

A painful chancre may appear at the site of inoculation of the parasite in some patients. The human African trypanosomiasis has two stages during its clinical course:
- **Stage I**, during which the parasite is found in the peripheral circulation but has not yet involved the central nervous system (CNS). During this stage, episodes of high-grade fever lasting several days occur that are separated by afebrile episodes. Pruritis and maculopapular

rashes are common. Malaise, headache, arthralgias, hepatosplenomegaly, oedema and tachycardia may occur.
- **Stage II**, begins when the parasites invade the CNS by crossing the blood–brain barrier. In this stage, neurologic manifestations appear including personality changes, daytime sleepiness with night time sleep disturbance, and progressive confusion. Extrapyramidal signs, ataxia may occur finally leading to coma and death.

The two forms of the disease (gambiense and rhodesiense) differ in several ways (Table 32.5).

Laboratory Diagnosis

Sample: Chancre fluid, blood, lymph node aspirate, bone marrow or CSF during the neurologic stage.

Microscopy

- **Wet mount:** Motile trypomastigotes are visible
- **Giemsa-stained smears:** Stained trypomastigotes are seen. To increase the sensitivity of parasite detection, serial specimens may be examined and samples may be concentrated before examination, by methods such as buffy coat examination, mini anion exchange columns for blood samples, centrifugation and examination of the sediment for CSF samples. A trypomastigote measures 14–33 μm in size, has a small kinetoplast at its posterior end, an undulating membrane, and a flagellum running along the undulating membrane, leaving the body at its anterior end (Fig. 32.13).

Serology

Detection of antibody by serology can be done but the sensitivity and specificity is highly variable and therefore, is of limited use in clinical practice but can be used for epidemiologic investigations.

Treatment

Pentamidine is the treatment of choice for first stage infection, though Suramin may also be used. For second stage disease, eflornithine either alone or in combination with nifurtimox is recommended for *T. b. gambiense* and melarsoprol is recommended for *T. b. rhodesiense*.

TABLE 32.5   Differences Between the Two Forms of Human African Trypanosomiasis

	Gambiense Form	Rhodesiense Form
Common name	West African sleeping sickness	East African sleeping sickness
Primary reservoir	Humans	Animals (antelope, cattle)
Epidemiology	Rural population	Rural population, hunters or tourists in game parks
Rate of disease progression	The disease progresses slowly and can last for months or years; the early symptoms are mild	The disease progresses rapidly and symptoms, such as fever, headache, malaise, lymphadenopathy appear within 1–2 weeks of bite of the tsetse fly; the untreated disease leads to death within a few months
CNS invasion by the parasites and neurological manifestations	Occurs usually after 1–2 years of infection	Occurs rapidly within a few weeks of infection
Parasitaemia	Low	High

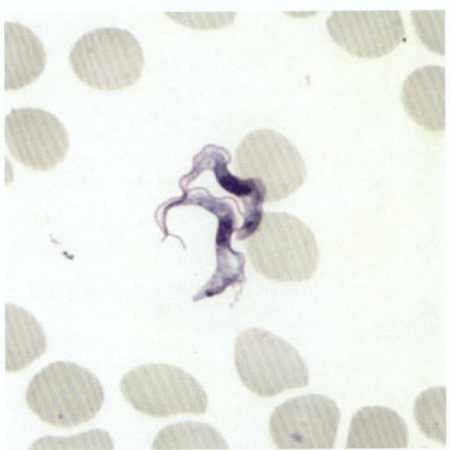

Fig. 32.13 *T. brucei* spp. in a Thin Blood Smear Stained With Giemsa. (*Source*: Centers for Disease Control and Prevention (CDC), DPDx - Laboratory Identification of Parasites of Public Health Concern, Parasite Biology, Leishmaniasis.)

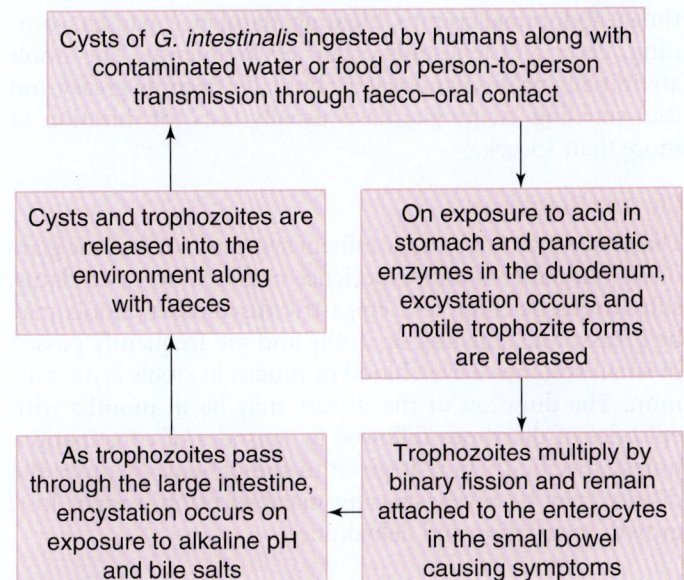

Flowchart 32.6 Life Cycle of *Giardia intestinalis*.

Prevention

Under the leadership of the World Health Organization, a coordination network was established in 2014 to eliminate the disease. The control program has a multipronged strategy that emphasises on effective surveillance, early diagnosis and management of cases and vector control.

LUMEN-DWELLING FLAGELLATES (INTESTINAL AND GENITAL)

GIARDIA INTESTINALIS

Introduction

Giardiasis is a zoonosis that is transmitted between humans and animals (e.g. dogs, cats, deer, cattle, and most importantly beavers). It is caused by a flagellated intestinal protozoan *Giardia*, which was discovered by Von Leeuwenhoek in 1681 in his own diarrhoeal stool. *Giardia intestinalis* (also known as *G. lamblia* or *G. duodenalis*) is the most common cause of epidemic and endemic diarrhoea worldwide. It is the most common intestinal parasite found in both the developing and developed countries.

Life Cycle

The life cycle of *Giardia intestinalis* is given in Flowchart 32.6.

Host: Giardia is monogenetic and requires only humans or animals (e.g. dogs, cats, deer, cattle and most importantly beavers) for the completion of life cycle. No intermediate host is required.

Developmental stages: The life cycle of *G. intestinalis* consists of two stages – the **trophozoite** (flagellated motile form) and the **cyst** (free-living form).

Transmission

Infection is acquired by faeco–oral transmission by ingestion of cysts along with contaminated water and food (when food gets contaminated with cysts after cooking). Person-to-person transmission is very common and may occur within families or in institutionalised persons like day care centres. Cysts are mainly responsible for transmission of infection because they are resistant to killing by routine water chlorination methods, can survive for months in fresh cold water and have a low infectious dose (ingestion of even 10 cysts can cause human infection). The cysts are however killed by heating, desiccation or prolonged exposure to faeces.

Epidemiology

Giardia has a worldwide distribution occurring in tropical and temperate regions. Children are more commonly affected than adults and epidemics are common in the day care centres. Others affected include elderly debilitated persons, people with poor hygiene and immunodeficient people. Faeco–oral transmission occurs in areas of poor hygiene.

Pathogenesis

Minimum infectious dose is 10–25 cysts. Trophozoites adhere to intestinal mucosa and cause disruption of brush border epithelium. This leads to increased permeability and decreased absorption. Malabsorption leads to steatorrhea (fats in stool), lactose intolerance and protein loosing enteropathy.

Clinical Manifestations

Most infections are asymptomatic, though during epidemics, symptomatic infections usually outnumber asymptomatic infections.

Asymptomatic Cyst Passers

It occurs in 5 %–15% of the people ingesting cysts.

Acute Giardiasis

The symptoms develop after an incubation period of 1–2 weeks. The patient usually presents with acute onset diar-

---

Flowchart 32.6 boxes:

Cysts of *G. intestinalis* ingested by humans along with contaminated water or food or person-to-person transmission through faeco–oral contact

On exposure to acid in stomach and pancreatic enzymes in the duodenum, excystation occurs and motile trophozite forms are released

Trophozoites multiply by binary fission and remain attached to the enterocytes in the small bowel causing symptoms

As trophozoites pass through the large intestine, encystation occurs on exposure to alkaline pH and bile salts

Cysts and trophozoites are released into the environment along with faeces

rhoea, abdominal cramps, bloating, flatulence, nausea, vomiting. Fever and tenesmus occur less frequently. The stools are initially profuse and watery but later become greasy and foul smelling. Acute giardiasis usually lasts for duration of more than 1 week.

Chronic Giardiasis

Diarrhoea may not be a prominent symptom of chronic giardiasis but patients may experience malaise, increased flatus, sulphurous belching and epigastric discomfort. Stools may be greasy, foul smelling or frothy and are frequently passed in small volumes. Fever, blood or mucus in stools is uncommon. The duration of the disease may be in months with episodes of diarrhoea followed by episodes of constipation, unless specific therapy is given to the patient. At times the disease may be severe causing malabsorption, weight loss, growth retardation and dehydration.

Unusual Manifestations

Unusual manifestations may include urticaria, arthritis, anterior uveitis and biliary tract disease. The disease may be severe in patients with hypogamaglobulinaemia, cystic fibrosis or other intestinal diseases and AIDS.

Laboratory Diagnosis

Sample required: Freshly passed stool/duodenal aspirates or duodenal biopsy.

Gross examination of stool: Blood, pus and mucus are usually not found.

Microscopic demonstration: Diagnosis is usually established by demonstration of cysts or trophozoites in wet mounts or iodine stained fresh stool samples. For stool samples preserved in 10% buffered formalin or polyvinyl alcohol, trichrome staining is done. The sensitivity of the procedure may be increased by concentration techniques (formalin-ether or zinc sulphate flotation technique) and by repeated sampling.

The cysts are smooth, thin walled, oval in shape, measuring 8–12 µm in size, contain two nuclei (immature cysts) or four nuclei (mature cysts) and fibrils (Fig. 32.14).

The trophozoites are pear shaped and measure 10–20 µm in length and 5–15 µm in width. It has a convex dorsal surface and a flat ventral surface containing sucking discs/adhesive discs. The transversely placed median bodies (tight collection of microtubules) and four pairs of flagella may also be seen (Fig. 32.15).

A direct fluorescence assay (DFA) may be used for detection of Giardia in stool. The antibodies tagged with fluorescent markers bind to the cysts, giving them an appearance of green, glowing ovoid bodies.

A stool antigen detection test is available that can simultaneously detect *E. histolytica* and *G. lamblia*.

Molecular tests may be used for subtyping *Giardia*.

Treatment

The drugs of choice include metronidazole or tinidazole.

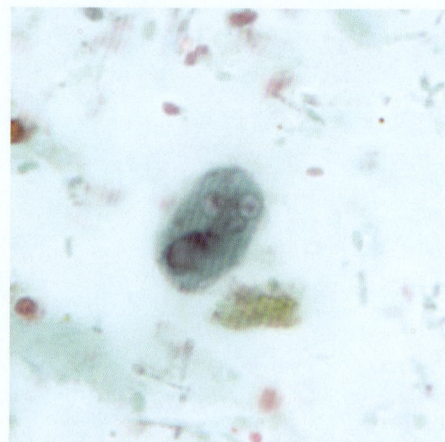

Fig. 32.14 *G. duodenalis* Cyst Stained With Trichrome. (*Source*: Centers for Disease Control and Prevention (CDC), DPDx - Laboratory Identification of Parasites of Public Health Concern, Parasite Biology, Giardiasis.)

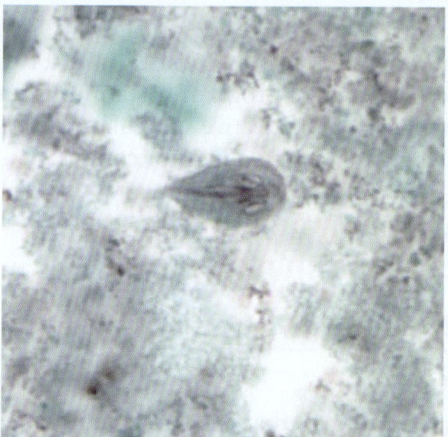

Fig. 32.15 *G. lamblia* Trophozoite Stained With Trichrome. (*Source*: Centers for Disease Control and Prevention (CDC), DPDx - Laboratory Identification of Parasites of Public Health Concern, Parasite Biology, Giardiasis.)

Prevention

The disease can be prevented by maintaining personal hygiene while caring for infected children and by boiling or filtering potentially contaminated water.

TRICHOMONAS VAGINALIS

Trichomoniasis is a very common sexually transmitted disease (STD) caused by the flagellated protozoan parasite *Trichomonas vaginalis* that can be cured easily. It is the most common cause of symptomatic vaginitis in women. In men, the infection is mostly asymptomatic, though in some cases it may cause urethritis. Besides *T. vaginalis*, other trichomonads include *Trichomonas tenax*, found in mouth (may cause periodontidis) and *Pentatrichomonas hominis*, found in the gastrointestinal tract.

Life Cycle

Host: Humans are the only natural host of *T. vaginalis.* The organism resides in the lumen and on the mucosal surfaces of the lower genital tract of females and of the urethra and prostrate of males.

Transmission

Transmission of trophozoites occurs primarily by sexual intercourse. The trophozoites multiply by binary fission. The cyst form does not exist.

Epidemiology

The parasite has a worldwide distribution. Its prevalence is highest among persons with high-risk sexual behaviour (e.g. having multiple sexual partners) and those with other STDs.

Clinical Manifestations

Infection by *T. vaginalis* in men is mostly asymptomatic but occasionally may cause epididymitis, prostatitis or urethritis. Infection in women is mostly symptomatic. After an incubation period of 5–28 days, vaginitis with a malodorous purulent yellow green discharge appears, along with vulvar erythema and itching, abdominal pain, vulvar and cervical lesions, dysuria and dyspareunia. Strawberry cervix (**colpitis macularis**) is a characteristic clinical sign observed during colposcopy.

Laboratory Diagnosis

Sample Required

Vaginal, urethral or prostatic secretions.

Microscopic Examination

Diagnosis is established by demonstrating actively motile trophozoites in wet mounts of the sample. This method has the benefit of providing immediate and rapid diagnosis but its sensitivity of only about 60% limits its use.

The trophozoites may also be seen in Giemsa-stained or Papanicolaou-stained smears of vaginal secretions. The trophozoites are pear shaped, measuring 7–30 μm in length, having four free flagella directed anteriorly and a flagellum directed posteriorly along the outer margin of the undulating membrane. It also has a large nucleus located at the anterior broad end, a small karyosome, an axostyle extending through the cell and many chromatin granules (Fig. 32.16).

Culture

The current gold standard for diagnosis is culture of the parasite, but the results are available only after 3–7 days.

Antigen Detection Tests

Immunochromatographic strip test that uses specific antibodies to detect Trichomonas protein antigens can be used.

Molecular Tests

PCR, PCR-based ELISA and transcription-mediated amplification (TMA) are available with good sensitivity and specificity. The other tests available include a dipstick ELISA and a RNA probe-based semi-automated system.

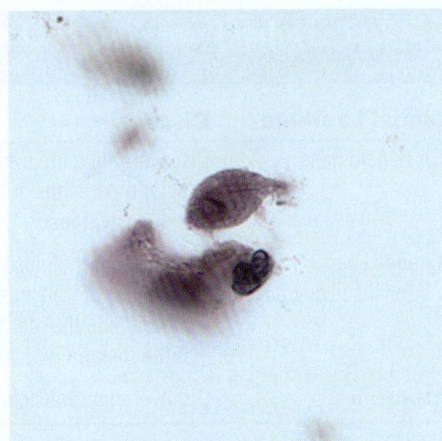

Fig. 32.16 Trophozoite of *T. vaginalis* in a Vaginal Smear, Stained With Giemsa. (*Source*: Centers for Disease Control and Prevention (CDC), DPDx - Laboratory Identification of Parasites of Public Health Concern, Parasite Biology, Trichomoniasis.)

Supportive Tests

Raised vaginal pH (>4.5) and positive whiff test (fishy odour when a drop of 10% KOH is added to vaginal discharge) supports the diagnosis.

Treatment

The treatment of choice is single dose of metronidazole or tinidazole taken orally. All the sexual partners must be treated simultaneously including asymptomatic males.

Prevention

Trichomoniasis can be prevented by refraining from high-risk sexual activities and by using latex condoms during each sexual activity.

OTHER FLAGELLATES

Other flagellates that usually are harmless commensals in human but can occasionally cause a disease include *P. hominis, C. mesnili, E. hominis, Retortamonas intestinalis, Dientamoeba fragilis* (commensals found in large intestine) and *T. tenax* (commensal found in mouth).

SPOROZOA

INTRODUCTION

This phylum is characterised by the presence of a specialised complex of apical organelles involved in host cell invasion. It includes pathogens, such as *Plasmodium, Babesia, Toxoplasma,* and *Cryptosporidium* species. Taxonomically they belong to subkingdom Neozoa, phylum Sporozoa, class Coccidea and three orders: Haemosporida (genus *Plasmodium*), Eimeriida (genera *Toxoplasma, Eimeria, Cryptosporidium, Cyclospora, Cystoisospora, Sarcocystis*), and Piroplasmida (genus *Babesia*) (refer to Chapter 31).

TABLE 32.6   Nobel Prize Winners for Malaria

Year	Nobel Laureate	Discovery
1902	Sir Ronald Ross	Discovered the life cycle of the malaria parasites in humans and mosquitoes
1907	Charles Louis Alphonse Laveran	Showed that the mosquito is the agent of transmission for malaria and the identification of the malaria parasite
2015	Youyou Tu	Discovered artemisinin

PLASMODIUM

Malaria is a life-threatening disease caused by protozoa of the genus *Plasmodium* that is transmitted by the bite of female *anopheles* mosquitoes. This was one of the most dreaded infections, which made the way for many scientists to win Noble Prize (Table 32.6). It is estimated that up to 40% of the world's population is at risk of acquiring malaria, with maximum cases occurring in the tropical regions of the world. There are five species of genus *Plasmodium* causing malaria in human namely *Plasmodium falciparum*, *Plasmodium vivax*, *Plasmodium malariae*, *Plasmodium ovale* and *Plasmodium knowlesi* (the monkey malaria parasite found in Southeast Asia, which is considered as zoonotic malaria). Of these, *P. falciparum* is responsible for most malaria-related deaths globally followed by *P. vivax*.

Life Cycle

Hosts: The life cycle of malaria parasite involves two hosts: female *anopheles* mosquitoes (*Anopheles culicifacies*, *Anopheles stephensi*, *Anopheles fluviatilis*, *Anopheles maculates*, *Anopheles philippinensis* etc.) (definitive host) and man (intermediate host). The sporogonic (sexual) cycle occurs in mosquito. Schizogony (asexual cycle) occurs in humans (Flowchart 32.7).

Transmission

Sporozoites (infective stage) are transmitted to humans by bite of infected female anopheles mosquito. Infection may also be transmitted rarely by blood transfusion or transplacental transmission.

Human Stages

In humans the asexual cycle occurs through the following stages – (1) exoerythocytic/pre-erythrocytic schizogony that occurs in the liver parenchymal cells, (2) the erythrocytic schizogony in red blood cells and (3) gametogony in blood vessels of internal organs, such as spleen and bone marrow.

Incubation period: It is the period between bite of an infected mosquito and the first appearance of symptoms. It

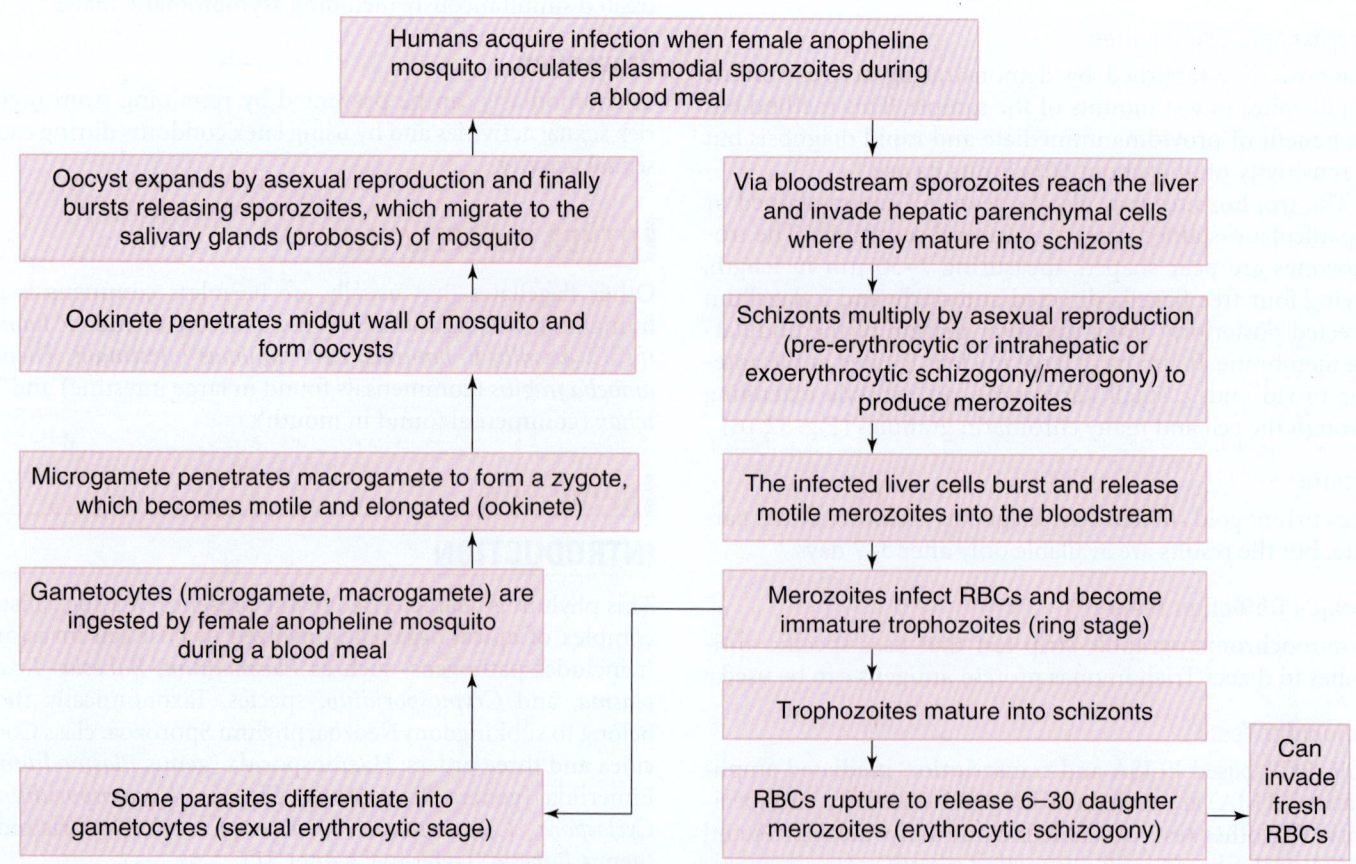

Flowchart 32.7 Life Cycle of *Plasmodium* spp.

varies from 7 to 30 days and depends on the species: *P. falciparum* has shorter incubation periods and *P. malariae* has longer incubation periods.

Prepatent period: The time interval between entry of the parasite in human and demonstration of ring forms in the peripheral blood is called prepatent period and depends on the species.

Pre-erythrocytic Schizogony

- The sporozoites leave the circulation and enter hepatocytes where they transform into trophozoites (feeding stage). The trophozoites undergo schizogony (multiplication) and transform into pre-erythrocytic schizonts containing several merozoites that are released in blood to start the erythrocytic schizogony. A single sporozoite may eventually form 10,000–30,000 merozoites. The duration of pre-erythrocytic schizogony varies from 5 to 15 days depending upon the species.
- In *P. vivax* and *P. ovale* infections, hypnozoites (dormant stage that do not divide immediately) can persist in liver for long periods. These hypnozoites are responsible for **relapses** that characterise infection with these species.

Erythrocytic Schizogony

- Merozoites enter the RBCs to form trophozoites. Early trophozoites are called the **ring form**, which have a signet ring appearance with a central vacuole and a peripheral thin rim of cytoplasm and a nucleus.
- **Haemozoin/malarial pigment:** In an infected RBC, the growing parasite rapidly consumes intracellular proteins especially haemoglobin and converts the potentially toxic haem to biologically inert haemozoin.
- **Amoeboid/late trophozoite form:** The ring form enlarges to become irregular and amoeboid. The parasite also alters the RBC membrane so that it becomes more irregular in shape, less deformable and more antigenic.
- **Erythrocytic schizont:** The late trophozoites get compact to become erythrocytic schizont, which has a big peripheral nucleus and malarial pigment scattered throughout cytoplasm.
- **Erythrocytic schizogony/merogony:** Erythrocytic schizonts undergo several divisions to produce 6–30 daughter merozoites arranged in a rosette.
- The symptoms in humans are caused when RBCs rupture to release the daughter merozoites, malarial pigment and toxins that stimulate the macrophages and other cells to secrete soluble factors including cytokines. These cytokines are responsible for fever, rigors and other clinical manifestations. The symptoms usually begin when the parasite density reaches about 50 per microliter of blood.
- **Sequestration**: *P. falciparum* infections induce the processes of cytoadherence, rosetting and agglutination. All these processes cause sequestration of RBCs containing mature forms of the parasite in vital organs like brain. The blood flow to these organs is thus compromised and conditions, such as cerebral malaria and blackwater fever (**malignant tertian malaria**) occur. Also, it is because of this sequestration that only the ring forms of the parasite

are seen in peripheral blood in case of *P. falciparum* infection. **Processes involved in sequestration are as follows**:
- **Cytoadherence:** Infected RBCs attach to the receptors on endothelium of venules and capillaries, thus blocking them.
- **Rosetting:** Infected RBCs attach to uninfected RBCs thereby forming rosettes.
- **Agglutination:** Infected RBCs attach to other parasitised RBCs.

Gametogony

- After few erythrocytic cycles some merozoites transform into gametocytes (sexual forms), which are released in peripheral blood. The male gametocyte is called microgametocyte and the female gametocyte, macrogametocyte.

Mosquito Stages

Extrinsic incubation period: It is the time required to complete the life cycle in mosquito and varies from 1 to 4 weeks.

Exflagellation: Microgametocyte divides into eight microgametes by exflagellation. Macrogametocytes mature to form macrogamete.

Mosquito cycle consists of the following sequential stages: zygote, ookinete (vermicular, elongated, motile form with an apical complex), oocyst (rounded form), and the infective form sporozoites (spindle-shaped released by rupture of mature oocyst).

Epidemiology

Malaria is endemic throughout most of the tropical countries of the world. According to the World Health Organization there were 216 million cases of malaria in 2016. Most of the global burden of malaria is borne by the WHO African Region countries: in 2016, 90% of malaria cases and 91% of malaria deaths occurred in these countries. India and 14 countries of sub-Saharan Africa accounted for 80% of the global disease burden. Between 2000 and 2010, the global malaria incidence and malaria specific mortality rates declined by 17 and 26%, respectively. Malaria mortality and morbidity has declined as a result of intensive **malaria control program** and several countries including Armenia, Morocco, Turmenistan and the United Arab Emirates have been certified as malaria free.

People at higher risk of acquiring malaria infection and developing severe disease include children less than 5 years of age, infants, pregnant women and patients with HIV/AIDS in regions of high malaria transmission. However an epidemiological shift has occurred towards males and adults in many malaria eliminating countries. The endemicity of malaria is very complex and depend upon several factors including man host interactions (agriculture, deforestation), parasite (different species with different sporogony cycle), vector (density, longevity, breeding sites), and climate (rainfall, temperature). Malaria endemicity in a population may be classified as per several indices (Box 32.5).

Host Immunity to Malaria

The initial host response to plasmodial infections is nonspecific. Spleen actively removes both parasitised and uninfected RBCs thereby resulting in decreased parasitaemia and anaemia. In asplenic individuals *P. falciparum* can rapidly cause high level parasitaemias.

Neonates, in the few first months of life, are relatively resistant to malaria. Maternal IgG and foetal haemoglobin may play a role in conferring this protection. Certain genetic disorders confer natural resistance to falciparum malaria, such as sickle cell disease, Haemoglobin C and E, hereditary ovalocytosis, the thalassaemia, and glucose-6-phosphate dehydrogenase deficiency. Duffy antigen negative individuals are resistant to *P. vivax* infection.

Premunition: The state of malaria infection without illness is called premunition. It occurs in areas with intense and stable transmission when asymptomatic parasitaemia is seen in adults and older children after repeated episodes of malaria.

Clinical Manifestations and Pathogenesis

The spectrum of clinical disease varies from asymptomatic or mild infection to severe disease and even death. If diagnosed early and treated correctly, malaria is a treatable disease.

Febrile Paroxysms

Fever occurs intermittently depending on the species. The classical malarial paroxysms consist of three stages: cold stage with cold and shivering; hot stage with fever, headache, vomiting and seizures (young children); and sweating stage when fever comes down with profuse sweating. The whole episode lasts 6–10 h, but is rarely seen. Other signs and symptoms may include anaemia, mild jaundice, splenomegaly and hepatomegaly.

Malignant Tertian Malaria

P. falciparum infections can be much more severe than the infections caused by other species. Severe malaria occurs as a result of metabolic acidosis combined with **microvascular sequestration** due to cytoadherence, rosetting and inflammatory cytokine production. Manifestations of severe malaria include the following:

- *Cerebral malaria*: It is characterised by coma and approximate death rates of 20% in adults and 15% in children. Abnormal behaviour, seizures, impaired consciousness or other neurologic abnormalities may also be seen.
- *Hypoglycaemia*: It is the common complication associated with poor prognosis. Children and pregnant women are more vulnerable to develop hypoglycaemia especially if treated with quinine.
- *Anaemia*: It occurs due to decreased production of RBCs and due to increased destruction of the RBCs by

(1) spleen, of both parasitised and non-parasitised RBCs due to decreased deformability and (2) during parasitic schizogony. Transfusions may be required in many cases.

- *Acute renal failure*: Acute tubular necrosis occurs due to sequestration and agglutination of RBCs in renal capillaries. It is more common in adults than in children.
- *Pulmonary oedema and respiratory distress*: It may occur due to: (1) local production of inflammatory cytokines due to sequestration of parasites in lung capillaries that increase capillary permeability; or (2) iatrogenic, by fluid overload or (3) secondary to acute renal failure.
- *Metabolic acidosis*: It occurs due to accumulation of organic aids like lactic acid.
- Abnormalities in blood coagulation.
- *Hyperparasitaemia*: More than 5% RBCs are infected by malarial parasites.
- *Black water fever*: It may occur in patients with severe malaria treated with amino-alcohol drugs, such as quinine, mefloquine and halofantrine. Haemoglobinuria (dark urine) and anaemia occurs due to massive haemolysis of RBCs in the blood stream.

Relapses of Malaria

Individuals after recovering from one episode of illness due to *P. vivax* or *P. ovale* infection may suffer additional episodes of malaria months or years later. This phenomenon occurs due to reactivation of hypnozoites (dormant sporozoites that do not enter pre-erythrocytic schizogony). Appropriate treatment may reduce the chance of relapse.

Recrudescence in Malaria

It is recurrence of infection with species that lack hypnozoites, that is *P. falciparum*, *P. malariae* and *P. knowlesi* as well as with *P. vivax* or *P. ovale*. It occurs due to persistent/drug-resistant blood stages of the parasite.

Other Manifestations in Malaria

- Malaria in pregnancy: It results in maternal morbidity and mortality, foetal mortality, low birth weight and intrauterine growth retardation.
- In children who survive cerebral malaria, neurologic defects may occasionally persist.
- Severe anaemia may occur due to recurrent infections with *P. falciparum*.
- Rarely splenic rupture may occur due to *P. vivax* infection.

Tropical Splenomegaly Syndrome/Hyperreactive Malarial Spenomegaly

It may occur in people living in tropical areas of Africa and Asia. It is attributed to an abnormal immunologic response to repeated malaria infections. It is characterised by enlarged spleen and liver, anaemia, abnormal immunologic findings including raised IgM, and increased susceptibility to other infections.

Nephrotic Syndrome

It occurs due to repeated or chronic infection with *P. malariae*.

Laboratory Diagnosis

Prompt diagnosis of malaria is essential to treat the patient appropriately. The diagnosis may be established by the following:

Microscopy

Microscopic examination of peripheral blood smear (PBS) is the gold standard test for laboratory diagnosis of malaria. The advantages are: (1) detection of parasite, (2) speciation and (3) parasite density.

Whole blood is collected from the patient and then smeared on slides as thick and thin films. They are stained with Giemsa (or other Romanovsky stain, such as Wright and Giemsa-Wright) and then observed under 100× oil immersion objective. The thick smears are used for detection of malaria parasite and for parasite quantification. The thin smear is used for species identification. Species identification is important since treatment depends on the infecting species. The characteristic findings of different malarial species are mentioned in Table 32.7.

Quantification of parasites: Quantification is essential to assess the severity of infection, monitoring response to

TABLE 32.7	Differences in Malarial Species			
	P. vivax	*P. ovale*	*P. falciparum*	*P. malariae*
Epidemiology				
Predominant species in	Central America, Southeast Asia and western Pacific	Africa	Tropical Africa, Southeast Asia and western Pacific	Found less commonly in most endemic areas
Asexual Cycle				
Hypnozoites (leading to relapses)	Present	Present	Absent	Absent
Repetition of asexual blood cycle and febrile paroxysms (h)	48 (tertian malaria)	48 (tertian malaria)	48 (tertian malaria)	72 (quartan malaria)
Diagnostic Points				
Forms visible in PBS	Early trophozoite (ring forms), late trophozoite, schizonts (Fig. 32.17) and gametocytes	Early trophozoite (ring forms), late trophozoite, schizonts and gametocytes	Only the ring forms and gametocytes are visible in PBS. Schizonts and late trophozoites are rarely seen (sequestered in microvasculature).	Early trophozoite (ring forms), late trophozoite, schizonts (Fig. 32.18) and gametocytes
Rings	Relatively larger (1/3rd of the size of RBC; Fig. 32.19)	Relatively larger (1/3rd of the size of RBC)	Rings are delicate (1/6th of the size of RBC) and multiple infection of RBC is common. Occasionally appliqué or accole forms are seen (rings lie at inner surface of the red cell membrane; Fig. 32.20)	
Haemozoin/malarial pigment	Numerous and yellowish brown in colour	Numerous and brown black in colour	Sparse and brown black in colour	Numerous and brown black in colour
Trophozoites	Amoeboid (Fig. 32.21)	Amoeboid		Band forms (Fig. 32.22)
Gametocytes	Gametocytes are spherical and almost occupy the RBC (Fig. 32.23A–B). Female gametocyte is larger than the male gametocyte and has a red, compact and eccentric nucleus whereas male gametocyte has a diffuse nucleus	Gametocytes are spherical and almost occupy the RBC. Female gametocyte is larger than the male gametocyte and has a red, compact and eccentric nucleus whereas male gametocyte has a diffuse nucleus	The gametocytes are crescentic or banana shaped (Fig. 32.24). The female gametocyte is larger than the male gametocyte and has a red, compact and central nucleus whereas male gametocyte has a diffuse nucleus	Gametocytes are spherical and almost occupy the RBC. Female gametocyte is larger than the male gametocyte and has a red, compact and eccentric nucleus whereas male gametocyte has a diffuse nucleus
Infected RBCs	RBCs are enlarged and rounded. Schuffner's dots (small red dots) are visible	RBCs are enlarged and rounded. Schuffner's dots (small red dots) are visible	Normal in size. Maurer's cleft (large red spots) may be visible	Normal in size. Ziemann's dots (small red spots) may be seen

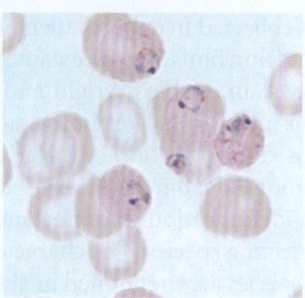

Fig. 32.17 Schizont of *P. vivax* in a Thick Blood Smear. (*Source*: Centers for Disease Control and Prevention (CDC), DPDx - Laboratory Identification of Parasites of Public Health Concern, Parasite Biology, Malaria.)

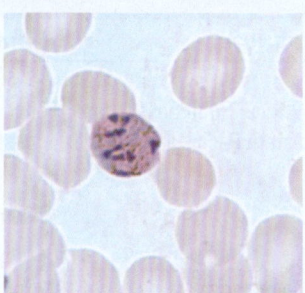

Fig. 32.18 Schizont of *P. malariae* in a Thin Blood Smear. (*Source*: Centers for Disease Control and Prevention (CDC), DPDx - Laboratory Identification of Parasites of Public Health Concern, Parasite Biology, Malaria.)

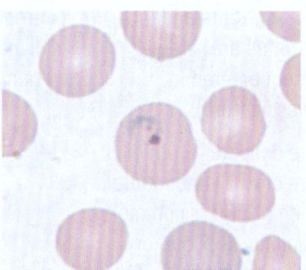

Fig. 32.19 Ring-Form Trophozoite of *P. vivax* in a Thin Blood Smear. (*Source*: Centers for Disease Control and Prevention (CDC), DPDx - Laboratory Identification of Parasites of Public Health Concern, Parasite Biology, Malaria.)

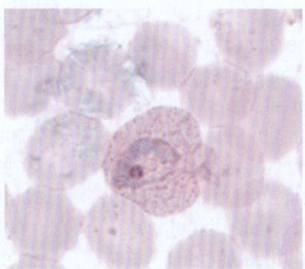

Fig. 32.20 Ring-Form Trophozoites of *P. falciparum* in a Thin Blood Smear, Exhibiting Maurer's Clefts. (*Source*: Centers for Disease Control and Prevention (CDC), DPDx - Laboratory Identification of Parasites of Public Health Concern, Parasite Biology, Malaria.)

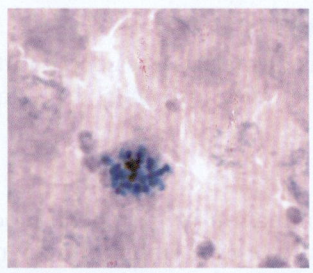

Fig. 32.21 Trophozoite of *P. vivax* in a Thin Blood Smear. Note the amoeboid appearance, Schüffner's dots and enlarged infected RBCs. (*Source*: Centers for Disease Control and Prevention (CDC), DPDx - Laboratory Identification of Parasites of Public Health Concern, Parasite Biology, Malaria.)

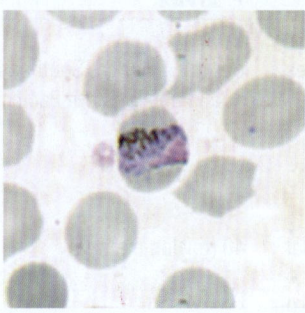

Fig. 32.22 Band-Form Trophozoite of *P. malariae* in a Thin Blood Smear. (*Source*: Centers for Disease Control and Prevention (CDC), DPDx - Laboratory Identification of Parasites of Public Health Concern, Parasite Biology, Malaria.)

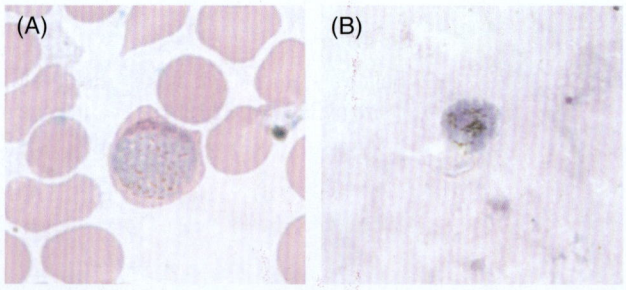

Fig. 32.23 (A) Macrogametocyte of *P. vivax* in a thin blood smear. Note the enlargement of the gametocytes compared to uninfected RBCs. (B) Gametocyte of *P. vivax* in a thick blood smear. (*Source*: Centers for Disease Control and Prevention (CDC), DPDx - Laboratory Identification of Parasites of Public Health Concern, Parasite Biology, Malaria.)

treatment and knowing the drug resistance. The RBCs are lysed in the thick blood smears; therefore, the parasite density may be calculated as number of parasites per 100 WBCs × total WBC count/100. In a thin smear it is calculated as number of parasites per 100 RBCs × total RBC count/100.

Quantitative Buffy Coat (QBC) Technique

This is a rapid and sensitive test for malaria that enhances the microscopic detection of parasites. Parasite DNA is stained with acridine orange in micro-haematocrit tubes and then

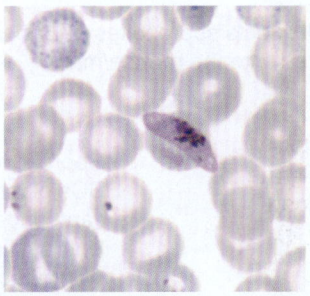

Fig. 32.24 Gametocyte of *P. falciparum* in a Thin Blood Smear. Ring-form trophozoites and an RBC exhibiting basophilic stippling (upper left) are also seen. (*Source*: Centers for Disease Control and Prevention (CDC), DPDx - Laboratory Identification of Parasites of Public Health Concern, Parasite Biology, Malaria.)

detected using epi-fluorescent microscopy. The nuclei of the parasite fluoresces bright green, and the cytoplasm appears yellow-orange. QBC is simple, reliable and user friendly technique, but has the disadvantage of being costlier than conventional microscopy, requires specialised instrumentation, and cannot determine the species or the parasite density.

Rapid Diagnostic Tests (RDTs)

These provide simple, quick, cost effective and accurate diagnosis of malaria. The principle of all RDTs is that malaria antigen is detected in patient's blood flowing along a membrane containing specific antimalaria antibodies. Depending on the manufacturer, the RDT may detect a *P. falciparum* specific protein, such as histidine-rich protein II (HRP-II) or pLDH (parasite lactate dehydrogenase) or pan-specific antigens that are common to all malarial species, for example aldolase, pan-malaria LDH. Its disadvantage are: (1) it does not differentiate between the non-*falciparum* species and (2) false positives may occur in some cases. It is recommended that RDT should always be accompanied with microscopy. Pf HRP-2-based kits may show positive results up to 3 weeks after successful treatment and parasite clearance.

Serology

Antibodies to malaria may not help in making diagnosis in a patient but can be used for screening blood before transfusion and for epidemiological surveys. The antibodies may be detected by indirect fluorescent antibody (IFA) test or by ELISA.

Molecular Diagnosis

The molecular tests offer rapid, sensitive and specific diagnosis of malaria. The different formats available are PCR technique, LAMP technique and mass spectrophotometry. The disadvantages are high cost, requirement of specialised instrumentation and trained manpower.

Treatment

- Drugs active against the parasite forms in blood include chloroquine, mefloquine, quinine, quinidine, atovaquone-

TABLE 32.8 Treatment Regimens for Malaria	
Uncomplicated Malaria	
P. vivax malaria	Chloroquine (25 g/kg divided over 3 days) + primaquine (0.25 mg/kg ×14 days)
P. falciparum malaria	ACT + single-dose primaquine (0.75 mg/kg) on day 2 ACT: an artemisinin derivative combined with a long-acting antimalarial (amodiaquine, lumefantrine, mefloquine, piperaquine or sulphadoxine–pyrimethamine)
Malaria in pregnancy	1st trimester: Quinine 2nd and 3rd trimester: ACT
Mixed infections	ACT + primaquine (if indicated)
Treatment Failure/Drug Resistance	
Early clinical failure, late clinical failure, late parasitological failure	Alternative ACT or quinine + doxycycline
Severe Malaria	
P. falciparum malaria	Parenteral artesunate or quinine or artemether or arteether
P. vivax malaria	Parenteral artesunate or quinine or artemether or arteether + primaquine

ACT, Artemisinin combination therapy.

proguanil, artemether-lumefantrine, doxycycline or clindamycin (in combination with quinine) and artesunate.
- Drug active against hypnozoites (dormant parasitic forms in liver) and thereby preventing relapse is primaquine. Primaquine is contraindicated in severe G6PD deficiency and during pregnancy.
- Monotherapy with oral artemisinin or its derivatives is banned in India. The treatment regimens for malaria are mentioned in Table 32.8.

Antimalarial Drug Resistance

Drug resistance in *P. falciparum* against chloroquine, sulphadoxine–pyrimethamine and mefloquine has been described from different geographical regions. The resistance arises as a result of mutations in *Plasmodium* species when drugs are used inadequately and irregularly, or due to poor compliance or host immunity. Degree of drug resistance can be assessed (Box 32.6).

Prevention

The **National Vector Borne Disease Control Programme** (NVBDCP), India follows a multipronged malaria control strategy, based on the following:
- Early case detection and prompt treatment.
- Vector control strategies: By residual spray with DDT, malathion; space application of pesticides; use of larvicides, for example mineral oil or Paris green or biological larvicides, for example Gambusia fish or *Bacillus thuringiensis*.

- Personal prophylactic protection: use of bednets, mosquito repellents and full-sleeved clothes.
- Community participation.
- Environmental management and source reduction.
- Monitoring and evaluation of program.

Besides the previously metioned program, other methods are as follows:

Chemoprophylaxis

It is recommended for travellers and military persons in malaria endemic areas. For short-term (<6 weeks) protection, doxycycline is preferred and for long-term (>6 weeks) protection, mefloquine is the drug of choice.

Vaccination Against Malaria

No vaccine is commercially available till date. More than 20 vaccine candidates at present are either in preclinical development or are being evaluated in clinical trials. RTS, S/AS01 has successfully completed Phase 3 clinical trial in seven countries in sub- Saharan Africa.

BABESIA

Babesiosis is a malaria-like illness that is tick borne and is caused by intraerythrocytic protozoan of the genus *Babesia*, order Piroplasmida. It consists of more than 100 species, though most of the human infections are caused by *Babesia microti (small rodent parasite), Babesia divergens, Babesia duncani (found in wild animals), and MO1 (currently unnamed agent)*. The natural reservoirs of *Babesia* are wild and domestic animals and humans are incidental hosts.

Life Cycle

The life cycle of *B. microti* is given in Flowchart 32.8.

Hosts: The life cycle of *B. microti* involves two hosts, a tick *Ixodes scapularis* (blacklegged ticks or deer ticks, definitive host) and *Peromysus leucopus* (white footed mouse, intermediate host). Humans are accidental and dead end hosts.

Transmission

Transmission of the parasite mainly occurs by bite of nymphal stage of the tick, most commonly during the warm months, in areas with grass, bushes or woods. Transmission may also occur via transfusion of infected blood or congenitally (from

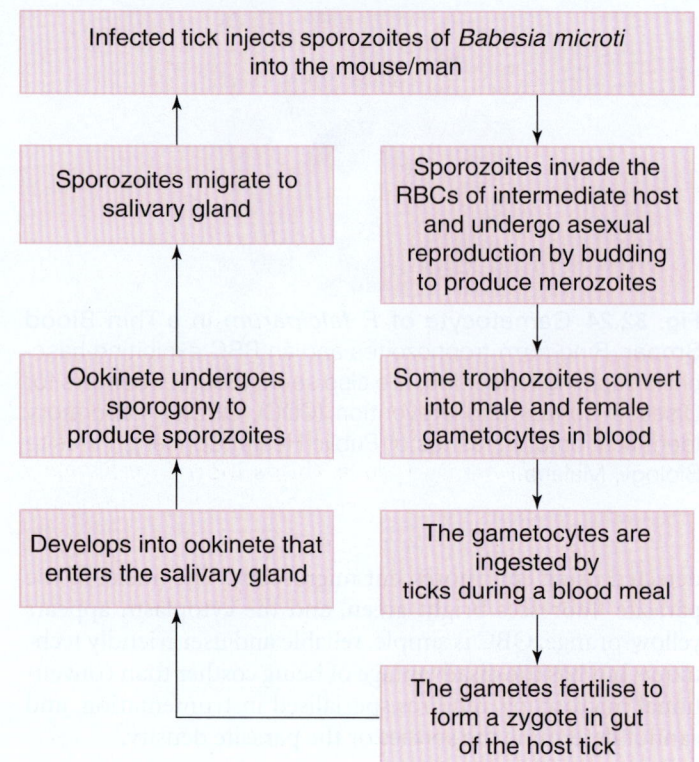

Flowchart 32.8 Life Cycle of *Babesia microti*.

infected mother to foetus). Transovarial or vertical transmission occurs in large *Babesia* species but not in small *Babesia* species (*B. microti*).

The incubation period is 1–6 weeks long. Clinical manifestations occur due to replication of the parasite in the human RBCs and rupture of these cells by the pathogen.

Epidemiology

B. microti is found in the northeastern and upper Midwest United States. Highly endemic areas include New York state, New Jersey, Wisconsin and Minnesota.

Clinical Manifestations

Many people are asymptomatic. Some people may experience non-specific symptoms, such as fever, fatigue, weakness, malaise, chills, sweat, headache and anorexia. Mild splenomegaly and hepatomegaly can occur and haemolytic anaemia and jaundice may also be seen due to destruction of the erythrocytes.

Severe babesiosis may develop in elderly patients, asplenic patients, immunocompromised patients or those with pre-existing serious health conditions. It is associated with parasitaemia levels of more than 10% or severe anaemia and may lead to acute respiratory distress syndrome, hypotension, haemolysis, thrombocytopenia, disseminated intravascular coagulation, congestive heart failure, renal failure and death.

Laboratory Diagnosis

- Diagnosis can be made by Giemsa-stained thick or thin blood smears. Several smears may need to be examined over many days. *Babesia* species appear as round or pear shaped

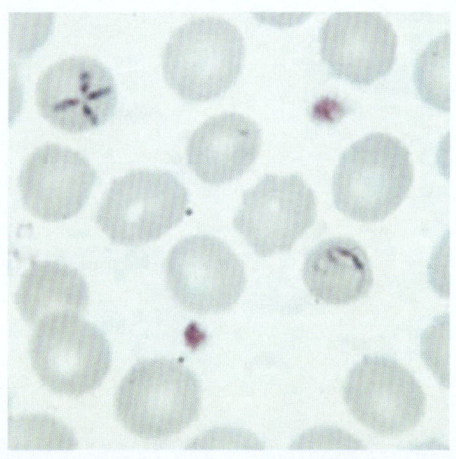

Fig. 32.25 *Babesia* spp. in a Thin Blood Smear. Note the tetrad form (Maltese cross) and amoeboid trophozoite. (*Source*: Centers for Disease Control and Prevention (CDC), DPDx - Laboratory Identification of Parasites of Public Health Concern, Parasite Biology, Babesiosis.)

organisms. These differ from the trophozoites of *P. falciparum* in that they do not produce brownish pigment (haemozoin), do not have schizonts and gametocytes, but rarely have tetrads of merozoites (**Maltese crosses**; Fig. 32.25), which if present is diagnostic of *B. microti* and *B. duncani*.

- Antibodies may be detected by indirect fluorescent antibody test (IFA) in individuals with very low levels of parasitaemia, and is used to confirm the diagnosis.
- Molecular tests may be used in cases where diagnosis is suspected and in cases where distinction cannot be made between *Babesia* and *P. falciparum*.

Treatment

Treatment is not required for most asymptomatic patients. For symptomatic patients the drugs recommended are combination therapy with atovaquone and azithromycin or with clindamycin and quinine.

Prevention

Chemoprophylaxis or vaccines are not available. Individuals in endemic areas should take personal prophylactic measures, for example applying tick repellents, covering the lower body completely and observe the skin after outdoor activities and remove ticks if any.

OPPORTUNISTIC COCCIDIAN PARASITES

They belong to phylum Sporozoa, class Coccidea, order Eimeriida and six genera (*Eimeria*, *Toxoplasma*, *Cryptosporidium*, *Cyclospora*, *Cystoisospora* and *Sarcocystis*).

TOXOPLASMA GONDII

Toxoplasmosis is caused by infection with obligate intracellular parasite *Toxoplasma gondii*. It uncommonly causes disease, though it infects a huge proportion of the world population. The disease may be life threatening in immunocompromised people and in children with congenital toxoplasma infection.

There are three major genotypes of *T. gondii* (types I to III), which differ in their geographic distribution and pathogenicity. Type III strains are more common in animals, while most human cases are caused by type II strains.

Life Cycle

The life cycle of *T. gondii* is given in Flowchart 32.9.

Hosts: The life cycle has two distinct hosts – the definitive host (domestic cats and other members of family Felidae) and the intermediate host (human, birds, rodents). In

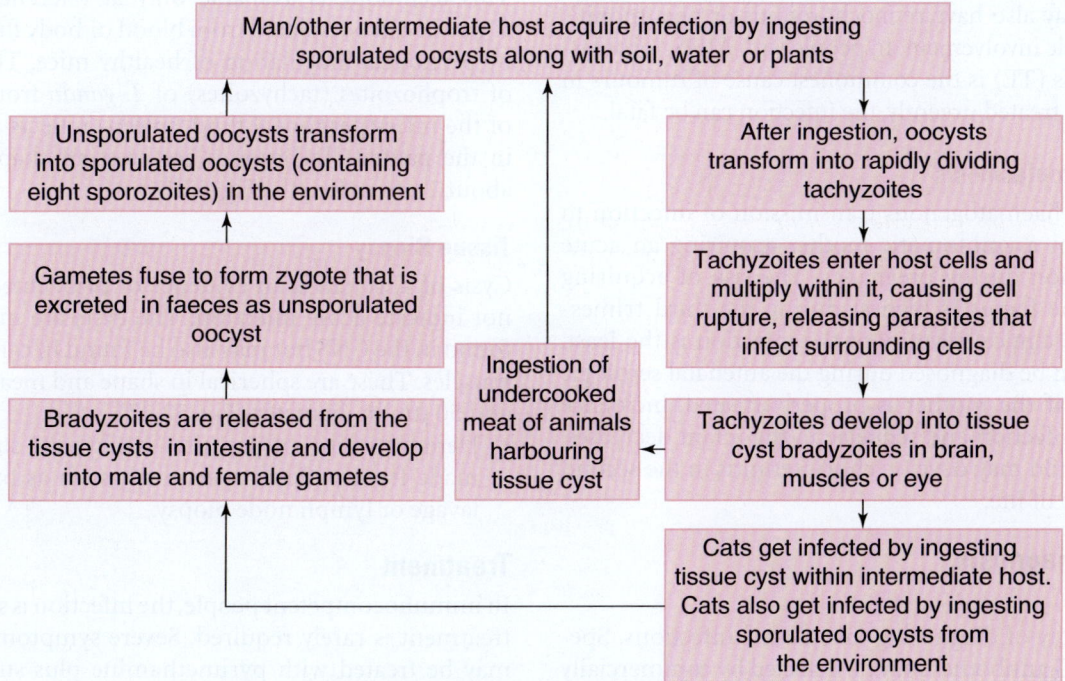

Flowchart 32.9 Life Cycle of *T. gondii*.

the intermediate hosts, the tissue cysts develop into various organs, but most commonly in the CNS and the muscle.

Transmission

Humans may acquire infection by (1) consuming improperly cooked meat of animals carrying tissue cysts; (2) consuming food or water contaminated with sporulated oocysts present in cat's faeces or in the environment; (3) vertical transmission (from mother to foetus), and (4) by blood transfusion or organ transplantation.

Epidemiology

T. gondii infection is a worldwide zoonosis. But prevalence is highest in areas with hot and humid climates. In many parts of the world (Europe and United States), the seroprevalence increases with age.

Clinical Manifestations

Infection in Immunocompetent Nonpregnant People

Acute toxoplasmosis is usually asymptomatic and self-limiting in people with intact immune systems. In symptomatic patients, lymphadenopathy is the most common symptom, which may be accompanied with fever, myalgia, headache and malaise. Rare complications, include myocarditis, encephalopathy and pneumonia. The acute infection usually resolves within weeks to months, though the parasite remains within the patient's body in the inactive state and can become reactivated during immunosuppressed states.

Infections in Immunocompromised People

The risk of acute toxoplasmosis is highest in AIDS patients and in patients receiving immunosuppressive therapy. Toxoplasma infection in these patients is either newly acquired or is due to reactivation of latent infection. In immunodeficient people the most common manifestation is CNS involvement, though they may also have retinochoroiditis or pneumonitis or other systemic involvement. In people with AIDS, toxoplasmic encephalitis (TE) is the commonest cause of tumours in the CNS. If not treated urgently, the infection can be fatal.

Congenital Toxoplasmosis

It results from haematogenous transmission of infection to the foetus from asymptomatic mother acquiring an acute primary infection during pregnancy. The risk of acquiring infection by the foetus is highest during the third trimester, though the clinical severity of the infection is the least. The mother can be diagnosed during the antenatal serologic screening and if the mother is properly treated, incidence of congenital infection and sequelae in the infant decreases. The infected child may develop chorioretinitis in the second or third decade of life.

Laboratory Diagnosis

Serology

It is the mainstay of diagnosing toxoplasma infections. Specific IgM or IgG antibodies may be detected by commercially available ELISA or indirect florescent antibody kits.

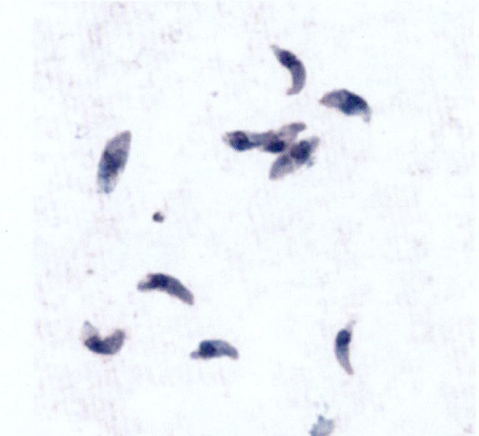

Fig. 32.26 *T. gondii* Tachyzoites Stained With Giemsa, From a Smear of Peritoneal Fluid Obtained From a Laboratory-Inoculated Mouse. (*Source*: Centers for Disease Control and Prevention (CDC), DPDx - Laboratory Identification of Parasites of Public Health Concern, Parasite Biology, Toxoplasmosis.)

IgG avidity testing can be used to differentiate recent from past infections. Avidity is low in recent infections and strong in past infections. This test is especially useful for estimating the risk of transmission from mother to foetus.

Sabin Feldman dye test is the gold standard test for detecting toxoplasma specific antibodies but is used only in reference laboratories.

Molecular Methods

It can detect the parasite DNA in various clinical samples, such as blood, body fluids and amniotic fluid.

Isolation of Parasite by Animal Inoculation

This technique is available only at reference laboratories. *T. gondii* can be isolated from blood or body fluids of patients by peritoneal inoculation of healthy mice. The observation of trophozoites (tachyzoites) of *T. gondii* from various sites of the mice (peritoneal fluid, spleen) reflects acute infection in the patient. Tachyzoites are crescent shaped, measuring about 7 μm in length (Fig. 32.26).

Tissue Biopsy

Cysts of *T. gondii* is an evidence of prior infection and does not indicate acute infection. The cysts are most commonly found in the CNS but may also be found in cardiac or skeletal muscles. These are spherical in shape and measure between 5 and 50 μm in diameter (Fig. 32.27).
- Demonstration of parasites (tachyzoite stage) in smears made from patients specimens, such as bronchoalveolar lavage or lymph node biopsy.

Treatment

In immunocompetent people, the infection is self-limited and treatment is rarely required. Severe symptomatic infections may be treated with pyrimethamine plus sulphadiazine or with clindamycin or trimethoprim plus sulphamethoxazole.

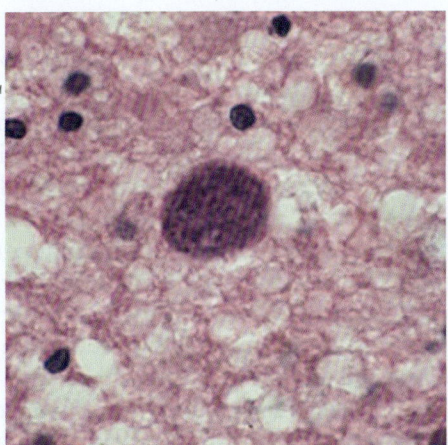

Fig. 32.27 *T. gondii* Cyst in Brain Tissue Stained With H&E. (*Source*: Centers for Disease Control and Prevention (CDC), DPDx - Laboratory Identification of Parasites of Public Health Concern, Parasite Biology, Toxoplasmosis.)

Foetal infections may be treated with spiramycin during early pregnancy and with pyrimethamine plus sulphadiazine during late pregnancy.

Prevention

By avoiding consuming undercooked meat, avoiding contact with oocyst-contaminated material, washing hands properly after gardening activities and washing fruits and vegetables thoroughly before consumption. Pregnant women should be screened by serology.

Persons with AIDS, seropositive for *T. gondii* and CD4$^+$ count <100/µl should receive chemoprophylaxis with trimethoprim-sulphamethoxazole.

CRYPTOSPORIDIA

Human cryptosporidiosis is a zoonotic disease caused by an intestinal coccidian parasite of the genus *Cryptosporidium* that causes self-limiting disease in immunocompetent people but may cause severe disease in immunodeficient people. Most human cases are caused by *Cryptosporidium hominis* and *Cryptosporidium parvum*, though 15 other species including *Cryptosporidium felis*, *Cryptosporidium canis* and *Cryptosporidium muris* are known to cause the disease.

Life Cycle

The life cycle of *Cryptosporidium* species is given in Flowchart 32.10.

The life cycle of the genus *Cryptosporidium* involves only one host (man or other animals). The infective stage is sporulated oocysts (containing four sporozoites), which are immediately infective when passed in faeces. The thick walled oocysts are hardy and can survive routine chlorination. Thin-walled oocysts can cause autoinfection. Incubation period is 2–10 days.

Transmission

The oocysts may be transmitted to humans by: (1) contaminated water (drinking or recreational water like swimming

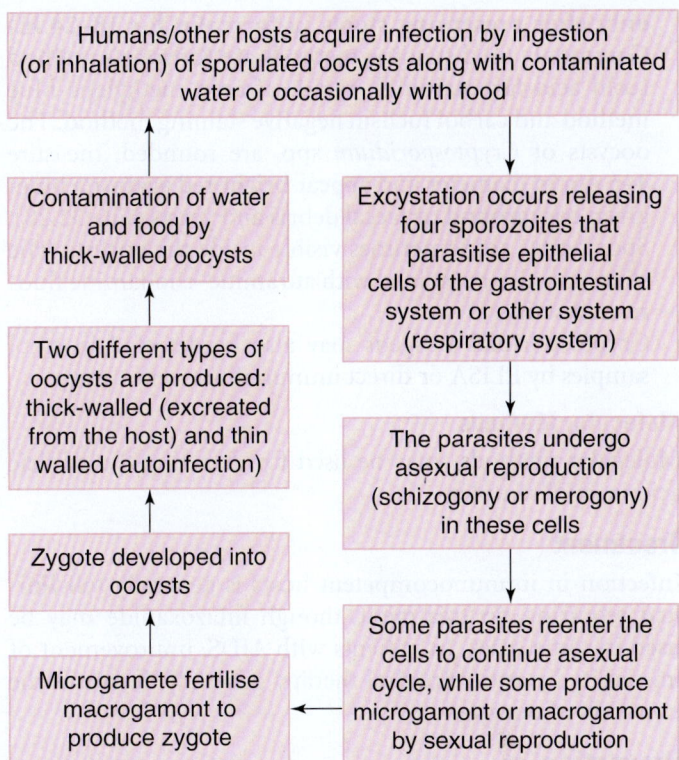

Flowchart 32.10 Life Cycle of *Cryptosporidium* spp.

pools); common source outbreaks can occur by contaminated pools or (2) by consumption of contaminated foods (e.g. chicken salad) and (3) person-to-person transmission may occur (e.g. household or day care centres).

Epidemiology

Cryptosporidium is present worldwide. The infection is most common in children under 5 years of age and in women of child bearing age who come in contact with infected children. In persons with AIDS, infection can occur at any age.

Clinical Manifestations

Asymptomatic infections may occur in immunocompetent and immunocompromised people. The most common symptom is watery diarrhoea. Other symptoms may include abdominal cramps, nausea, vomiting, dehydration and weight loss. Symptoms usually subside within 1–2 weeks. Patients with AIDS or other immunocompromised conditions may develop chronic or severe diarrhoea, weight loss, wasting, abdominal pain, biliary tract or respiratory tract involvement.

Diagnosis

Sample required: Stool

Microscopic Examination

• Diagnosis is usually made by microscopic examination of stool preserved in 10% buffered formalin and stained with Ziehl–Neelsen-modified acid-fast stain. (**Since oocysts in unpreserved stools remain viable and infectious for a long time, samples should be handled cautiously and** only after preserving stools in formalin for 18–24 h.) Commonly used staining methods are Kinyoun's method (cold acid-fast staining), rapid safranin methylene blue method and carbol fuchsin negative staining method. The oocysts of *Cryptosporidium* spp. are rounded, measure 4–5 µm in diameter and appear bright red against a blue-green background of faecal debris and yeasts (Fig. 32.28). Sporozoites are sometimes visible inside the oocysts. The slides can also be stained with auramine–rhodamine fluorescent stain.

- Cryptosporidial antigens may also be detected in stool samples by ELISA or direct immunofluorescence assay.

Molecular Methods

Molecular methods may be used for detecting the nucleic acid in stool samples.

Treatment

Infection in immunocompetent hosts is generally self-limited requiring no treatment, though nitazoxanide may be used in some cases. In patients with AIDS, improvement of immunity by anti-retroviral therapy generally decreases the excretion of oocysts.

Prevention

The exposure to oocysts in human and animal faeces should be minimised. Water should be filtered before consumption.

CYCLOSPORA

Cyclosporiasis is a diarrhoeal illness caused by the coccidian parasite *Cyclospora cayetanensis*. *C. cayetanensis* is the only species known to infect humans.

Life Cycle and Transmission

Host: Man is the only known host.

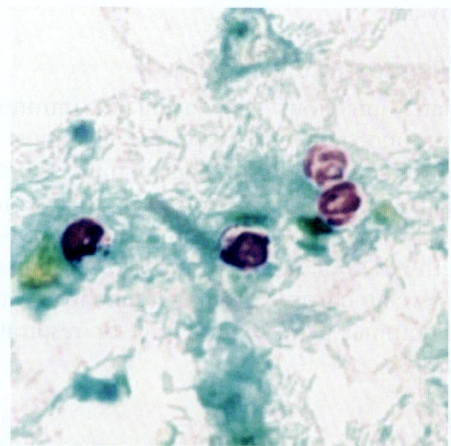

Fig. 32.28 *Cryptosporidium parvum* Oocysts Stained With Modified Acid-Fast. Against a blue-green background, the oocysts stand out in a bright red stain. Sporozoites are visible inside the two oocysts to the right. (*Source*: Centers for Disease Control and Prevention (CDC), DPDx - Laboratory Identification of Parasites of Public Health Concern, Parasite Biology, Cryptosporidiosis.)

Transmission to humans occurs by ingestion of sporulated oocysts (that develop after exposure to environment for 7–10 days) along with contaminated water or different types of fresh produce, such as raspberries, snow peas and basil. In the gastrointestinal tract the oocysts excyst releasing sporozoites that invade the epithelial cells of the small intestine. The parasite undergoes asexual reproduction in these cells and continues the cycle endogenously and also undergoes sexual reproduction leading to the formation of zygotes. The zygotes transform into oocysts, which in turn are released in faeces. The oocysts are non-sporulated and thus noninfective when passed freshly in faeces and require 7–10 days at ambient temperature to develop into sporulated oocysts (sporulated oocyst contains two sporocysts each containing two sporozoites). This is in contrast to *Cryptosporidium* spp. where sporulated oocysts are excreted in faeces.

Epidemiology

C. cayetanensis is found worldwide, but is more common in the tropical and subtropical countries.

Clinical Manifestations

Some patients may be asymptomatic. The most common manifestations include watery diarrhoea, weight loss, loss of appetite, bloating, flatulence, cramping, nausea and fatigue. The disease may be self-limited, may have a waxing and waning course or in untreated cases may persist for several weeks to months. Rarely, the infection may lead to extraintestinal manifestations, such as Reiter's syndrome or Guillain–Barre syndrome.

Diagnosis

The diagnosis can be made by microscopically examining stool samples that are either fresh or preserved in 10% formalin. Since the oocysts are shed intermittently in stool, the sensitivity of the test can be increased by examining 2–3 stool samples from a patient and by concentrating stool samples by techniques, such as formalin ethyl acetate sedimentation technique. The sediments may be examined as follows:

- A wet mount or after staining with modified acid-fast stain, where oocysts are seen as spherical bodies measuring 8–10 µm in diameter (Fig. 32.29).
- The oocysts of *Cyclospora* spp. are **autofluorescent** and therefore their detection can be enhanced by UV fluorescence microscopy or by differential interference contrast (DIC; Fig. 32.30).
- The sediments can also be stained by modified acid-fast stain or by modified safranin stain.
- Molecular methods like PCR may also be used for making the diagnosis.
- Histopathologic examination of intestine may also help in diagnosis.

Treatment

Trimethoprim–sulphamethoxazole for 7–10 days is the treatment of choice. A longer duration of therapy may be required for HIV patients.

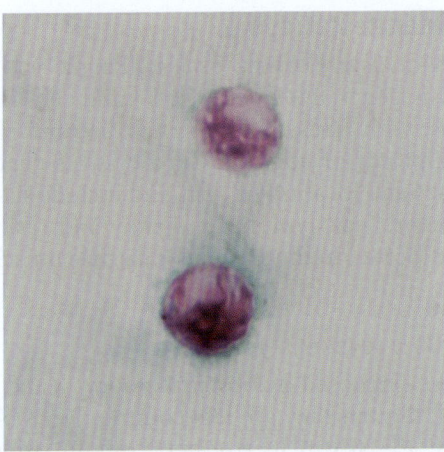

Fig. 32.29 Oocysts of *Cyclospora* spp. Stained With Modified Acid-Fast Stain. (*Source*: Centers for Disease Control and Prevention (CDC), DPDx - Laboratory Identification of Parasites of Public Health Concern, Parasite Biology, Cyclosporiasis.)

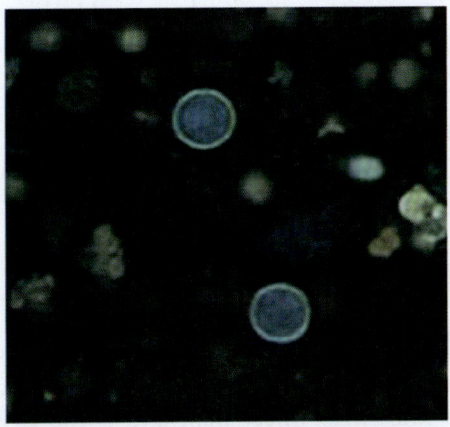

Fig. 32.30 Oocysts of *Cyclospora* spp. Showing Autoflorescence Under UV Florescent Microscope. (*Source*: Centers for Disease Control and Prevention (CDC), DPDx - Laboratory Identification of Parasites of Public Health Concern, Parasites - Cyclosporiasis (Cyclospora Infection).)

CYSTOISOSPORA (FORMERLY CALLED ISOSPORA)

Cystoisosporiasis (formerly called isosporiasis) caused by the least common intestinal coccidian parasite, *Cystoisospora belli* (formerly called *Isospora belli*). The genus *Cystoisospora* is closely related to *Toxoplasma*, *Cyclospora* and *Cryptosporidium*.

Life Cycle

Hosts: Humans are the only hosts known. Infection in man occurs by ingestion of oocysts (containing sporocysts) along with contaminated food or water. The sporocysts excyst releasing sporozoites in the small intestine that enter the enterocytes and begin asexual reproduction (schizogony producing schizonts). The schizonts rupture releasing the merozoites that enter fresh epithelial cells and continue schizogony. With time sexual reproduction occurs resulting in production of male and female gametocytes that fuse to form oocyst. The oocysts are immature and contains one sporoblast (rarely two) when passed freshly in stool. In the environment, the sporoblast divides to form two sporoblasts, which secrete a cyst wall around them to from sporocyst. The resulting oocyst is now infective with two sporocysts and four sporozoites.

Epidemiology

Cystoisospora is found all over the world but is predominant in tropical and subtropical regions of the world. It rarely causes infection in immunocompetent people, though is common in persons with AIDS. Cases have also been reported from patients with malignancies or with transplants.

Clinical Manifestations

Acute infections in immunocompetent persons may present with fever, nonbloody diarrhoea, abdominal pain, and can last for weeks to months resulting in weight loss. In immunosuppressed patients diarrhoea may be severe.

Diagnosis

The diagnosis can be made by microscopically examining stool samples that are either fresh or preserved in 10% formalin. Since the oocysts are shed intermittently in stool, the sensitivity of the test can be increased by examining 2–3 stool samples from a patient and by concentrating stool samples by techniques, such as formalin ethyl acetate sedimentation technique. The sediments can then be examined as follows:
- A wet mount where oocysts are seen as large, ellipsoidal bodies measuring 25–30 μm in diameter.
- UV fluorescence microscopy or by DIC, since the oocysts of *Cystoisospora* spp. are autofluorescent.
- Modified acid-fast stained or modified safranin-stained smears (Fig. 32.31).

Molecular methods like PCR may also be used for making the diagnosis.

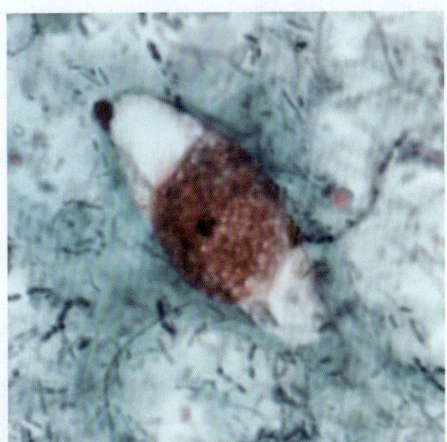

Fig. 32.31 Immature Oocyst of *C. belli* Stained With Safranin, Containing a Single Sporoblast. (*Source*: Centers for Disease Control and Prevention (CDC), DPDx - Laboratory Identification of Parasites of Public Health Concern, Parasite Biology, Cystoisosporiasis.)

Treatment

Trimethoprim–sulphamethoxazole for 7–10 days is the treatment of choice. A longer duration of therapy may be required for HIV patients. Second-line drug is ciprofloxacin.

SARCOCYSTIS

Sarcocystosis is caused by zoonotic protozoa, *Sarcocystis*. Most of the human infections are caused by *Sarcocystis hominis* and *Sarcocystis suihominis*.

Life Cycle

The life cycle of *Sarcocystis* is given in Flowchart 32.11.

Hosts: In contrast to most of the other coccidian parasites *Sarcocystis* has two hosts – definitive host (human/carnivores) and intermediate host (herbivores).

Intestinal Sarcocystosis

It is caused by *S. hominis* and *S. suihominis* which use humans as the definitive hosts. It is transmitted by raw or under-cooked beef or pork containing sarcocysts (tissue cysts).

Muscular Sarcocystosis

It is caused by non-human *Sarcocystis* spp. like *Sarcocystis lindemanni*, where humans act as intermediate accidental host. It is transmitted by infective sporocysts ingested along with contaminated food or water.

Epidemiology

Sarcocystis is worldwide in distribution, although most human cases are reported from the tropical and subtropical regions particularly in Southeast Asia.

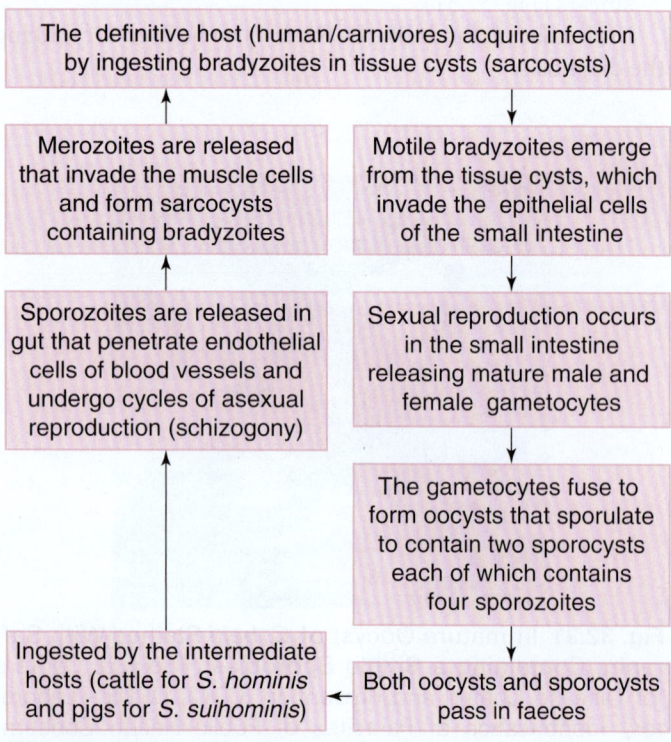

The definitive host (human/carnivores) acquire infection by ingesting bradyzoites in tissue cysts (sarcocysts)

↑

Merozoites are released that invade the muscle cells and form sarcocysts containing bradyzoites

Motile bradyzoites emerge from the tissue cysts, which invade the epithelial cells of the small intestine

↑

Sporozoites are released in gut that penetrate endothelial cells of blood vessels and undergo cycles of asexual reproduction (schizogony)

Sexual reproduction occurs in the small intestine releasing mature male and female gametocytes

↓

The gametocytes fuse to form oocysts that sporulate to contain two sporocysts each of which contains four sporozoites

Ingested by the intermediate hosts (cattle for *S. hominis* and pigs for *S. suihominis*)

Both oocysts and sporocysts pass in faeces

Flowchart 32.11 Life Cycle of *Sarcocystis*.

Clinical Manifestations

Most individuals are asymptomatic. People with intestinal sarcocystosis generally experience a mild self-limiting gastroenteritis, with nausea, abdominal discomfort and diarrhoea. Symptoms of muscular sarcocystosis depend on the number, size and location of cysts and include subcutaneous nodules, arthralgias, lymphadenopathy and rash. The cysts are usually located in the skeletal or cardiac muscles.

Diagnosis

Intestinal Sarcocystosis

- It can be diagnosed by observing oocysts and sporocysts in stool, though they are less commonly visible. Oocysts measure 15–20 μm in size and contain two sporocysts each of which contain four sporozoites.
- The oocysts of *Sarcocytis* spp. are autofluorescent and therefore their detection can be enhanced by UV fluorescence microscopy or by DIC.

Muscular Sarcocystosis

Muscular Sarcocystosis can be diagnosed by biopsy of muscle tissue stained with H&E. The sarcocysts contain several bradyzoites (Fig. 32.32).

Treatment

Currently no specific treatment or prophylaxis of sarcocystosis is available, though albendazole was efficacious in one case.

CILIOPHORA AND CHROMISTA

BALANTIDIAISIS

Balantidiasis is caused by *Balantidium coli*, the only ciliate protozoan known to infect man and is the largest protozoan parasite of man. The genus *Balantidium* belongs to the phylum Ciliophora, class Litostomatea, order Vestibuliferida [refer Chapter 31, Molecular Classification of Protozoa (2000)].

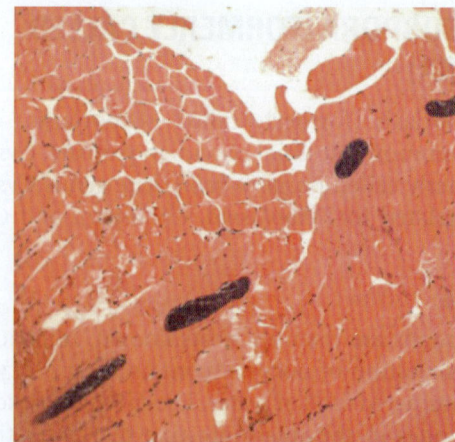

Fig. 32.32 Sarcocysts of *Sarcocystis* spp. in Muscle Tissue Stained With H&E. (*Source*: Centers for Disease Control and Prevention (CDC), DPDx - Laboratory Identification of Parasites of Public Health Concern, Parasite Biology, *Sarcocystosis*.)

Life Cycle

Host: Pigs are the main reservoirs of infection, though rodents and other non-human primates can also act as animal reservoir. It completes its life cycle in a single host.

Transmission

Humans acquire infection by ingesting cysts of *B. coli* along with contaminated food and water.

In humans, the cysts excyst in small intestine to form trophozoites. The motile trophozoites (with the help of cilia) migrate to the lumen of large intestine where they multiply by binary fission (asexual cycle) and conjugation (sexual cycle). Some trophozoites multiply in the colon wall. Some trophozoites convert into infective cysts that are passed in faeces and survive well in the external environment.

Epidemiology

B. coli has a worldwide distribution. Human infections mostly occur in areas where food and water get contaminated with excreta of pigs or where pigs are raised. Human to human transmission of infection also occurs.

Clinical Manifestations

Infection is mostly asymptomatic. Clinical manifestations may include persistent diarrhoea, abdominal pain and weight loss. Severe symptoms, such as fulminant colitis and intestinal perforation may occur in debilitated patients.

Diagnosis

- *Stool examination*: *B. coli* can be diagnosed by observing motile trophozoites in fresh stool samples. Repeated samples are required because trophozoites are excreted intermittently. The trophozoites are large in size, measuring 40–200 µm, have cilia on their surface, have a bean-shaped macronucleus, a micronucleus, and a cytostome.
- Trophozoites may also be observed in tissues obtained by endoscopy. The trophozoites may also be stained with H&E stain in tissue sections (Fig. 32.33).

Treatment

Tetracycline is the drug of choice. Metronidazole or iodoquinol may also be used.

Prevention

Transmission may be controlled by preventing contamination of food or water with pig and human faeces.

CHROMISTA

Blastocystosis

Blastocystosis is caused by *Blastocystis hominis*, an organism with controversial taxonomy and pathogenicity. Molecular analysis of the ssrURNA gene places *B. hominis* in a diverse group of protists including diatoms, brown algae, water

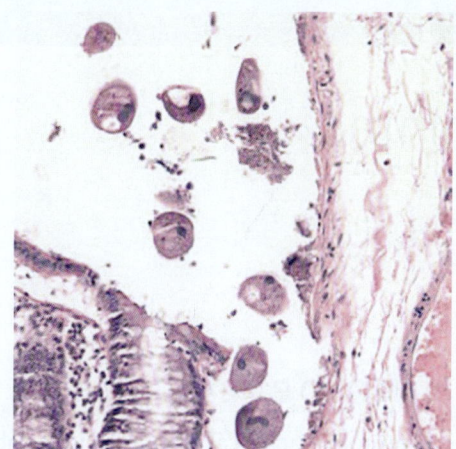

Fig. 32.33 Trophozoites of *B. coli* in Biopsy From Intestine Stained With H&E. (*Source*: Michael Abbey/Science Source.

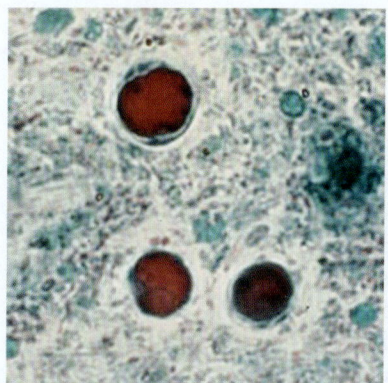

Fig. 32.34 *B. hominis* Cyst-Like Forms Stained With Trichrome. (*Source*: OMICS International: Blastocystis Hominis Infection. https://www.omicsonline.org/india/blastocystis-hominis-infection-peer-reviewed-pdf-ppt-articles/)

moulds etc. called the stramenophiles. Some scientists place it in the kingdom Chromista.

The organism is ubiquitous with a worldwide distribution. The life cycle of *B. hominis* is still not clear though four major morphologic forms have been observed: granular, vacuolar, amoeboid and cystic forms. Transmission of infection is thought to be by thick walled cysts via faeco–oral route by consumption of contaminated food and water. The pathogenicity of *B. hominis* is also debatable. The infection may be asymptomatic or symptomatic with watery diarrhoea, abdominal pain, flatulence and bloating. Diagnosis is based on microscopic examination of stool by wet mount or preferably by permanent stains (e.g. trichrome stain). Spherical cysts measuring 5–30 µm in size are visible and contain a large vacuole surrounded by a thin layer of cytoplasm (Fig. 32.34). Treatment is usually not required. Symptomatic infections may be treated with metronidazole, trimethoprim/sulphamethoxazole or nitazoxanide.

33

Helminths

Parul Jain

LEARNING OBJECTIVES

Life cycle, epidemiology, clinical manifestations, laboratory diagnosis, treatment and prevention of helminths affecting humans:

- **Intestinal nematodes**
  - *Ascaris lumbricoides* (Ascariasis)
  - *Trichuris trichiura* (Trichuriasis)
  - Hookworm infection (*Ancyclostoma duodenale, Necator americanus*)
  - *Strongyloides stercoralis*
  - *Enterobius vermicularis* (Enterobiasis)
- **Tissue nematodes**
  - *Trichinella* spp. (Trichinellosis/Trichinosis)
  - *Dracunculus medinensis* (Dracunculiasis)
  - Filariasis
    - Lymphatic filariasis
    - Tropical pulmonary eosinophilia
    - Loa loa (Loiasis)

- Onchocerca volvulus (Onchocerciasis)
- Mansonella (Mansonellosis)
- **Intestinal cestodes**
  - *Taenia saginata* and *Taenia solium* (Taeniasis)
  - Hymenolepis (Hymenolepiasis)
  - *Diphyllobothrium latum* (Diphyllobothriasis)
- **Invasive cestodes**
  - *Echinococcus granulosus* and *Echinococcus multilocularis* (Echinococcosis)
  - *Taenia multiceps, T. serialis* (Coenurosis)
- **Common trematodes**
  - Schistosomes (Schistosomiasis)
  - Clonorchis and Opisthorchis
  - Fasciola (Fascioliasis)
  - *Fasciolopsis buski* and *Heterophyes hetrophyes*
  - *Paragonimus westermani* (Paragonimiasis)

INTESTINAL NEMATODES

INTRODUCTION

More than a billion people worldwide harbour nematodes in the intestine. These constitute a major public health problem in developing countries with poor sanitation. The nematodes have cylindrical bodies covered with a smooth cuticle and have a body cavity and hence are called roundworms. Nematodes are broadly classified into intestinal nematodes and tissue nematodes, depending on the habitat of the adult worms. The adult worms range from 1 mm to several centimetres in size and have separate sexes. Intestinal nematodes commonly affecting man are roundworms (*Ascaris lumbricoides),* hookworms (*Ancyclostoma duodenale* and *Necator americanus*), *Strongyloides stercoralis, Trichuris trichiura* (whipworms) and *Enterobius vermicularis* (pinworms). Depending upon the species, transmission occurs by ingestion of cysts or larvae or by penetration of intact skin by larvae. The **soil transmitted helminths** (hookworms, whipworms and *Ascaris)* account for a major

burden of disease worldwide. These are more common in moist and warm areas with poor sanitation. The adult worms inhabit the intestine and eggs get deposited in the soil when a person defecates in open. Eggs of *Ascaris* and whipworms become infective in soil and are transmitted by faeco-oral route. Eggs of hookworms are not infective. They transform into larvae, which get transmitted by skin invasion. The salient features of medically important intestinal nematodes are mentioned in Table 33.1.

ASCARIS LUMBRICOIDES (ASCARIASIS)

A. lumbricoides is the **most common helminthic infection** and is the **largest intestinal nematode** of humans. Females measure up to 40 cm and are longer than males. It is the **most common human parasite**.

Life Cycle (Flowchart 33.1)

A. lumbricoides needs a single host to complete its life cycle. The eggs are non-infective when freshly passed in stool and require 1–2 weeks in soil for the development. Adult worms can live up to 1–2 years.

TABLE 33.1 **Features of Medically Important Intestinal Nematodes**

Features	Ascaris lumbricoides	Ancyclostoma duodenale, Necator americanus	Strongyloides stercoralis	Trichuris trichiura	Enterobius vermicularis
Common name	**Round worm**	**Hookworm**		**Whipworm**	**Pinworm**
Epidemiology	Worldwide	Warm, humid areas	Warm, humid areas	Worldwide	Worldwide
Transmission	Oral by ingestion of infective eggs	Percutaneous, by penetration of skin by filariform larvae	Percutaneous, by penetration of skin by filariform larvae, autoinfection	Oral by ingestion of infective eggs	Oral by ingestion of infective eggs
Location of adult worm in human gastrointestinal tract	Jejunum (lumen)	Jejunum (mucosa)	Duodenum, jejunum (mucosa)	Caecum and colon (mucosa)	Caecum, appendix, colon (lumen)
Migrate through lungs	Yes	Yes	Yes	No	No
Longevity of adult worms	1–2 years	6–8 years (*A. duodenale*) 2–5 years (*N. americanus*)	Not known due to autoinfection	Several years	2 months
Prepatent period*	2–3 months	6–8 weeks	18–28 days	3 months	1 month
Laboratory diagnosis	Demonstration of eggs in stool	Demonstration of eggs in stool	Identification of rhabditiform larvae in stool	Demonstration of eggs in stool	Demonstration of eggs on perianal skin
Treatment	• Mebendazole • Albendazole • Pyrantel • Ivermectin	• Mebendazole • Albendazole • Levamisole	• Ivermectin • Albendazole • Thiabendazole	• Mebendazole • Albendazole	• Mebendazole • Albendazole • Pyrantel • Ivermectin

*Time period between the entry of infectious form in human body to the appearance of eggs in faeces.

Epidemiology

Ascariasis is common in hot and humid regions and in places with poor sanitation. In endemic areas, malnourished children are most commonly affected because of hand to mouth transmission while playing in soil contaminated with human faeces. The whole community is at risk through dust and contaminated fruits and vegetables.

Clinical Manifestations

Infection is generally asymptomatic or produces mild symptoms such as dyspepsia, abdominal discomfort, nausea or loss of appetite. Heavy infections can lead to malnutrition in children. Some patients may experience hypersensitivity response during larval migration through lungs.

In severe cases, **Loeffler syndrome** may occur, which includes eosinophilic pneumonia with transient patchy infiltrates on chest X-rays. Heavy infections may cause intestinal obstruction, obstruction of biliary and pancreatic ducts and appendix. An adult worm may migrate up the oesophagus and may be expelled orally.

Laboratory Diagnosis

Sample Required

Freshly passed stool.

Microscopic Examination

1. Diagnosis is made by microscopic detection of fertilised, unfertilised or decorticated eggs in smears from stool samples (Fig. 33.1A–C).
   a. The fertilised eggs are bile stained, 50–75 μm in size, are rounded in shape and have a thick shell with a mammillated (rough, bumpy) albuminous coat.
   b. Unfertilised eggs are larger (90 × 55 μm) in size, elongated than the fertilised eggs and have thinner outer shell. They contain mass of disorganised, highly refractile granules.
   c. Decorticated eggs are fertilised or unfertilised eggs that have lost the outer layer.
2. Diagnosis can also be made by identifying an adult worm passed in stool or through mouth or nose. The worm is large in size and has an unsegmented cream coloured cuticle (Fig. 33.2).
3. Larvae may be found in sputum or gastric aspirates during the migratory phase.
4. Ultrasound and endoscopic retrograde cholangiopancreatography (ERCP) can be used for diagnosing worms in biliary tree and pancreatic duct.

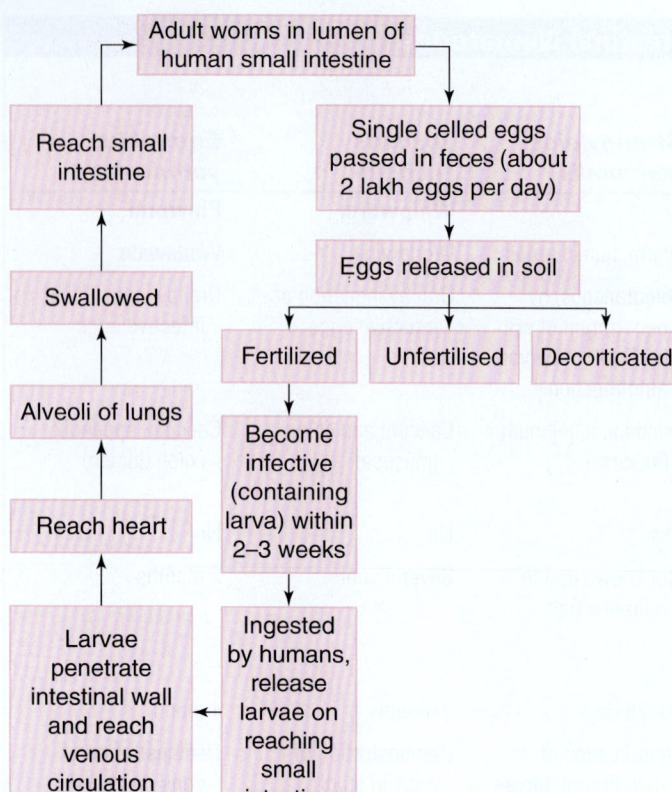

Flowchart 33.1 Life Cycle of *Ascaris lumbricoides*.

Treatment

Drugs of choice are albendazole and mebendazole (antihelminthic medications) for 1–3 days. Single dose of levamisole or piperazine may also be used.

Prevention

Infections may be prevented by washing hands properly with soap and water before handling food, after defecation or playing in soil. Vegetables should be thoroughly washed, peeled or cooked before consumption. Transmission can also be prevented by effective sewage systems and by not defecating outdoors.

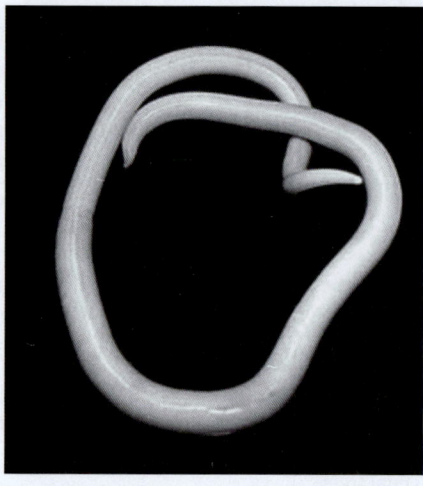

Fig. 33.2 Adult Female *A. lumbricoides.* (*Source*: Centers for Disease Control and Prevention (CDC), DPDx - Laboratory Identification of Parasites of Public Health Concern, Parasite Biology, Ascariasis.)

TRICHURIS TRICHIURA (TRICHURIASIS)

Trichuriasis occurs primarily in children of low socio-economic status in tropical and subtropical regions of the world. Since *Trichuris* and *Ascaris* have the common geographic distribution, simultaneous infection by both the worms is not uncommon. No tissue migratory phase occurs in the life cycle of *Trichuris* and the infection remains limited to the gastrointestinal tract. This is in contrast to *Ascaris* where tissue migration occurs. Adult *Trichuris trichura* has a characteristic whip-like shape and is therefore commonly known as the whipworm. It buries its anterior half into the intestinal mucosa and feeds on the tissue fluids.

Life Cycle (Flowchart 33.2)

T. trichiura needs a single host to complete its life cycle. The eggs are non-infective when freshly passed in stool and requires 1–2 weeks in soil for the development.

Epidemiology

It is the third most common nematode affecting humans and is common in areas with humid tropical climate and with

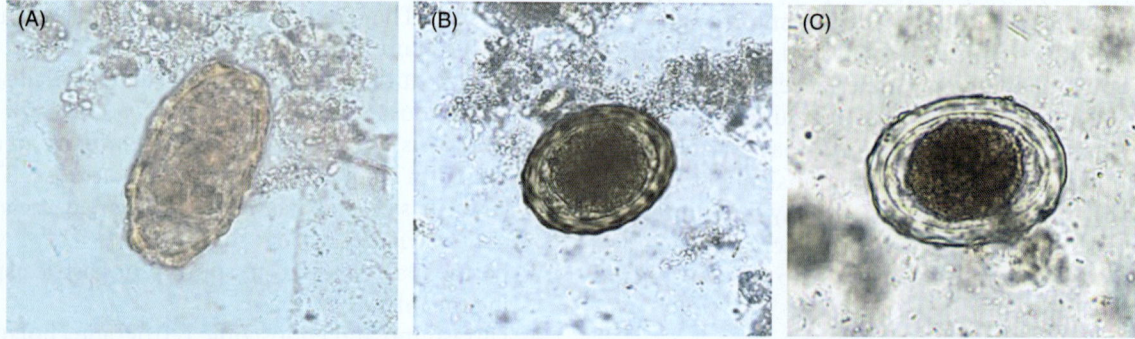

Fig. 33.1 (A) Unfertilised egg of *A. lumbricoides* in an unstained wet mount of stool. (B) Fertilised egg of *A. lumbricoides* in an unstained wet mount of stool. (C) *A. lumbricoides* decorticated, fertile egg in wet mounts, 200× magnification. (*Source*: Centers for Disease Control and Prevention (CDC), DPDx - Laboratory Identification of Parasites of Public Health Concern, Parasite Biology, Ascariasis.)

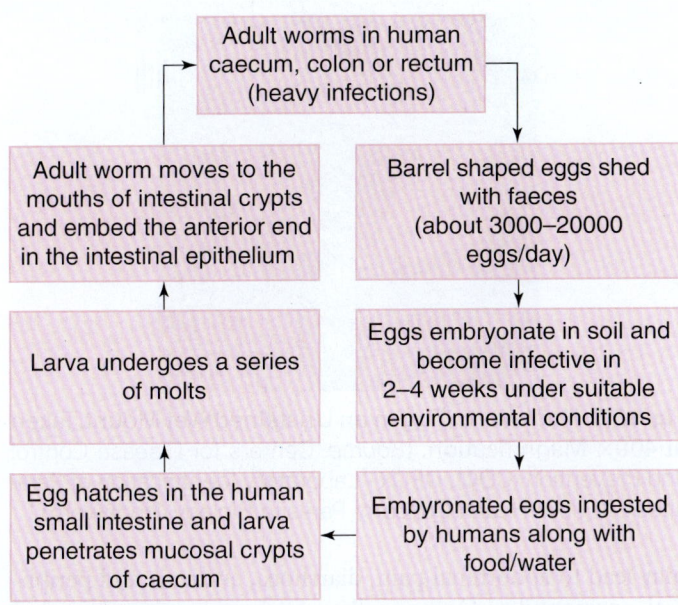

Flowchart 33.2 Life Cycle of *Trichuris trichiura*.

poor sanitation. Transmission occurs by faeco-oral route, that is, through contaminated hands or improperly washed or cooked vegetables and fruits.

Clinical Manifestations

People with light infection usually have no symptoms. People with heavy infections may experience abdominal pain, anorexia, stool mixed with blood or mucus or even rectal prolapse. Anaemia, growth retardation or vitamin A deficiency may occur in children with heavy infection. Peripheral eosinophilia occurs occasionally.

Laboratory Diagnosis

Sample Required

Freshly passed stool.

Microscopic Examination

Diagnosis may be made by microscopic detection of bile stained barrel-shaped eggs having bipolar hyaline plugs and a smooth shell. These measure 50–70 µm in length and 25–35 µm in width (Fig. 33.3).

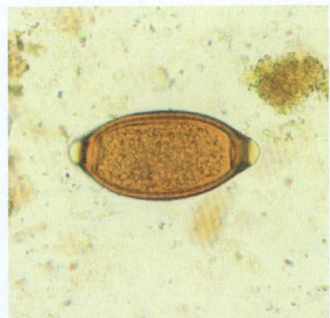

Fig. 33.3 Eggs of *T. trichiura* in an Iodine-Stained Wet Mount of Stool. (*Source*: Centers for Disease Control and Prevention (CDC), DPDx - Laboratory Identification of Parasites of Public Health Concern, Parasite Biology, Trichuriasis.)

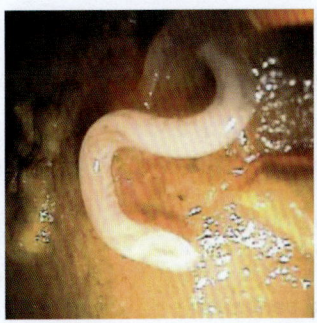

Fig. 33.4 Posterior End of an Adult *T. trichiura*, Taken During a Colonoscopy. (*Source*: Centers for Disease Control and Prevention (CDC), DPDx - Laboratory Identification of Parasites of Public Health Concern, Parasite Biology, Trichuriasis.)

Colonoscopic Examination

Elongate whip-like adult worm measuring 3–7 cm with a thin long anterior end and a thick posterior end may be seen at colonoscopy or on mucosa of the prolapsed rectum (Fig. 33.4) (Remember that it buries its anterior half into the intestinal mucosa).

Treatment

Albendazole/Mebendazole for 3 days is the treatment of choice. A stool test may be repeated after completion of treatment.

Prevention

Infections may be prevented by washing hands properly with soap and water before handling food, after defecation or playing in soil. Vegetables should be thoroughly washed, peeled or cooked before consumption. Transmission can also be prevented by effective sewage systems and by not defecating outdoors.

HOOKWORM INFECTION (*ANCYCLOSTOMA DUODENALE, NECATOR AMERICANUS*)

Human hookworm infection is caused by two species of nematodes, *A. duodenale* and *N. americanus* that attach to the small intestinal mucosa and feed on blood and interstitial fluid. After Ascariasis, these are the **second most common human helminthic infections**. Hookworms are important causes of morbidity in underdeveloped countries especially in children, causing iron deficiency anaemia and hypoproteinemia. *Ancylostoma caninum* (canine hookworm) and *Ancylostoma braziliense* (dog hookworm) are animal hookworm species, which may cause 'creeping eruptions' or 'cutaneous larva migrans'.

Life Cycle (Flowchart 33.3)

Host

Human. Human hookworms require only human hosts for completing their life cycle.

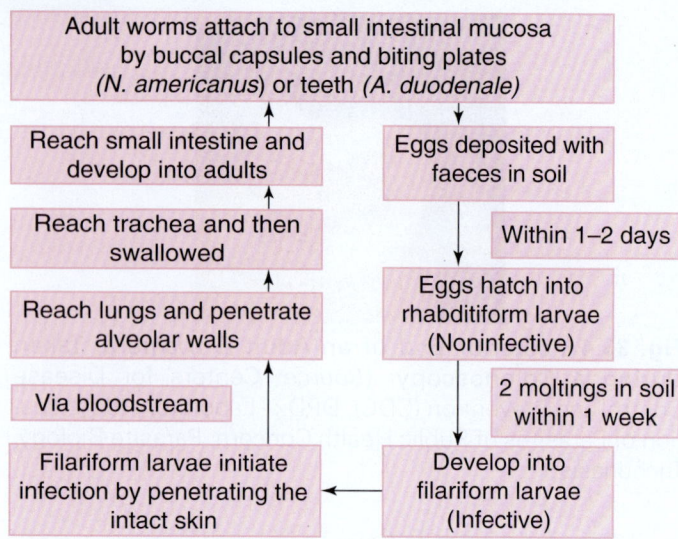

Flowchart 33.3 Life Cycle of Hookworms.

Non-human species. *A. braziliense* and *A. caninum* have dogs and cats, respectively as the definite hosts. Their life cycle is similar to that of the human species. Occasionally the filariform larvae of these species present in soil penetrate human skin, where the larvae cannot mature further but wander aimlessly within the epidermis. Sometimes these larvae move several centimetres a day and are known as **creeping eruptions** or **cutaneous larva migrans**.

Epidemiology
- *A. duodenale* predominates in the Middle East, North Africa and southern Europe and is therefore commonly known as the old world hookworm. *N. americanus* predominates in north and south America and in Australia and is therefore known as the new world hookworm. Both the species are found in Africa, Asia and America.
- These infections are common worldwide but are most common in areas with warm and humid climates and with poor sanitation (open defecation or where night soil is used as a fertiliser).

Transmission
These are transmitted mainly by walking barefoot on contaminated soil when the infective filariform larva penetrates the intact human skin. *A. duodenale* may also be transmitted by ingestion of larvae and by transmammary route. Heavy infections may occur in both children and adults (in contrast to infection by Ascaris and whipworm where heavy infections occur only in children), though are more common in children.

Clinical Manifestations
Most hookworm infections are asymptomatic. The major manifestation of hookworm infection in children is iron deficiency anaemia and hypoproteinemia, which may retard their physical and cognitive development. Other symptoms include itching and localised maculopapular rash (ground itch) at the site of skin penetration. In previously sensitised hosts, serpiginous tracks of subcutaneous migration may occur. Heavy infection

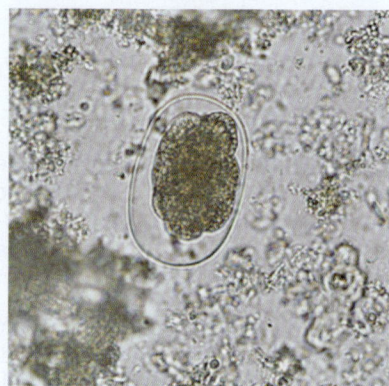

Fig. 33.5 Hookworm Egg in an Unstained Wet Mount, Taken at 400× Magnification. (*Source*: Centers for Disease Control and Prevention (CDC), DPDx - Laboratory Identification of Parasites of Public Health Concern, Parasite Biology, Hookworm.)

may lead to abdominal pain, diarrhoea, anorexia and peripheral eosinophilia. Occasionally transient pneumonitis may occur during larval migration through lungs. Zoonotic infections with animal hookworms may result in intensely pruritic serpiginous track in the upper dermis. Rarely the larvae may cause eosinophilic enteritis or unilateral subacute retinitis.

Laboratory Diagnosis
Sample Required
Freshly passed stool.

Microscopic examination. Diagnosis is made by microscopic detection of eggs in stool, but the eggs of two species (*N. americanus* and *A. duodenale*) cannot be differentiated microscopically. Eggs are colourless (not bile stained), oval in shape measuring about 40×70 μm in size, has a thin outer shell, and contains 4–8 segmented embryo when passed fresh in stool (Fig. 33.5).

Macroscopic examination. Adult worms or rhabditiform larvae may be observed macroscopically in stool.
1. **Rhabditiform larvae** from the eggs may appear in stool samples that are not fresh (when there is a delay in processing of stool sample). These larvae are 250–300 μm long and may be identified by a long narrow buccal canal and flask-shaped muscular oesophagus (Fig. 33.6). These must be differentiated from larvae of *S. stercoralis* (Table 33.2).

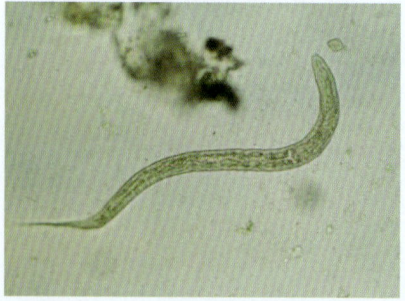

Fig. 33.6 Hookworm Rhabditiform Larva (Wet Preparation). (*Source*: Centers for Disease Control and Prevention (CDC), DPDx - Laboratory Identification of Parasites of Public Health Concern, Parasite Biology, Trichuriasis.)

TABLE 33.2 Methods for Differentiating Larvae of *S. stercoralis* and Hookworm

	S. stercoralis	Hookworm
Larva	Larva currens	Larva migrans/creeping eruptions
Morphology of rhabditiform larvae	Short buccal cavity, prominent genital primordium	Long narrow buccal canal, absent/small genital primordium
Morphology of filariform larvae	Long oesophagus and a notched tail	Short oesophagus and a pointed tail
Furrows on agar plate culture	Larvae show whip-like movement	Larvae show snake-like gliding

TABLE 33.3 Morphological Differences Between *Ancyclostoma duodenale* and *Necator americanus*

A. duodenale	*N. americanus*
Commonly known as old world hookworm	Commonly known as new world hookworm
Larger with 'C' shaped body curvature	Smaller with 'S' shaped body curvature
Prominent buccal capsule with two pairs of ventral teeth	Smaller buccal capsule with semilunar ventral cutting plates
Vulva opens at the junction of middle and posterior one third	Vulva opens in front of the middle portion
Copulatory bursa has a tridigitate dorsal ray	Copulatory bursa has a bidigitate dorsal ray
Spicules are not fused, bristle-like	Spicules are fused, barbed tip

2. **Adult worms** may sometimes be seen in stool sample specially if collected after treatment. Adult worms measure about 12 mm in length and are covered with a creamy white cuticle. Males are smaller than females. Anterior end of both the species is bent dorsally like a hook (hence called hookworms) and has a large buccal capsule. At the posterior end, the males have a copulatory bursa with two spicules.

Species of human hookworms can be differentiated by examining the adult worms expelled after the treatment. Species can be differentiated morphologically (Table 33.3 and Fig. 33.7) or by polymerase chain reaction (PCR) on stool samples.

Diagnosis of creeping eruptions is a clinical diagnosis and is based on finding red raised intensely pruritic serpiginous tracks usually on lower legs or feet. No blood test is available for its diagnosis.

Treatment

Drug of choice is albendazole or mebendazole or flubendazole. Levamisole or pyrantel may also be used. Cutaneous larva migrans is usually a self-limiting disease since the larvae die within 4–6 weeks in human host but may require treatment with antiparasitic drugs that reduce symptoms and prevent bacterial infections.

STRONGYLOIDES STERCORALIS

The genus *Strongyloides* consists of more than 40 species but *S. stercoralis* is the only species infective to man. It is one of the **smallest known nematode** infectious to man (larvae measure less than 2 mm long). *S. stercoralis* is unique among the helminths because of its ability of **autoinfection**, that is, to replicate in the human host. Therefore, in immunocompetent hosts, it can persist for decades while remaining asymptomatic and in immunocompromised hosts; it can cause invasive, disseminated infection that can be fatal. Eggs are produced by **parthenogenesis** (a form of asexual

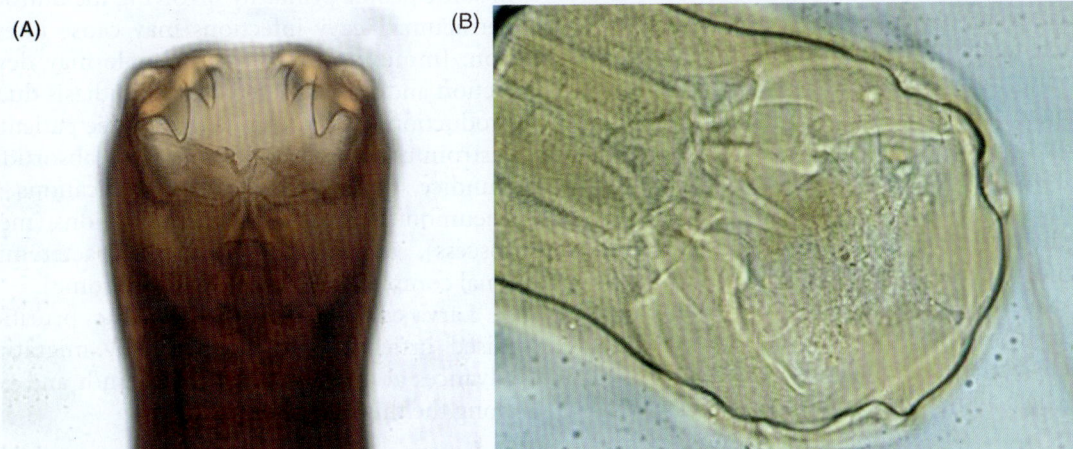

Fig. 33.7 (A) Apical view of the mouth of *Ancyclostoma duodenale* showing the two prominent pointed ventral teeth on each side. (B) Apical view of the mouth of *N. americanus* showing the rounded ventral cutting plate. (*Source*: Centers for Disease Control and Prevention (CDC), DPDx - Laboratory Identification of Parasites of Public Health Concern, Parasite Biology, Ancylostomiasis.)

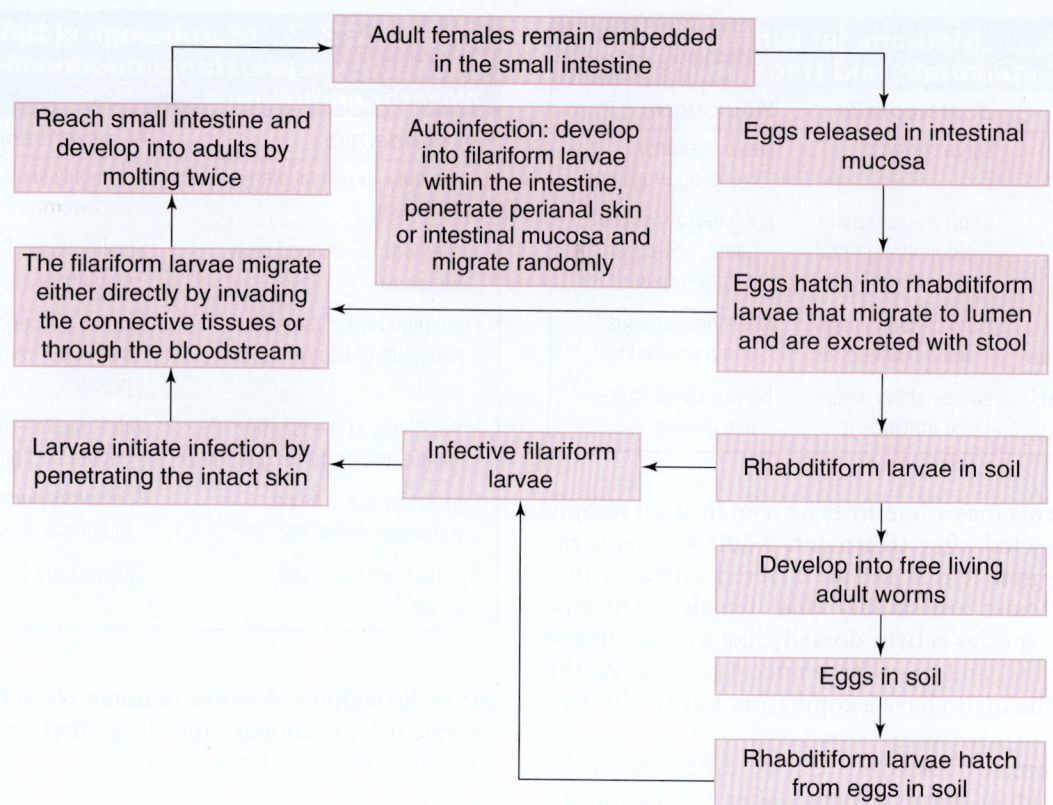

Flowchart 33.4 Life Cycle of *Strongyloides stercoralis*.

reproduction in which development of embryos occur from ovum without fertilisation) since adult parasitic male forms do not exist.

Life Cycle (Flowchart 33.4)

Adult worms live embedded in the human small intestinal mucosa where they release their eggs. The eggs hatch into the rhabditiform larvae (non-infectious forms). These larvae make their way to the intestinal lumen and are released into the soil along with faeces. In the soil, the rhabditiform larvae develop into either the infectious filariform larvae after undergoing several molts or free-living adults that produce eggs, which hatch and produce rhabditiform larvae and finally develop into filariform larvae. These infectious larvae on coming in contact with the host penetrate the skin or mucous membrane to enter the blood stream. The larvae migrate to the lungs where they break through the alveoli, ascend the bronchial tree, reach the trachea and are swallowed to reach the intestine.

Few rhabditiform larvae develop into filariform larvae within the intestine, which penetrate the colonic mucosa or the perianal skin to re-enter the circulation. Thus, the life cycle is complete without leaving the host. This is called autoinfection, which allows the worm to persist in a host for decades.

Epidemiology

The parasite is endemic in patchy areas of tropics, the subtropics and the warm temperate regions.

Transmission

S. stercoralis is transmitted mainly by walking barefoot on the soil contaminated with human waste or sewage, when the infective larva penetrates the intact skin. Occasionally, person-to-person transmission has also been observed.

Clinical Manifestations

Most patients are asymptomatic or may have mild symptoms such as abdominal cramps, indigestion, anorexia, nausea, diarrhoea and weight loss. Recurrent urticaria may develop in some people primarily involving the buttocks, thighs and perineum. Heavy infections may cause intestinal obstruction. Immunocompromised people may develop hyperinfection and disseminated strongyloidiasis due to unchecked production of filariform larvae. These patients may develop gastrointestinal complications (malabsortion, obstructive jaundice, ileus), pulmonary complications (haemorrhage, pneumonitis), neurologic complications (meningitis, brain abscess), vascular complications (bacteremia) and rarely renal complications (nephritic syndrome).

'Larva currens' is pathognomonic, pruritic, serpiginous, raised urticarial rash produced by migrating larvae that advances at a speed of up to 10 cm/h and typically occurs along the thigh and buttocks.

Laboratory Diagnosis

Sample Required

Freshly passed stool/duodenojejunal aspirate are the samples of choice. For disseminated strongyloidiasis, other samples

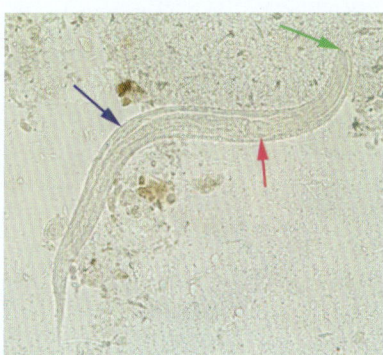

Fig. 33.8 Rhabditiform Larva of *S. stercoralis* in an Unstained Wet Mount of Stool. Notice the prominent genital primordium (*blue arrow*), rhabditoid esophagus (*red arrow*) and short buccal canal (*green arrow*). (*Source*: Centers for Disease Control and Prevention (CDC), DPDx - Laboratory Identification of Parasites of Public Health Concern, Parasite Biology, *Strongyloides stercoralis*.)

such as sputum, bronchoalveolar lavage should also be looked for in addition to the stool sample.

Microscopic Examination

The diagnosis is usually made by microscopic detection of rhabditiform larvae in faeces (Fig. 33.8) or duodenojejunal aspirate (more sensitive). The rhabditiform larvae are slender measuring about 250 µm in length and are fast moving. The sensitivity of stool examination may be increased by repeated stool examinations, by specialised stool concentration techniques or by nutrient agar plate culture (Table 33.3).

Colonoscopy

Female adult worms (male parasitic forms do not exist) may also be seen colonoscopically. They are 30–50 mm in length and 2–3 mm in diameter and have a cylindrical oesophagus. The worms can also be observed in biopsy specimens taken endoscopically (Fig. 33.9).

Serologic Diagnosis

Can be made by the detection of specific tests, though extensive cross reactivity exists with hookworm, schistosomes and filariae.

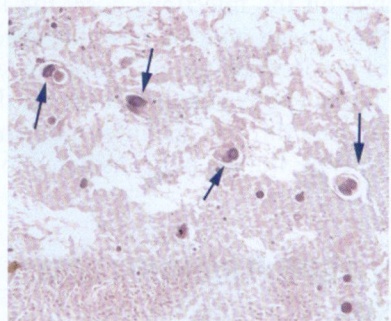

Fig. 33.9 Cross-Section of Female *S. stercoralis* (*blue arrows*) in Small Intestine Tissue (H&E; 200×). (*Source*: Centers for Disease Control and Prevention (CDC), DPDx - Laboratory Identification of Parasites of Public Health Concern, Parasite Biology, *Strongyloides stercoralis*.)

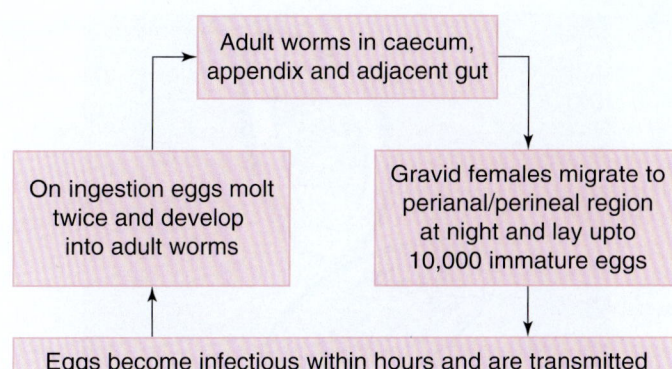

Flowchart 33.5 Life Cycle of *Enterobius vermicularis*.

Treatment

The drug of choice is ivermectin for 1–2 days. Albendazole can be used as an alternative drug. For disseminated infections, ivermectin is continued until stool or sputum samples are negative for 2 weeks.

Prevention

The best way of prevention is not to walk barefoot on the soil or on areas contaminated with human faeces.

ENTEROBIUS VERMICULARIS (ENTEROBIASIS)

E. vermicularis is commonly known as human pinworm, threadworm or seatworm. Humans are the only host of *E. vermicularis*.

Life Cycle (Flowchart 33.5)

Epidemiology

It is prevalent throughout the world. The infection is particularly common in children and institutionalised people. Mode of transmission of infection is by hand to mouth passage, person-to-person transmission can occur through contaminated clothes or bed linen, sexual transmission can occur through oro-anal contact. Retroinfection may also occur, in which the newly hatched larvae may migrate back into the anus (Box 33.1).

Clinical Manifestations

Most cases are asymptomatic. The most common symptom is perianal pruritis, which is worse at night because

---

> **BOX 33.1   Unique Feature of *Enterobius vermicularis***
>
> *E. vermicularis* is commonly found in developed countries and in the temperate regions, which is in contrast to other human pathogenic helminths.

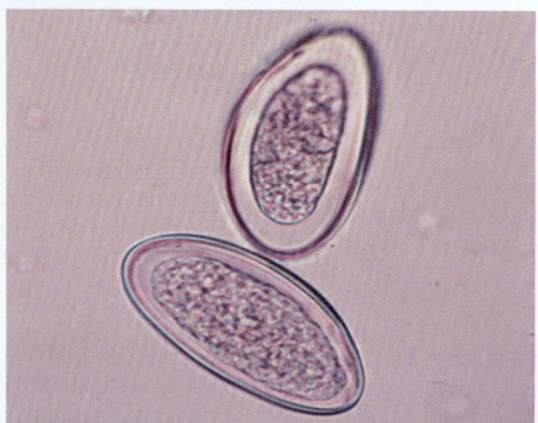

Fig. 33.10 Eggs of *E. vermicularis* in a Wet Mount. (*Source*: Centers for Disease Control and Prevention (CDC), DPDx - Laboratory Identification of Parasites of Public Health Concern, Parasite Biology, Enterobiasis.)

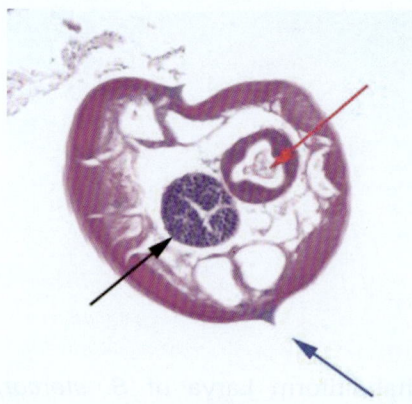

Fig. 33.11 Cross-Section of a Male *E. vermicularis* From Tissue, Stained With H&E. Notice the presence of the alae (*blue arrow*), intestine (*red arrow*) and testis (*black arrow*). (*Source*: Centers for Disease Control and Prevention (CDC), DPDx - Laboratory Identification of Parasites of Public Health Concern, Parasite Biology, Enterobiasis.)

of migration of adult females to the anus. Bacterial superinfections and excoriation may occur in the perianal region due to intense scratching. In females, genital tract infection may also occur.

Laboratory Diagnosis

Examination of faeces is not helpful for establishing the diagnosis of pinworm.

Scotch Tape Test

For making the diagnosis, clear adhesive cellulose tape may be applied to the perianal skin early in the morning (before any washing is done). The tape is then examined under the microscope for detection of pinworm eggs. Lactophenol cotton blue may be added to facilitate the detection of eggs. The procedure should be done on three consecutive mornings to increase the sensitivity of the test.

The eggs are elongate oval, flattened on one side (D-shaped) and measure $55 \times 25\ \mu m$. When shed, these eggs are partially embryonated (Fig. 33.10).

Examination of Adult Worms

Adult worms may sometimes be found in the perianal area, or found during ano-rectal or vaginal examinations (Fig. 33.11)

Treatment

Drug of choice is mebendazole/albendazole/pyrantel pamoate single dose followed by another single dose after 15 days. In institutional outbreaks or in households when more than one member is infected, all the members should be treated simultaneously.

Prevention

Washing hands with soap and warm water after defecation and before handling food is important in breaking the chain of disease transmission. Fingernails should be cut and hygiene should be maintained.

PREVENTION OF INTESTINAL NEMATODES

Infection by intestinal nematodes can be prevented by hygiene maintenance. Emphasis should be laid on washing hands with soap and water after defecation, after touching or handling animals, or handling manure or soil contaminated with human or animal faeces. Fruits and vegetables must be thoroughly washed, peeled or cooked before consuming. Sewage disposal systems should be made effective.

TISSUE NEMATODES

INTRODUCTION

Tissue nematodes are widely distributed throughout the world though their impact is greatest in the developing or under-developed nations of the tropical and subtropical region. All of them are zoonotic infections and are acquired by accidental exposure to animal nematodes. The disease severity depends on the infectious dose, the immunological status of the patient, and the infecting species. The tissue nematodes reside in tissues or lympho-haematogenous system of the human body and thus cause the disease. Some nematodes may pass through the intestine temporarily as a part of their life cycle, but are not considered to be intestinal nematodes.

TRICHINELLA SPP. (TRICHINELLOSIS/TRICHINOSIS)

Trichinellosis is an important public health hazard. It is also damaging to porcine meat production and food safety. The infection is common in carnivorous animals (wild boars) or omnivorous animals (pig, horses).

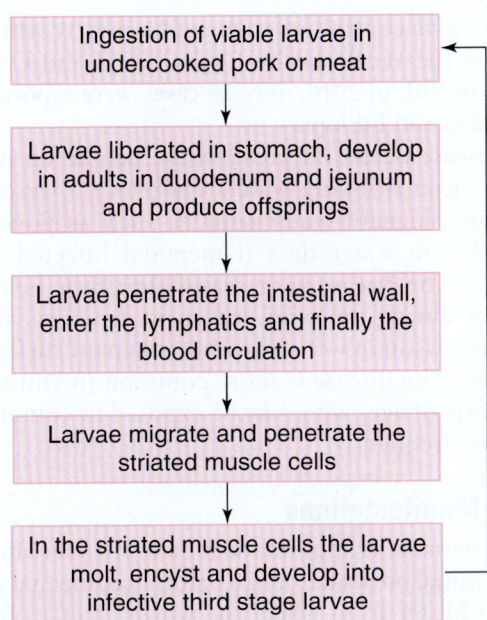

Ingestion of viable larvae in undercooked pork or meat

↓

Larvae liberated in stomach, develop in adults in duodenum and jejunum and produce offsprings

↓

Larvae penetrate the intestinal wall, enter the lymphatics and finally the blood circulation

↓

Larvae migrate and penetrate the striated muscle cells

↓

In the striated muscle cells the larvae molt, encyst and develop into infective third stage larvae

Flowchart 33.6 Life Cycle of *Trichinella spiralis*.

Life Cycle (Flowchart 33.6)

- The life cycle comprises of two generations of *Trichinella* spp. in the same host. Host may be humans or other mammals including pigs, cats, dogs, foxes and rats.
- Both adult and larval stages of *Trichinella* occur in a single host.

Epidemiology

Several species of *Trichinella* can cause human infection, but *Trichinella spiralis* is the most common species and is distributed worldwide. Other species such as *T. pseudospiralis*, *T. murrelli* are found usually in wild animals in various parts of the world. It is a zoonotic infection.

Transmission

Humans get infected by ingesting raw or improperly cooked meat of the infected animals (pigs, cats, dogs, foxes, rats) containing cysts of *Trichinella* in the muscle tissue.

Clinical Manifestations

Symptoms of Trichinellosis vary.

Abdominal symptoms such as nausea, vomiting, diarrhoea or abdominal discomfort may occur 1–2 days after infection. Fever, weakness, eosinophilia, pruritis, periorbital and facial oedema, urticarial rash, conjunctival and subungual haemorrhages may occur 2–8 weeks after ingestion of the contaminated meat as a result of systemic hypersensitivity reaction to larval migration and muscle invasion. Myocarditis, encephalitis or pneumonitis may occur in patients with severe infections. Symptoms of myositis such as myalgia, muscle oedema, and weakness may develop at the onset of larval encystment in muscles.

Diagnosis

Serology

Diagnosis of Trichinellosis may be established by detection of specific antibodies by ELISA. IgG antibodies can be detected about 3 weeks after ingestion of infective larvae and may be detectable for 10 years or more following infection. The drawback is that these antibodies may cross react with other parasitic infections.

Muscle Biopsies

Muscle biopsies may be performed infrequently, for molecular or histological identification of the *Trichinella* species. Squash preparations are made from muscle biopsy and are stained with H&E. Characteristic appearance is a coiled larva in a cyst within a striated muscle cell, surrounded by a nurse cell (derived from the host); and stichosome, which is a column of large rectangular cells present along the oesophagus of the larvae (Figs. 33.12 and 33.13).

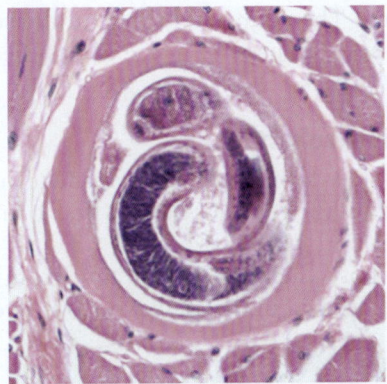

Fig. 33.12 Encysted Larvae of *Trichinella* in a Muscle Biopsy (H&E; 400×). (*Source*: Centers for Disease Control and Prevention (CDC), DPDx - Laboratory Identification of Parasites of Public Health Concern, Parasite Biology, Trichinellosis.)

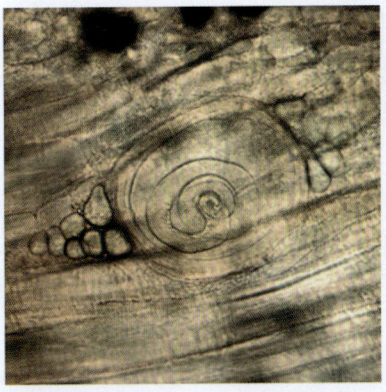

Fig. 33.13 *Trichinella* Larva in Pressed Bear Meat, Partially Digested With Pepsin. (*Source*: Centers for Disease Control and Prevention (CDC), DPDx - Laboratory Identification of Parasites of Public Health Concern, Parasite Biology, Trichinellosis.)

Treatment

Albendazole or mebendazole may be used for treatment. In patients where larvae have established in muscle cells, prolonged or repeated courses of treatment may be required. In such patients, serial monitoring of complete blood counts is necessary.

Prevention

Infection with *T. spiralis* may be prevented by cooking meat properly until pink flesh is not visible or by freezing meat at −15°C for 3 weeks.

DRACUNCULUS MEDINENSIS (DRACUNCULIASIS/GUINEA WORM DISEASE)

The name is derived from Latin language meaning 'little dragon from Medina'. Dracunculiasis is a crippling parasitic disease that was prevalent in Asia and parts of the Middle East but is now about to be eradicated. It is the **largest nematode** measuring up to 1.2 m and is commonly known as 'Guinea worm'. The adult worm lives in the subcutaneous tissue of the affected person, hence is classified as a tissue nematode.

Life Cycle (Flowchart 33.7)

Definitive host is human and intermediate host is water fleas/copepods of the genus *Cyclops*. It has no known animal reservoir.

In human host, the approximate time from entry of the larvae to development of the adult worms is 1 year.

Epidemiology

In 1980s more than 3.5 million people were infected by Guinea worm disease in 20 endemic countries. But because of the intensive control programs of Guinea Worm Disease (GWD) by the World health Organisation and Centre for Disease Control, in 2016, only 25 cases were reported from Chad, Sudan and Ethiopia.

The disease occurs in areas where people do not have access to clean drinking water. Therefore, they consume water from stagnant water sources such as ponds, contaminated with water fleas (copepods) infected with *D. medinensis*. The disease has a seasonal pattern; being more prominent during the rainy season in dry areas and during the dry season in wet areas when stagnant surface water is available. The disease is more common in young adults (15–45 years of age) who may be exposed to contaminated water more frequently.

Clinical Manifestations

A patient remains asymptomatic for about 1 year after infection. The initial presentation is a painful papule that enlarges to form a blister from where the worm emerges from the skin. Local erythema, fever, pruritis may accompany the blister. The usual site of appearance of worms is the lower limb but worms may emerge from any part of the body including head, neck and genitalia. Wound complications may include local cellulitis, abscess, sepsis, septic arthritis, joint deformities or tetanus due to secondary bacterial infection of the worm's exit site. Chronic cases may develop encapsulation of the adult worm that can cause chronic pain for up to 18 months.

Diagnosis

Appearance of the skin blister (Fig. 33.14) and adult worm are sufficient for diagnosis. Adult worms are white and filamentous, and may measure up to 1 m in length (Fig. 33.15).

Treatment

Antihelminthic drugs are not effective against *D. medinensis*. Treatment includes gentle traction of the worm over a period of several days using a small stick, application of wet compresses or immersion of the affected part in water that encourages further worm emergence, topical antibiotics and administration of anti-inflammatory agents.

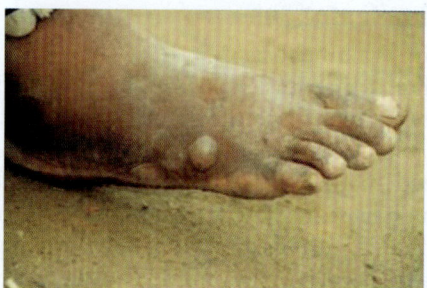

Fig. 33.14 The Female Guinea Worm Induces a Painful Blister. (*Source*: Centers for Disease Control and Prevention (CDC), DPDx - Laboratory Identification of Parasites of Public Health Concern, Parasite Biology, Dracunculiasis.)

Water fleas infected with L3 larvae enter a person who drinks contaminated and unfiltered water

The water fleas become infective for humans over 2 weeks containing L3 larvae

Water fleas are killed in acidic environment of stomach and larvae are released

The free swimming larvae infect the water fleas where they undergo 2 molts

Larvae penetrate wall of the host's stomach and small intestine, where they mature and reproduce

When skin is immersed in water the female worm induces a local blister through which it releases large number of L1 larvae

Gravid female worm migrate throughout the host's body and reaches the skin surface particularly of the lower limb

Flowchart 33.7 Life Cycle of *Dracunculus medinensis*.

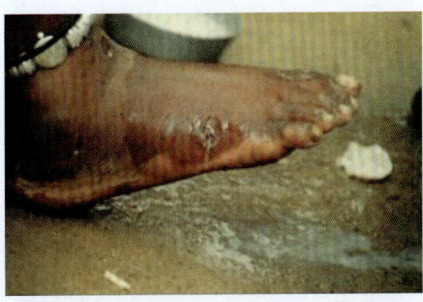

Fig. 33.15 After Rupture of the Blister, the Worm Emerges as a Whitish Filament in the Centre of a Painful Ulcer Which is Often Secondarily Infected. (*Source*: Centers for Disease Control and Prevention (CDC), DPDx - Laboratory Identification of Parasites of Public Health Concern, Parasite Biology, Dracunculiasis.)

Prevention

The disease may be prevented by filtration of drinking water, application of larvicide for vector control, provision of clean drinking water and health education aiming for behavioural change in people.

FILARIASIS

Filarial parasites are transmitted to humans by blood-feeding insects. They affect the lymphatic system or the subcutaneous tissues. Both the adult and microfilarial forms are found in man. Though hundreds of filarial diseases are known, four filarial diseases caused by eight species are known in humans (Table 33.4). Of these, lymphatic filariasis is considered to be a major form of disability worldwide by the World Health Organisation because of the permanent disfigurement associated with the disease. The microfilaria of these eight species may be sheathed or unsheathed.

W. bancrofti, B. malayi (Lymphatic Filariasis)

Lymphatic filariasis (elephantiasis) is a neglected tropical disease causing significant socio-economic burden and is a leading cause of permanent disability. Person-to-person transmission occurs by infected mosquito bites. Three different parasites can cause filariasis: *Wuchereria bancrofti* (90% of cases), *Brugia malayi* (most of the remaining 10% cases) and *Brugia timori*. The adult worms lodge in the human lymphatic vessels producing millions of microfilariae and disrupt the normal function of the lymphatic system.

Life Cycle (Flowchart 33.8)

Host. Definitive host is man and intermediate host is mosquito belonging to the genera *Culex, Anopheles, Aedes* or *Mansonia* depending on the geographical location.

Periodicity/circadian rhythm. The microfilariae have a nocturnal periodicity, that is, they are found in the peripheral blood during the night and in the lung capillaries during the day.

Epidemiology

W. bancrofti is the most widely distributed filarial worm of man contributing to 90% cases worldwide. It has no animal reservoir and human is the only definite host. *B. malayi* is prevalent in south and southeast Asia and can also infect primates in addition to humans.

Both the species have two forms (adults and microfilaria). The periodicity of microfilariaemia depends on species of the parasite, vector and geographical area. It may be nocturnal (microfilaria increase in blood during night and are rarely found during the day) or subperiodic (microfilaria increase in blood in the afternoon but are found at all times).

Clinical Manifestations

Asymptomatic infections. Some people may lodge millions of microfilariae in blood and adult worms in the lymphatics and can still be asymptomatic. These patients may have compromised lymphatic functions. They are an important reservoir of infection for others.

Acute adenolymphangitis. Recurrent episodes of filarial fevers involving the skin, lymph nodes and lymphatic vessels may occur in adolescence and early childhood, which manifests as fever, swelling of upper or lower extremities or the male genitalia. These may be due to direct damage caused by adult worms, inflammatory damage caused by host's immune response to the worms, or due to bacterial skin infections that may result from hampered body immunity as a result of lymphatic damage.

Elephantiasis. May occur in chronic cases when lymphatic obstruction and permanent changes occur. Pitting oedema, hyperkeratosis and hyperplastic changes occur in the skin that lead to increase in size of the affected extremity, which looks like that of an elephant (Fig. 33.19). In males, it can lead to hydrocoele, scrotal lymphoedema and scrotal elephantiasis. In females, breasts may also be affected.

Chyluria. Occasionally, patient may complain of intermittent milky white urine (chyluria), which occurs due to rupture of renal lymphatics.

Diagnosis

Sample required. Whole blood collected at night between 10:00 p.m. and 2:00 a.m. (because microfilaria in India have nocturnal periodicity) is taken as the sample. Blood sample may also be collected 30–60 min after giving a single dose of diethylcarbamazine (DEC) to the patient.

Test required. Thick and thin blood smears stained with Giemsa or haematoxylin and eosin. Knott's concentration technique may be used to improve the sensitivity of detection.

Microscopic examination. Microfilaria measures 290 × 6–7 µm in size, with a rounded head, a gently curved body, a pointed tail and are covered with a sheath. Somatic nuclei are seen as granules. In microfilaria of *W. bancrofti*, nuclei do not extend to tip of the tail, while in *B. malayi*, the tail contains two distinct nuclei (Figs. 33.16 and 33.17).

Histology. Microfilaria and adult worms may also be found in the tissue sections (Figs. 33.18).

TABLE 33.4 Characteristics of Filarial Diseases

Disease	Lymphatic Filariasis			Loiasis	Onchocerciasis	Mansonellosis		
Organism	*W. bancrofti*	*B. malayi*	*B. timori*	*Loa loa*	*O. volvulus*	*M. ozzardi*	*M. perstans*	*M. streptocerca*
Vector	Mosquitoes (*Anopheles, Culex, Aedes* spp.)	Mosquitoes (*Anopheles, Culex, Aedes* spp.)	Mosquitoes (*Anopheles, Culex, Aedes* spp.)	Deer flies (*Chrysops* spp.)	Black flies (*Simulum* spp.)	Midges (*Culicoides* spp.)		
Endemic areas	Southeast Asia, South America, Africa, West Pacific	Southeast Asia, South America, Africa, West Pacific	Southeast Asia, South America, Africa, West Pacific	West and Central Africa	Africa, central and South America	West and central Africa, South and central America		
Habitat of adults	Lymphatics	Lymphatics	Lymphatics	Subcutaneous tissues	Subcutaneous tissues	Subcutaneous tissues	Mesenteries and connective tissues of abdominal organs	Dermis
Habitat of microfilaria	Blood	Blood	Blood	Blood	Skin	Blood	Blood	Skin
Periodicity of microfilaria	Nocturnal	Nocturnal	Nocturnal	Diurnal	—	Aperiodic	Aperiodic	—
Sheath over microfilaria	Present	Present	Present	Present	Absent	Absent	Absent	Absent
Posterior end (Arrangement of nuclei in the tip of tail)	No nuclei in tip of tail	Two distinct nuclei in tip of tail	Two distinct nuclei in tip of tail	Nuclei extend to tip of tail	No nuclei to tip of tail, typically flexed	No nuclei to tip of tail	Nuclei extending to tip of tail	Nuclei extending to tip of hooked tail

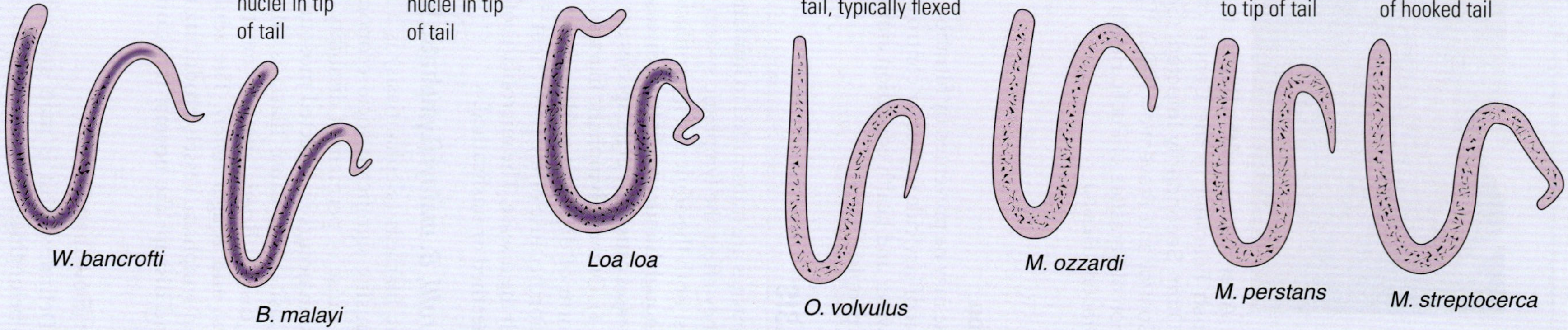

W. bancrofti

B. malayi

Loa loa

O. volvulus

M. ozzardi

M. perstans

M. streptocerca

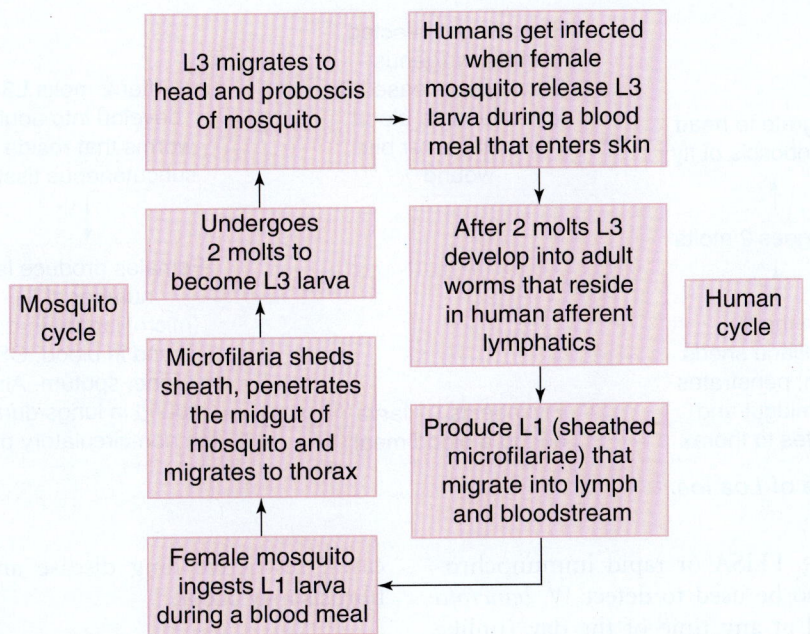

Flowchart 33.8 Life Cycle of Lymphatic Filarial Worms.

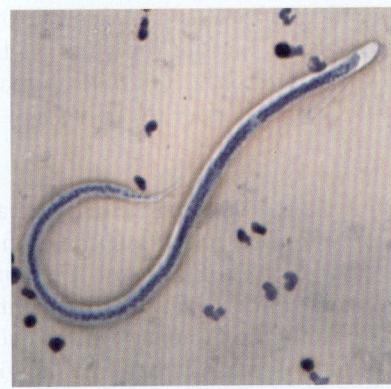

Fig. 33.16 Microfilariae of *W. bancrofti* in Thick Blood Smears Stained With Giemsa. (*Source*: Centers for Disease Control and Prevention (CDC), DPDx - Laboratory Identification of Parasites of Public Health Concern, Parasite Biology, Lymphatic Filariasis.)

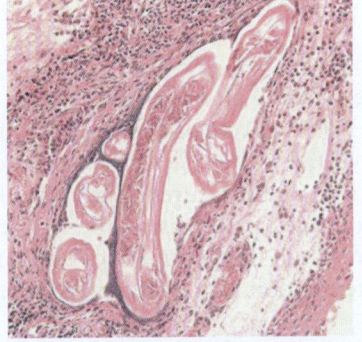

Fig. 33.18 Sections of Adults of *Brugia* spp. From a Lymph Node Stained With Haematoxylin and Eosin (H&E 200×). (*Source*: Taken at Magnification. Centers for Disease Control and Prevention (CDC), DPDx - Laboratory Identification of Parasites of Public Health Concern, Parasite Biology, Lymphatic Filariasis.)

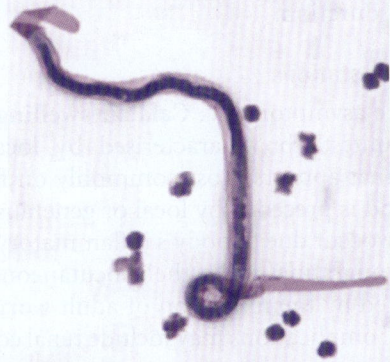

Fig. 33.17 Microfilaria of *B. malayi* in a Thick Blood Smear, Stained With Giemsa. (*Source*: Centers for Disease Control and Prevention (CDC), DPDx - Laboratory Identification of Parasites of Public Health Concern, Parasite Biology, Lymphatic Filariasis.)

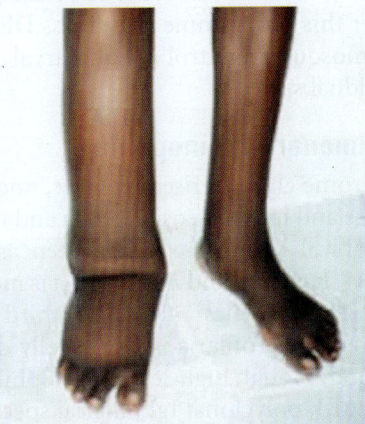

Fig. 33.19 Patient With Lymphedema. (*Source*: Centers for Disease Control and Prevention (CDC), DPDx - Laboratory Identification of Parasites of Public Health Concern, Parasites – Lymphatic Filariasis; Disease.)

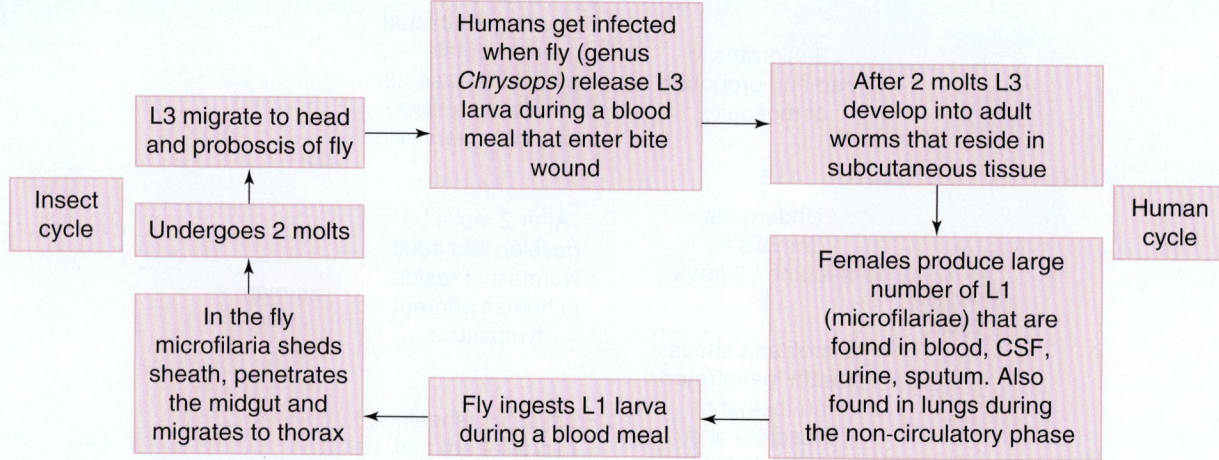

Flowchart 33.9 Life Cycle of *Loa loa*.

Serological techniques. ELISA or rapid immunochromatographic tests may also be used to detect *W. bancrofti* antigens that can be used at any time of the day (unlike microscopy which is sensitive when the blood is collected at night).

PCRs. PCRs may also be used.

Treatment
- Single dose of DEC alone or albendazole with DEC or ivermectin is 99% effective in killing microfilaria in blood.
- Rigorous hygiene of the affected part (to prevent bacterial superinfection) and measures to improve the lymph flow play a major role in decreasing the frequency of filarial fevers and result in improvement of elephantiasis.
- Surgical intervention may be required for hydrocoeles.

Prevention
- The World Health Organisation launched the Global Programme to Eliminate Lymphatic Filariasis (GPELF) in the year 2000, which has two components: transmission interruption and morbidity control.
- The National Filarial Control Programme was launched in India by the National Vector Borne Disease Control Programme (NVBDCP) in 1955. The major interventions done under this programme were mass DEC administration and mosquito control by antilarval measures and indoor residual spray.

Tropical Pulmonary Eosinophilia
TPE is a syndrome characterised by fever, nocturnal wheezing, weight loss and marked eosinophilia and is seen in India and other tropical countries. It has been associated with infections by *W. bancrofti* and *B. malayi*. It is most commonly seen in males (male:females= 4:1) in the third decade of life. In this condition, microfilaria is not usually detected in the peripheral blood, though high-level eosinophilia (eosinophil count> 3000/μL), polyclonal IgE, filarial specific antibodies are found in blood of patient. Adult worms may be detected in the lymph nodes. Treatment consists of DEC for 3 weeks. Retreatment may be required in cases of relapse (20% cases) and steroids may be required in severe cases. In untreated

cases, restrictive lung disease and progressive interstitial fibrosis may occur.

Loa loa (Loiasis)
The nematode *Loa loa* is also known as African eye worm. The parasite is endemic in West and Central Africa because the insect vector of the parasite (deerflies) is found in the rain forests of that region. The adult worm may localise in the conjunctiva of eye and hence the name.

Life Cycle (Flowchart 33.9)
Host. Definitive host is human in which the parasite develops into an adult, reproduce and produce microfilariae. Intermediate host/vector are flies of the genus *Chrysops* (deerflies) in which microfilariae undergo few developmental stages.

Periodicity. The microfilariae have a diurnal periodicity, that is, they are found in the peripheral blood during the day but are found in the lung capillaries at night.

Epidemiology
The parasite is found in Central and West Africa, especially in the rain forests. The infection may be acquired by people living in endemic areas and have a prolonged exposure to infected vectors, or by the people who have repeated exposures of short duration.

Clinical Manifestations
Most cases are asymptomatic. **Calabar swelling** is the most common manifestation characterised by local, non-tender swelling that appears most commonly on face, and the extremities and is preceded by local or generalised pain and itching. These occur due to body's inflammatory response to adult worms migrating under the subcutaneous tissue. **Eye worm** is caused by the migration of adult worm across the conjunctiva. Complications may include renal complications and encephalitis.

Diagnosis
- By identification of microfilaria in thick blood film prepared from the blood sample collected during the day (between 10:00 a.m. and 2:00 p.m.). The microfilaria

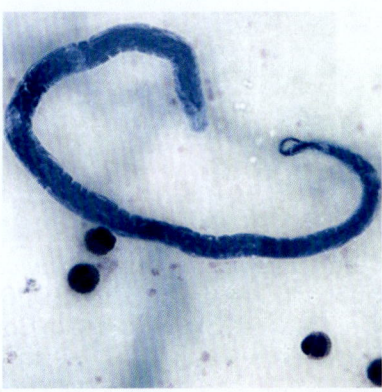

Fig. 33.20 Microfilaria of *Loa loa* in Thick Blood Smear, Stained With Giemsa. (*Source*: Centers for Disease Control and Prevention (CDC), DPDx - Laboratory Identification of Parasites of Public Health Concern, Parasite Biology, *Loa loa*.)

are sheathed, measuring about 300 μm in size, and have nuclei that are continuous from tip to tail (Fig. 33.20).
- By identification of the adult worm after its removal from the eye or under the skin. Adult females measure 40–70 × 0.5 mm and males measure 30–34 × 0.3 mm. Adult worm may also be seen in eye.
- Serologic tests detecting the specific antibodies may also be done.

Treatment

DEC is the treatment of choice that kills both the adult forms and microfilariae. Albendazole may be used in non-responding cases.

Prevention

DEC 300 mg taken once a week, avoiding contact with vector, using insect repellents can help prevent the disease.

Onchocerca volvulus (Onchocerciasis)

Onchocerciasis or river blindness is caused by the filarial worm *Onchocerca vovulus* that is transmitted by blood-feeding black flies (genus *Simulium).* The average life span of adults is 9 years and can extend up to 18 years.

Life Cycle

The life cycle of *Onchocerca vovulus* is similar to that of *Loa loa* except that the vector for the former is black flies (genus *Simulium*). In humans, the adult worms are found in the subcutaneous nodules where they produce microfilariae. These microfilariae are found in skin and lymphatics but may occasionally be found in blood, sputum or urine.

Epidemiology

The disease is endemic in tropical areas of West Africa, South America and in Yemen in the Middle East. The risk is higher in people living near the streams or rivers in areas having the *Simulum* black flies. Usually, several bites of the vector are required for the infection to occur. Therefore, residents and long-term travellers are at risk of acquiring the disease.

Clinical Manifestations

Many people are asymptomatic. Many people experience intense pruritis, skin rashes and skin nodules. Long-term damage may result in '**leopard skin**' or '**cigarette paper appearance**'. In eye, the inflammatory response to the dead microfilaria causes punctate keratitis initially, which in the absence of treatment can permanently damage the cornea, resulting in **river blindness**. Symptoms in *O. vovulus* infection are caused by the immune response to dead and decaying microfilaria. This is in contrast to lymphatic filariasis where the immune response is directed against the adult worms.

Diagnosis

Sample required. Skin snip or snip from eye is the sample of choice. Total six skin snips are collected from different parts of the body and are dipped in physiological saline.
Microscopic examination. Skin snips are stained with H&E and are examined for microfilaria (Fig. 33.21).
- If microfilariae are not visible, diagnosis can be established by PCR.
- Adult worms may be identified in the skin nodules removed during surgery (Fig. 33.22). The worms are thread-like measuring 300–400 × 0.3 mm in size.
- In the eye, microfilaria can be identified by slit lamp examination.
- Specific antibodies may also be detected by immunological tests, though they are not the tests of choice.

Treatment

As per WHO, the treatment of onchocerciasis requires ivermectin at least once yearly for 10–15 years.

Prevention

No vaccine or chemoprophylaxis exists against *O. volvulus*. The disease may be prevented by using personal protective measures against the insect bites, by using vector control methods and mass scale treatment with ivermectin.

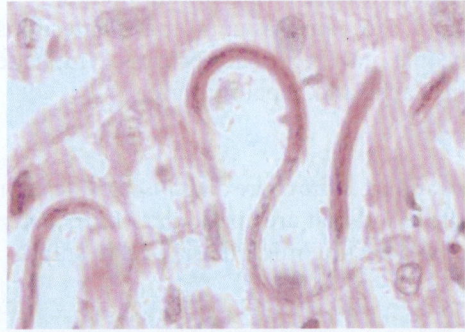

Fig. 33.21 Microfilariae of *O. volvulus* From a Skin Nodule of a Patient From Zambia, Stained With H&E. (*Source*: Image taken at 1000× oil magnification. Centers for Disease Control and Prevention (CDC), DPDx - Laboratory Identification of Parasites of Public Health Concern, Parasite Biology, Onchocerciasis.)

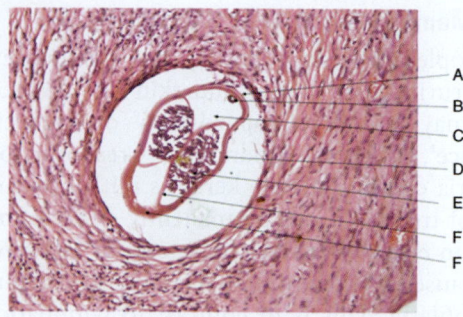

Fig. 33.22 A Photomicrograph Showing the Micro-Anatomy of a Cross-Sectional View of a Female *Onchocerca volvulus* in a Paraffin Section (H&E 200×). A, cuticle; B, intestine; C, pseudo-coelomic cavity; D, hypodermis; E, genital tract with microfilaria; F, genital tract wall; G, muscle of worm body. (*Source: Am J Biomed Life Sci* 2016;4(3):35–40.)

Mansonella (Mansonellosis)

Mansonellosis is caused by three species of the genus *Mansonella*: *M. ozzardi*, *M. perstans* and *M. streptocerca*. The parasites are transmitted by bites of infected blood-feeding midges.

Life Cycle

Vector is midges of genus *Culicoides* or black flies of genus *Simulium*. Life cycle is similar to that of *Loa loa*.

Epidemiology

M. ozzardi is endemic in South and Central America and some Caribbean islands. *M. perstans* is found in Central and North Africa, South America and the Caribbean. *M. streptocerca* is endemic in Africa.

Clinical Manifestations

- *M. streptocerca* can cause asymptomatic infections, though some people may experience pruritis, skin rashes and pigmentation.
- *M. perstans* can cause pruritis, angioedema, fever, headache or urticaria. This filariasis is often accompanied with eosinophilia and rise in specific antibodies,
- *M. ozzardi* can cause asymptomatic infections, though some people may experience pruritis, lymphadenopathy, fever, headache and hepatomegaly.

Diagnosis

M. perstans and *M. ozzardi* are diagnosed by detecting microfilaria in the blood that are aperiodic. *M. streptocerca* can be diagnosed by detecting microfilaria in skin snips.

Treatment

M. perstans and *M. streptocerca* can be treated with DEC. For *M. ozzardi*, ivermectin may be used for treatment.

Differential Diagnosis of Filariasis (Table 33.4)

The diagnosis of filarial worms depends principally on the identification of microfilariae (larval stage) because adults

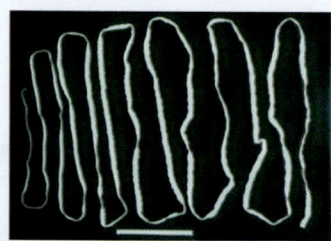

Fig. 33.23 Adult *Taenia saginata*. (*Source*: Centers for Disease Control and Prevention (CDC), DPDx - Laboratory Identification of Parasites of Public Health Concern, Parasite Biology, Taeniasis.)

are usually inaccessible for identification since they inhabit different tissues or organs of the human body. Microfilaria is serpentine in shape, with several nuclei along their body and may be sheathed or unsheathed (may or may not be covered with a membrane). Several anatomical landmarks exist in microfilariae, which may be used for species identification: cephalic space, nerve ring, excretory pore, excretory cell and anal pore. For establishing the diagnosis, following points should be taken into consideration: periodicity, location of microfilariae, sheath present in microfilariae and arrangement of nuclei in the tip of tail of microfilariae.

CESTODES

INTRODUCTION

Cestodes or tapeworms are ribbon shaped, segmented hermaphrodite worms, which have no alimentary canal or a body cavity. They have a scolex with hooks or sucking cups through which they attach to the intestinal mucosa. The scolex is followed by narrow neck, which forms the progottids/segments and the strobila at successively later stages of the development. These worms are oviparous (egg laying). They need two hosts to complete their life cycle: definitive host for sexual maturity and intermediate host for larval development. Exceptions are *Diphyllobothrium* spp. (requires three hosts: two intermediate hosts, cyclops and fish and one definitive host, man) and *Hymenolepis nana* (requires only human host). The characteristics of medically important cestodes are listed in Table 33.5.

Humans can be the definitive host (*Taenia saginata*, *Diphyllobothrium* spp., *H. nana*), harbouring the sexually mature adult worms or can be the intermediate host (*Echinococcus granulosus*, *Coenurus* spp.), with larval stage parasites. The adult worms are found in the small intestine of man and the larvae in any extraintestinal location producing cysts and cause clinical symptoms depending on the location, size and number of cysts.

CLASSIFICATION

Cestodes can be classified according to their habitat in humans into intestinal cestodes (adult worms inhabit intestine) and invasive/somatic/tissue cestodes (larvae inhabit human organs/tissues; Flowchart 33.10).

TABLE 33.5 Characteristics of Medically Important Cestodes

Organism	Common name	Definitive Host	Intermediate Host	Source of Transmission	Stage of Development in Humans
Taenia saginata	Beef tapeworm	Human	Cattle or other herbivores	Cysts in beef	Adult tapeworm
Taenia solium (Taeniasis)	Pork tapeworm	Human	Pigs	Cysts in pork	Adult tapeworm
T. solium (Cysticercus cellulosae)		Human	Human	Eggs in faeces of infected persons	Cysticerci
Echinococcus granulosus	Hydatid cyst disease/ dog tapeworm	Foxes, dogs and other carnivores	Sheep, cattle, human, camels and horses	Eggs from infected dogs/other canines	Cysts containing larvae
Echinococcus multilocularis	Alveolar cyst disease	Foxes, dogs and other carnivores	Mice/other rodents	Eggs from infected dogs/other canines	Cysts containing larvae
Hymenolepis nana	Dwarf tapeworm	Humans	None	Infected humans	Adult tapeworm, cysticercoid
Diphyllobothrium latum	Fish tapeworm or broad tapeworm	Man or other mammals such as cats, wolves, foxes	1st: freshwater crustacean (Copepod), 2nd: freshwater small fish (minnows)	Plerocercoid cysts in freshwater fishes	Adult tapeworm

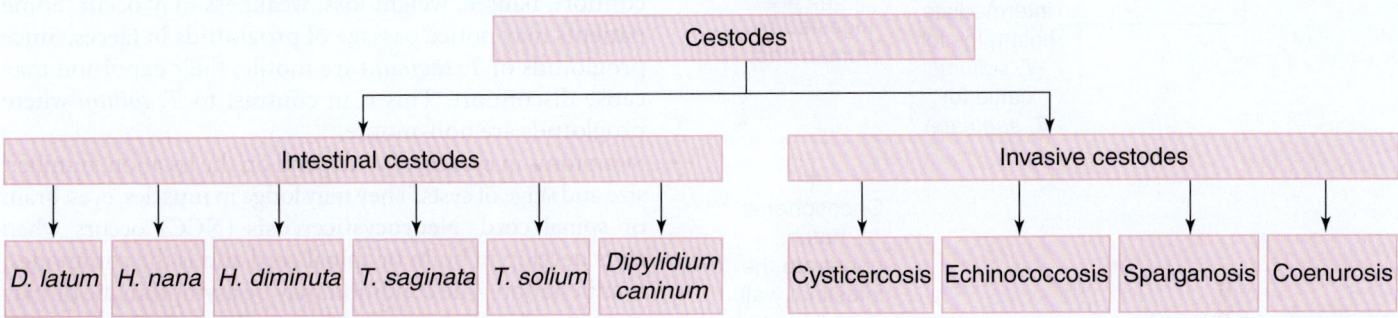

Flowchart 33.10 Classification of Cestodes.

TAENIA SAGINATA AND TAENIA SOLIUM (TAENIASIS)

These cestodes are acquired when measly (containing cysts) uncooked or improperly cooked meat (beef for *T. saginata* and pork for *T. solium*) is consumed. Therefore, *T. solium* and *T. saginata* are commonly known as the pork tapeworm and beef tapeworm, respectively. Human is the definitive host for *T. saginata* and can act as the definitive (intestinal taeniasis) or intermediate host (cysticercosis) for *T. solium*.

Morphology

Difference in the morphology of adult *T. solium* and *T. saginata* is mentioned in Table 33.6. An adult *T. saginata* worm is shown in Fig. 33.23.

Life Cycle (Flowchart 33.11)

Host

For *T. saginata*, intermediate host is cattle and definitive host is man. For *T. solium*, when larval forms present in pork are

TABLE 33.6 Morphological Differences Between Adult *T. solium* and *T. saginata*

	Adult *T. solium*	Adult *T. saginata*
Length	2–3 m in length	5–10 m in length (largest species of genus *Taenia*)
Scolex	Globular with 4 suckers and 2 rows of alternating small and large hooks (armed tapeworm)	Quadrate scolex with 4 circular suckers but without any hooklets or rostellum (unarmed tapeworm)
Neck	Measures 5–10 mm in length	Narrow and long
Proglottids	>1000 in number	1000–2000 in number
Gravid segments	• Usually expelled in chains of 4–5 • Non-motile • 8–12 uterine branches	• Usually expelled singly • Motile • 15–30 uterine branches

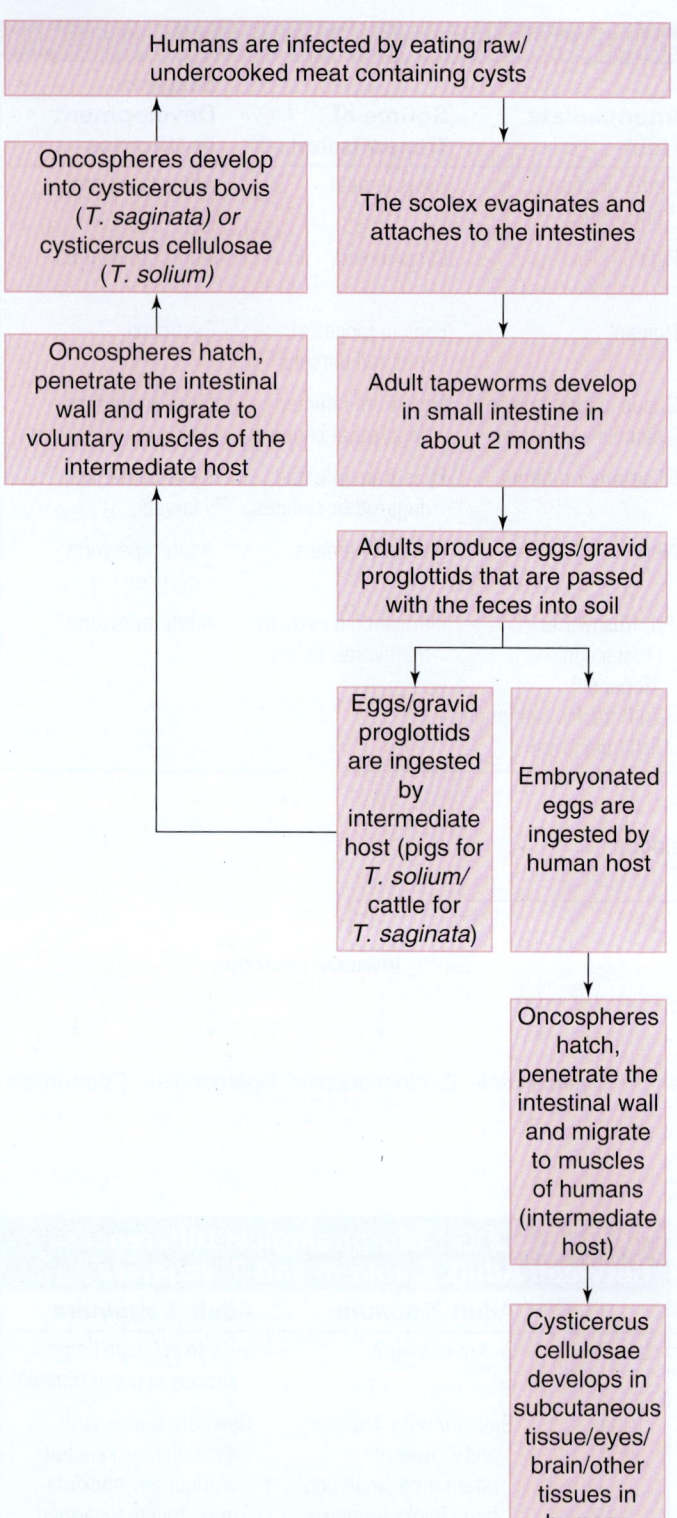

Flowchart 33.11 Life Cycle of *T. solium/T. saginata*.

Fig. 33.24 Egg of *Taenia* spp. (*Source*: Centers for Disease Control and Prevention (CDC), DPDx - Laboratory Identification of Parasites of Public Health Concern, Parasite Biology, Taeniasis.)

Middle East countries, and the latter in Asia, sub-Saharan Africa and Latin America.

Clinical Manifestations

Taeniasis

- Most patients are asymptomatic. Mild abdominal discomfort, nausea, weight loss, weakness may occur. Some patients may notice passage of proglottids in faeces. Since proglottids of *T. saginata* are motile, their expulsion may cause discomfort. This is in contrast to *T. solium* where proglottids are non-motile.
- Symptoms in cysticercosis depend on the location, number, size and stage of cysts. They may lodge in muscles, eyes, brain or spinal cord. **Neurocysticercosis (NCC)** occurs when cysts develop in brain or spinal cord and is the commonest manifestation. The condition may be asymptomatic or may present as headaches and seizures or as signs of increased intracranial pressure such as ataxia, confusion, dizziness, etc. Cysts in muscles may cause tender lumps under the skin.

Laboratory Diagnosis

Diagnosis may be made by demonstration of eggs (Fig. 33.24) or proglottids in stool (Fig. 33.25). Several stool samples collected on different days may be required because eggs are released irregularly in stool.

Macroscopic Examination of Stool

White segments against yellow stool are easily visible. The scolex may be visible, if the sample is taken after administering an antihelminthic agent.

Microscopic Examination

1. Eggs are bile stained, golden yellow in colour, measure 30–40 μm in size, contain an oncosphere 14–20 μm in size with three pairs of hooklets and a striated outer shell (Fig. 33.24). The eggs do not float in saturated salt solution and therefore sedimentation techniques are used to isolate tapeworm eggs. Eggs of both the species are undistinguishable.
2. Proglottids: Each gravid segment of *T. saginata* has 15–30 uterine branches but that of *T. solium* has 8–12

ingested, pig acts as the intermediate host and man as the definitive host and taeniasis (intestinal infection with adult tapeworm) develops. If eggs are consumed as a result of auto-infection or along with the contaminated food or water, man acts as the intermediate host and cysticercosis develops.

Epidemiology

Both *T. saginata* and *T. solium* are distributed worldwide but the former is more common in sub-Saharan Africa and

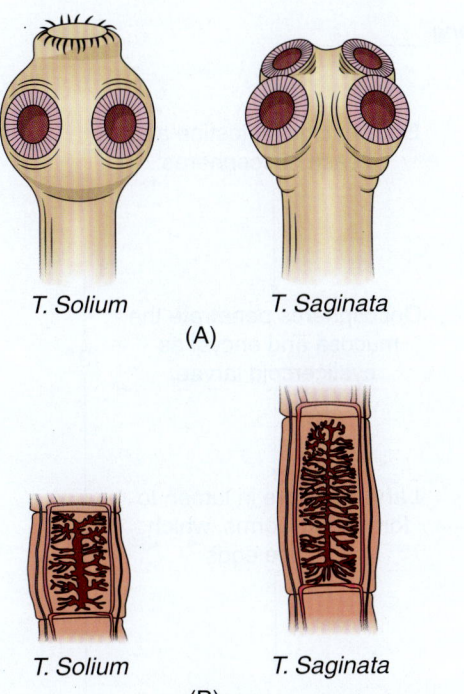

Fig. 33.25 (A) Scolices and (B) Gravid Proglottids of *T. solium* and *T. saginata*.

uterine branches (useful for species; Fig. 33.25 and Table 33.6).

Imaging

For NCC, CT or MRI scan is essential, which shows cysticerci, calcified cysts, solid nodules or ring-enhancing lesions. Fundoscopic examination of eye may also visualise the parasite.

Histopathology

The histopathology of biopsy of a subcutaneous nodule shows cysticercus cellulosae, the larval stage of *T. solium* in the muscles of pig or humans. It is an opalescent ellipsoidal body measuring 8–10 × 5 mm. The cyst lies parallel to the muscle fibre along its long axis. At the site of invaginated scolex, a milky white spot is found. The cyst is filled with salt and albuminous material rich fluid.

A highly sensitive and specific immunoblot assay is also available.

Treatment

Tapeworm infection can be treated with praziquantel or niclosamide. For cysticercosis outside the CNS, surgical removal may also be needed. For NCC, treatment with antihelminthic agents (albendazole, praziquantel or niclosamide), steroids, anticonvulsant agents or surgical removal may be needed.

Prevention

Maintaining proper hygiene can help prevent the faeco-oral transmission of eggs. Transmission in the community may be controlled by identification and treatment of tapeworm carriers.

HYMENOLEPIS (HYMENOLEPIASIS)

Hymenolepiasis is caused by two tapeworm species: *H. nana* and *H. diminuta*. *H. nana* is also known as the dwarf tapeworm since it is the **smallest tapeworm** known to infect man (adults measure 15–40 mm in length). It is the **only tapeworm known that can be transmitted from human to human** and is the **most common human cestode**.

H. diminuta is also known as the **rat tapeworm** as it primarily affects the rodents, and humans are incidental hosts. The adults measure 20–60 cm in length.

Life Cycle (Flowchart 33.11)

Host

Rodents are definitive hosts and insects such as fleas, grain beetles or others are intermediate hosts. Though insects may act as the intermediate host, *H. nana* may also complete its life cycle without it.

Transmission

H. nana can be transmitted to humans by (1) ingesting parasitised insects (intermediate hosts), which get infected while feeding on the contaminated droppings of rodents (definitive hosts), (2) human-to-human by oro-anal transmission or (3) autoinfection, when eggs hatch within the gut initiating the second generation of infection without exiting from a single host (Flowchart 33.12). Both adult and larval forms are found in humans.

H. diminuta can be transmitted to humans by accidentally ingesting the insects in food (uncooked cereals containing insect larvae) or directly from the environment.

Epidemiology

H. nana is the most common cestode infection of humans and has worldwide distribution, though is more prevalent in countries with low levels of hygiene or sanitation. It is most commonly found in school-going children. *H. diminuta* is commonly found in various parts of the world.

Clinical Manifestations

Usually, the infection is asymptomatic. Intense infections with *H. nana* may cause abdominal pain, anorexia and diarrhoea.

Laboratory Diagnosis

1. By examination of direct (unstained) wet mount of formol-ether concentrated stool sample containing eggs. The sensitivity may be increased by multiple stool examinations.
2. Eggs of *H. nana* are spherical measuring 30–50 μm in size, and have a characteristic double membrane. They contain an oncosphere bearing three pairs of hooklets and surrounded by a membrane. Within the space between the oncosphere and outer shell, thread-like polar filaments arise (**distinguishes *H. nana* from *H. diminuta*, which does not have these polar filaments**; Fig. 33.26).
3. Adult worms and proglottids are rarely seen in stool sample.

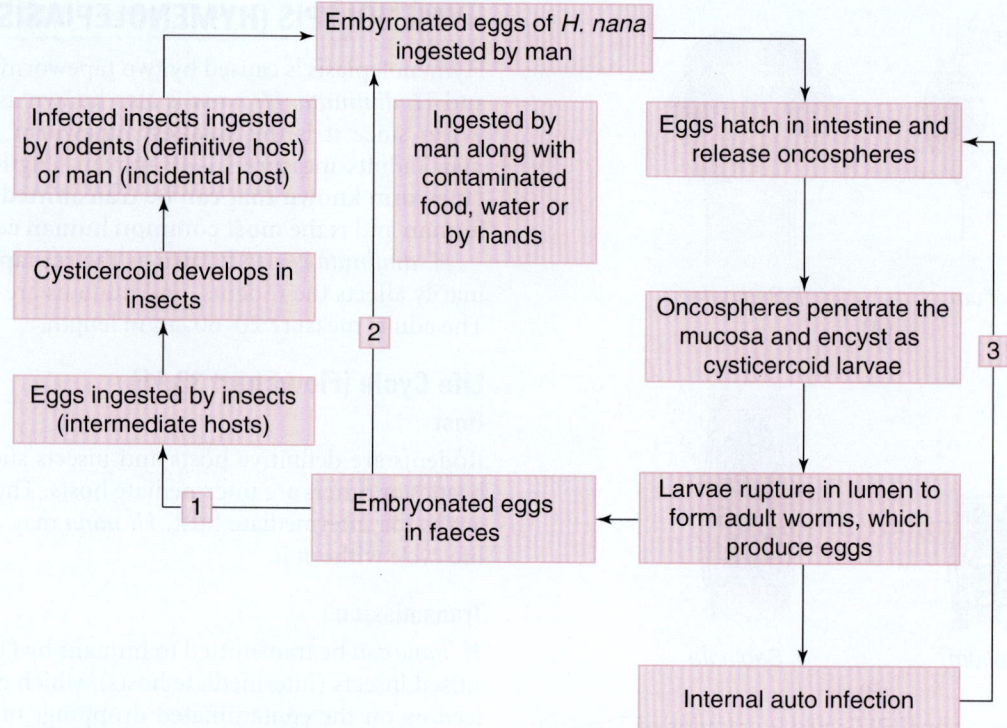

Flowchart 33.12 Life Cycle of *Hymenolepis nana*.

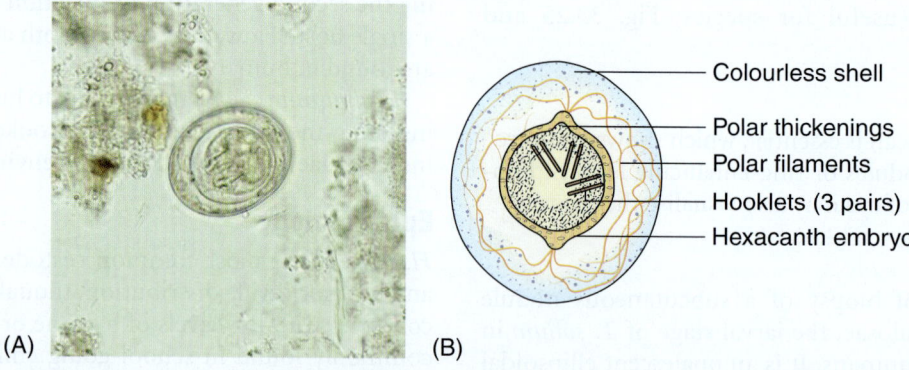

(A)   (B)

Fig. 33.26 Eggs of *H. nana*. (*Source*: Centers for Disease Control and Prevention (CDC), DPDx - Laboratory Identification of Parasites of Public Health Concern, Parasite Biology, Hymenolepiasis.)

Treatment

The infection can be treated with praziquantel or niclosamide. Prolonged therapy (for 5–7 days) is required to eliminate emerging tapeworms.

Prevention

The transmission can be prevented by proper hygiene and by improving environmental sanitation.

DIPHYLLOBOTHRIASIS (*DIPHYLLOBOTHRIUM LATUM*)

Diphyllobothriasis is a disease caused by pseudophyllidean cestode *D. latum*. It is also known as the **fish tapeworm** or **broad tapeworm** and is the **largest tapeworm affecting humans** (may be up to 25 m long) carrying 3000–5000 proglottids. Other species that can affect humans infre-

quently include: *D. cordatum, D. ursi, D. dendriticum, D. pacificum*, etc. Members of this order mainly parasitise fish, birds and fish-eating mammals. The adults attach to human intestine by two bothria or bilateral sucking grooves on their scolex. Their proglottids are flattened dorsoventrally.

Life Cycle

D. latum requires two intermediate hosts (1st: freshwater crustacean such as copepod, 2nd: freshwater small fish-like minnows) and a definitive host (man or other mammals such as cats, wolves, foxes) to complete its life cycle (Flowchart 33.13). It has three larval stages, first stage larva is called coracidia, second stage larva: procercoid and third stage larva: plerocercoid. Humans acquire infection by eating raw or undercooked fish containing plerocercoid larvae (sparganum) in their musculature.

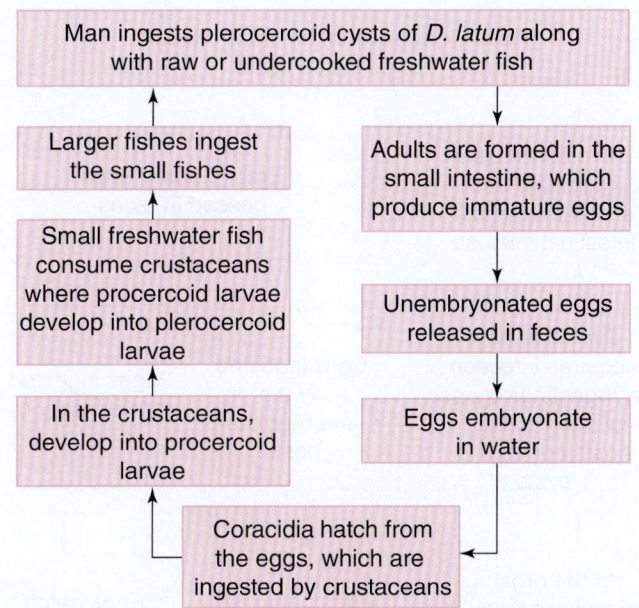

Flowchart 33.13 Life Cycle of *Diphyllobothrium latum*.

Epidemiology

The parasite has a worldwide distribution, though it is endemic in lakes of deltas of the Northern Hemisphere, Central Africa and South America.

Clinical Manifestations

Infection is mostly asymptomatic. Symptoms may include weakness, dizziness, diarrhoea, intermittent abdominal pain and weight loss. Vitamin B12 deficiency leading to megaloblastic anaemia and neurologic sequelae may occur in some individuals.

Laboratory Diagnosis

1. By detecting characteristic eggs in stool, which measure 45–65 μm in size, yellowish-brown in colour, are undeveloped when passed in stool, have a single shell and a small knob-like operculum at one end (Fig. 33.27).
2. Diagnosis may also be established by the recovery of proglottids in stool that has a characteristic central rosette-shaped uterus.

Treatment

Single course of praziquantel or niclosamide is required for treatment. Vitamin B12 deficiency may be treated by parenteral vitamin B12 therapy.

Prevention

The disease may be prevented by heating the fish to 54°C for 5 min or by freezing it at −18°C for 24 h. Placing fish for long periods in high salt concentration also kills the eggs.

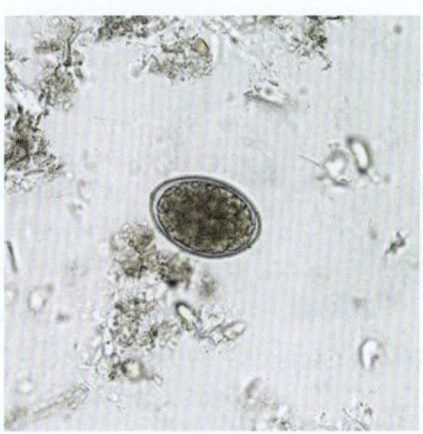

Fig. 33.27 Eggs of *D. latum* in an Unstained Wet Mount. (*Source*: Centers for Disease Control and Prevention (CDC), DPDx - Laboratory Identification of Parasites of Public Health Concern, Parasite Biology, Diphyllobothriasis.)

ECHINOCOCCUS GRANULOSUS AND *ECHINOCOCCUS MULTILOCULARIS* (ECHINOCOCCOSIS)

Echinococcosis is caused by larvae of tiny tapeworm of the genus *Echinococcus*. Two main forms of the disease are seen in humans: cystic echinococcosis/hydatidosis/unilocular cyst disease (caused by *E. granulosus*), and alveolar echinococcosis (caused by *E. multilocularis)*. Both forms of the disease and especially alveolar echinococcosis are chronic life-threatening diseases with a high mortality and morbidity in the absence of proper management.

Morphology

Adult tapeworms are about 5 mm long and have three proglottids: immature, mature and gravid (Fig. 33.28). The scolex has four suckers and a rostellum with hooks. Hydatid cysts are fluid-filled unilocular cysts (have only one bladder), have an external laminated layer and an inner germinal layer. The inner side of the germinal layer gives rise to daughter

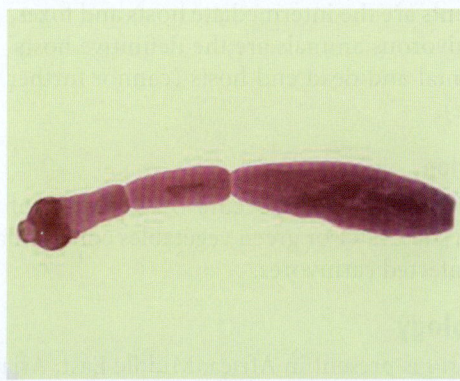

Fig. 33.28 Adult *Echinococcus granulosus*. (*Source*: Centers for Disease Control and Prevention (CDC), DPDx - Laboratory Identification of Parasites of Public Health Concern, Parasite Biology, Echinococcosis.)

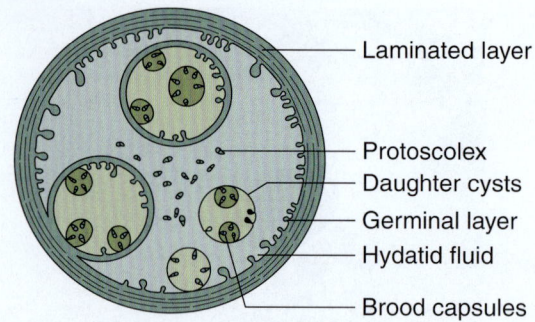

Fig. 33.29 Morphology of a Hydatid Cyst.

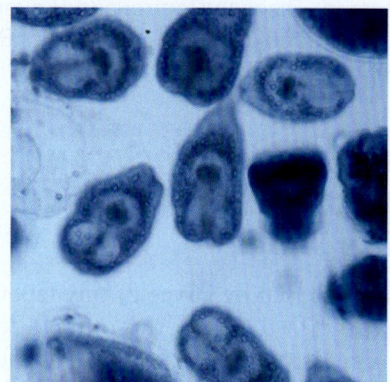

Fig. 33.30 *Protoscoleces* Liberated from a Hydatid Cyst. (*Source*: Centers for Disease Control and Prevention (CDC), DPDx - Laboratory Identification of Parasites of Public Health Concern, Parasite Biology, Echinococcosis.)

cysts and form the protoscolices (new larvae; Figs. 33.29 and 33.30). All these are suspended in a hydatid fluid.

Life Cycle (Flowchart 33.14)

Host

For *E. granulosus*, dogs and other carnivorous animals are definitive hosts; and sheep, cattle, humans, camels and horses are intermediate hosts. For *E. multilocularis*, mice/other rodents are the intermediate hosts and foxes, dogs and other carnivorous animals are the definitive hosts. Humans are accidental and dead end hosts (cannot further transmit the disease).

Transmission

Transmission to humans occurs by ingestion of the eggs along with soil, water or green vegetables contaminated with faeces of infected carnivores.

Epidemiology

E. granulosus is present in Africa, Middle East, Asia, Central and South America and the Mediterranean region. Hydatid disease occurs most commonly in people involved with sheep rearing. *E. multilocularis* is found worldwide but is more prevalent in Asia, North America and Europe.

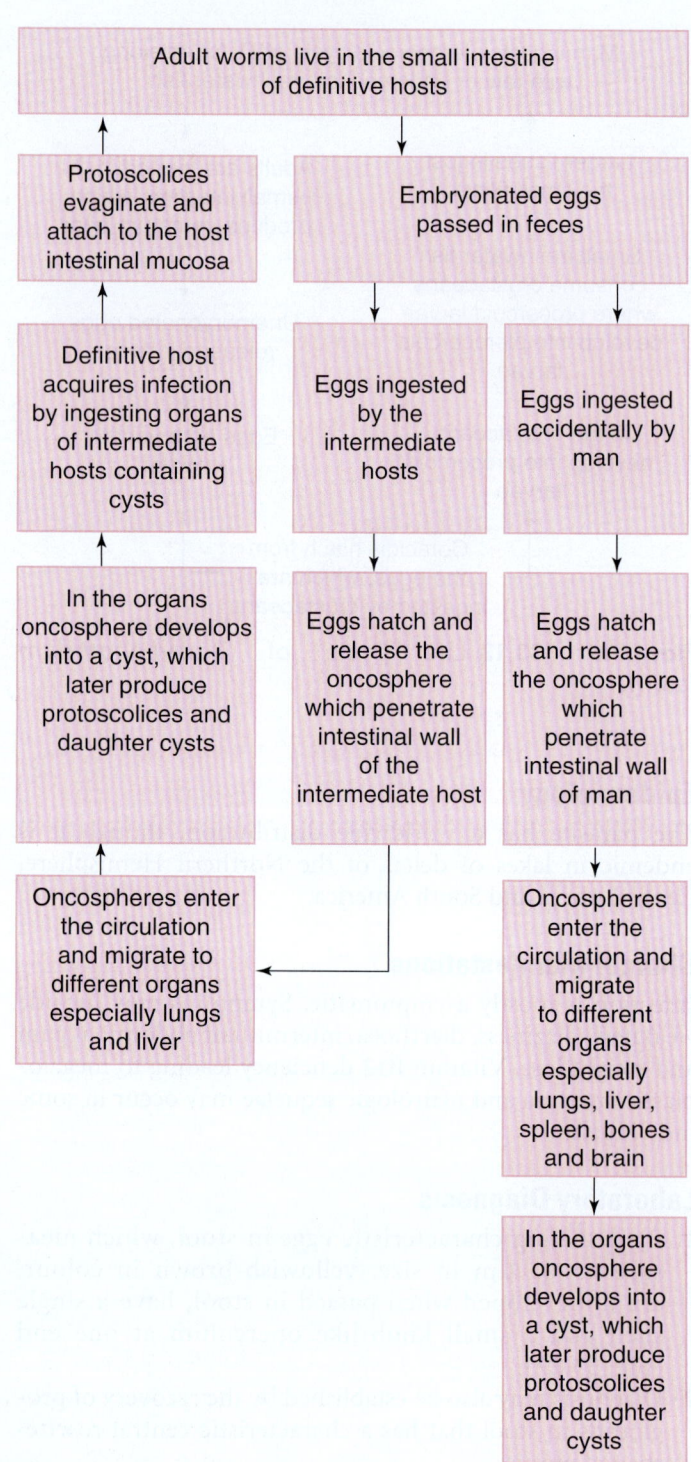

Flowchart 33.14 Life Cycle of Genus *Echinococcus*.

Clinical Manifestations

The echinococcal cysts usually do not produce any symptoms. But they may cause space-occupying effect on the affected organ once these gradually growing cysts containing the new parasites become large enough. The most common locations of these cysts are lungs and liver but may also be found in spleen, kidneys, bones, heart and the central nervous system. Symptoms depend on the organ involved, for example abdominal pain, jaundice may occur in hepatic involvement, haemoptysis and chest pain may occur in pulmonary involvement. If cyst ruptures due to trauma or surgery, the patient may experience anaphylactic reactions of varying severity due to the release of cystic fluid.

The larvae of *E. multilocularis* typically present as slow-growing parasitic hepatic tumour and may also involve the lungs and brain. The vesicles cause progressive destruction of liver and invade the surrounding tissue causing weight loss, pain and malaise. The disease may even progress to hepatic failure and death.

Laboratory Diagnosis

Imaging

Cysts in the human body may be diagnosed by imaging studies such as ultrasonography, CT scan and MRI, which show well-defined cysts.

Serology

The diagnosis is supported by serologic tests that detect specific IgG antibodies by ELISA or indirect haemagglutination tests. Negative serology does not rule out echinococcosis because the antibody response depends on the location of the cyst and integrity of the cyst wall (Box 33.2). For *E. granulosus*, the diagnosis may be confirmed by an immunoblot assay that demonstrates 'Arc 5'.

Examination of Cyst Fluid

In seronegative patients with imaging suggestive of echinococcosis, fine needle aspiration biopsy of the cyst may be done under ultrasound guidance. The presence of numerous hooklets, brood capsules and protoscolices (invaginated or evaginated) in the aspirate is diagnostic of echinococcal disease.

Histology. Diagnosis may also be made by the histology of the cyst wall after surgical removal.

Treatment

Treatment of hydatid disease depends on the location, size and manifestations of the cyst: surgery, albendazole therapy, cyst puncture, PAIR (percutaneous aspiration, injection of scolicidal agents such as 95% ethanol or hypertonic saline and reaspiration) are the different modalities of treatment available. *E. multilocularis* is more difficult to treat and requires complete surgical resection and/or prolonged albendazole therapy.

BOX 33.2 **False-Positive and False-Negative Serology for Echinococcosis**

False-negative antibody responses:
- Occur in 10% cases of liver cysts and 40% lungs cyst
- In calcified cysts or cysts of brain or eye

False-positive reactions: May occur in persons with cancer, chronic immune diseases and with other helminthic infections.

Prevention

By controlling transmission of the parasite using measures such as proper hand washing with soap and water after handling dogs and avoiding contact with wild animals.

TAENIA MULTICEPS, T. SERIALIS (COENUROSIS)

Coenurosis is caused by larval stage of cestodes *T. multiceps* and *T. serialis*. These cestodes are widely distributed, though clinical cases have been reported mostly from Africa, Europe, Canada and the United States. Definitive hosts are dogs and other canines for *T. multiceps* and dogs and foxes for *T. serialis*. Intermediate hosts are horses, sheep, cattle, goats, rabbits and rodents. Adult worms remain attached to the intestinal wall of the definite hosts. Eggs or gravid proglottids are shed in faeces, which are then ingested by the intermediate host. Humans accidentally get infected by ingesting the eggs along with contaminated food or water. These eggs hatch in the intestine releasing oncospheres that circulate in blood and lodge in different parts of the body where they develop into coenuri. The usual location of coenuri of *T. multiceps* is brain and eyes and those of *T. serialis* is the subcutaneous tissue. The inner surface of the cyst wall gives rise to multiple scolices (in contrast to cysticercus which contains a single protoscolex). Clinical manifestations depend on the location of the coenuri and may cause neurologic symptoms or may mimic lymphomas and lipomas. The coenurus may rarely cause serious damage or even death. These may be diagnosed by observing coenuri in autopsy or biopsy specimens. Surgery is the usual modality of treatment for intracranial and intraocular coenurosis. Antiparasitics such as praziquentel may also be used.

TREMATODES

INTRODUCTION

Trematodes are flatworms belonging to the phylum *Platyhelminths*, which have a leaf-like, dorsoventrally flattened, bilaterally symmetrical, unsegmented body. They have a head with suckers but no hooks and have a rudimentary alimentary canal. In contrast, the cestodes have a ribbon-like segmented body, their head contains suckers with hooks and do not have an alimentary canal. The trematodes are similar to the cestodes in that they do not have a body cavity and are hermaphrodites (except *Schistosoma* spp.) that have separate male and female worms. The trematodes require a definite host (mammal/human) for sexual reproduction and either one or two intermediate host for asexual multiplication of larvae. Humans can acquire infection either by direct penetration of the skin by larvae or by ingestion of eggs. The characteristics of medically important trematodes are mentioned in Table 33.7.

TABLE 33.7   Characteristics of Medically Important Trematodes

	Organism	Common Name	Definitive Host	Intermediate Host	Source of Transmission to Humans
Blood flukes	*Schistosoma* spp.	Bilharzia	Man, other primates	Freshwater snail	Penetration of skin by cercariae
Liver flukes	*Clonorchis sinensis*	Chinese liver fluke disease, oriental fluke	Dogs, man, fish-eating carnivores	1st: Freshwater snail 2nd: Fish	Metacercariae in fish
	Opisthorchis viverrini	Southeast Asian liver fluke	Cats, dogs	1st: Freshwater snail 2nd: Fish	Metacercariae in fish
	Fasciola hepatica	Common liver fluke/ sheep liver fluke	Sheep/cattle/man	1st: Freshwater snail 2nd: Aquatic plants (watercress)	Metacercariae on aquatic plants
Intestinal flukes	*Fasciolopsis buski*	Giant intestinal fluke	Pigs/humans	1st: Freshwater snail 2nd: Aquatic plants	Metacercariae on aquatic plants
	Heterophyes hetrophyes		Fish-eating mammals and birds	1st: Freshwater snail 2nd: Fish	Metacercariae in fish
Lung flukes	*Paragonimus westermani*	Oriental lung fluke	Wild and domestic felines, man	1st: Freshwater snail 2nd: Freshwater crustaceans (crabs, crayfish)	Metacercariae in crustaceans

SCHISTOSOMA (SCHISTOSOMIASIS; BOX 33.3)

The blood fluke/trematode belonging to the genus *Scistosoma* causes an acute and chronic disease called Schistosomiasis/Bilharzia. The disease ranks second in terms of impact of disease (next only to malaria) and is a neglected tropical disease. Human infection is caused mainly by five species, the **intestinal species** (residing in mesentric and portal blood vessels): *S. mansoni, S. japonicum, S. mekongi* and *S. intercalatum* and the **urogenital species** (living in vesical veins): *S. haematobium*. Humans become infected when skin comes in contact with water contaminated with cercariae of Schistosomes. Eggs appear in faeces or urine 4–6 weeks after cerariae penetration. The average life span of adults is 3–5 years but can survive for 30 years or more.

Morphology

Adult flukes are elongate tubular worms measuring 1–2 cm in length and 0.3–0.6 mm in width. Males are shorter and stouter than females and have a gynaecophoral canal or longitudinal cleft in which the female worm lives (Fig. 33.31).

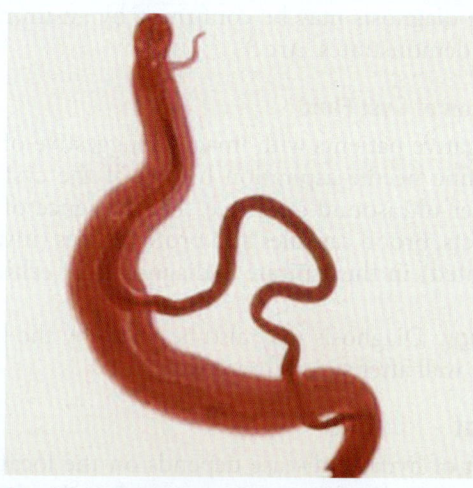

Fig. 33.31 Adult Male and Female Schistosomes. (*Source*: Centers for Disease Control and Prevention (CDC), DPDx - Laboratory Identification of Parasites of Public Health Concern, Parasite Biology, Schistosomiasis Infection.)

Life Cycle

Host

Schistosomes are digenetic with man or other primates as the definitive hosts and freshwater snails as the intermediate hosts.

They have five different developmental stages: Eggs, miracidia, sporocysts, cercariae and adult worms (Flowchart 33.15).

Epidemiology

Schistosomiasis is an important disease in several parts of the world, particularly in places with poor sanitation. *S. mansoni* is

BOX 33.3   Unique Features of Schistosomes

Schistosomes differ from other trematodes in having the following:
1. Separate sexes of adults
2. Transmission through skin penetration by larvae
3. Requiring only one intermediate host for completing its life cycle
4. They are the only trematodes that reside in the blood vessels of warm-blooded animals

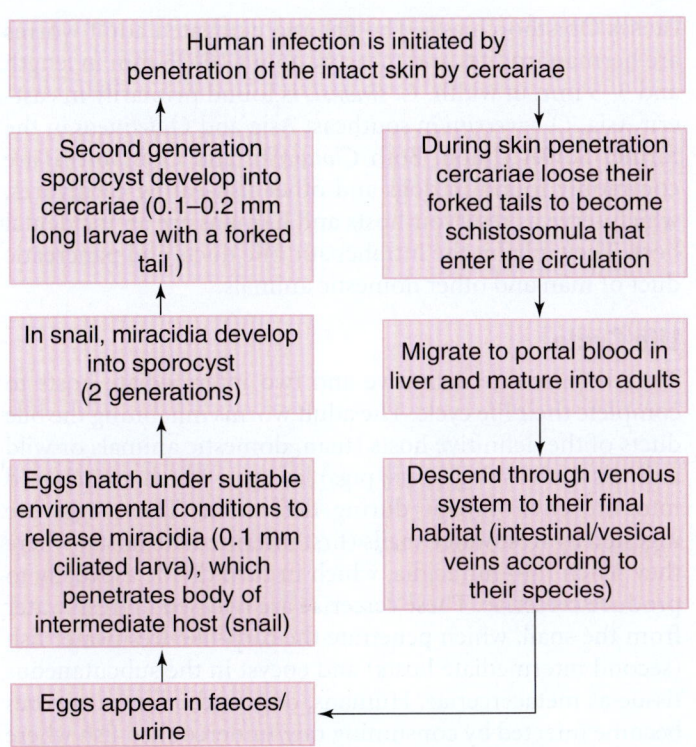

Human infection is initiated by penetration of the intact skin by cercariae

Second generation sporocyst develop into cercariae (0.1–0.2 mm long larvae with a forked tail)

During skin penetration cercariae loose their forked tails to become schistosomula that enter the circulation

In snail, miracidia develop into sporocyst (2 generations)

Migrate to portal blood in liver and mature into adults

Eggs hatch under suitable environmental conditions to release miracidia (0.1 mm ciliated larva), which penetrates body of intermediate host (snail)

Descend through venous system to their final habitat (intestinal/vesical veins according to their species)

Eggs appear in faeces/ urine

Flowchart 33.15 Life Cycle of Schistosomes.

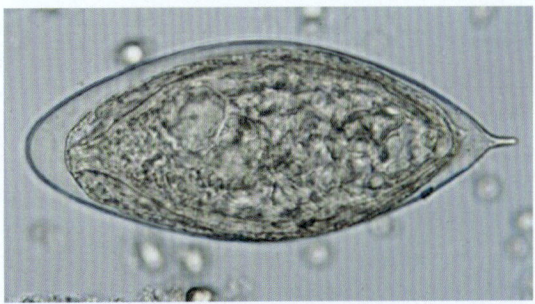

Fig. 33.32 Egg of *S. haematobium* in a Wet Mount of Urine Concentrates, Showing the Characteristic Terminal Spine. (*Source*: Centers for Disease Control and Prevention (CDC), DPDx - Laboratory Identification of Parasites of Public Health Concern, Parasite Biology, Schistosomiasis Infection.)

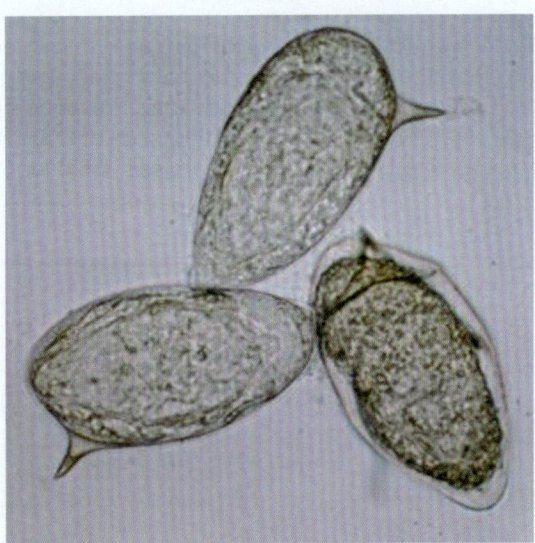

Fig. 33.33 Eggs of *S. mansoni* in an Unstained Wet Mount Showing the Characteristic Lateral Spine. (*Source*: Centers for Disease Control and Prevention (CDC), DPDx - Laboratory Identification of Parasites of Public Health Concern, Parasite Biology, Schistosomiasis Infection.)

found in Africa, South America and the Caribbean; *S. haematobium* is found in Africa and the Middle east; *S. japonicum* in Indonesia, China and southeast Asia; *S. mekongi* in Cambodia and Laos and *S. intercalatum* in Central and West Africa. School age children are most commonly affected by the disease.

Clinical Manifestations

Symptoms depend on parasitologic, host, associated infections, nutritional and environmental factors.

- Swimmers's itch (an itchy maculopapular rash) may appear at the site of cercarial invasion.
- *Acute schistosomiasis or Katayama fever:* Characterised by fever, hepatosplenomegaly and generalised lymphadenopathy that may develop during worm maturation.
- *Intestinal schistosomiasis:* May cause abdominal pain, diarrhoea and blood in stool. In chronic cases, hepatomegaly, ascites and portal hypertension may develop.
- *Urogenital schistosomiasis:* Classically presents with haematuria. It may also lead to fibrosis of bladder, ureter and kidney or even may cause bladder cancer. It may also cause genital lesions or infertility.
- In children, it may cause anaemia, stunting and reduced ability to learn.

Laboratory Diagnosis

Microscopy

Laboratory diagnosis can be done by detection of parasite eggs in faeces or urine. Eggs of schistosomes are round to oval in shape, operculate (hinged at one end) and contains miracidium (developing embryonic larva).

1. *S. haematobium* (in urine) produces oval eggs (140 × 40 µm) with a sharp terminal spine (Fig. 33.32)
2. *S. mansoni* (in faeces) produces oval eggs (150 × 45 µm) with sharp lateral spine and a thin shell (Fig. 33.33)
3. *S. japonicum* (in faeces) forms smaller rounded eggs (100 × 50 µm) with a small lateral spine or hook-like structure (Fig. 33.34)
4. *S. mekongi* egg (in faeces) is similar to that of *S. japonicum* but smaller (65 µm) and *S. intercalatum* egg (in faeces) is similar to that of *S. haematobium* but larger (190 µm) (Fig. 33.35)

Kato Katz Technique

Used for quantification of egg output by using 20–50 mg of faeces or filtering a standard volume of urine through a nucleopore membrane. Counts more than 400 eggs/g of faeces or 10 mL of urine, are heavy infections associated with high risk of complications.

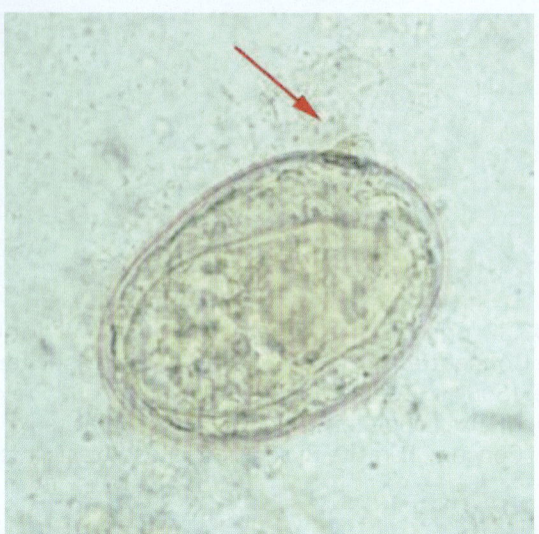

Fig. 33.34 Egg of *S. japonicum* in an Unstained Wet Mount. Note the small, inconspicuous spine *(red arrow)*. (*Source*: Centers for Disease Control and Prevention (CDC), DPDx - Laboratory Identification of Parasites of Public Health Concern, Parasite Biology, Schistosomiasis Infection.)

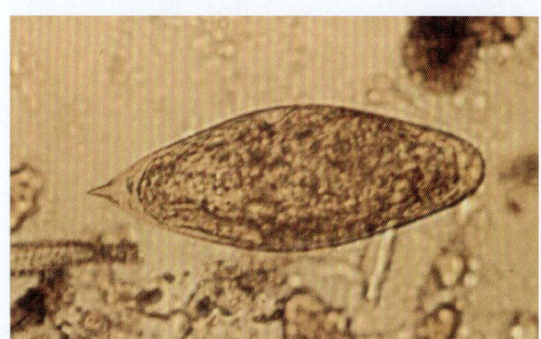

Fig. 33.35 Egg of *S. intercalatum* in a Wet Mount. (*Source*: Centers for Disease Control and Prevention (CDC), DPDx - Laboratory Identification of Parasites of Public Health Concern, Parasite Biology, Schistosomiasis Infection.)

Serological Tests

These may be used in non-endemic areas, but these tests do not distinguish between present and past infection.

Treatment

Drug of choice for schistosomiasis is praziquantel.

Prevention

The World Health Organisation strategy for Shistosomiasis control is based on large-scale preventive chemotherapy (praziquantel therapy to targeted population), improved sanitation, public awareness, snail control and access to safe water.

CLONORCHIS AND OPISTHORCHIS

Clonorchis sinensis, also called **Chinese liver fluke** or **oriental fluke** causes Clonorchiasis. *Opisthorchis viverrini* or *O. felineus*

causes Opisthorchiasis. The flat and elongated adult worms are hermaphrodites and measure about 10–25 mm in length and 3–5 mm in width. *C. sinensis* is found primarily in eastern Asia, *O. viverrini* in southeast Asia and *O. felineus* in the former Soviet Union. Both *Clonorchis* and *Opisthorchis* are commonly found in dogs and other fish-eating carnivores, which serve as reservoir hosts and humans are an incidental host. They inhabit the intrahepatic bile ducts and pancreatic duct of man and other domestic animals.

Life Cycle

They require one definitive and two intermediate hosts to complete their life cycle. The adult worms inhabiting the bile ducts of the definitive hosts (man, domestic animals or wild animals such as dogs, cats, pigs) lay eggs, which are released into the environment during defecation. These eggs are ingested by freshwater snails (first intermediate hosts) where they hatch into miracidia, which in turn divide asexually to produce cerceriae. These cerceriae are released in fresh water from the snail, which penetrate the carp-like freshwater fish (second intermediate hosts) and encyst in the subcutaneous tissue as metacerceriae. Humans, dogs and other carnivores become infected by consuming raw/undercooked fish where the metacerceriae excyst in the duodenum and young worms migrate to the bile ducts.

Adult flukes can live in the bile ducts for up to 30 years. Infection does not lead to protective immunity and reinfections may occur.

Clinical Manifestations

Acute infections may be asymptomatic but chronic severe infections may cause right upper quadrant pain, anorexia, weight loss, cholangitis, biliary obstruction or even cholangiocarcinoma.

Laboratory Diagnosis

Stool Microscopy

Diagnosis is made by detecting eggs in stool that are yellow-brown in colour, are operculated, measure 30×14 μm in size and are embryonated (Fig. 33.36). Eggs of both *C. sinensis* and *O. viverrini* are morphologically indistinguishable and are also similar to those of *Heterophyes* and *Metagonimus*. The intensity of infection may be estimated by Kato-Katz method.

Adult Worms

Diagnosis can also be made by identification of adult worms during surgery. Adult *C. sinensis* worms are dorsoventrally flattened, elongated, lancet-shaped, measure 10–20 mm in length and 3–5 mm in width (Fig. 33.37). Adult worms of *O. viverrini* are similar except that it is smaller (8–12 mm in length).

Others

Immunological tests that can detect specific antibodies in blood, or antigens in stool or blood are being developed. Molecular techniques are still under development.

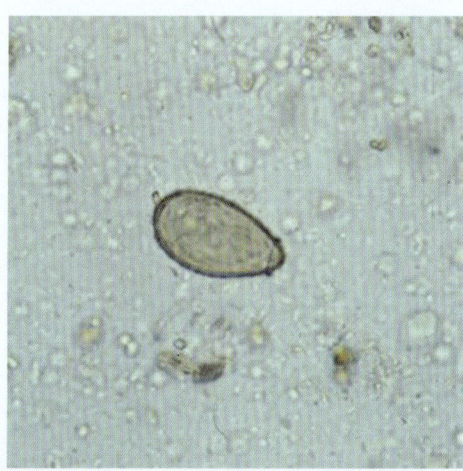

Fig. 33.36 Eggs of *C. sinensis*. (*Source*: Centers for Disease Control and Prevention (CDC), DPDx - Laboratory Identification of Parasites of Public Health Concern, Parasite Biology, Clonorchiasis.)

Fig. 33.37 Adult Worm of *C. sinensis*. (*Source:* Wikipedia. https://en.wikipedia.org/wiki/Clonorchis_sinensis.)

Treatment

Praziquantel is the drug of choice.

Prevention

It can be done by preventive chemotherapy with praziquantel. Primary prevention can be done by avoiding consumption of raw or undercooked fish (the parasite in fish gets killed by adequately cooking it or by freezing it).

FASCIOLA (FASCIOLIASIS)

Fasciola hepatica and *Fasciola gigantica* are the two principal species of the genus *Fasciola* that infect man causing fascioliasis. *F. hepatica* is also known as the **sheep liver fluke** or the **common liver fluke**. Fascioliasis is a zoonotic disease present worldwide, but is especially prevalent in sheep-raising countries. The adult worms reside in common and hepatic bile ducts of sheep and cattle/man (definitive hosts) for as long as 10 years.

Morphology

The adult worms are flat, brown and leaf-shaped measuring 20–30 × 13 mm (*F. hepatica*) and 25–75 × 12 mm (*F. gigantica*) and are hermaphrodites.

Life Cycle

Humans/sheep or other herbivores (definitive host) acquire infection by (1) ingestion of metacercariae attached to water plants such as watercress or water mint, or (2) by consumption of contaminated water or (3) by consuming salad vegetables washed with such water. In the definitive host, the metacercariae excyst, penetrate the intestinal wall, migrate through the peritoneum and reach the liver capsule. The larvae invade the liver tissue and reach the hepatic ducts where they mature into adults. The adult worms produce eggs that are passed along with bile and faeces into the environment.

In freshwater bodies, these eggs hatch into miracidia, which penetrate snail (intermediate host), undergo asexual reproduction in snails to produce sporocysts, which further transform into two generations of rediae. These rediae finally convert to cercariae, which are released in water. The cercariae infect the aquatic plants and encyst as metacercariae, which is the infective stage.

Epidemiology

The worms are more common in sheep-raising countries. Human cases have been reported from Europe, Africa, Australia, Far East and South America.

Clinical Manifestations

Clinical symptoms depend on the stage of disease. The acute phase coincides with the migration of immature worms through the intestine, peritoneum and liver tissue and includes symptoms such as nausea, fever, abdominal pain and skin rashes.

Chronic phase occurs when the worms reach the bile ducts and start producing eggs. This stage is usually subclinical associated with eosinophilia, but may lead to bile duct obstruction, biliary cirrhosis, pancreatitis, bacterial superinfections. **Fascioliasis has not been associated with hepatic carcinoma.**

Laboratory Diagnosis

Diagnosis may be established by demonstration of eggs in faeces: Eggs are large (140 × 75 μm in size), oval, yellowish-brown, operculate (Fig. 33.38). Repeated stool examinations may be required for diagnosis.

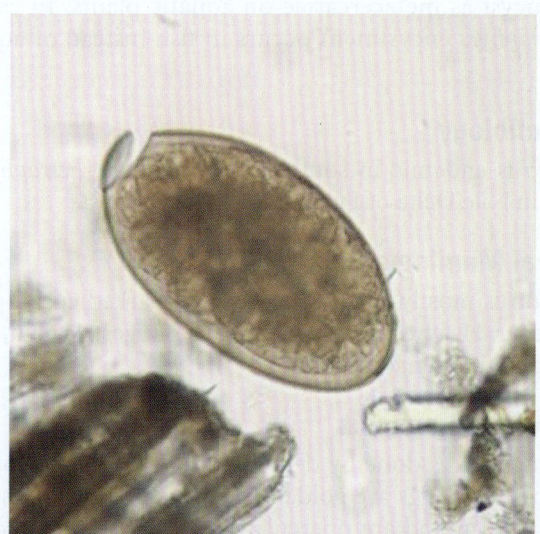

Fig. 33.38 Eggs of *F. hepatica*. (*Source*: Centers for Disease Control and Prevention (CDC), DPDx - Laboratory Identification of Parasites of Public Health Concern, Parasite Biology, Fascioliasis.)

Serologic tests that can detect specific antibodies in serum, or antigens in stool or serum can be used. Molecular techniques are still under development.

Treatment

Triclabendazole is the only drug approved by the WHO for treatment of Fascioliasis.

Prevention

The disease may be prevented by preventive chemotherapy by Triclabendazole and by reducing transmission rates by adopting measures such as health education, snail control, and avoiding consumption of raw water plants.

FASCIOLOPSIS BUSKI AND HETEROPHYES HETROPHYES

F. buski is commonly known as **giant intestinal fluke** since it is the largest intestinal fluke infecting man. The adult worms are thick and fleshy measuring about 2×7 cm. *Heterophyes hetrophyes* measure <2 mm in length.

Life Cycle

Humans or pigs (definitive host) acquire infection by ingestion of metacercariae attached to waterplants like watercress or water mint, or by consumption of contaminated water or consuming salad vegetables washed with such water. On ingestion the metacercariae excyst and attach to the small intestine, where they develop into adult worms. The eggs produced by the adult worms pass along with faeces into the environment. In freshwater bodies these eggs embryonate, hatch into miracidia, which penetrate snail (intermediate host). Within the snail the miracidia undergo many developmental stages including sporocysts, redia, and finally cercariae, which are released from the snail and later encyst as metacercariae on aquatic plants. In case of *H. hetrophyes* encystment occurs in fish instead of aquatic plants.

Epidemiology

F. buski is endemic in southeast Asia and *H. hetrophyes* is found in Nile Delta of Egypt and in the Far East.

Clinical Manifestations

Infection is mostly subclinical though marked eosinophilia may occur. Heavy infections may cause epigastric pain, diarrhoea and duodenal ulcerations.

Laboratory Diagnosis

The diagnosis is done by demonstration of eggs in sputum or faeces. The eggs are golden-brown in colour, measuring 135 × 80 μm and are operculated (Fig. 33.39). Serological tests may also be used.

Treatment

It is by Praziquantel.

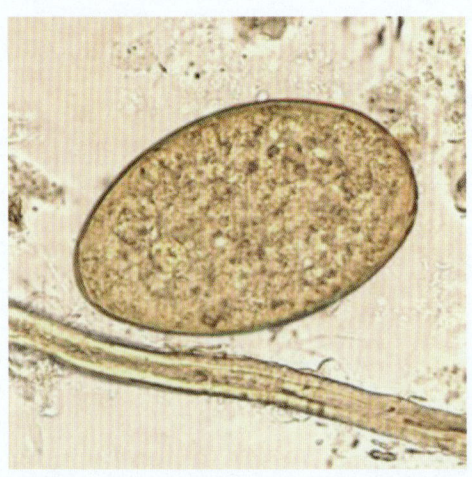

Fig. 33.39 Egg of *F. buski* in an Unstained Wet Mount. (*Source*: Centers for Disease Control and Prevention (CDC), DPDx - Laboratory Identification of Parasites of Public Health Concern, Parasite Biology, Fasciolopsiasis.)

PARAGONIMUS (PARAGONIMIASIS)

The most important species of *Paragonimus* infecting the lungs of mammals is *Paragonimus westermani* (the **oriental lung fluke**). The adult worms (measuring 7 × 4 mm) reside in the lung parenchyma of mammals in an encapsulated cystic cavity. Occasionally, they may be found in the CNS (**cerebral paragonimiasis**) or abdominal cavity.

Life Cycle

P. westermani needs a definitive host (dog, man or cat) and two intermediate hosts (snail and crab/cray fishes) to complete their life cycle. Humans acquire infection by consuming uncooked or improperly cooked crustaceans (e.g. crabs or crayfish) containing metacercariae of the parasite. Metacercariae excyst in the duodenum, penetrate the intestinal wall, traverse through the peritoneal cavity, abdominal wall and diaphragm; and finally reach the lungs where they develop into adult worms. The worms produce eggs, which are coughed up and released in sputum or are swallowed and released in faeces. In freshwater bodies, these eggs embryonate, hatch into miracidia, which penetrate snail (intermediate host) undergo several developmental stages including sporocysts, redia and finally cercariae, which are released and invade the second intermediate hosts (crustacean such as crab) where they encyst as metacercariae.

Epidemiology

It is endemic in many countries of the world except North America and Europe.

Clinical Manifestations

Most patients remain asymptomatic. Acute phase (occurring 2–15 days after ingestion of metacercariae) may consist of abdominal pain, fever, chest pain, urticaria and eosinophilia. Heavy infections may lead to pulmonary calcifications, nodules, pleural effusion and pneumothorax. The patients

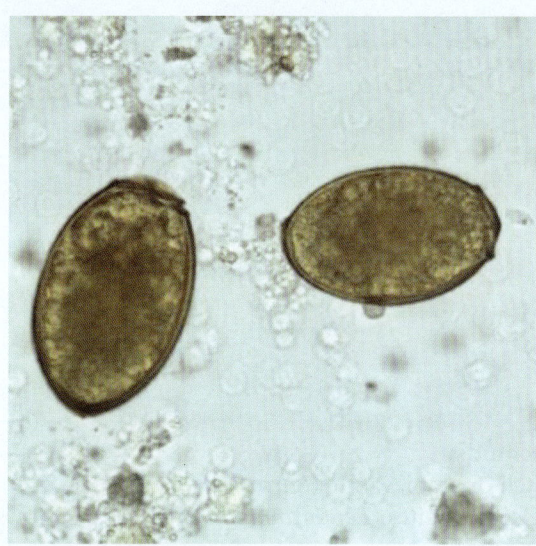

Fig. 33.40 Egg of *P. westermani* in Sputum. (*Source*: Centers for Disease Control and Prevention (CDC), DPDx - Laboratory Identification of Parasites of Public Health Concern, Parasite Biology, Paragonimiasis.)

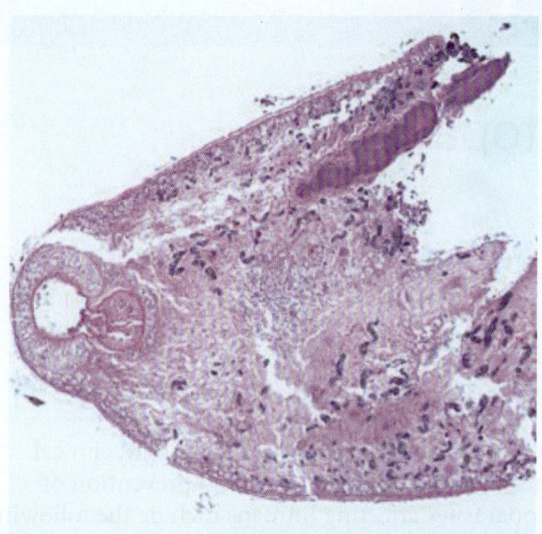

Fig. 33.41 Adult of *Paragonimus* spp., Taken From a Lung Biopsy Specimen Stained With Hematoxylin and Eosin (**H&E**). (*Source*: Centers for Disease Control and Prevention (CDC), DPDx - Laboratory Identification of Parasites of Public Health Concern, Parasite Biology, Paragonimiasis.)

usually present as productive cough with brownish-blood tinged sputum with an offensive fishy odour that may mimic tuberculosis or bronchitis. Extrapulmonary paragonimiasis includes cerebral and cutaneous paragonimiasis.

Laboratory Diagnosis

By demonstration of eggs in sputum or faeces. The eggs are golden-brown in colour, measuring 120 × 60 μm and are operculated (Fig. 33.40). The adult may be seen in a lung biopsy specimen (Fig. 33.41). Serological tests detecting specific antibodies may also be used.

Treatment

Drug of choice is Praziquantel.

34

Ectoparasites

Parul Jain

LEARNING OBJECTIVES

Medically important species, epidemiology, clinical manifestations, management and prevention of ectoparasites affecting humans include the following:
- **Insects:**
  - Lice (Pediculosis)
  - Fleas (*Ctenocephalides* spp., *Xenopsylla cheopis*, *Tunga penetrans*)
  - Flies (*Dermatobia hominis*, *Cordylobia* species, *Sarcophaga* species)
- **Arachnids:**
  - Itch or scabies mite (*Sarcoptes scabiei*)
  - Chigger mites (*Leptotrombidium* species, *Liponyssoides* species)
  - Ticks (*Hyalomma* spp., *Dermacentor* spp., *Ixodes* species)
- Other medically important arthropods

INTRODUCTION

The term 'ecto' means outer. By definition, ectoparasites are the parasites found on the skin or in the superficial layers of the skin. Arthropods and helminths can live as ectoparasites. Ectoparasites of human importance fall into two main groups: insects (lice, fleas and flies) and arachnids (mites and ticks). Ectoparasites may cause disease per se or can act as a vector for transmission of other pathogens (e.g. infected larval trombiculid mites transmit *Orientia tsutsugamushi*, the cause of scrub typhus). The effects caused by ectoparasites on human health may include direct injury, anaemia, detrimental immune reactions (hypersensitivity, anaphylaxis etc.), irritability, dermatitis, skin necrosis, secondary infections, focal haemorrhages, blockage of orifices (ears, nares, orbit), inoculation of toxins (tick paralysis) or exsanguination. The medically important ectoparasites, their effects on humans, management and prevention are summarized in Table 34.1.

Other medically important arthropods: Other arthropods that can act as vectors for disease transmission but do not act as ectoparasites include mosquito, housefly, sandfly and tsetse fly of class Insecta; and cyclops of class Crustacea (Table 34.2).

TABLE 34.1 Medically Important Species, Epidemiology, Clinical Manifestations, Management and Prevention of Ectoparasites Affecting Humans

Ectoparasites	Medically Important Species	Epidemiology	Clinical Manifestations	Management	Prevention
Insects					
Lice (pediculosis)	• *Pediculus humanus* var. *corporis* (body louse) • *P. humanus* var. *capitis* (head louse) • *Phthirus pubis* (crab louse or pubic louse)	• Head louse is the commonest lice infestation affecting mainly school-aged children • Body louse infestation is associated with low socioeconomic status and poor hygiene • Spread commonly by close person-to-person contact • Pubic louse is also transmitted during sexual contact • Only body louse can act as a vector for several bacterial diseases	• Relapsing fever (*Borrelia recurrentis*) • Epidemic typhus (*Rickettsia prowazekii*) • Itching depending on the site of infestation: scalp (head louse), trunk (body louse) and genital area (pubic louse)	• Diagnosis may be established by demonstrating live nymph or adult louse or nits in their precise ecological niches • Topical application of permethrin or pyrethrins • 0.5% Malathion may be used for chronic infections • Oral ivermectin, trimethoprim—sulphamethoxazole may be used for severe infestations • Benzyl alcohol, levamisole, lindane, petrolatum may be used	• Maintaining personal hygiene • For body louse infestation, additionally bathing, laundering clothes and bedding and drying are helpful
Fleas	• Dog and cat fleas (*Ctenocephalides* spp.) • Rat flea (*X. cheopis*) • Chigoe/jigger sand flea (*T. penetrans*)	• Chigoe fleas are endemic in tropical regions of Africa and America	• Hypersensitivity reactions, such as urticaria, erythematous papules or vesicles at the site of bite • Tungiasis is characterised by a subcutaneous papule with a central black dot; resembles folliculitis, impetigo, paronychia. Complications may include tetanus, septicaemia, osteomyelitis, autoamputation of fingers or toes • Rat flea can also transmit bubonic plague (*Yersinia pestis*) and murine typhus (*Rickettsia typhi*) • Cat flea can also transmit flea-borne spotted fever (*Rickettsia felis*) and cat scratch fever (*Bartonella henselae*)	• Antipruritics and histamines • For tungiasis, removal of embedded fleas with sterile needles or scalpels, tetanus vaccination, oral or topical antibiotics • Specific treatment required for transmitted diseases	• Wearing shoes • Insecticidal treatment of dogs, cats and floors
Flies	• *D. hominis* • *Cordylobia* spp. • *Sarcophaga* spp.	• Endemic in America and Africa	• Myiasis (infestation by maggots/flea larvae) which may be cutaneous (wound infestation), subcutaneous (furuncular) or cavitary (invading orbits, nostrils or ear canals)	• Surgical debridement and antibiotic therapy for cavitary myiasis • Occlusive coating by petroleum jelly or surgical extraction of larvae followed by antibiotics for furuncular myiasis	• Using insect repellents • Control of animal larval infestations and safe disposal of caracasses

(Continued)

TABLE 34.1   Medically Important Species, Epidemiology, Clinical Manifestations, Management and Prevention of Ectoparasites Affecting Humans (Cont.)

Ectoparasites	Medically Important Species	Epidemiology	Clinical Manifestations	Management	Prevention
Arachnids					
Itch or scabies mite	• *S. scabiei*	• Occurs in all ethnic and socioeconomic groups • Transmission occurs by direct intimate contact	• Intense nocturnal itching at hairless areas, such as between the fingers, elbows, knees, wrists, penis and breasts • Burrows, small papules or pustules may be present • Crusted scabies mimics psoriasis • Atypical forms include Norwegian scabies (mite hyperinfestation due to immunodeficiency states or mental illness)	*Diagnosis:* • Symmetric polymorphic itchy skin lesions at typical locations • Microscopic demonstration of mite, eggs or faecal pellets from skin scrapings or biopsies *Treatment:* Permethrin cream locally and ivermectin orally	• Adequate treatment of infested patients and all household members • Maintaining good personal hygiene
Chigger mites	• *Leptotrombidium* spp. • *Liponyssoides* spp.	• Distributed worldwide • Only the larvae are ectoparasites and adults are free living • Larvae pierce the skin of the host, produce stylostome and feed on host tissue juices	May transmit diseases: • Scrub typhus (*O. tsutsugamushi*) • Rickettsial pox (*Rickettsia akari*) • Scrub itch • Antigenic reactions, such as allergy, asthma and rhinitis • Papules and vesicular eruptions	• Specific diagnostic tests and antibiotic therapy for transmitted diseases • For dermatitis: History of exposure to the mite's source/ identification of mites • Treatment: warm water soaks, antihistaminics or local steroid application	• Environmental spraying of pyrethrin- and pyrethroid-containing insecticides • Rodent control

Ticks	• *Hyalomma* spp.
	• *Dermacentor* spp.
	• *Ixodes* spp.
		• Distributed worldwide
		• Outstanding vectors of pathogenic agents
		• Transmit pathogens during blood feeding on animals/humans
		• Transmission of microorganisms among ticks:
		  • Transovarian (vertical transmission to offsprings)
		  • Trans-stadial (carrier state transmitted among all generational growth stages)
			Tick-borne diseases:
			• Lyme's disease (*Borrelia burgdorferi*)
			• Tularaemia (*Francisella tularensis*)
			• Ehrlichiosis (*Ehrlichia canis*)
			• Q fever (*Coxiella burnetii*)
			• Indian tick typhus (*Rickettsia conorii*)
			• Kyasanur forest disease (Group B Toganvirus)
			• Crimean–Congo haemorrhagic fever (Norovirus)
			Tick paralysis: Acute ataxia and ascending paralysis in specific regions and seasons
				• Depends on the pathogenic microorganism transmitted (see respective sections)	• Immunisation against tick-borne viral infections
					• Personal protective measures
					• Integrated tick control strategies

TABLE 34.2	Arthropods (Not Ectoparasite) which Transmit Diseases to Human
Arthropods	**Examples of Diseases Transmitted**
Mosquito	Malaria, filaria, yellow fever, dengue fever, chikungunya, Japanese encephalitis, other flaviviruses
Housefly	Typhoid fever, paratyphoid fever, cholera
Sandfly	Kala-azar, oriental sore, sandfly fever, oroya fever
Tsetse fly	Sleeping sickness
Cockroach	Enteric pathogens
Cyclops	Dracunculiasis, diphyllobothriasis
Crabs, crayfish	Paragonimiasis

35

Introduction to Fungal Pathogens

Suruchi Shukla

LEARNING OBJECTIVES

- Characteristics, classification and laboratory diagnosis of fungi
- Historical milestones in mycology
- Basics of antifungal therapy

INTRODUCTION

Fungi are achlorophyllous eukaryotic organisms, which multiply sexually as well as asexually by the production of spores. They are somatically composed of **chitin** and cannot produce their own food, hence called **heterotrophs**. Fungus is a Latin word that means **mushroom**. Mycology is a branch of biological science that deals with the study of fungi. The diseases caused by fungi are known as **mycoses** (myco' means fungi).

In the era of antimicrobials, corticosteroids, and immuno-suppressive therapy, prosthetic devices, and organ transplantation, invasive mycoses have started to develop new clinical aspects, occurring with high frequency and severity. Many of the opportunistic fungal species are now causing invasive infections and are now hospital-acquired. Fungi are ubiquitous and there are millions of different fungal species on Earth, but the common ones in the environment causes fungal disease.

HISTORICAL MILESTONES OF MEDICAL MYCOLOGY

The historical milestones of mycology are listed in Table 35.1.

CHARACTERISTICS OF FUNGI

1. Unlike bacteria, fungus cell contain nuclei with **chromosomes**.
2. Fungi are **heterotrophs**, as they cannot photosynthesise.
3. Fungi absorb food, as they are **osmotrophic**.
4. Fungal **hyphae** exude enzymes and absorb food at their growing tips.
5. Fungi reproduce by the means of **spores**.
6. Fungal diseases can pose threat to public health:
   - **Opportunistic infections**: For example cryptococcosis and aspergillosis are major threats in immunocompromised patients such as cancer patients,

	TABLE 35.1 Historical Milestones of Mycology	
	Scientist/Time Period	**Discovery in the Field of Medical Mycology**
1	**Atharva Veda** in India (about 2000–1000 BC)	Foot anthill was mentioned
2	Hippocrates (460–377 BC) and Gallen (AD 130–200)	Described disease aphthae albae also called as oral thrush
3	Robin (460~370 BC)	Oidium (Candida) was identified as agent of oral thrush
4	Hooke (1665)	First illustrations of microfungi such as Mucor
5	Antonie van Leeuwenhoek (1680)	First one to observe yeasts microscopically
6	Gruby (1843)	Described an important agent of tinea capitis, which he named as *Microsporum audouinii*
7	Sabouraud (1864–1938)	Studied dermatophytes
8	Alexander Fleming (1928)	Discovered Penicillin from *Penicillium notatum*
9	*Pneumocystis carinii* (PCP) (1976)	PCP was proposed as *P. jirovecii*

transplant recipients, patients on immunosuppressant and people with HIV/AIDS often succumb to fungal infections.

- **Nosocomial infections**: Advancements in the health-care practices like usage of prosthesis and devices provides opportunities of fungal biofilm formation. Fungi multiply inside the biofilm and become drug resistant. Candida is the leading cause of hospital-acquired infection.
- **Community-acquired infections**: Dimorphic fungi such as histoplasmosis, blastomycosis and coccidioidomycosis that live in the environment (soil and tree decay) in specific geographic areas can cause community acquired infections.

CLASSIFICATION OF FUNGAL PATHOGEN

Five kingdom classification of living organisms were described by Whitaker (1969):

1. Monera: All prokaryotic organisms such as bacteria and cynobacteria.
2. Protista: Eukaryotic organisms such as algae, lower fungi and protozoa.
3. **Fungi: Multicellular eukaryotes.**
4. Plantae: Multicellular plants with eukaryotic cells.
5. Animalia: Multicellular animals with eukarotic cells.

Classification Based on Modes of Existence

- **Commensalism**: Fungus neither gets benefit nor harm in host–parasite relationship
- **Mutualism**: Fungus and host gets mutual benefits from host–parasite relationship

- **Parasitism**: Fungus gets benefit and host is harmed by host–parasite relationship

Classification Based on Spectrum of Adaptability

- **Pathogenic fungi**: Dimorphic fungus invade tissue and causes systemic infection
- **Opportunistic fungi**: Weakened immunity of host helps to establish fungal infection
- **Toxigenic fungi**: Disease is caused due to ingestion of food contaminated with preformed toxins
- **Allergenic fungi**: Fungal species act as allergens

Classification Based on Primary Site of Involvement (Flowchart 35.1)

- **Superficial, mucocutaneous and cutaneous mycoses**
- **Deep and systemic mycoses**
- **Subcutaneous mycoses**

Classification Based on Fungal Morphology (Flowchart 35.1)

They are eukaryotes broadly divided into two main groups:

- *Yeasts*: Unicellular fungi, which reproduce by asexual, process known as budding
- *Moulds*: Fungal spores germinate and grow into slender tubular thread-like structure called as hyphae

Classification Based on Their Sexual Spore Formation (Flowchart 35.2)

Fungi can be classified according to mode of reproduction:

- Sexually reproducing fungi: Phycomycetes/ascomycetes/zygomycetes
- Fungi where sexual reproduction is not seen: Deuteromycetes also called as fungi imperfect

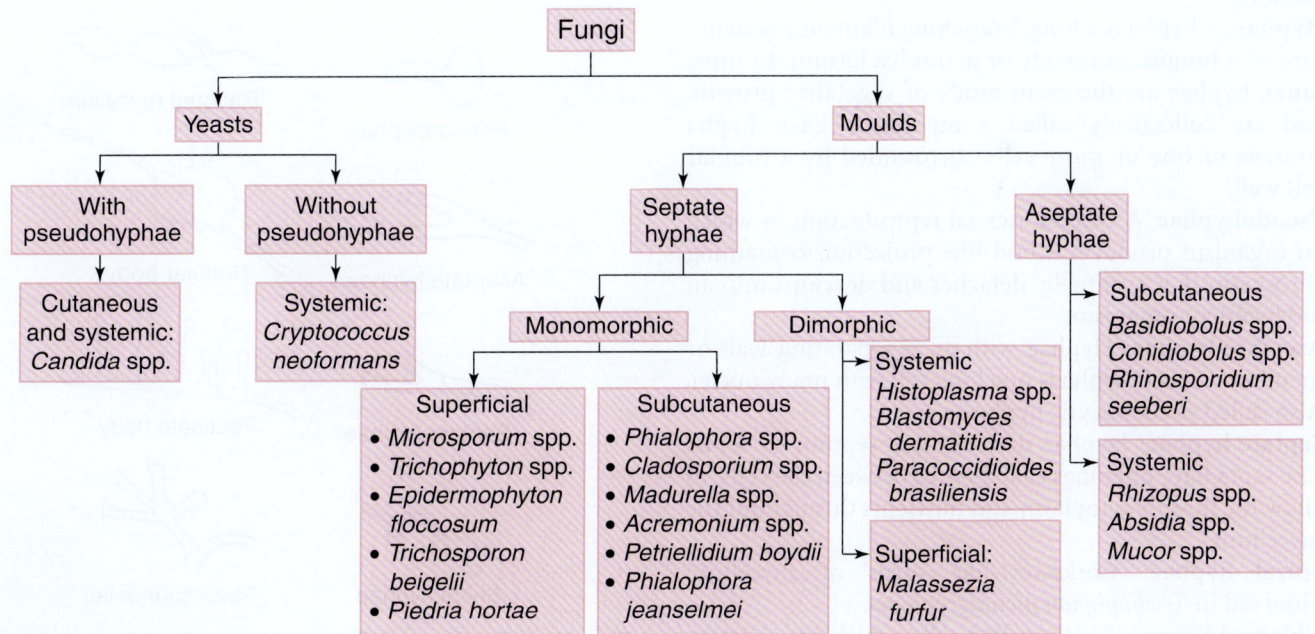

Flowchart 35.1 Fungal Classification Based on Morphology and Primary Site of Involvement.

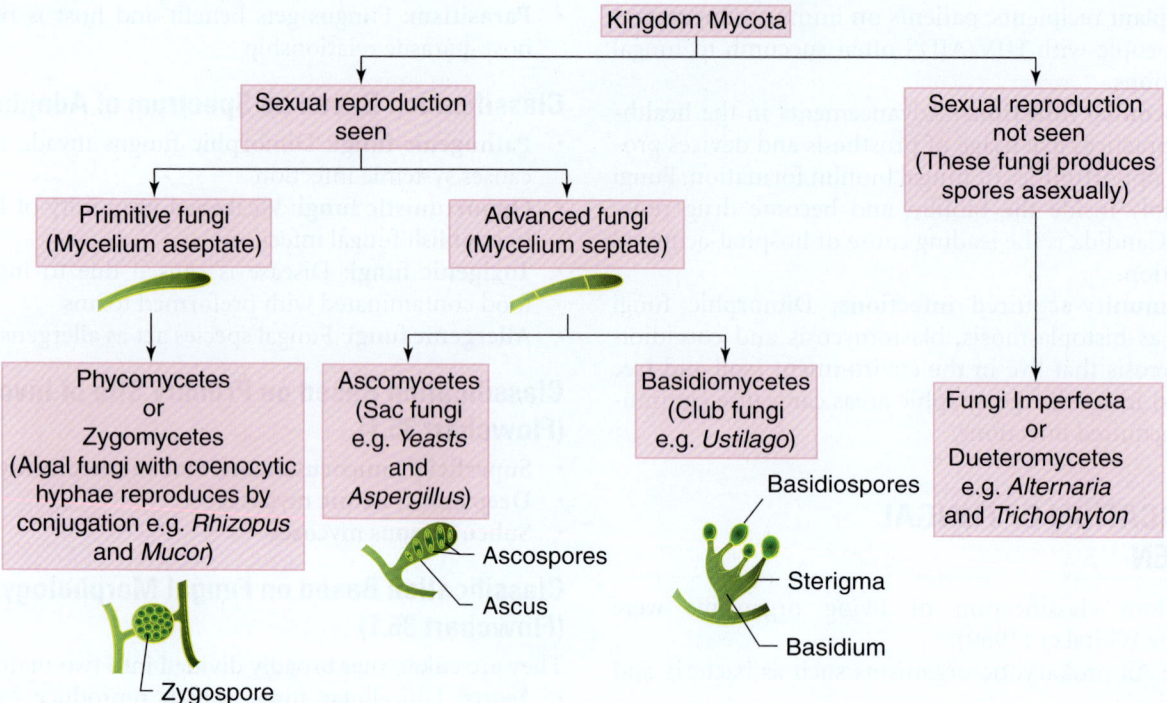

Flowchart 35.2 Fungal Classification Based on Reproductive Methods.

FUNGAL VEGETATIVE STRUCTURES (Fig. 35.1)

- **Yeast cell**: A general term including single-celled, usually rounded fungi that are produced by budding; some yeasts transform to a mycelial stage under certain environmental conditions, while others remain single-celled. They are fermenters of carbohydrates, and a few are pathogenic for humans.
- **Budding yeast cell**: A short chain of cells that results from a lack of separation of daughter cells following budding.
- **Hyphae**: A hypha is a long, branching filamentous structure of a fungus, oomycete or actinobacterium. In most fungi, hyphae are the main mode of vegetative growth, and are collectively called a mycelium. Each hypha consists of one or more cells surrounded by a tubular cell wall.
- **Pseudohyphae**: A type of asexual reproduction in which an organism produces a bud-like projection containing chromatin that eventually detaches and develops into an independent organism.
- **Aseptate hyphae**: Hyphae without a separating wall or membrane. Each hypha is one long cell with many nuclei. Also called as coenocytic hyphae.
- **Septate hyphae**: Hyphae divided by a septum or septa. The septa have openings called pores between the cells, to allow the flow of cytoplasm and nutrients throughout the mycelium.
- **Spiral hyphae**: Corkscrew-like turns of mycelium observed in *Trichophyton mentagrophytes*.
- **Chlamydospores**: Thick-walled cells, which are resistant to adverse conditions found commonly in *Candida* spp.

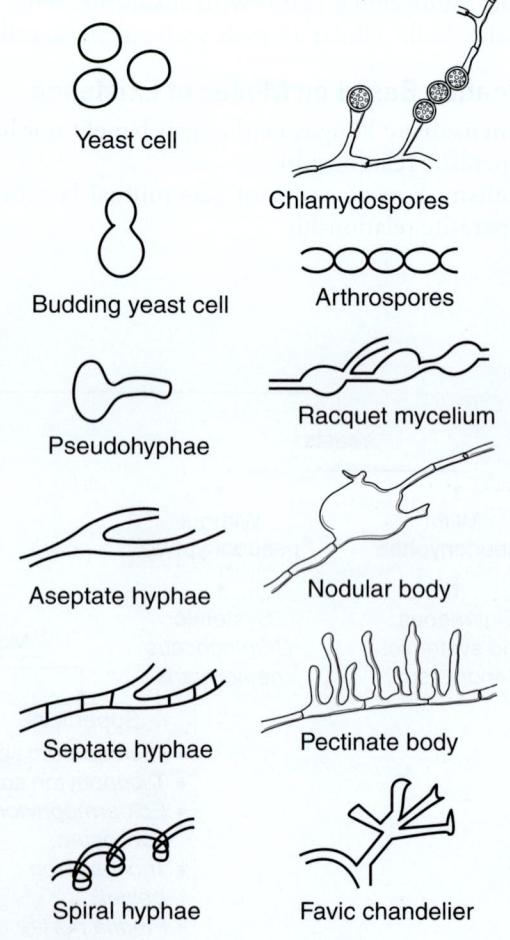

Fig. 35.1 Shapes of Fungal Vegetative Structures and Yeast Forms.

- **Arthrospores**: Small cuboidal or rectangular spores joined by septa in the hyphae, appear to be like a chain and are most commonly seen in *Coccidioides* spp.
- **Racquet mycelium**: Regular enlargement of one end of each segment with large and small ends being in opposition, most commonly seen in dermatophytes.
- **Nodular body**: Closely twisted hyphae observed in dermatophytes.
- **Pectinate body**: Unilateral, short, irregular projections formed at one side of hypha that gives appearance of broken comb.
- **Favic chandeliers**: Short, multiple branches appearing at the end of hyphae seen in *Trichophyton*.

LABORATORY DIAGNOSIS

Clinical history of occupation, living condition, contact with pets and previous intake of drug treatment is very important in diagnosing the fungal infection. The laboratory diagnosis of fungal infection is made by microscopy, culture serology, molecular assays, and skin tests.

Specimens

Clinical material from the site of active lesion is sent for laboratory examination. Blood sample is needed in case of suspected invasive fungal disease.

Microscopy

Mostly **potassium hydroxide** (10%–20% KOH), with or without the fluorescent stain **calcofluor white** stained wet mount prepared from clinical specimen is required for direct fungal element visualisation. In some laboratories and procedures, the periodic acid-Schiff or Grocott's methamine silver stains are used. **Indian ink preparation** is used for negative staining of capsulated yeast cells, for example *Cryptococcus*.

FUNGAL CULTURE

- All fungi are basically aerobic organisms.
- Fungi commonly associated with disease grow well at temperatures between 25°C and 30°C.
- Culture is preferably done in a humid environment
- Incubation period of 4–6 weeks is required.
- Conversion of mould to yeast form of dimorphic fungi occurs at 35–37°C temperature in humid environment.
- A primary medium for the isolation is **Sabouraud's dextrose agar** (SDA) with cycloheximide and chloramphenicol, which has an acidic (pH 5.4) or neutral pH.
- Other culture media used for fungal isolation are dermatophyte test medium (DTM), brain heart infusion broth (BHI), blood agar, Czapek-Dox medium, bird seed agar and corn meal agar.
- Colony characteristics: are best noted when cultures are grown on the Petri dishes rather than in tubes. Gross characters observed on SDA include the following:
  - Growth rate
  - Colour of the surface of mycelia growth: reverse and observe
  - Colour due to diffusible pigment
  - Texture of the colony growth: glabrous/waxy/powdery/granular/suede-like, velvety/downy/fluffy
  - Topography of the growth: flat/cerebriform/crateriform
- **Microscopic morphology of culture growth:** Microscopic morphology can be studied in **teased out** mounts, slide cultures or pressure-sensitive scotch-tape preparations. They are examined for colony morphology, micro and macroconidia and the other structures such as foot cells and vesicle. These preparations may be mounted in **lactophenol cotton blue**. Specialised media that may be necessary to **stimulate sporulation** include cornmeal agar, potato flake agar and potato glucose agar.

OTHER SPECIAL DIRECT EXAMINATION

Wood's Lamp Examination

It is a device used to diagnose superficial fungal infections by exposing the affected skin directly under Wood's lamp. Wood's glass is made up of barium silicate and 9% nickel oxide. On transmission of 365 nm UV rays, a characteristic flourescence is produced by microbial agents. *Microsporum* spp. mainly fluorescence bright green and *Trichophyton* spp. fluorescence dull green whereas *Malassezia furfur* produce golden yellow fluorescence.

FUNGAL SEROLOGICAL TESTING

Serological tests are useful in the diagnosis of systemic and disseminated fungal diseases. Various tests such as agglutination, immunodiffusion, complement fixation and ELISA are commercially available for use in the diagnosis of fungal diseases. The latex agglutination tests for cryptococcal capsular antigen, for use in diagnosing cryptococcal meningitis and detection of glycomanan, for diagnosis of hyphal fungi, are reliable fungal serological tests.

MOLECULAR TESTING

Molecular assays help in the rapid identification of fungi in clinical specimens. Polymerase chain reaction (PCR) and sequencing, aid in speciation of the fungi. The use of molecular methods helps to trace and investigate the epidemiology and environmental sources of fungi that infect immunocompromised patients. Various DNA-based molecular assays used in medical mycology are mentioned in Table 35.2.

ANTIFUNGAL AGENTS

Antifungal susceptibility testing is conducted to identify the most appropriate antifungal treatment. It is standardised as per the Clinical and Laboratory Standards Institute (CLSI) Guidelines. CLSI document includes minimum inhibitory concentration (MIC) references for antifungal susceptibility testing of various antifungal drugs such as amphotericin B, flucytosine, ketoconazole, fluconazole, itraconazole and the new triazoles (ravuconazole, voriconazole, posaconazole). The method is based on visual reading of MIC (MIC, µg/mL) values. Methods to test are disc diffusion test, E-test and Vitek -2 automated system.

Antifungal agents are usually classified on the basis of their mechanism of action on the targeted fungal cell sites

TABLE 35.2   General DNA-Based Methods in Medical Mycology

Method	Application		
	Strain Identification, Molecular Epidemiology, Population Genetics	Species Identification	Phylogenetics and Systematic Classification
Electrophoretic karyotype	✓		
RFLP	✓	✓	
Southern hybridisation	✓	✓	
RAPD, PCR fingerprint, AFLP	✓	✓	
Microsatellites	✓	✓	
Microarrays	✓	✓	
SNP, MLST, DNA sequencing	✓	✓	✓

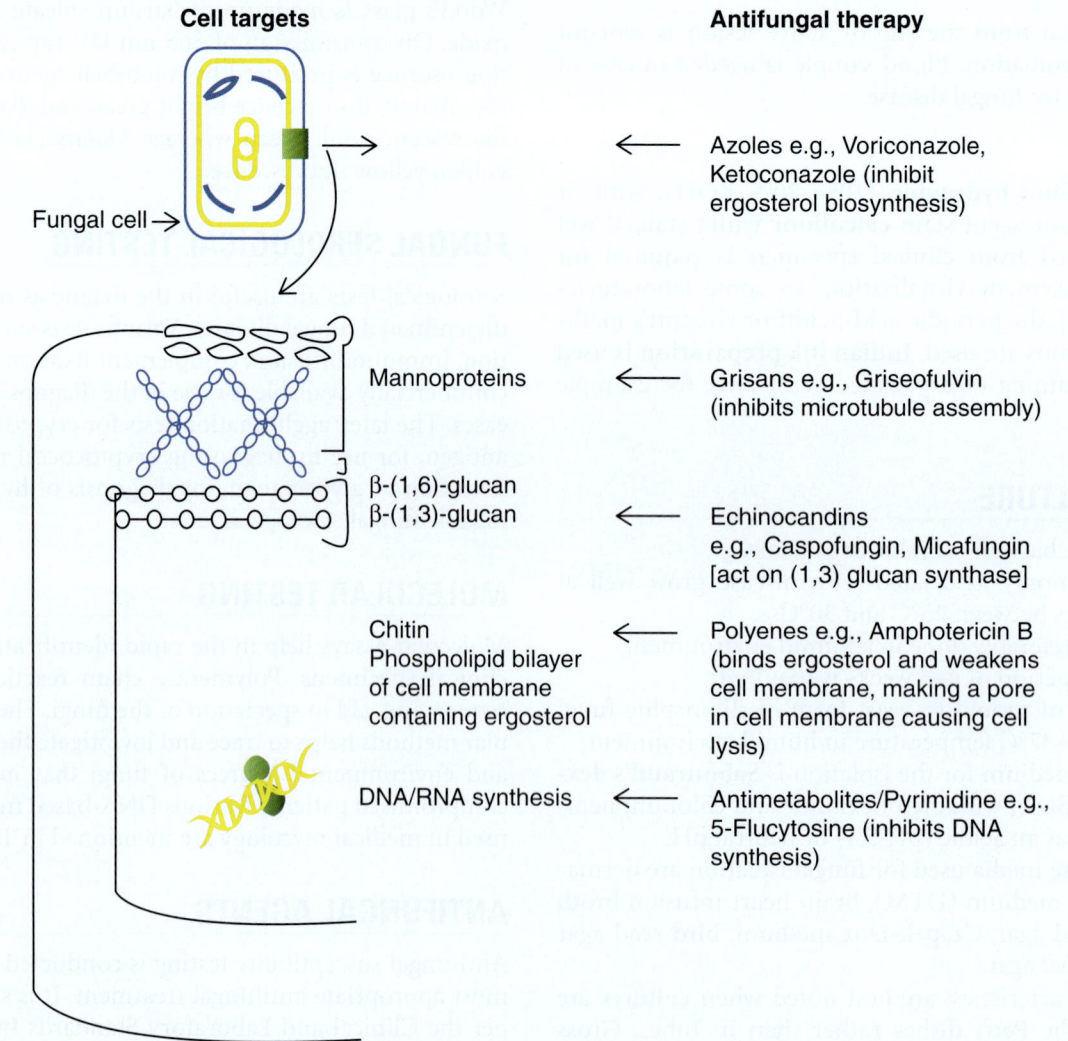

Cell targets

Fungal cell →

Mannoproteins

β-(1,6)-glucan
β-(1,3)-glucan

Chitin
Phospholipid bilayer of cell membrane containing ergosterol

DNA/RNA synthesis

Antifungal therapy

← Azoles e.g., Voriconazole, Ketoconazole (inhibit ergosterol biosynthesis)

← Grisans e.g., Griseofulvin (inhibits microtubule assembly)

← Echinocandins e.g., Caspofungin, Micafungin [act on (1,3) glucan synthase]

← Polyenes e.g., Amphotericin B (binds ergosterol and weakens cell membrane, making a pore in cell membrane causing cell lysis)

← Antimetabolites/Pyrimidine e.g., 5-Flucytosine (inhibits DNA synthesis)

Fig. 35.2 Antifungal Agents and Their Site of Action.

(Fig. 35.2). They can also be classified on the basis of their application as follows;

- **Topical antifungal compounds**
  - Clotrimazole
  - Miconazole
  - Benzoic acid
  - Nystatin
  - Ciclopirox
  - Potassium iodide

- **Systemic antifungal compounds used for the treatment of superficial fungal infection**
  - Griseofulvin
  - Terbenafine
- **Systemic antifungal compounds used for the treatment of deep fungal infection**
  - Azoles
  - Amphotericin B
  - Echinocandins
  - 5-Flucytosine

36

Fungi Causing Superficial and Cutaneous Infections

Suruchi Shukla

LEARNING OBJECTIVES

Aetiological agents, clinical manifestations, laboratory diagnosis, and therapy of the following:
- Piedra
- Tinea nigra
- *Malassezia* spp.
- Dermatophytes

INTRODUCTION

Superficial fungi invade epithelium of keratinised human tissue such as skin, hair and nail. Superficial Mycoses is caused by the following fungal agents: *Trichosporon beigelli, Piedraia hortae, Hortaea werneckii, Malassezia* spp. and *dermatophytes*. Detailed description of different groups of superficial mycosis is illustrated in Table 36.1. There may be no pathology elicited by their presence, and patients may be unaware of their fungal diseases and report to physician only for cosmetic reasons. The warm and moist climate is a major predisposing factor. Superficial mycosis present with typical clinical features and diagnosis can be confirmed by direct microscopy or culture of affected clinical specimen. Superficial mycosis can be treated as per the site and severity of the infection either by topical azoles or with longer periods of oral antifungal.

PIEDRA

Introduction

Piedra, also known as tinea nodosa, is a fungal infection of the hair shaft. It is characterised by the presence of firm, irregular nodules along the shaft axis which are composed of cemented fungal elements. It is of two types, that is, white and black piedra.

Aetiological Agent

White piedra is caused by *T. beigelii*, and black piedra is caused by *P. hortae*.

Clinical Presentation

Generally, it does not cause discomfort to the patient. The diseased hair shaft gets weaken and may rupture or break.

White piedra has soft nodules and can be easily pulled off the hair shaft. Hair of the beard, eyelashes, eyebrows, scalp, groin and axilla is commonly affected.

Black piedra is macroscopically visible colonisation of the shaft of scalp hair. They are hard, black, gritty nodules, consisting of dense fungal stromata thicker at one end. The nodules cannot be pulled off the hair shaft.

Laboratory Diagnosis

Direct microscopy of infected hair in a 10% potassium hydroxide (KOH) will help in identifying both types of piedra. Oval and rectangular arthroconidia are present in *T. beigelii* whereas asci and ascospores are seen in *P. hortae*. Both grow on Sabouraud's dextrose agar (SDA) culture media at 25°C. *T. beigelii* colonies are dry pustular, with a broad, deeply fissured creamy white colonies with radial furrows. *P. hortae* produces small, dark brown to black colored colonies having folded velvety texture. The reverse of the colony appears black due to reddish brown diffusible pigment.

Therapy

The disorder can be controlled by shaving or cutting of infected hair. Topical fungicides like clotrimazole cream, benzoic acid, salicylic acid or 5% ammoniated mercury ointment can be used.

TINEA NIGRA

Aetiological Agent

Tinea nigra is caused by the single species *H. werneckii*. It is a tropical disease and common in Central and South America, Africa and Asia.

TABLE 36.1 Description of Different Types of Superficial and Cutaneous Mycosis

Cutaneous Mycosis	Aetiological Agent	Clinical Manifestation	Direct Microscopy	Culture	Diagram
Black piedra	*Piedraia hortae*	Hard, black, gritty nodules, consisting of dense fungal stromata thicker at one end. The nodules cannot be pulled off the hair shaft	Septate hyphae with asci, ascospores and intercalary chlamydoconidia are seen	On SDA/25°C: Colonies are small, folded, velvety in texture and dark brown to black in colour. The reverse of the colony appears black due to reddish brown diffusible pigment	Hard, black nodule
White piedra	*Trichosporon beigelii*	Soft nodules and can be easily pulled off the hair shaft	Septate hyphae with oval and rectangular arthroconidia	On SDA/25°C: dry pustular, with a broad, deeply fissured creamy white colonies with radial furrows	Soft, white nodule
Tinea nigra	*Hortaea werneckii*	Brown to black painless lesions on palm and sole	Brown multiple-branched, septate hyphae with budding cell	Colonies on oatmeal agar are restricted, smooth and slimy with an oily, glistening olivaceous-black colour. On microscopic examination of fungal mycelia growth, budding cells are subhyphae, 0–1 septate with production of daughter cells by annellidie conidiogenesis from their poles	Hyaline septate hyphae with annellides
Pityriasis versicolour or tinea versicolour	*Malassezia furfur* and *Malassezia ovalis*	Fawn, yellow brown or dark brown maculopapular scaling sheet like pattern over chest, trunk or abdomen	Dimorphism seen. Clusters of yeast cells. 4–8 µm in size, round, budding with short septate branched fungal hyphae—spaghetti and meat appearance	SDA media covered with olive oil is used as culture media for lipophilic yeasts. Colonies are creamy yellow in color with smooth to slightly wrinkled texture	Spaghetti and meat appearance
Tinea infection/ Ringworm	Tinea capitis/ Tinea corporis/ Tinea barbae and so on	Irregular rings with inflammatory borders with some clearing in the central areas of the lesion	Multiple thin, branched septate hyphae	SDA with cycloheximide and chloramphenicol and the dermatophyte test medium (DTM) used. Colonies produce pigmentation and have characteristic macroconidia and microconidia	Microconidia / Macroconidia

Clinical Presentation

Brown to black painless lesions are formed, mainly on the skin of the palm but occasionally on the sole of the foot. There is macular pigmentation with sharp and darker margins. The fungus is present in the form of melanised, ellipsoidal cells, which often have a thick median cross-wall.

Laboratory Diagnosis

Direct Mount

Brown, multiple-branched, septate hyphae with budding cell are observed in 10% KOH mount of infected skin scrapping.

Fungal Culture

Colonies on oatmeal agar are restricted, smooth and slimy with an oily, glistening olivaceous-black colour. On microscopic examination of fungal mycelia growth, the budding cells are subhyaline, 0–1 septate with production of daughter cells by annellidic conidiogenesis from their poles.

Therapy

The fungus can easily be removed by keratinalotic compounds such as Whitfield's ointment.

MALASSEZIA INFECTION

Introduction

Malassezia infection is also called pityriasis versicolour or tinea versicolour. It is a chronic asymptomatic infection involving stratum corneum of skin.

Aetiological Agent

M. furfur and *Malassezia ovalis* both are lipophilic yeasts and part of normal flora of skin.

Clinical Presentation

The infection can be confused with vitiligo or pityriasis rosea, seborrheic dermatitis and other pigmentary skin lesions. They typically present with fawn, yellow brown or dark brown patches scaling sheet like pattern over chest, trunk or abdomen. Pityriasis versicolour are tiny maculopapular lesions coalescing to form scales of various shades. They are non-inflammatory lesions seldom causing redness and pruritis. Folliculitis is more severe form where infection involves hair follicles and sebaceous glands leading to painful inflammation.

Laboratory Diagnosis

Direct 10% KOH mount of skin scrapping shows both pathognomonic dimorphic yeast and hyphae. It appears like clusters of yeast cells of 4–8 µm in size, round, budding with short septate, branched fungal hyphae. It is also called as **spaghetti and meat** appearance or **banana and grapes** appearance.

Woods lamp examination of the affected skin shows yellow fluorescence.

SDA media covered with olive oil is used, culture media for lipophilic yeasts (see Chapter 35). The colonies are creamy yellow in colour with smooth to slightly wrinkled texture. On microscopic examination a 3–6 µm oval to spherical budding yeast cells is seen. Hyphae are not seen in culture.

Therapy

The fungus can control and get cured by keratinolytic compounds such as Whitfield's ointment or salicylic acid, but recurrences are commonly observed.

DERMATOPHYTOSES

Introduction

Raymond Sabouraud, a French dermatologist, in 1890, established that the ringworm is caused by fungi, the dermatophytes. The **dermatophytes** are closely related keratinophilic fungi causing dermatophytosis (ringworm or tinea). They invade the superficial protective epithelium of body like skin, hair and nails, and destroy keratin. Dermatophytes belong to family *Arthrodermataceae* of the order *Onygenales* with three genera *Epidermophyton*, *Microsporum* and *Trichophyton*. The infection prevalence depends on environmental condition and personal hygiene of the susceptible individuals. Animal (zoophilic) and soil (geophilic) are source of infection to man. Fungi exclusively affecting humans are known as anthropophilic fungi.

Aetiological Agent

Tinea capitis is the most common form of dermatophytic infection worldwide. Aetiological agents and common site affected by them are mentioned in Table 36.2.

Pathogenesis

Dermatophytes invade the stratum corneum of all types of keratinised tissue like skin, hair and nail. The degree of inflammation produced in the lesions depends primarily on the immunological response of the host and the environmental conditions (warm, moist conditions are known to favour infection). Conditions with weaken immune system like diabetes, malignancies and collagen vascular diseases act as predisposing factors for dermatophyte infections.

IgE antibodies through histamine activity modulation play a major role in the suppression of cell-mediated immunity (CMI) evoked by fungal antigens.

Clinical Presentation

Infections caused by dermatophytes (ringworm, tineas) are clinically classified on the basis of the location of the lesions on the body. The invading dermatophyte forms lesions with irregular rings, with central clearing and bordered inflammation. The infection is named according to anatomical site involved after the Latin word *tinea*, as shown in Table 36.2. Few of them are described as follows.

Tinea Corporis

It is a dermatophyte infection of the glabrous (non-hairy) skin involving stratum corneum caused by *Trichophyton* and

TABLE 36.2 Dermatophyte Infections, Affected Area on Body and Aetiological Agents

Disease	Affected Area of Human Body	Dermatophytes Causing Infection
Tinea capitis	Scalp, eyebrows, eyelashes	*Microsporum* spp. and *Trichophyton* spp.
Tinea favosa	Honeycomb like crust around infected follicles of scalp	*T. schoenleinii*
Tinea barbae	Beard and moustache	*T. mentagrophytes, T. rubrum, T. violaceum*
Tinea corporis	Skin	*T. rubrum, T. mentagrophytes, M. audouinii*
Tinea imbricate	Concentric rings of scaling of skin	*T. concentricum*
Tinea cruris	Groin	*Epidermophyton floccosum, T. rubrum*
Tinea pedis	Feet involves infection of the interdigital webs and soles	*T. rubrum, T. mentagrophytes, E. floccosum*
Tinea manuum	Hand	*T. rubrum, E. floccosum, T. mentagrophytes*
Tinea unguium	Nail	*T. rubrum, T. mentagrophytes*

Microsporum. It is also called as **tinea glabrata circinata**. Disease is characterised by erythematous scaly lesion with sharp margins and raised borders. Infection of the hair follicles can lead to a deep dermal inflammatory reaction. This pustular, elevated lesion is called as **Majocchi's** dermatophytic granuloma. Infection is transmitted by direct contact with infected individual or animal and fomites such as clothes and furniture. Clinically it is difficult to get differentiated with pityriasis versicolour, pityriasis rosea, lichen planus, contact dermatitis and fixed drug reaction.

Tinea Cruris (Jock Itch)

The infection of the inguinal area involves the groin perianal and perineal areas, often involving the upper thighs. Tinea cruris is mostly present in men with an underlying predisposing factor such as long-term use of tight-fitting garments. It is also seen in other areas such as under pendulous breasts, axilla and around the umbilicus of obese patients. It is frequently caused in adults by *Trichophyton rubrum* and *Epidermophyton floccosum*. Lesions are well delineated, erythematous and scaly, bilateral extending to the inner thighs, waist and buttock.

Tinea Capitis

This ringworm involves infection of the scalp hairs and intervening skin. Infection is often accompanied by secondary bacterial infection resulting in scarring and alopecia. Ectothrix and endothrix are two patterns of hair invasion. **Ectothrix** of scalp is most commonly caused by *Microsporum audouinii*. In ectothrix, the fungi invade the surface of hair shaft and make hair brittle and fragile, thus it breaks of leading to partial alopecia. **Endothrix** of scalp is most commonly caused by *Trichophyton schoenleinii* and *Trichophyton tonsurans*. Endothrix fungi invade into the hair shaft and severely damage the hair leading to total alopecia with black dot appearance.

Tinea Pedis

It is a superficial infection of feet involving toe webs and soles. It is also called as athlete's foot. It occurs due to wearing of shoes and sweaty socks. Three dermatophytes majorly cause

tinea pedis include *T. rubrum, Trichophyton mentagrophytes* and *E. floccosum*. Lesions are of four types:

1. Chronic intertriginous type is the most common form affecting areas between fourth and fifth toes and third and fourth toes. Skin becomes white, macerated and have foul odour.
2. Chronic papulosquamous hyperkeratotic type involves whole foot.
3. Subacute or vesicular type evokes "id" reaction.
4. Acute ulcerative type is often complicated by secondary bacterial reaction.

Tinea Unguium

It is invasion of nail plates by *T. rubrum* and *T. mentagrophytes*. Initially, small yellow-whitish spots are observed on nails which spreads to base of nail and persists for many years. Nails become brittle, friable and thick with black discolouration that later may show gross distortion. It may asymmetrically involve single nail. It should be differentiated from Candidal onychomycosis which symmetrically involves distal portion of several nails without gross distortion. Tinea unguium form of dermatophytosis is most resistant to treatment.

Tinea Incognito

It is steroid modified tinea which occurs due to misuse of corticosteroids in combination with topical antimycotic drugs for the treatment of dermatophytosis.

"ID" Reaction

It is dermatophytid reaction in which secondary eruption occurs on skin in sensitive tinea patients because of allergenic products from the primary site of infection. It may also develop after antimycotic therapy. Itching is the only symptom due to absorbed fungal products.

Laboratory Diagnosis

The history of occupation, living condition, contacts with pets and previous intake of drug treatment is important in diagnosing tinea patients. Clinical material from site of active lesion is sent for laboratory examination. The typical

TABLE 36.3   Microscopic Morphology of Different Dermatophytes

Genus	Macroconidia	Microconidia
Microsporum	Usually present and more numerous than microconidia. Fusiform shaped	Usually present; pyriform or clavate; typically borne singly along the hypha
Trichophyton	Often absent or less numerous than microconidia. Smooth pencil-shaped	Usually more numerous than macroconidia clavate shaped in clusters or singly along the hypha
Epidermophyton	Smooth-walled; broadly clavate; singly or in banana-like clusters	Absent

specimens submitted for the diagnosis of the dermatophyte infection are skin scrapings, hair stubs and nail clippings or scrapings.

Wood's Lamp Examination

The Wood's lamp is used for the detection of tinea infection based on the fact that some dermatophyte species produce characteristic fluorescence under UV light (see Chapter 35). *Microsporum* spp. mainly fluoresce bright-green, and *Trichophyton* spp. fluoresce dull green. *M. furfur* produces golden yellow colour.

Direct Microscopic Examination

Dermatophytes appears as multiple, thin, branched septate hyphae on direct visualization of wet or stained clinical specimen.

Fungal Culture

A primary medium for the isolation of dermatophytes commonly used is SDA with cycloheximide and chloramphenicol and the dermatophyte test medium (DTM).

Identification is based on its gross colonial morphology on SDA and on its microscopic morphology (Table 36.3). *T. rubrum* produces red pigment on culture media with teardrop microconidia. *T. mentagrophytes* are white cottony colonies with cluster of cigar-shaped microconidia. *T. schoenleinii* has smooth waxy colonies, and on

microscopic examination hyphal swelling and favic chandelier are seen.

In Vitro Hair Perforation Test

This test is also helpful in identifying dermatophyte species invading hair. The test is positive when dermatophytes show wedge-shaped perforation of hair. Dermatophyte species showing positive in vitro hair perforation test are *T. mentagrophytes* and *Microsporum canis* whereas no hair perforation (negative test) is seen with species like *T. rubrum* and *Microsporum equinum*.

Urease Test

The ability to hydrolyse urea in either an agar or a broth medium aids in the distinction of *T. rubrum* (urea-positive). Both Christensen's urea agar and broth media may be used. The test is considered negative if there is no colour change from straw to reddish purple within 7 days at 23–30°C.

Therapy

Antifungal agents are available to treat the dermatophytoses. This includes oral fluconazole, itraconazole, naftifine, griseofulvin and terbinafine. Treatment of tinea capitis requires systemic antifungal therapy. Terbinafine and griseofulvin are both effective against *T. tonsurans* and are FDA-approved for this indication in children. Topical antifungal therapy is usually effective for tinea corporis.

37

Fungi Causing Subcutaneous Infections and Mycetoma

Suruchi Shukla

LEARNING OBJECTIVES

- Aetiology, disease, clinical features, laboratory diagnosis, therapy of fungi causing subcutaneous infection including *Sporothrix schenckii*, chromoblastoma, phaeoid fungi, *Rhinosporodium seeberi* and *Mycotic mycetes*

INTRODUCTION

Subcutaneous fungi have limited invasive ability and gain entrance to the body by traumatic implant. They may take several years to produce noticeable disease as they require time to adapt to the tissue environment. All the organisms are soil saprophytes of regional epidemiology. Various fungi causing subcutaneous mycosis are described in Table 37.1.

SPOROTHRIX SCHENCKII

S. schenckii causes a subcutaneous or systemic fungal infection known as sporotrichosis. The disease is also commonly called as 'Rose handler's disease'. The fungi enter the skin, that is dermis and subcutis through a cut or puncture wound. The systemic form of disease conjointly presents as pneumonitis, arthritis or meningitis.

Laboratory Diagnosis

Clinical samples such as skin scrapings, aspiration fluid, biopsy material, tissue and pus are collected from infected lesion. On direct microscopy a characteristic 3–5 μm elongated **cigar-shaped yeast cells also called as asteroid body** are seen. The histopathological methenamine silver-stained biopsy specimen shows fungi surrounded by eosinophilic refractile halo, known as **Splendore-Hoeppli** phenomenon. The fungi generally isolated in culture are disregarded as laboratory contaminant. *S. schenckii* is a dimorphic fungus found in both yeast and hyphae forms (Table 37.1). Mycelial strands are intertwined to give appearance of **twisted ropes** with **flower-like pattern** of sporulation, as shown in Fig. 37.1. Latex agglutination using peptide-L-rhamno–D-mannan antigen is helpful in diagnosing systemic disease. The sporotrichin skin test has been documented but not used to diagnose the infection.

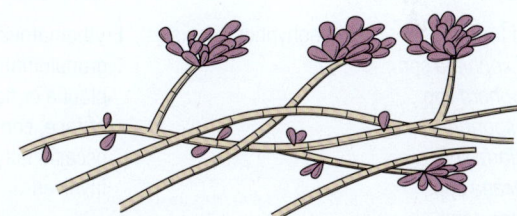

Fig. 37.1 Mycelial Strands of *Sporothrix schenckii*: Twisted Ropes With Flower-Like Pattern of Sporulation.

Therapy

Saturated solution of potassium iodide (SSKI) is the drug of choice. SSKI is given orally 250 mg in three divided doses over a period of 2–4 months. Oral azoles such as fluconazole (400 mg/day), ketoconazole and itraconazole are used for long-term treatment of systemic sporotrichosis.

CHROMOBLASTOMA

Chromoblastoma are group of dematiaceous fungi causing subcutaneous mycosis commonly called as chromoblastomycosis. They are also known as **chromomycosis, verrucous dermatitis and phaeosporotrichosis**. The disease is mainly prevalent in tropical or subtropical climate areas of America, Asia and Africa. From Asia, the chromoblastomas have been frequently reported from Japan, Sri Lanka and India. The infection is not uncommon in India; several case reports are from the sub-Himalayan belt and western and eastern coasts of India. The fungi, as seen in other subcutaneous mycoses, gain entrance through the skin by traumatic implantation. A skin lesion varies from small nodules to large papillary-like eruptions. Chromoblastoma pathognomically depicts sclerotic cells or muriform cells on tissue biopsies.

TABLE 37.1	Features of Fungi Causing Subcutaneous Mycosis				
Fungal Agents Causing Subcutaneous Mycosis	**Disease**	**Clinical Features**	**Direct Microscopy**	**Culture**	**Treatment**
Sporothrix schenckii	Sporotrichosis *Gardener's disease Rose handler's disease*	Pustule or ulcers with nodules, most commonly on hands	Cigar-shaped elongated yeast	Dimorphic fungi. Cigar-shaped yeast cells and on temperature conversion hyphae with daisy like conidia	Potassium Iodide for local application and itraconozole
Loboa loboi	Lobomycosis 'Lacazia'	Subcutaneous nodular keloids	Obligatory parasitic fungi	Spherical thick walled yeast cells (5–12 µm diameter) present in chains	Surgery with antifungal therapy
Chromoblastoma, e.g., *Fonasecaea pedrosoi*, *Cladosporium carona*	Chromoblastomycosis	Verrucous or cauliflower hyperkeratotis lesions with microabscess present on both hands and feet	Muriform/ sclerotic bodies	Dematiaceous fungi: On culture black-olivaeous coloured colonies of 2–4 µ in size	Surgery with antifungal therapy
Phaeoid fungi, e.g., *Exophiala* spp., *Phialophora* spp., *Cladosporium* spp., *Curvularia* spp., *Fonsecaea* spp., *Alternaria* spp.	Phaeohyphomycosis	Erythematous granulomatous plaque or nodules of face, cornea, eye occasionally brain involved	Brown septate hyphae	Dematiaceous fungi: On culture black-olive coloured colonies of 2–4 µ in size	Surgery with antifungal therapy
Rhinosporodium seeberi	Rhinosporidiosis	Granulomatous polyp in the nasal cavities seen. Rarely it is present on vagina, anus, penis and ear	Obligate parasitic fungi	Spherules with endospores seen	Surgery
Zygomycetes, e.g., *Conidiobolus coronatus* and *Basidiobolus ranarum*	Entomophthoromyco-sis/Subcutaneous zygomycosis	Painless subcutaneous nodules involving entire limb	Short, broad fragmented hyphae	Difficult to isolate	Surgery and azoles
Mycotic mycetes, e.g., *Madurella mycetomatis*, *Pseudallescheria boydii*	Maduramycosis (fungal) or *Madura foot*	Localised skin abscess with discharging sinuses due to granulomatous infection of dermis and subcutaneous tissue	White/green/ yellow pigmented grain	Blood agar, BHI agar or SDA are used as culture media and incubated at 37°C both aerobically and anaerobically. Identification depends on colony morphology, conidia types and sugar assimilation patterns	Antifungal drugs with antibiotics and surgery

Aetiological Agents

It is caused by soil-inhabiting dematiaceous fungi. They are light brown, pigmented organisms. They are found in decaying woods and forest litter. Agents causing chromoblastomycosis are as follows:

- *Phialophora verrucosa*
- *Fonasecaea pedrosoi*
- *Cladophialophora carrionii*
- *Fonasecaea compacta*
- *Rhinocladiella aquaspera*

Clinical Features

The infection is more prevalent in middle-aged males engaged in agriculture activities. The infection remains localised to site of trauma or puncture wound. It is slow developing polymorphic *hyperkeratosis* and *pseudoepithe-*

liomatous hyperplasic nodular, verrucose, tumoural and plaque skin lesion. The lesion is 1–3 cm raised above the skin, typically manifesting like cauliflower nodules. There is no apparent discomfort to patient. Secondary bacterial infection leads to considerable lymph stasis leading to elephantiasis. There is no bone or muscle invasion or fistula formation which is commonly seen in mycetoma. Rarely verrucous lesion may appear on chest, abdomen or trunk due to haematogenous spread.

Laboratory Diagnosis

The chromoblastomycosis should be differentiated from similar skin lesions observed in blastomycosis, lobomycosis, protothecosis, sporotrichosis, leishmaniasis, Hansen's disease and tertiary syphilis. The demonstration of sclerotic bodies and fungal culture are essential modes to diagnose CBM.

Direct Microscopy

The clinical samples like skin scraping, crusts, aspirated debris and excised material can be examined in 10% potassium hydroxide mount. Chromoblastoma typically presents as **medlar** and **sclerotic bodies.** The sclerotic bodies are predominantly seen in the sample; they are thick walled, brown, pigmented with planate division in cluster or chain, as shown in Fig. 37.2. On histopathological examination, the fungi are observed as pigmented yeasts like medlar bodies or 'copper pennies'. Special stains of biopsy material, such as periodic acid Schiff and Gömöri methenamine silver, are used to demonstrate the fungal elements if needed.

Fungal Culture

Sabouraud's dextrose agar is culture media recommended for these fungal agents. Culture should be kept at 25°C for at least 6 weeks as they are slow growers. Their colonies are floccuse and black in colour. They exhibit more than one pattern of sporulation due to pleomorphism. On microscopic examination of culture isolates, simple stalk called as conidiophores and conidia in different morphological arrangement present over conidiophores. Three distinct types of conidia sporulation are exhibited by chromoblastoma, that is *Cladosporium* (disjuntor), *Rhinocladiella* (bottle-brush) and *Phialophora* (flower in a vase; Fig. 37.3).

Therapy

Chromoblastomycosis is very difficult to cure. The therapeutic modalities are cryotherapy, thermotherapy, laser and surgery along with antifungal drugs. The primary treatments of

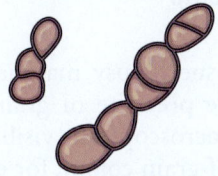

Fig. 37.2 Sclerotic Bodies of Chromoblastoma.

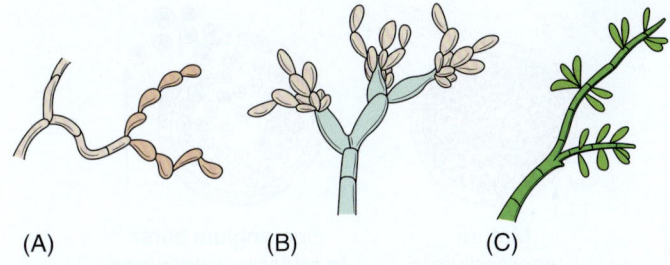

Fig. 37.3 Types of Sporulation in Chromoblastoma: (A) *Cladosporium*-Like Disjunctor, (B) *Phialophora*-Like Flower in a Vase, and (C) *Rhinocladiella*-Like Bottle Brush.

choice are itraconazole, an antifungal azole, given orally, with or without flucytosine.

PHAEOID FUNGI

The phaeoid are group of dematiaceous fungi causing the disease phaeohyphomycosis. The disease was introduced by Ajello in 1974. They are present in the environment as soil saprotrophs and plant pathogen. They cause both subcutaneous and systemic mycosis. On the basis of site and depth of infection, phaeoid were classified by Rippon in 1973 into three following types:

- Superficial: Black piedra and Tinea nigra
- Cutaneous: Dermatomycosis and onychomycosis and mycotic keratosis
- Subcutaneous: Invasive, systemic and cerebral

Aetiological Agents

Genera causing phaeohyphomycosis are as follows:

- *Bipolaris* spp.
- *Curvularia* spp.
- *Alternaria* spp.
- *Exophiala* spp.
- *Cladophialophora* spp.
- *Exserohilum* spp.
- *Phialophora* spp.
- *Wangiella* spp.

Clinical Presentation and Laboratory Diagnosis

Subcutaneous solitary ulcers preceding trauma are the commonest type of lesion caused by phaeohyphomycotics. They comprise vast array of opportunistic fungal infections characterised by **melanised** (brown melanin pigment in cell walls) fungal elements in tissue. Brain abscess is caused by *Cladophialophora bantiana*. Subcutaneous and intramuscular lesions are caused by *Exophiala jeanselmei* and *Exophiala dermatitidis*.

The diagnosis is established on biopsy tissue samples through histopathological examination and culture. This pigmented fungal cell wall can be detected by **Masson-Fontana** stain on direct microscopic examination. Culture will initially yield black yeasts but later forming mycelia during the course of incubation. Sclerotic bodies are not found

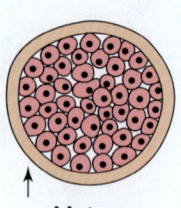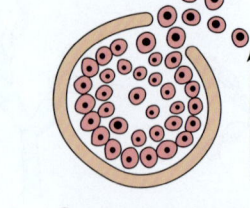

Mature sporangium — Sporangium burst to release endospores

Fig. 37.4 Rhinosporodium: Sporangium With Endospores.

unlike chromoblastomycosis. Radiological examination of affected site helps to visualise the systemic lesions.

Therapy

Systemic phaeohyphomycotic infections are difficult to treat. Surgery and antifungal like azoles are combined to treat the infection.

RHINOSPORIDIUM SEEBERI

R. seeberi causes a chronic granulomatous disease of the mucocutaneous tissue called as rhinosporidiosis. It was previously considered to be a fungus, hence was classified as a fungal disease. It is now considered to be a protest classified under mesomycetozoea. The disease is endemic in South India, Sri Lanka, South America and Africa. It is not a contagious disease, but exogenous source of infection is also not known. Animal to man transmission is not proven, but incidences of disease are more common in people who are habitually bathing with domestic animals in stagnant pools and lakes.

Clinical Manifestation and Laboratory Diagnosis

It is characterised by polyposis of the **nasal cavity**, conjunctiva and other body sites like vagina, penis and anus. The nose is most commonly affected site. Tissue becomes friable and large polyp or wart-like lesion is formed. Clinically wart is red and bleeds on touch. Histopathology examination of affected tissue is helpful in diagnosis of infection. HPE section reveals **sporangium with endospores**, as shown in Fig. 37.4.

Therapy

Surgical debridement and dapsone therapy is mainstay of treatment. Recurrence of infection is a characteristic of rhinosporidiosis.

MYCETOMA

Mycetoma causes a chronic, destructive morbid inflammatory disease usually of the foot but any part of the body can be affected commonly known as maduramycosis or **Madura foot**. Mycetoma was delineated in the history in 1694; however, it was first rumoured in the mid-19th century within the Indian town of Madura, and hence was initially called

Madura foot. Godfrey first documented a case of mycetoma in **Madras, India**. However, the term 'Mycetoma' (fungal tumour) was proposed by Carter, who established the fungal aetiology of this disorder. This malady primarily affects poorer individual in rural regions of continents Latin America, Africa and Asia that are located near Earth's equator and have dry climates.

In developing countries young men aged between 20 and 40 years are commonly affected by the mycetes. Individual belonging to low socio-economic stature and field labourer are the worst affected. Usually, some predisposing condition may be associated such as poor general health, diabetes and malnutrition, and this may lead to a more invasive and widespread infection.

Aetiological Agents

Mycetoma is caused by **bacteria (actinomycetoma)** or **fungi (eumycetoma)** found in soil and water. Actinomycotic mycetoma is caused by aerobic species of actinomycetes belonging to the genera *Nocardia*, *Streptomyces* and *Actinomadura* with *Nocardia brasiliensis*, *Actinomadura madurae*, *Actinomadura pelletieri* and *Streptomyces somaliensis* being most common. *Madurella mycetomatis* is the most common cause of eumycotic mycetoma. Clinically, the different mycetoma aetiology produces grains of different colours.

Pathogenesis

The organism is usually implanted after a penetrating injury while performing agricultural work barefoot or through pre-existing abrasions. Complement-dependent chemotaxis of polymorphonuclear leukocytes has been shown to be induced by both fungal and actinomycotic antigens in vitro. Epithelioid granuloma is formed. T-cell responses also seem to play an important part in the development of mycetoma.

Clinical Presentation

The disease depicts characteristic **triad** of painless dermal mass, multiple sinuses and granular discharge from the sinuses. It is usually noncontiguous and spreads to involve the skin, deep structures and bone resulting in destruction, deformity and loss of function, which may be fatal. Mycetoma commonly involves the extremities, back and striated muscle region.

As mycetoma is painless and slowly progressing, many patients are diagnosed late with advanced infection, hence amputation may be the only treatment available. Superadded bacterial infection is common, and lesions may lead to pain, disability and fatal septicaemia.

Laboratory Diagnosis
Direct Microscopy

Pus, exudates or tissue biopsy material from patient may be examined for the presence of grains. These grains are 0.5–2 mm in size, macroscopically visible and fungus can be identified on basis of grain colour, for example Actinomadura pelletieri grains are **red** in colour, *Nocardia* grains are **white**, *Madurella* and Phialophora grains are **black** in colour and Streptomyces are **yellow** in colour.

Fungal Culture

A slide culture (growing the bacteria or fungi in the laboratory) can determine the specific type of bacteria or fungus causing the infection. Deep tissue biopsy is required for culture as it is not contaminated. Blood agar, BHI agar or Sabouraud's dextrose agar may be used as culture media and incubated at 37°C both aerobically and anaerobically. Identification depends on colony morphology, conidia types and sugar assimilation patterns.

Molecular Assay

PCR and DNA sequencing provide rapid diagnosis.

Radiodiagnosis

An imaging test such as an X-ray or ultrasound to diagnose mycetoma may help to see the muscle and bone damage.

Therapy

Eumycetoma, that is bacterial infection, is treated by long-term antibiotics, whereas mycotic mycetoma is treated by combined surgical excision and antifungal drugs. Preventive measures like wearing of shoes might prevent injuries that cause mycetoma.

38

Yeasts Causing Human Infections

Suruchi Shukla

LEARNING OBJECTIVES

- History, pathogenesis, virulence factors, clinical features, laboratory diagnosis and therapy of *Candida* spp.
- History, classification, pathogenesis, virulence factors, clinical features, laboratory diagnosis and treatment of *Cryptococcus* spp.

INTRODUCTION

Yeasts are single-cell fungi (eukaryotic organisms) found globally in soil, on plants especially in sugar-containing parts like nectar and fruits. Most yeast belong to the phylum Ascomycota. Only a few belong to order Basidiomycota. Most yeasts reproduce asexually by budding, but a few yeasts multiply by binary fission. Some yeast form true hyphae such as *Trichosporon* and in contrast some yeast has true polysaccharide capsule, for example, *Cryptococcus* and *Rhodotorula*. Some yeast like *Saccharomyces cerevisiae* are useful industrially and are used to make bread, beer and wine. In contrast some yeast like *Candida* and *Cryptococcus* are pathogenic to human and animals. The yeasts causing human infection are mentioned as follows:

- *Candida* spp.
- *Cryptococccus* spp.
- *Torulopsis*
- *Trichosporon beigelii*
- *Geotrichum candidum*
- *Saccharomyces cerevisiae*
- *Rhodotorula* spp.

This chapter describes two most common yeasts, that is, *Candida* spp. and *Cryptococcus* spp. whereas other yeasts are discussed in Chapter 40.

CANDIDA SPP.

Candida is the most common cause of fungal infection of human being; its infection is known as candidiasis or moniliasis. There are more than 150 anamorphic species of *Candida* (yeast-like fungi). *Candida* belongs to phylum **Fungi imperfecti**, order **Moniliales**, family **Cryptococcaceae**. *Candida albicans* is the representative species that has two serotypes A and B based on its cell wall component, that is, mannan. *Candida* species are part of normal flora of human beings commonly found on the skin, gastrointestinal tract and female genital tract. Varity of factors predispose to candidal infection like alteration of microbial flora of the body or lowering of host resistance. *C. albicans* is the most common cause of the superficial and deep fungal infection. *Candida* spp. is the leading cause of **blood stream** and **nosocomial (hospital acquired)** infection. Non-judicious use of antibiotics further leads to increased incidence of candidiasis. It is commonly associated with AIDS. Candidiasis is caused by *C. albicans* and other *Candida* species (non-*C. albicans* spp.) like *C. tropicalis*, *C. glabrata*, *C. parapsilosis*, *C. krusei*, *C. kefyr* and *C. dubliniensis*.

History

Important landmarks in history of candidiasis are as following:

- Oral thrush was first described in the 4th century BC by **Rosen von Rosenstein** and **Underwood**.
- *C. albicans* was initially named as *Oidium albicans* and *Monilia albican*.
- Torulopsis glabrata was first isolated from grapes in 1894 by Berlese.
- First case of Torulopsis infection in humans was described by Anderson in 1917.

Pathogenesis and Virulence Factors

C. albicans has several virulence factors:

- *Toxins*
- *Enzymes*
- *Adhesin*
- *Complement receptors*
- *Phenotypic switching*
- *Antigens present in cell wall and cytoplasm*

These virulence factors contribute to invasion and disease manifestation. Enzymes like phospholipases present at hyphal tip help in invasion of yeast. Adhesin majorly aids in biofilm formation under which yeast colonies escape phagocytosis and colonise.

Clinical Features

Candidiasis presents in two forms:

1. **Infectious diseases**
   a. **Mucocutaneous manifestation**
      Oral thrush/somatitis/glossitis/cheilitis/esophagitis/gastritis/vulvovaginitis/balanitis/ocular candidiasis.
   b. **Cutaneous manifestation**
      Intertriginous/generalised/paronychia/onycho-mycosis/diaper rashes.
   c. **Systemic manifestation**
      Urinary tract infection/endocarditis/pulmonary candidiasis/meningitis/arthritis/osteomyelitis/endophthalmitis.
   d. Candida is able to form **biofilms** on medical devices such as intravenous lines, catheters, prosthetic heart valves, stents and prosthetic joints. Biofilms make the organism resistant to antimicrobial treatment and host immune system, thus the organism is difficult to clear.
2. **Allergic diseases**
   Candidids/eczema/gastritis/asthma.

Laboratory Diagnosis (Flowchart 38.1)

Clinical samples required for laboratory diagnosis: Nail/hair/skin/tissue/blood/sputum/urine/vaginal swab/oral swab/cereberospinal fluid.

Direct Examination

- KOH mount/calcofluor white stain/stained tissue section: Budding yeast cells are seen.

Fungal Culture

- Sabouraud's dextrose agar: At 37°C smooth, creamy, white pasty colonies with yeasty odour within 48 h of incubation.
- Blood cultures are done for invasive yeast infection.
- **Phenotypic media to differentiate *C. albicans* from non-*C. albicans* spp.**
  - **Corn meal agar:** At 25°C typical **chlamydospore** and **psuedohyphae** formed (Fig. 38.1). Psuedohyphae are long projections extending out of yeast cells with constriction at the point of attachment.
  - **Tetrazolium reduction medium:** It is an agar-based differential media showing cream coloured colonies of *C. albicans*.
  - **CHROM agar:** It is a selective and differential agar-based media. *C. albicans* colonies are light green/bluish green in colour.
  - **Germ tube test (Reynolds Braude phenomenon):** On treating Candida culture isolate with sheep or human serum and incubating at 37°C for 2–4 h; **germ tube** formation occurs. Germ tube is a long tube like projection from yeast cells without constriction at the point of attachment (Fig. 38.1).
  - **Sugar assimilation and fermentation test:** Sugars used for the phenotypic identification of *Candida* spp. are

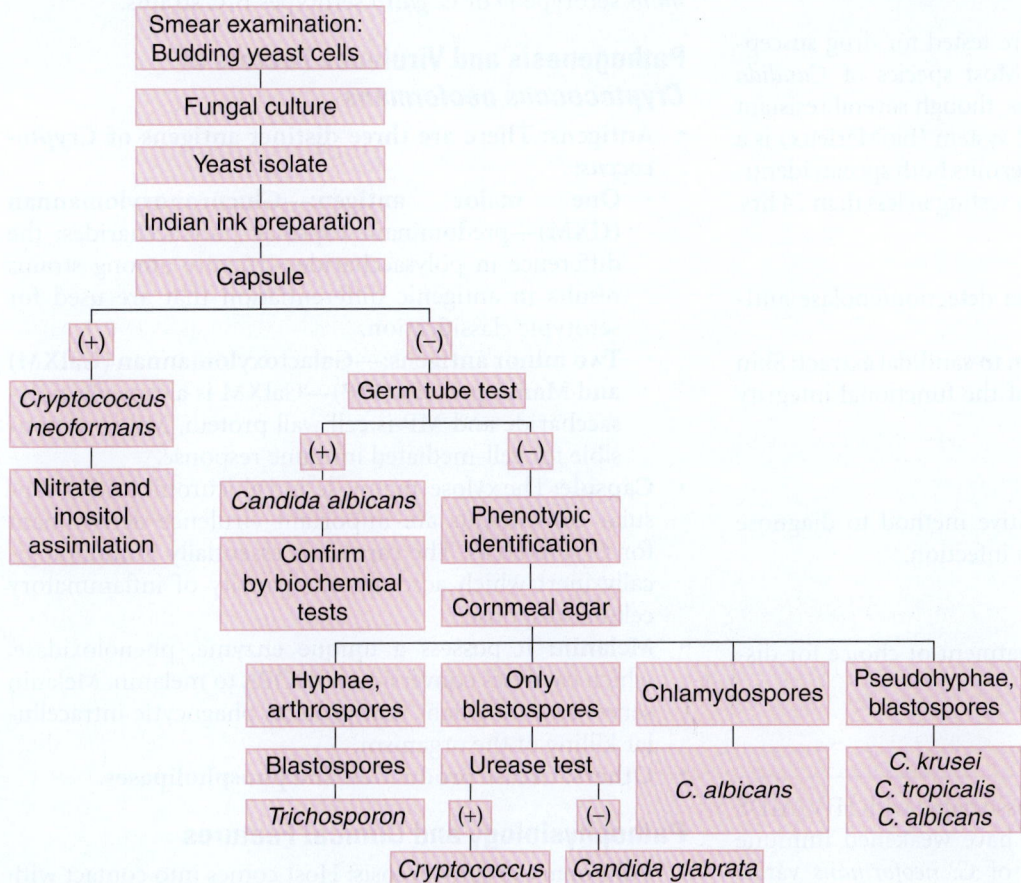

Flowchart 38.1 Laboratory Diagnosis of Yeast Like Fungi.

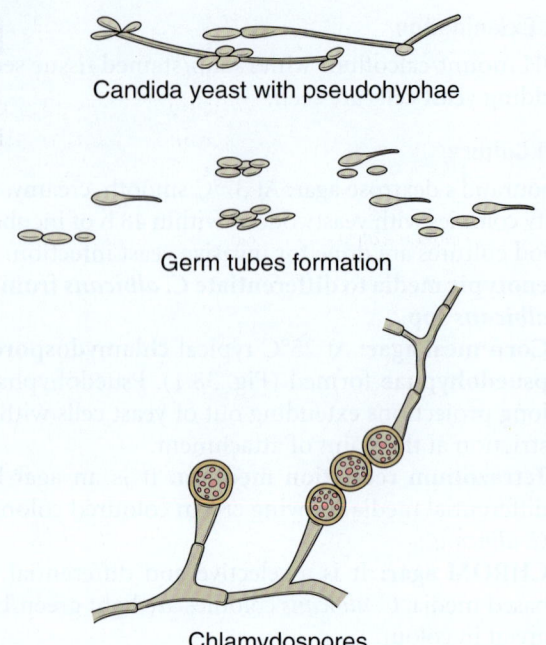

Candida yeast with pseudohyphae

Germ tubes formation

Chlamydospores

Fig. 38.1 Yeast Formations.

glucose, maltose, sucrose, lactose, galactose and trehalose. A commercially available **API 20c** analytic profile index system incorporates 20 different carbohydrate for yeast identification by carbohydrate assimilation method.

Antifungal Sensitivity

Amphotericin B and azole drugs are tested for drug susceptibility against candidial isolates. Most species of *Candida* are sensitive to both the drug classes, though several resistant strains have emerged. The VITEK 2 system (bioMérieux) is a rapid fully automated system that permits both species identification and antifungal susceptibility testing in less than 24 hrs.

Serology Test

- Latex agglutination/β-1,3-glucan detection/enolase antigen test.
- Delayed hypersensitivity reaction to candidal extract: Skin testing is used as an indicator of the functional integrity of cell-mediated immunity.

Molecular Assay

Real-time PCR is rapid and sensitive method to diagnose yeast in disseminated blood stream infection.

Therapy

Amphotericin B and azoles are treatment of choice for disseminated candidiasis.

CRYPTOCOCCUS SPP.

C. neoformans infections are often associated with HIV/AIDS and mostly occur in people who have weakened immune system. The clinical presentation of *C. neoformans* varies from a localised skin lesion or asymptomatic colonisation

of lung to a widely dissemination extrapulmonary lesions at sites like central nervous system. Brain is the most common site affected after the lung. The patient clinically presents with headache, meningitis, altered mental status, seizures and raised intracranial pressure. Focal lesions, that is cryptococcoma, are more common by newer emerging opportunistic cryptococcal species like *C. gatti*. Cryptococcal species causing infection are as follows:

- **C. neoformans**
- **Non-C. neoformans** spp. like *C. albidus* and *C. laurentii*.

History

Human cryptococcosis was described in 1894 by Busse and Buschke. It was named cryptococcosis by Benhem's. It was isolated from pigeon droppings and soil by Emmons in 1955. It is also inhabitant of decayed wood in trunk hollows of a wide spectrum of host trees like Eucalyptus.

Classification

Cryptococcus is ubiquitously distributed in the environment. It belongs to genus *Basidiomycetes*. Only two cryptococcal species are commonly known to cause human disease, that is, *C. neoformans* and *C. gattii*. Based on structural differences in the polysaccharide capsule, the genus was further classified into three varieties, five serotypes and eight molecular subtypes (Table 38.1). Ninety-five per cent of cyptococcal infections are caused by *C. neoformans* serotype A, whereas the remaining 4%–5% of infections are caused by *C. neoformans* serotype D or *C. gattii* serotypes B/C strains.

Pathogenesis and Virulence Factors of *Cryptococcus neoformans*

- **Antigens:** There are three distinct antigens of *Cryptococcus*:
  - **One major antigen:—Glucuronoxylomannan (GXM)**—predominant capsular polysaccharides; the difference in polysaccharide structure among strains results in antigenic differentiation that are used for serotypic classification.
  - **Two minor antigens:—Galactoxylomannan (GalXM) and Mannoprotein (MP)**—GalXM is a capsular polysaccharide and MP is cell wall protein. MP is responsible for cell-mediated immune response.
- **Capsule:** The xylose, mannose and glucuronic acid of capsular constituents are important virulence determinant for *Cryptococcus*. The capsule is essentially immunologically inert which accounts for paucity of inflammatory cellular response.
- **Melanin:** It possess a unique enzyme, phenoloxidase, which catalyses conversion of DOPA to melanin. Melanin serves as antioxidant that protects phagocytic intracellular killing of the organism.
- **Others: Urease production and phospholipases.**

Pathophysiology and Clinical Features

- Pulmonary cryptococcosis: Host comes into contact with cryptococci from the environment through inhalation of

TABLE 38.1   Pathogenic Classification of *Cryptococcus* spp.

Serotype	Species and Variety	Molecular Genotypes	Geographical Distribution	Shape of Basidio-spores	Association Found	Utilise Glycine as Sole Nitrogen Source	Urease Sensitivity to Chelating Agents EDTA
A	*C. neoformans* var. *grubii*	VN I–II	Worldwide	Round	Pigeon excreta droppings and tree hollows of variety of tree species	No	Resistant
B C	*C. gattii*	VG I–IV	Tropical and subtropical countries like Brazil, Australia, Southeast Asia, and central Africa	Ellipsoidal/rod-shaped oval	Eucalyptus tree	Yes	Yes
D	*C. neoformans* var. *neoformans*	VN IV	European countries	Round	Pigeon excreta droppings and tree hollows of variety of tree species	No	Resistant
AD	*C. neoformans*	VN III	—	Round	—	No	—

yeasts. Typically after entering lungs, the organism disseminates rapidly to establish CNS infection. Patient has overwhelming cryptococcal pneumonia with adult respiratory distress syndrome later developing meningitis.

- Disseminated infection: Fever, malaise and chest pain are most common clinical symptoms. Weight loss, cough and haemoptysis, headache and meningitis or space occupying lesion in brain may present according to severity.
- Skin lesion: *C. neoformans* can produce variety of skin lesions. It may appear as purpura, papules, vesicles, nodules, tumour, abscess and granuloma.
- Eye involvement: Ocular involvement is not rare. Patient may develop ocular palsy, endophthalmitis, optic neuritis and may lead to rapid vision loss also.
- The primary defence mechanisms against cryptococcal infection are helper T cell. They release cytokines including tumour necrosis factor (TNF), interferon-γ and interleukin-2, leading to granuloma formation.
- Immune reconstitution inflammatory syndrome (IRIS): Cryptococcal IRIS may present as a clinical deterioration, paradoxical relapse or new presentation of subclinical cryptococcal disease following initiation of antiretroviral therapy (ART) and is believed to be caused by recovery of Cryptococcus-specific immune responses.

Laboratory Diagnosis (Flowchart 38.1)

- Definitive diagnosis can be made by isolation of *Cryptococcus* form clinical specimen and/or direct detection of encapsulated yeast by Indian ink preparation from body fluids, as shown in Fig. 38.2.
- Fungal culture is done on Sabouraud dextrose agar with antibiotics inoculated at 25–37°C. Yeast like smooth,

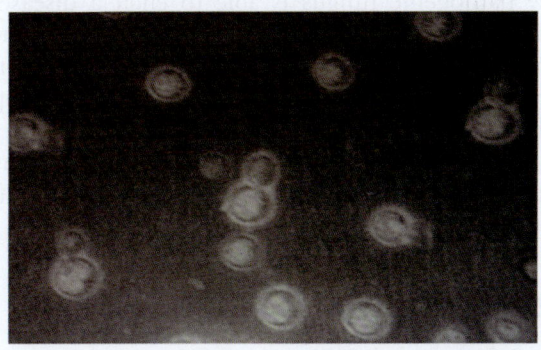

Fig. 38.2 Encapsulated Budding Yeast Cells (Indian Ink).

mucoid cream coloured colonies are seen. Special culture media like Niger seed agar and Sunflower seed agar can be used for identification of *Cryptococcus*. *C. neoformans* var. neoformans can hydrolyse urea.

- Serology for cryptococcal polysaccharide capsular antigen and histopathological specimens are other methods to identify the agent. Cryptococcal antigen titre (CALAS test) is a **latex Agglutination System for the Detection of Cryptococcal Antigen in Serum and CSF.**
- Animal pathogenicity: Cryptococcal isolate is intracerebrally or intaperitoneally inoculated into the mice. Fatal infection develops in the mice and encapsulated yeast cells can be demonstrated in CSF.
- Molecular PCRs are rapid methods to diagnose the infection.

Therapy

Amphotericin B (in conventional or liposomal formulation) combined with flucytosine remains the induction therapy of choice. In HIV/AIDS patient's antiretroviral therapy is initiated along with antifungal therapy.

39

Dimorphic Fungi Causing Human Infections

Suruchi Shukla

LEARNING OBJECTIVES

- Geographical distribution, pathogenesis, clinical features, laboratory diagnosis and therapy of dimorphic fungi: *Paracoccidioides, Coccidioides, Histoplasma, Blastomycetes*

INTRODUCTION

The systemic diseases caused by fungi fall into two categories: (1) infections caused by true pathogenic fungi and (2) group of infections caused by opportunistic fungi.

True pathogenic fungus is able to elicit a disease in a normal human host and is an accidental phenomenon. The infections are either completely asymptomatic or may be chronic or residual, leading to granulomatous response resembling tuberculosis. True pathogenic fungi are also called dimorphic as they exhibit a morphological transition from mycelia to the parasitic or yeast-like form in infected tissue (see Table 39.1).

PARACOCCIDIOIDES

Paracoccidioides brasiliensis is a thermally dimorphic fungi causing paracoccidioidomycosis disease. It is acute, subacute or chronic granulomatous systemic fungal infection. It primarily affects lungs and then subsequently disseminates to oral mucosa, skin, lymph nodes and sometimes to internal organs.

Geographical Distribution

Disease is endemic in Latin America extending from Mexico to Brazil and Ecuador, also called as **South American Blastomycosis or Lutz's mycosis.**

Pathogenesis and Clinical Features

Soil is considered to be the source of infection. Infection is acquired by inhalation. Children of both the sexes may get infected; however, disease is common in men aged between 20 to 50 years. The fungal spores are inhaled and affect the lungs and later spread to the oral, nasal and gastrointestinal tract mucosa, skin and lymph nodes leading to ulcerative granulomas. After entry into the lungs, organisms get converted into yeast form. It gets phagocytosed by mononuclear cells and is found in cytoplasm of giant cells in the tissues. Upon chronicity of infection granulomatous lesions develop.

The paracoccidioidomycosis infection are of subclinical type and detected only by skin test positivity. The disease is slowly progressive and has three clinical forms:
- Acute or juvenile form
- Chronic or adult form
- Sequelae or quiescent or latent form

Laboratory Diagnosis

Direct Microscopy

On KOH/Calcofluor white wet mount or Grocott-Gomori's methenamine silver stained tissue section; a characteristic multipolar budding yeast cell of 2–10 μm resembling **Mariner's wheel or Pilot wheel** is observed, as shown in Fig. 39.1. It also gives an appearance of **Mickey Mouse.**

Fungal Culture

Yeast-like creamy-white colonies are seen at 37°C on Sabouraud's dextrose agar medium, which after few weeks of incubation become wrinkled and rough. Mycelial growth is observed at room temperature. On microscopic examination multiple budding yeast cells of 3–30 μm in size are seen that are oval, globose or pyriform shaped. Daughter yeast cells, giving the appearance of a Mariner's wheel, surround the spherical mother cell. The hyphae measure around 1–2 μm with single-celled conidia. Numerous intercalary chlamydospores are also observed.

Skin Test

Intradermal injection of paracoccidioidin antigen at local site evokes inflammation of skin and thus induration is measured.

Yeast cells resembling
Mariner's wheel

Fig. 39.1 Yeasts Cells of *Paracoccidioides brasiliensis*.

TABLE 39.1 Important Features of Dimorphic Fungi

Aetiological Agent	Systemic Mycosis	Endemic Zone	Direct Microscopy	Culture at 37°C	Culture at Room Temperature/25°C
Paracoccidioides, e.g., Paracoccidioides brasiliensis	Paracoccidiomycosis commonly called South American Blastomycosis or **Lutz's mycosis**	Latin America	Multipolar budding yeast cell of 2–10 µm resembling Mariner's wheel or Pilot wheel (Fig. 39.1) are observed	Yeast-like creamy-white colonies. On microscopy, multiple budding yeast cells of 3–30 µm in size and oval, globose to pyriform shaped; appearing like Mariner's wheel	The hyphae measure around 1–2 µm with single-celled conidia. Numerous intercalary chlamydospores are also observed
Coccidioides, e.g., Coccidioides immitis	Coccidiodomycosis called as San Joaquin Valley fever/desert fever	North, central and South America	Double refractile thick-walled globular spherules of 30–60 µm diameter size	The mature colony morphology are flat, fluffy-white, gray, or brownish-black in colour. After 2-3 weeks of incubation, mycelial growth is seen even at 37°C	Grows as molds with a characteristic microscopic sporulation pattern of alternating arthroconidia
Histoplasma capsulatum	Histoplasmosis commonly called Darling's disease, Ohio valley disease	USA	Small budding, intracellular, 2–5 µm yeast cells surrounded by a halo in tissue sections (Fig. 39.3)	Budding yeast cells with narrow neck between mother and daughter cells	Large (8–14 µm) sunflower-like tuberculate macroconidia and small teardrop-like micrconidia seen
Blastomycetes, e.g., Blastomyces dermatitidis	Blastomycosis commonly called Gilchrist's disease/ Chicago disease	North America	Double-contoured thick-walled, multinucleated giant yeast cells with broad-base budding daughter cells (Fig. 39.4)	Creamy white yeast-like colonies; yeast cell with figure of eight morphology	Fluffy white growth; filamentous septate hyphae with round or oval conidia

Serological Tests

Antibodies against circulating antigen are measured by ELISA or counter-immunoelectrophoresis.

Therapy

It is treated by long-term antifungal therapy of azoles.

COCCIDIOIDES

Coccidioides immitis is a thermally dimorphic fungi causing coccidioidomycosis. It is the oldest mycosis of the major human fungal disease reported by Alejandro Posadas. It is also called as **San Joaquin Valley fever**/desert fever/desert rheumatism/kok-see after the name of the valley of California. The infections caused by *Coccidioides* are usually subacute and self-limiting. They primarily affect respiratory system of humans and animals and rarely extra-pulmonary dissemination occurs in immunocompromised individuals to skin, bone and meninges.

Geographical Distribution

The disease is caused by *Coccidioides immitis* and is endemic in North, Central and South America. It is soil-dwelling dimorphic fungi that primarily affect the lungs.

Pathogenesis and Clinical Features

The life cycle of *C. immitis* involves development of hyphae, arthroconidia and spherules (Fig. 39.2). Mycelium to spherule conversion is not temperature-dependent. Fungi enter the lung through inhalation of arthroconidia leading to pyogenic inflammatory response later forming granulomas and calcification. The clinical features can be described into the following headings:

- Pulmonary infection: Lung Infiltrates and fibrosis are observed.
- Disseminated disease: The infection disseminates to extrapulmonary site such as skin, bones, and meninges.
- Chronic meningitis: Meningitis involving basilar portion of brain

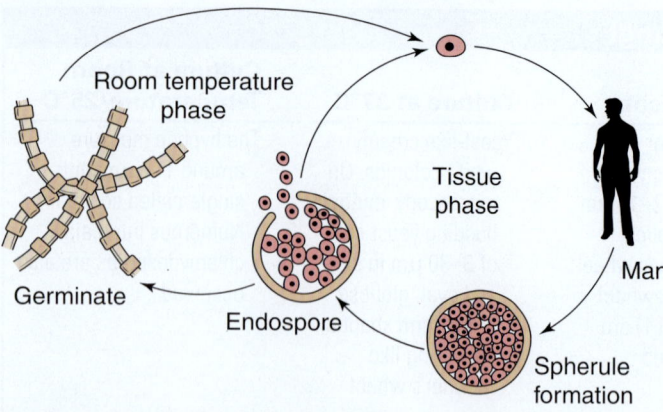

Fig. 39.2 Life Cycle of *C. immitis*.

- Infection associated with AIDS patients: It is one of the most common opportunistic infection among HIV-infected patients

Laboratory Diagnosis

Clinical Samples

Sputum, gastric lavage, bronchoalveolar lavage pus, cerebrospinal fluid.

Direct Examination

On KOH mount of clinical specimen, double refractile thick-walled globular spherules of 30–60 μm diameters are observed. The spherules are also seen in histopathological stained slides.

Fungal Culture

The infected material is cultured on SDA or brain heart infusion agar. They are **biosafety level 3** contaminant and the cultures should never be sniffed. At 25°C, downy suede-like greyish-white colonies are seen, which develop abundant aerial mycelia by 10 days. They form alternate, thick-walled barrel-shaped arthroconidia. The culture of *Coccidioides* species differs from other dimorphic fungi in that it grows as moulds at 25°C as well as 37°C under standard laboratory conditions.

Serological Tests and Immunodiagnosis

Skin tests are performed by intradermally inoculating coccidioidin antigen. A positive reaction is an induration zone of >5 mm measured at 24–48 h. Antibodies are measured via latex particle agglutination test and complement fixation test.

Molecular Assay

The organism can be phenotypically identified via nucleic acid amplification tests.

Radiological Diagnosis

On chest X-ray hilar lymphadenopathy or coin like consolidation is seen in lungs.

Therapy

Most patients recover without treatment and develop life-long immunity. Patients with severe form of disease are treated by ketoconazole, fluconazole or itraconazole.

HISTOPLASMA

Histoplasma capsulatum causes *histoplasmosis disease*. Darling who also established the genus *Histoplasma* in 1906 coined the name *Histoplasma capsulatum*. It is also called **Darling's disease**, Ohio valley disease or Tingo Maria fever. Approximately 95% of cases are inapparent, subclinical or completely benign. Acute or chronic disseminated disease develops that has a poor prognosis. They are intracellular uninucleate organism involving reticuloendothelial system. Its growth is associated with the presence of guano and debris of birds and bats.

Geographical Distribution

Histoplasma disease is reported throughout the world; however, it is endemic in the areas bordering Ohio and Mississippi rivers, that is central and eastern American states.

Aetiological Agent

Histoplasma capsulatum is the causative agent. They are thermally dimorphic fungi with three varieties:

   H.c var. *capsulatum* causing classic histoplasmosis in human

   H.c. var. *duboisii* causing African histoplasmosis in human

   H.c. *farciminosum* causing epizootic histoplasmosis in animals

Pathogenesis and Clinical Features

Lungs are the primary sites of infection. Infection is usually acquired via inhalation of spores followed by haematogenous spread to various parts of the body. Later, disease becomes chronic similar to chronic tuberculosis leaving residual calcification in the lung and sometimes in spleen. The disease rarely becomes progressive and fatal. It has four clinical varieties:

- **Acute pulmonary histoplasmosis:** Majority of immunocompetent individuals remain asymptomatic. Few symptomatic patients manifest influenza-like illness, malaise, fever, sore throat, dry cough and dyspnea. Radiological chest X-ray shows small-scattered pulmonary infiltrate, which later gets calcified.
- **Chronic pulmonary histoplasmosis:** It is the slowly progressive form of histoplasmosis. Typical pulmonary lesions are seen with central necrotic areas. Patients present with haemoptysis and apical pulmonary cavities.
- **Cutaneous and mucocutaneous histoplasmosis:** Petechial and ecchymotic purpura is usually seen on abdominal wall and thorax.
- **Disseminated histoplasmosis:** This form is associated with HIV. It is manifested as fever, weight loss, deteriorated general condition, leucopenia, anaemia, lymphadenopathy and hepatosplenomegaly.

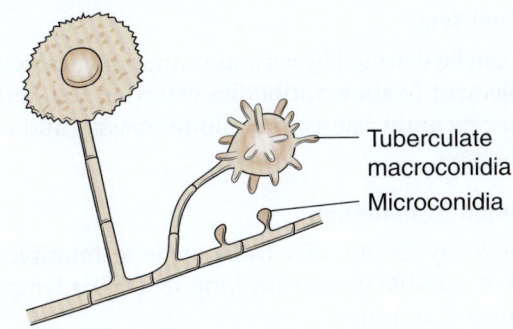

Fig. 39.3 Conidia of *Histoplasma capsulatum*.

Histoplasma capsulatum var. *duboisii* or African histoplasmosis primarily affects cutaneous, subcutaneous and osseous tissue. Lungs are not affected and rarely reticuloendothelium are involved.

Laboratory Diagnosis

Clinical Specimen

Sputum, bronchial lavage, blood, bone marrow, scrapings and biopsies from lesion and lymphnodes.

Direct Microscopy

Infected tissues are stained with Geimsa or Wright stains. Intracelluar and extracellular yeasts are observed.

Culture

Fungal culture is done on Sabouraud's dextrose agar medium. Media is incubated at two temperatures for depicting thermoconversion of dimorphic forms, that is yeast and mycelial; yeast form is seen at 37°C and mycelia grows at 25°C. *H. capsulatum* produces both macroconidia and microconidia. Macroconidia are 8–15 µm in size, globose with **tuberculate** or finger-like cell wall and microconidia are 2–4 µm in size with smooth cell wall (Fig. 39.3). The **budding yeast cells** are characteristically seen in clusters within phagocytic cells, including histiocytes and other macrophages, as well as monocytes.

Serology

Antibody titre is measured during the course of disease. It has diagnostic and prognostic value as antibodies decline when infection is inactive and increase in progressive disease.

Skin Test

Delayed hypersensitivity can be elicited with histoplasmin antigen.

Animal Pathogenicity

Mouse, hamsters, guinea pig, dog and rabbit are used as animal models. It is an important method for primary isolation of organism from clinical samples. Animal acquires infection through normal route, that is by inhalation of airborne conidia.

Molecular Assay

Real time-based assays are available for identification of histoplasma.

Radiological Examination

On chest X-ray old lesions appear as coin shaped along with calcified nodules.

Therapy

Amphotericin B is the mainstay of treatment for severe disseminated disease. Oral azoles are used for less severe infections.

BLASTOMYCETES

Blastomyces dermatitidis causes a chronic granulomatous and suppurative disease, that is blastomycosis. Its perfect or teleomorphic state is known as *Ajellomyces dermatitidis*. The lesion is commonly seen in adult male in any part of body, but with a marked predilection for the skin and lungs. The fungi are naturally found on decaying wood, debris and soil.

Geographical Distribution

It is endemic in North America hence known by the name of **Chicago disease**. Scientist T. Casper Gilchrist described this disease, and thus a name was designated to honour its discoverer as **Gilchrist's disease**.

Pathogenesis and Clinical Features

Lung is primarily affected through inhalation of fungal spores. Two clinical forms are seen in blastomycosis, that is primary pulmonary and chronic cutaneous. Later, infection disseminates to organs via bloodstream and multiple abscesses form in different parts of body. With development of immunity, inflammatory pyogranulomatous reaction occurs at the pulmonary site. Mixed neutrophilic and mononuclear cell response is distinctive of blastomycosis. Acutely infected patient presents with self-limiting pneumonia. In chronic cases it mimics tuberculosis. Cutaneous dermatitidis is the commonest presentation, and hence, the name of the species *dermatitidis* is derived from it. The initial lesion is a papule around which secondary nodule develops and coalesce leading to large elevated ulcerative lesions.

Laboratory Diagnosis

Clinical Specimen

Sputum, bronchial lavage, pus, blood, bone marrow, scrapings and biopsies from lesion and lymphnodes.

Direct Microscopy

A characteristic double-contoured, thick-walled, multinucleated giant yeast cell with **broad-based** budding daughter cells are observed. The most sensitive method to detect organism is in Gomori's methenamine silver-stained tissue histopathological sections. It shows broad-based yeasts along with granulomatous inflammatory response known as Splendore–Hoeppli phenomenon.

Fungal Culture

Dimorphism is depicted in fungal culture; on SDA at 37°C creamy white yeast-like colonies are seen at an early stage and later hyphal projections develop on surface and finally entire surface becomes fluffy white within 2 weeks of incubation.

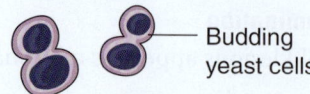
Budding yeast cells

Fig. 39.4 Broad-Based Budding Yeast Cells of *Blastomyces dermatitidis*.

On microscopy large (7–20 µm) spherical budding yeast cells are seen that have thick double-contoured wall. Each yeast cell carries only a single broad-based bud easily recognisable as *figure of eight* morphology, as shown in Fig. 39.4. At 22–25°C filamentous septate hyphae with round or oval conidia are seen. Thick-walled chlamydospores of 7–18 µm may also be observed in older cultures.

Serological Test

Antigen can be detected by various immunological tests such as complement fixation antibodies detection test, immunodiffusion precipitin bands, radioimmunoassay and enzyme immunoassays.

Radiological Examination

On chest X-ray lesions vary from single to multiple round densities of variable sizes in any lung area. Hilar lymph node enlargement is common.

Therapy

Most of blastomycosis infections require antifungal therapy; amphotericin B and azoles are the drugs of choice.

40

Opportunistic Fungal Infections and Miscellaneous Fungal Diseases

Suruchi Shukla

LEARNING OBJECTIVES

- Clinical features, laboratory diagnosis and therapy of opportunistic fungi:
  - Hyaline moulds infections
  - Pigmented moulds (phaeoid fungi) infections
  - Yeast-like fungi infections
- Mycotoxin
- Mushroom poisoning
- Fungal allergic diseases
- Miscellaneous fungal infections

OPPORTUNISTIC FUNGAL INFECTIONS

INTRODUCTION

Opportunistic fungi cause systemic or invasive diseases and often manifest acutely with rapidly progressive pneumonia or fungaemia or manifestations of extrapulmonary dissemination. Opportunistic fungi are usually not pathogenic except in immunocompromised conditions such as diabetes mellitus, AIDS, azotemia, lymphoma, leukaemia, burns, long ICU admission, long-term steroid therapy and poor nutrition.

The factors predisposing for opportunistic fungal infection may be classified into two types:

1. **Intrinsic factors: Dependent on the host**
   a. Alterations of the immune system
   b. Functional defects in chemotaxis
   c. Acquired immunodeficiency syndrome
2. **Extrinsic: Iatrogenic factors**
   a. Antibiotics use
   b. Immunosuppressants use
   c. Intravenous drug abuse
   d. Desferrioxamine therapy

There are various modes of transmission of infection by opportunistic fungi. The common modes of transmission are listed as follows:

- Inhalation of spores
- Percutaneous inoculation in cutaneous and subcutaneous infections
- Mucosal penetration by normal commensal of gastrointestinal tract such as *Candida* spp.
- Ingestion of a toxin in contaminated food or drink

Fungal agents causing typical opportunistic systemic fungal infections are listed in Box 40.1.

BOX 40.1 Fungi Causing Opportunistic Infections

- Hyaline moulds
  - *Aspergillus* spp.
  - *Zygomycetes* spp.
  - *Penicillium marneffei*
  - *Fusarium* spp.
  - *Acremonium* spp.
  - *Scedosporium* spp.
- Phaeoid fungi: *Cladophialophora bantiana/Chaetomium* spp.
- Yeast-like fungi: *Candida* spp. other than *Candida albicans/Trichosporon* spp./*Geotrichum* spp./*Rhodotorula rubra/Saccharomyces cerevisiae*
- *Pneumocystis jirovecii*

HYALINE MOULDS

ASPERGILLUS SPP.

Aspergillus is ubiquitous organism present in soil, dust and decomposing organic matter. Aspergillosis is broadly defined as a group of diseases in which members of the genus *Aspergillus* are involved. *Aspergillus* is present everywhere in the environment, and individuals with compromised immune condition are prone to this disease. **Micheli** in 1729 defined genus *Aspergillus*. **Raper and Fennel**, in 1965, classified *Aspergillus* in 151 species in 18 different groups.

Characteristics of Genus *Aspergillus*

- Genus comprises of more than 185 species, including *A. terreus, A. glaucus, A. nidulans, A. oryzae* and *A. clavatus*.
- About 95% of all infection is caused by three species: *Aspergillus fumigatus, Aspergillus flavus* and *Aspergillus*

TABLE 40.1 Test to Diagnose Aspergillosis

Clinical Manifestation	Direct Microscopic Examination	Fungal Culture	Serological Tests	Skin Test
Allergic bronchopulmonary aspergillosis (ABPA)	Positive	Positive	Strongly positive	Strongly positive
Aspergilloma	Positive	Positive	Very strongly positive	Negative
Invasive aspergillosis	Positive	Positive	May be positive	Negative

niger. Among them *A. fumigatus* causes majority of invasive and non-invasive aspergillosis.

- They are anamorphic (asexual) filamentous organisms which reproduce by means of asexual spores termed conidia. *Aspergillus* is characterised by **septate hyphae** from which nonseptate conidiophores arise. Hyphae branched regularly at **about 45-degree angle**. Aspergilli produce conidia in a basipetal fashion in chains (youngest conidium at base). The conidiophores are hyphal form which enlarges at its apex to form swollen vesicle. The basal cell from where conidia arise is called phialide. There can be one (uniseriate) or two (biseriate) layers of basal cells. The stem holding the vesicle is called stipe. Some of the species like *A. fumigatus* are uniseriate while others like *A. niger*, flavus and terreus are biseriate. Phialides may cover whole of vesicle (*A. niger* and *A. flavus*) or only upper 2/3 of vesicle's surface (*A. fumigatus* and *A. terreus*). Pigmentation of conidia determines the colour of the colony.
- They are thermotolerant, that is able to grow at 40°C and higher temperature.
- Proteolytic and elastase enzyme are used as virulence mechanism by *Aspergillus* spp. for invasion into the human tissue.
- *Aspergillus* spp. synthesise mycotoxins, including aflatoxin.

Clinical Manifestation

The organism primarily infects lungs and later may disseminate to other body organs. Various clinical forms are as follows:

1. *Pulmonary aspergillosis*
   a. **Allergic bronchopulmonary aspergillosis (ABPA):** Conidia of *Aspergillus* are small in size <5 μm and has aerodynamic properties, thus bypass upper respiratory tract defences and reach distal portion of lung, causing allergic reactions.
   b. **Extrinsic allergic alveolitis:** It can cause asthma, eosinophilia and recurrent pulmonary infiltrates.
   c. **Aspergilloma:** It can also develop localised infection in on old tuberculous cavities called as aspergilloma or fungal ball. They are also associated with cystic fibrosis.
2. *Disseminated infection*: *Aspergillus* infection can spread to sites like eye, bones, ear, heart, gastrointestinal tract, skin, liver and spleen.
3. *Central nervous system infection*: It occurs due to haematogenous spread of infection from the lungs or direct spread from nasal sinuses. Patients present with altered behaviour, confusion and delirium. On computed tomography head imaging, the brain lesion appears like a halo or crescent.

4. *Cutaneous aspergillosis*: It may be primary or secondary and occurs due to immunosuppression. Skin lesion starts with discrete papules which become pustular later.
5. *Paranasal sinuses aspergillosis*: Maxillary sinus is the most common sinus involved.
6. *Iatrogenic aspergillosis*: *Aspergillus* is present in the environment and therefore contaminates hospital ward and supplies. They gain entry to susceptible patients via instrumentations.

Laboratory Diagnosis

The high index of suspicion, microscopy, culture isolation, serological assay, skin test and radiological imagings are the criteria for definite diagnosis of aspergillosis (Table 40.1).

Direct Microscopic Examination

The specimen like exudates, sputum, bronchoalveolar lavage (BAL) or tissue biopsy is examined for fungal elements. Ten per cent potassium hydroxide mount demonstrates hyaline septate hyphae with dichotomous branching of *Aspergillus* spp. Tissue specimens are stained with calcofluor-white stain. Fluorescent antibody techniques are also used to demonstrate septate hyphae.

Fungal Culture

The *Aspergillus* can be grown in culture from blood, CSF, bone marrow or other organs (brain, kidney and liver). Aspergilli are identified on the basis of colony morphology (on Malt Extract Agar or Czapek Dox Agar at 37°C and 25°C). Species of genus *Aspergillus* can be distinguished on the basis of *colony morphology*. The microscopic details of the growth can be demonstrated by scotch-tape or tease mount *methodology* (Table 40.2). As drug resistance of *Aspergillus* spp. is common nowadays, in vitro drug susceptibility testing is recommended using European Committee on Antimicrobial Susceptibility Testing (EUCAST) and the Clinical Laboratory Standard Institute (CLSI) as reference testing methods. Disk diffusion is the most attractive methodology to perform in vitro drug susceptibility. Another method is E test which is easy to perform and use on a daily basis, yet it is expensive.

Serology

Invasive aspergillosis can be detected by enzyme-linked immunosorbent assay (ELISA). Patients' sera are tested for antibodies against galactomannan polysaccharides. ß-1-3 Glucan Assay (G-test, Glucatell, Fungitell) test helps in the diagnosis of invasive fungal infection. Cross-reactivity and lack of consistent results are major drawbacks of this assay.

TABLE 40.2 Morphological Details of *Aspergillus* spp.

Aspergillus spp.	Colony Morphology	Phialides Arrangement	Conidiophores	Diagrammatic Representation
A. fumigatus	Velvety, greenish blue colonies with reverse white to tan	It has small, conical-shaped terminal vesicle with uniseriate phialides covering only upper two-third of vesicle head	Short, smooth and erect	
A. flavus	Velvety yellow green colonies with reverse golden brown	Uniseriate and biseriate both forms, covering the entire vesicle	Spiny and rough of variable length	
A. niger	Wooly black colonies with white to yellow colour	Biseriate arrangement in which philaides are borne on metulae. Rough, dark, black conidia covers the entire vesicle in radiate form	Long and smooth	
A. terreus	Velvety and Cinnamon brown colonies with reverse brown	Biseriate arrangement. Conidia's are globose and ellipsoidal	Short and smooth	

Molecular Assay

Aspergillus spp. can be rapidly diagnosed by real-time PCR from clinical samples such as BAL and blood.

Radiodiagnosis

Chest X-ray and CT scan are important adjuncts to diagnosis.

Treatment

Voriconazole is the most effective agent against isolates. The recommended treatments for invasive aspergillosis are lipoidal Amphotericin B and azoles like itraconazole and voriconozole. Corticosteroids may also be helpful invasive aspergillosis. People who have severe aspergillosis may need surgery.

ZYGOMYCETES

Zygomycetes are group of fungi causing mucormycosis (previously called zygomycosis). These fungi live throughout the environment, particularly in soil and in association with decaying organic matter, such as leaves, compost piles or rotten wood. They belong to phylum zygomycota which are further classified in two orders, that is Mucorales and Entomophthorales. *Rhizopus* spp., *Mucor* spp., *Absidia* spp., *Apophysomyces* spp. and *Saksenaea* spp. are classified under order Mucorales, whereas *Conidiobolus* spp. and *Basidiobolus* spp. are grouped under Entomophthorales. They grow primarily as filaments of long cells called hyphae. Unlike the so-called 'higher fungi' comprising the *Ascomycota* and

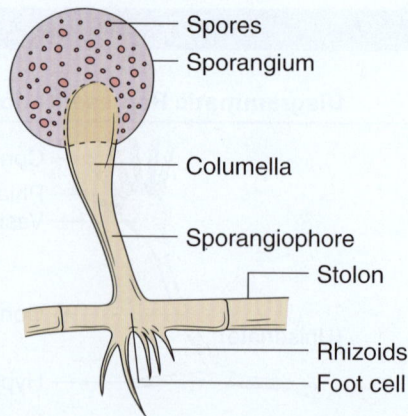

Fig. 40.1 Morphological Features of *Zygomycetes*.

Basidiomycota which produce regularly septate hyphae, most *Zygomycota* form hyphae which are generally coenocytic because they lack cross walls or septa. Kurchenmeister in 1855 described zygomycosis in patient and Lichtheim in 1884 established pathogenecity of mucorales.

Zygomycetes reproduce asexually by the formation of sporangia and sporangiospores as shown in Fig. 40.1. They also sexually reproduce with help of thick walled zygospore. *Zygomycetes* have characteristic **broad aseptate thick walled coenocytic hyphae** with irregular wide angle/obtuse branching. They have predilection to invade arterial vessels causing embolisation and necrosis of the surrounding tissue. In non-trauma cases, it usually begins in the nose and paranasal sinuses after inhaling fungal spores from the air and is one of the most rapidly spreading fungal infections in humans. It enters the skin through a cut, scrape, burn or other type of skin trauma. Mucormycosis mainly affects people with weakened immune systems and can occur in nearly any part of the body. Etiological agents causing zygomycosis or mucormycosis are given in Box 40.2.

Clinical Manifestation

The infection is acquired by inhalation of fungal spores, which are present in the environment. Lungs and sinuses are the most common site infected by the spores. This form of mucormycosis is most common in people with uncontrolled diabetes. Fungal spores may be ingested to gastrointestinal

> ### BOX 40.2   Fungal Pathogens Causing Zygomycosis
>
> - *Rhizopus oryzae* (*R. arrhizus*)
> - *Rhizopus microspores*
> - *Mucor racemosus*
> - *Rhizomucor pusillus*
> - *Cunninghamella bertholletiae*
> - *Apophysomyces elegans*
> - *Saksenaea vasiformis*
> - *Cokeromyces recurvatus*
> - *Lichtheimia* (formerly *Absidia*) *corymbifera*
> - *Conidiobolus coronatus*
> - *Basidiobolus ranarum*

tract in children and young adults. The infection may disseminate to brain and organs such as the spleen, heart, and skin. Clinically, mucormycosis manifests as several types:

- *Rhinocerebral (sinus and brain) mucormycosis*
- *Lung or pulmonary mucormycosis*
- *Gastrointestinal mucormycosis*
- *Skin or cutaneous mucormycosis*
- *Disseminated mucormycosis*

Laboratory Diagnosis

Direct Microscopy

Specimens like biopsy, scrapings and aspirates can be directly examined in KOH mount. Histopathological stained (H&E, GMS, PAS) tissue sections are screened for aseptate hyphae.

Fungal Culture

Culture in SDA at 37°C and 25°C is sufficient to isolate the fungus. The viability of the fungi is lost during crushing of tissue; therefore, tissue should be handled without grounding. Most fungi can be identified on routine lactophenol cotton blue (LCB) mounts. As they are ubiquitous fungi, they are often disregarded as culture contaminants. Fungal isolation may be reconfirmed on another sample also. The morphological details are described in Table 40.3.

Serology

Detection of β-glucan in patient's serum is helpful in diagnosis of systemic zygomycosis.

Radiodiagnosis

Radiographic findings of infection are often suggestive and helpful in delineating the extent of disease. The *Zygomycetes* invade blood vessels and therefore tissue gets infracted.

Animal Pathogenicity

Mice and rabbits are used to define mechanism of pathogenicity of zygomycosis.

Therapy

Surgical debridement and antifungal is the mainstay of therapy. Antifungal therapies, such as Amphotericin B, are initiated as early as possible. Novel azoles like isavuconazonium sulphate (isavuconazole) has been recently approved for the treatment of invasive mucormycosis.

PENICILLIUM SPP.

The organism is present in the environment and grows on various substrates such as bread, jam, fruits and cheese. It has been reported from Europe, Australia, Thailand and also from northeastern state of India. It is an emerging clinical entity and associated with later stages of HIV infection. It is **thermally dimorphic** facultative intracellular fungal organism that exists as both mycelia and yeast-like forms. Some of the *Penicillium* spp. causes opportunistic mycosis, among them ***Penicillium marneffei*** is most commonly associated with **HIV**-infected patients.

TABLE 40.3 Morphological Features of Different Agents Causing Zygomycosis

Genus	Culture	Rhizoids and Stolon	Sporangiophores and Conidiophores	Columella/ Sporangia	Diagrammatic Representation
Mucor	Colonies are very fast growing, cottony to fluffy, white to yellow, becoming dark-grey, with the development of sporangia	Absent	They are erect, simple or branched	Large terminal, globose to spherical multispored sporangia, without apophyses and with well-developed columellae	Sporangia; Columella
Rhizopus	Colonies are fast growing and cover an agar surface with a dense cottony growth that is at first white becoming grey or yellowish brown with sporulation	Present/ Rhizoids are pigmented and nodal in origin	They arise singly or in groups from nodes directly above the rhizoids. They are globose to ovoid, one-celled, hyaline to brown and striate in many species	Columella often collapses to form an umbrella-like structure	Spores; Sporangio-phore; Stolon; Rhizoids
Absidia	Rapid growing, flat, woolly to cottony and olive grey colonies. The reverse side is uncoloured as there is no pigment production	Present/ Internodal rhizoids	They are branched and arise in groups of 2–5 at the internodes	Sporangia are found in the sporangium and are released to the surrounding when the sporangium ruptures	Sporangium; Sporangio-phore; Rhizoids
Rhizomucor	Cottony colonies grows very rapidly, fill the Petri dish. From the front the colour of the colony is white initially and turns grey to yellowish brown in time. The reverse is white to pale	Present/Single or weak branched rhizoids	They do not arising between rhizoids unlike Absidia	Sporangia are found in the sporangium and are released to the surrounding when the sporangium ruptures	Sporangio-phore; Rhizoids
Cunninghamella	Colonies are very fast growing, white at first, but becoming dark grey and powdery with sporangiola development	Present	Sporangiophore end in globose or pyriform-shaped vesicles from which several one-celled globose to echinulate sporangia develop on swollen denticle	Columellae subglobose to pyriform, with globose sporangiola	Sporangium; Columella
Saksenaea	Colonies are fast growing, downy, white with no reverse pigment	Present/ Simple, darkly pigmented rhizoids	They are small, oblong and are discharged through the neck following the dissolution of an apical mucilaginous plug	Flask-shaped sporangia with columellae	Sporangiospores; Flask-shaped sporangium; Columellal; Sporangiophore; Septate hyphae; Rhizoids

(Continued)

TABLE 40.3   Morphological Features of Different Agents Causing Zygomycosis (*Cont.*)

Genus	Culture	Rhizoids and Stolon	Sporangiophores and Conidiophores	Columella/ Sporangia	Diagrammatic Representation
Apophysomyces	Colonies are fast growing, whitish with scarce aerial mycelium and no reverse pigment	Present	They are unbranched, straight or curved, slightly tapering towards the apex	Columellae are hemispherical in shape and the apophyes are distinctly funnel or bell-shaped	Spores, Sporangium, Columella, Apophysis, Sporangiophore, Stolon, Foot cell, Rhizoids
Conidiobolus	Colonies are rapidly growing flat, cream-coloured, glabrous becoming rapidly folded and covered by a fine, powdery, white surface mycelium and conidiophores. The colour of the colony may become tan to brown with age	Absent	Conidiophores are simple forming solitary, terminal conidia which are spherical, single-celled and have a prominent papilla	Conidia produces hair-like appendages, called villae. Conjugation beak is absent	Conidia, Hyphae
Basidiobolus	Colonies are moderately fast growing at 30°C, flat, yellowish-grey to creamy-grey, glabrous, becoming radially folded and covered by a fine, powdery, white surface mycelium	Absent	Large vegetative hypae forming numerous round smooth, thick-walled zygospores that have two closely appressed beak-like appendages	Primary conidia are globose, one-celled, solitary and are forcibly discharged from a zygospores. Conjugation beaks are prominent feature	Conidia with prominent conjugation beak

Pathogenesis and Clinical Features

Fungus primarily involves reticuloendothelial system and then disseminates to involve multiple organs. The organism clinically manifests as characteristic generalised **Molluscum contagiosum-like** vesiculo-papular lesions of various sizes often at face and chest. Histopathologically, three distinct patterns are observed: granulomatous reaction, suppurative reaction and necrotising reaction.

Laboratory Diagnosis

They are common airborne contaminants of culture media in the laboratory. On Sabouraud's dextrose agar, the *Penicillium* spp. develops bluish green colonies with white border and powdery surface. One of the *Penillium* spp., that is **P. marneffei**, **produces red diffusible pigment** in culture. On microscopy examination, yeast cell of 2–4 mm in diameter size and thin, filamentous, hyaline septate hyphae with branched conidiophores and two rows of sterigmata bearing chains of spores are observed, as shown in Fig. 40.2.

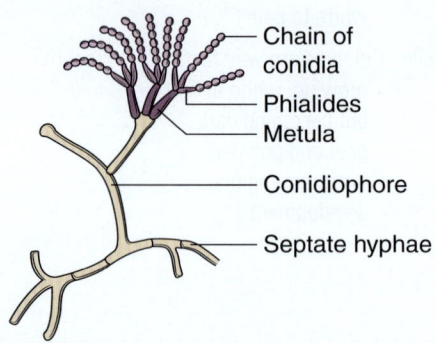

Fig. 40.2 *Penicillium* spp.

Treatment

Penicillosis can be treated by Amphotericin B and oral itraconazole.

PSEUDALLESCHERIA BOYDII

It is the perfect stage of *Scedosporium apiospermum*. It is an ascomycetes belonging to family *Microascaceae* of order

Microascales. It is a saprophyte frequently found in soil and environment. It causes severe systemic infection in immunocompromised host such as granulocytopenic patients. The organism is capable to cause sinusitis, endophthalmitis, pneumonia and meningitis. Skin lesion can spread to bones and joints. The fungus is histopathological analogous to *Aspergillus* spp.

The histopathological examination of specimen collected from active lesion reveals hyaline septate hyphae of 2–7 μm in diameter with an acute angle branching within the abscess and with granulomatous reaction. On SDA media, their colonies appears cottony fluffy molds initially whitish in color later turns brownish grey with age. The reverse colony is black. On microscopic examination, septate hyphae with oval- or club-shaped, pale brown conidia are seen which produces elliptical, sperm-shaped, single-celled conidia borne at tip of condiophores. The drug of choice is itraconazole. Surgical excision of localised lesion is required for complete successful treatment.

FUSARIUM SPP.

The fungus is found in soil. The common species involved in the human infections are *F. solani*, *F. oxysporum* and *F. verticilloides*. They are emerging fungal pathogens causing morbidity and mortality in patients with malignancies and in those who received bone marrow transplantation. The infection is acquired through inhalation of aerosolised conidia. *Fusarium* spp. has three types of pathological features: infection caused by mycotoxin, locally invasive lesion and disseminated infections.

Ecthyma-like skin lesions are initially seen which later develops into cellulitis and fasciitis. The fungus disseminates to the sinuses, brain, lungs, eye and blood stream.

The diagnosis depends on high index of suspicion in neutropenic patients. The final diagnosis relies on blood culture or tissue biopsy specimen examination. On microscopy, a hyaline septate acute-angled dichotomous-branched hypha with vascular invasion is seen. Surgical excision and Amphotericin B are administered for successful outcome of the patients.

ACREMONIUM SPP.

It is a saprophyte found in decaying vegetation, food stuff and soil. It causes human infections of eyes, skin, nails and foot. Immunocompromised individuals manifest as arthritis, osteomyelitis, peritonitis, endocarditis, pneumonia, cerebritis and subcutaneous infection. It is also diagnosed on blood culture and histopathological specimens. On direct microscopy of clinical specimen, a septate hyaline hypha is seen with vascular invasion like *Fusarium* spp. On culture the colonies of *Acremonium* spp. are often fluffy white colonies later turning grey. The conidiophores are absent and phialidic conidiogenous cells are erect, gradually tapering towards apex and arising singly from creeping hyphae. Antimicrobial resistance is recently seen with the *Acremonium*

isolates also. Now newer azoles like voriconozole are active against *Acremonium* spp.

PAECILOMYCES SPP.

It is generally a soil saprophytes and occasionally airborne contaminant. Although uncommon, invasive infection is observed in patients with neutropenia, HIV and organ transplant recipient. It usually presents with superficial infection of skin. The pathogen enters through indwelling catheters leading to systemic infection of sinusitis, soft tissue and bone. The pathogen is similar to *Penicillium* species. The organism grows on SDA producing white, flat, powdery colonies with red-brown reverse. As culture matures **lilac colour** appears and later darkens to violet colour. On microscopic examination characteristic elongated and tapered phialides with elliptical conidia. Amphotericin B and azoles are used in invasive infections.

PIGMENTED MOULDS (PHAEOID FUNGI)

Infection caused by dematiaceous fungi is known as phaeohyphomycoses (discussed in Chapter 37). They are widely present in soil and decaying vegetation and woods. These are ubiquitous saprophytes causing human disease. **Cladophialophora bantiana** is a neurotropic opportunistic phaeoid fungi causing brain abscess via hematogenous spread. Cerebral mycosis caused by fungi belonging to genus **Chaetomium** is another emerging group of opportunistic phaeoid pathogens in intravenous drug abusers and renal transplant patients.

Clinical manifestation varies from localised skin lesion to invasive and systemic infection. Invasive form is most severe form of the disease.

In laboratory it is diagnosed on blood culture and histopathological examination of tissue specimens. On direct KOH wet mount examination, **pigmented and dark brown hyaline mycelial** strands are seen. Special stain such as **Masson-Fontana** may be used to elicit the presence of melanin in the hyphae. On histopathological examination, fungal hyphae of 3–4 μm diameter in variable size and shape within the fibrous cyst cell of 25 μm in diameter are seen. On culture slow growing spreading olive grey to brown colonies are seen. On microscopic examination of culture isolate, a single-celled blastoconidia without darkly pigmented hila, borne on long and sparsely branched chains, is seen. Radiological CT/MRI imaging of brain aids in localisation and diagnosis of cerebral lesion. Deep-seated localised cerebral lesion shows **double ring sign** on contrast enhancement imaging.

Invasive infection requires both surgical excision and antifungal therapy. Amphotericin B and oral flucytosine are successfully used for treatment of cerebral abscess.

YEAST-LIKE FUNGI

These organisms are present in the environment or food or water and can also be normal human microbial flora. When the individual's immunity is weakened, these organisms get opportunity to colonise by invading the human defence barrier and cause severe infection.

CANDIDA SPP. OTHER THAN CANDIDA ALBICANS

They are now emerging as major cause of blood stream infection. The commonly reported species are *Candida glabrata, Candida krusei, Candida guillierimondii, Candida dubliniensis* and *Candida rugosa.* These species are rare but now are observed in nosocomial clusters. They are resistant to one or more group of antifungal agents.

TRICHOSPORON SPP.

The genus *Trichosporon* is inhabitant of soil and part of human skin flora. The fungus is an established entity as white peidra caused by *Trichosporon beigelli* (described in Chapter 36). ***Trichosporon asahii*** and ***Trichosporon mucoides*** **cause deep-seated invasive infection**. *T. cutaneum, T. inkin* and *T. ovoides* cause superficial skin, hair and nail infection and are casually grouped as ***T. beigelli*-complex**.

It causes catheter-associated fungemia in leucopenic patients and gains entrance to the blood stream, thus causing blood stream infection. The patient clinically presents with fever and pneumonia. The disease disseminates to other body parts like spleen, liver, kidney, bone marrow, brain and eye.

The fungus is isolated from blood cultures. On Sabouraud's dextrose agar there are cream coloured wrinkled yeast like colonies. On direct microscopy, 3–8 μm diameter, pleomorphic septate hyaline yeast-like cells with arthrospores and blastospores are seen. On biochemical analysis, urease test is positive in *Trichosporon* spp.

Antifungal susceptibility to Amphotericin B and azoles is variable for deep-seated trichosporonosis infection, and thus drug resistance makes this organism difficult to treat.

RHODOTORULA SPP.

It is found in environment, cheese and milk products, and commensal of skin. *Rhodotorula* includes species *R. mucilaginosa, R. glutinis, R. rubra* and *R. minuta. R. rubra* is emerging as an important opportunistic pathogen especially in patients with indwelling catheter. It causes CNS infection, blood stream infection, eye infection and peritonitis.

Rhodotorula spp. produces red glistening mucoid colonies on SDA medium. On microscopic examination a budding yeast cells similar to *Cryptococccus* spp. except for the presence of **carotinoid pigment.** *Rhodotorula* species are separated from cryptoccaceae by their lack of inositol fermentation, urease and pigment production. Amphotericin B has excellent activity against *Rhodotorula* spp.

GEOTRICHUM SPP.

It is a saprophytic fungi present as microbial commensal of oral, gastrointestinal tract, bronchi, lungs and genitourinary tract. The species causing geotrichosis is *G. candidum*, formally called as *Blastoschizomyces capitatus.* It causes superficial infection of skin, oral and gastrointestinal mucosa. Its pulmonary infection clinically mimics tuberculosis.

Geotrichosis is diagnosed by culture and biochemical reactions of the clinical isolates. At 25°C the organism grows as whitish dry, smooth or hairy colonies, and at 37°C it grows like slow growing mould. On microscopic examination, hyaline hyphae with 4–8 μm rectangular arthroconidia are seen. The arthroconidia produces hyphal extension from one corner, giving appearance of **hockey-stick appearance**. Antifungal therapy is sometimes difficult as decreased antifungal susceptibility to Amphotericin B and azoles has been reported for *Geotrichum.*

SACCHAROMYCES CEREVISIAE

This fungus is also known as **baker's or brewer's yeast** used commercially to ferment beer, wine and mash. It is found in soil, ripe fruit and sewage and also as microbial commensal of skin. *S. cerevisiae* is clinically significant pathogen causing oral thrush, vulvovaginitis and fungemia. On microscopy, the oval budding yeast cells with short rudimentary pseudohyphae are seen. Amphotericin B and azoles are used in the treatment of systemic fungal infection.

PNEUMOCYSTIS JIROVECII

The organism was previously called as ***Pneumocystis carinii.*** It is yeast-like fungus of the genus *Pneumocystis.* It is important human pathogen associated with HIV and bone marrow transplant patients. It was first identified both in human and rats, therefore called *P. carinii*, and was believed to be a *protozoan.* The organism was first identified in pneumonia patient in 1952 by **Otto Jerovec.** Afterwards the organism was reclassified as *fungus* in 1976, and the nomenclature changed as *P. jirovecii.*

Life Cycle

The organism is extra-cellular fungi and all the stages are found in lungs only. The trophozoite is the vegetative state of the organism. The trophozoites are multilobed, amoeboid-shaped single-celled structure, later the cell wall thickens.

Clinical Feature

The organism causes atypical interstitial pneumonitis of the lung.

Laboratory Diagnosis

As the organism cannot be cultured, therefore its identification is difficult. Clinical samples like sputum, lung aspirate or bronchoalveolar lavage are stained with methenamine silver dye. On direct microscopic examination, the thick walled trophozoite is identified on the stained specimen.

TABLE 40.4 Fungi and Actinomycetes Producing Allergens

Clinical Disease	Fungi	Allergic Disease	Source of Allergy
Allergic respiratory fungal diseases	Thermoactinomyces saccharii	Bagassosis	Sugar cane
	T. vulgaris, Faenia rectivirgula	Farmer's lung	Stocked hay
	Cryptostroma coricale	Maple bark stripper's lung	Maple tree bark
	Penicillium casei	Cheese washer's disease	Cheese
	Aspergillus clavatus	Maltster's lung	Barley malt
	Aspergillus fumigatus, A. flavus, A. niger	Allergic bronchopulmonary aspergillosis (ABPA)	
	Alternaria spp.	Wood pulp worker's disease	Wood pulp
	T. vulgaris, T. candidus	Humidifier lung	Humidifier air conditioner
Fungal rhinosinusitis	Aspergillus flavus, A. fumigatus, Pseudallescheria boydii, Fusarium spp., Paecilomyces spp., Trichoderma spp. Mucor spp., Rhizopus spp., Rhizomucor spp., Apophysomyces elegans, Curvularia spp., Alternaria spp., Cladosporium spp., Bipolaris spp.	Acute fulminant invasive fungal sinusitis/chronic invasive fungal sinusitis/granulomatous invasive fungal sinusitis/fungal ball	

Treatment

Trimethoprim-sulphamethoxozole is the drug of choice. Pentamidine or dapsone can also be used in drug-resistant patients.

FUNGI CAUSING ALLERGIC DISEASES

Numerous fungi produce allergens. Allergens lead to hypersensitivity reactions like type I and type III. Hypersensitivity to inhaled fungal conidia manifests allergic bronchopulmonary diseases. There are three main stages of clinical manifestation: acute, subacute and chronic stages. The fungi and actinomycetes producing allergens are listed in Table 40.4. The clinicians diagnose the disease on the basis of high suspicion index. The final diagnosis requires demonstrating the fungi on direct KOH microscopy and/or histopathological specimens. On radiological imaging, a characteristic polypoidal growth and **eosinophilic mucin** are present. The treatment usually requires surgical debridement.

FUNGI PRODUCING TOXINS

A substantial number of fungus produces toxins which are exclusively utilised to produce antibacterial antibiotics and food-related products, whereas some fungi found naturally in food and produce secondary metabolites known as mycotoxins. Ingestion of mycotoxin producing food may lead to wide range of disease called as **mycotoxicoses**, while the study of mycotoxins and mycotoxicoses is known as mycotoxicology. Similarly, a clinical entity, **mycetismus** or **mycetism** or **mushroom poisoning,** produced while ingesting toxins along with fungus. Medically significant fungi causing mycotoxins and mushrooms poisoning are shown in Tables 40.5 and 40.6. The mycotoxins can be detected by

TABLE 40.5 Mycotoxin

Types of Mycotoxins	Agent	Food Source	Clinical Manifestation
Aflatoxins	Aspergillus flavus, A. nomius, Penicillium puberulum	Nuts, maize	Aflatoxicosis, hepatoma, hepatitis, Reye's syndrome
Ochratoxins	Aspergillus ochraceus, A. niger, Penicillium verrucosum	Cereals, coffee-beans, bread	Nehropathies
Trichothecenes	Fusarium graminearum	Maize and sorghum	Human toxicosis infects alimentary tract, biological warfare as yellow rain
Cyclopiazonic acid	Aspergillus flavus, A. versicolor, A. oryzae, Penicillium verrucosum	Groundnut and corn, meat	Kodua poisoning and co-contaminant
Patulin	Penicillium griseofulvum	—	Viral like common cold in man and animals
Fumonisins	Fusarium moniliforme	Maize	Fatal disease of horse, pigs and rats

TABLE 40.6 Mushroom Poisoning

Types of Mushroom Poisoning	Agent	Food Source	Clinical Manifestation
Ergot alkaloid poisoning	*Claviceps purpurea*	Rye flour	St. Antony's fire-adrenergic blockage causes burning sensation due to marked vasoconstriction
Muscarine poisoning	*Inocybe fastigiata, I. napipes*	Food material	Cholinergic effect
Coprine poisoning	*Coprine atrementarius*	Cream butter sauce	Dilation of vessels leads to unpleasant taste and hot flash sensation
Ibotenic acid, Muscimol and muscazone	*Amanita pantherina, A. muscaria*	Edible mushroom	Abdominal pain, vomiting, diarrhoea
Cyclopeptides	*Amanita phalloides, A. verna*	Toadstools	Hepatocellular failure

TABLE 40.7 Fungi Causing Oculomycoses

Fungi Causing Keratomycosis			Fungi Causing Endophthalmitis
Hyaline Fungi	**Phaeoid Fungi**	**Yeast-Like Fungi**	
Aspergillus fumigatus, A. flavus, Fusarium solani, F. oxysporum, Penicillium spp., *Acremonium* spp., *Paecilomyces* spp., *Pseudallescheria boydii*	*Curvalaria* spp., *Alternaria* spp., *Bipolaris* spp., *Cladosporium* spp.	*Candida albicans, Candia tropicalis, Candida krusei*	*Candida albicans, C. glabrata, C. tropicalis, Cryptococcus neoformans, Fusarium* spp., *Curvularia* spp., *Sprothrix schenckii, Aspergillus* spp., *Histoplasma capsulatum, Coccidiodes* spp.

physiological methods like spectrophotometery, HPLC and ELISA. The best way is to prevent these food poisoning and toxicity is by maintaining good food safety regulation laws and legislature about highest permissible levels of toxins in food. Intoxicated patients are decontaminated and detoxified. Only symptomatic and supportive therapy is given as no specific antidote is available.

MISCELLANEOUS FUNGAL INFECTIONS

OCULOMYCOSES

The fungi are important cause of ocular infection. The filamentous fungi are responsible for one-third of infectious keratitis. Among them **Aspergillus genus** is the most common etiological agent of all types ocular infection. Other fungal pathogens are also listed in Table 40.7. In keratomycosis (cornea infection), patients complaint of gritty sensation, foreign body sensation, burning, discomfort and photophobia. Later hyphopyon and fungal corneal ulcers are observed. In fungal endophthalmitis, patient complaint of severe pain due to progressive granulomatous uveitis, retinitis and vitreous abscess and if not timely treated may suffer from complete loss of vision. These fungus are diagnosed on KOH wet mount and culture. The topical antifungals are mainstay of therapy such as 5% Natamycin, Amphotericin B eye drop and itraconazole eye cream. In extensive damage of the eyes, surgical debridement is also considered.

OTOMYCOSIS

Otomycosis is subacute or chronic superficial infection of external auditory canal. The agent causing ear infection are **Aspergillus spp.** and **Candida spp.** The patients complaint of itching, irritation, pain and discomfort with feeling of ear blockage. Otoscopic examination reveals greenish-black fluffy growth in the ear canal in case of otomycosis caused by *Aspergillus* and cheesy white material in case of candidal infection. KOH wet mount helps to demonstrate the pathogenic fungi. The topical antifungals, such as Nystatin and econozole are effective in ear infection.

Laboratory Approach to Common Clinical Syndromes

41

Laboratory Approach to Common Clinical Syndromes

Amita Jain

LEARNING OBJECTIVES

- Definitions, categories, pathogenesis, laboratory diagnosis, testing algorithm of:
  - Pyrexia of unknown origin
  - Respiratory tract infections
  - Urinary tract infections
  - Meningitis, encephalitis and brain abscess

- Gastrointestinal infections and food poisoning
- Sexually transmitted infections and reproductive tract infections
- Skin and soft tissue infections
- Prevention of Hospital Acquired Illness and Blood Born infections

PYREXIA OF UNKNOWN ORIGIN

DEFINITION

The pyrexia of unknown origin (PUO) is also known as fever of unknown origin (FUO). PUO is defined as; 'a person presenting with fever more than 38.3°C on several occasions; lasting more than 3 weeks, when a diagnosis is not made despite 1 week of inpatient investigation'.

CATEGORIES

Based on the potential aetiology, four categories of PUO are described in Table 41.1.

COMMON AETIOLOGIES OF PUO/FUO

1. **Infections:** Most common cause contributing to 25%–50% cases.

TABLE 41.1 Categories of PUO

Categories of PUO	Definition	Common Aetiologies
Classic	• Temperature >38.3°C (100.9°F) • Duration of >3 weeks • Patient undergoes either at least 3 outpatient visits with 3 days in hospital for clinical evaluation or 1 week of logical and intensive outpatient testing without clarification of the cause of fever	Infection, malignancy, collagen vascular disease
Nosocomial	• Temperature >38.3°C • Patient hospitalised ≥24 h and does not manifest an obvious source of infection that could have been present before or at the time of admission • At least 3 days of clinical evaluation	*Clostridium difficile* enterocolitis, drug-induced, pulmonary embolism, septic thrombophlebitis, sinusitis
Immune deficient (neutropenic)	• Recurrent fever with temperature >38.3°C • Neutrophil count ≤500 per mm³ • Evaluation of at least 3 days including negative cultures after 48 h	Opportunistic bacterial infections, aspergillosis, candidiasis, herpes virus
HIV-associated	• Temperature >38.3°C • Duration of >4 weeks for outpatients, >3 days for inpatients • HIV infection confirmed • Diagnosis remain uncertain after appropriate evaluation	Cytomegalovirus, Mycobacterium avium–intracellulare complex, *Pneumocystis carinii* pneumonia, drug-induced, Kaposi's sarcoma, lymphoma. Non-infectious causes are less common and include lymphomas, Kaposi's sarcoma, and drug-induced fever

Note: Geographic considerations are especially important in determining the aetiology of PUO in patients with HIV.
PUO = pyrexia of unknown origin.
(*Source:* Adapted from Benett JE, Dolin R, Blaser MJ. Fever of unknown origin. Mandell, Douglas, and Bennett's principles and practice of infectious diseases. 8th ed. pp. 779–790 [Chapter 55].)

2. **Connective tissue disorders:** Contribute to 10%–20% cases and include SLE, rheumatoid arthritis, vasculitis, giant cell arteritis etc.
3. **Neoplasms:** Contribute to 5%–35% cases and include lymphoma, leukaemia, renal cell carcinoma, hepatocellular carcinoma and metastasis.
4. **Miscellaneous causes:** Contribute to 15%–25% cases and include drug reactions, deep venous thrombosis, sarcoidosis, inflammatory bowel disease and unidentified causes.

Common Infectious Causes of PUO in Indian Continent

- Tuberculosis (especially extrapulmonary)
- Abdominal abscesses
- Pelvic abscesses
- Dental abscesses
- Endocarditis
- Osteomyelitis
- Sinusitis
- Cytomegalovirus infection
- Epstein–Barr virus infection
- HIV infection
- Lyme disease
- Prostatitis
- Sinusitis

LABORATORY DIAGNOSIS

1. **Blood culture:** It is the most common test to detect the presence of bacteria or yeasts in the blood and to guide treatment, when a person is having symptoms of sepsis. Two or more blood cultures are typically ordered and collected as consecutive samples. Obtain two separate samples initially. After 24–36 h, two more samples should be obtained. Yield in less than four cultures is minimal. The volume of blood is critical because the concentration of organism in many of bloodstream infections is low. In adults, the recommended volume is 8–10 mL in each bottle. In infants and children, the concentration of microorganisms during bacteraemia is higher than in adults; therefore, less blood is required for culture. Recommended volume of blood for children is 1–3 mL in each bottle.
2. **Others:** A complete blood count (CBC), chemistry panel or urine, sputum or cerebrospinal fluid (CSF) culture is usually ordered along with blood culture to determine the source of the original infection.

DIAGNOSTIC WORKUP

An algorithm for diagnostic management of a case of PUO is shown in Flowchart 41.1.

RESPIRATORY TRACT INFECTIONS

DEFINITIONS

The respiratory tract infection is an infectious disease of the upper or lower respiratory tract. It includes the following:

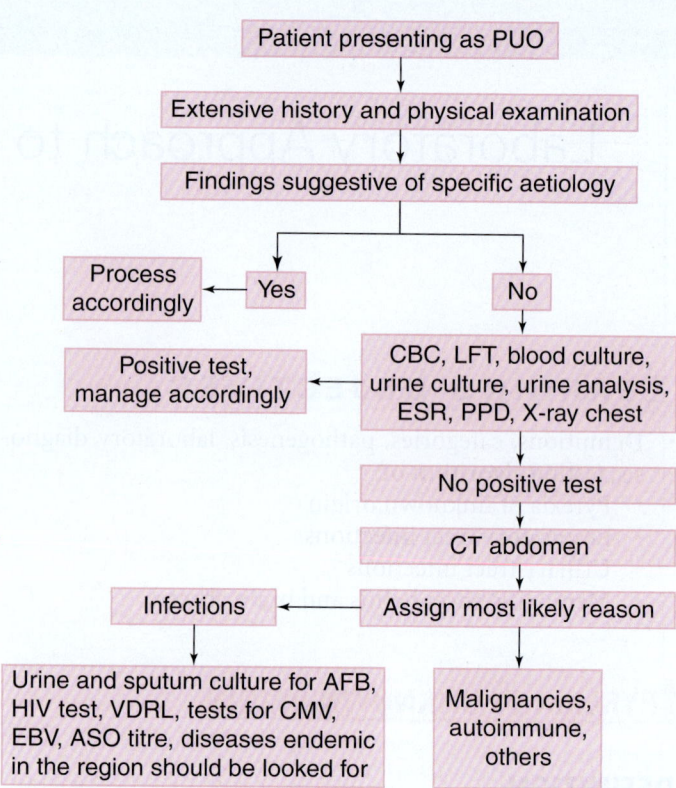

Flowchart 41.1 Algorithm for Diagnostic Management of a Case of PUO.

1. **Upper respiratory tract infections (URTI):** URTI is usually named according to the sites involved and includes common cold, laryngitis, pharyngitis, rhinitis, sinusitis, laryngotracheobronchitis (croup) and tonsillitis. The infection can also spread to sinuses and middle ear.
2. **Lower respiratory tract infections (LRTI):** LRTI include bronchitis, bronchiolitis, pneumonia and tracheitis.

UPPER RESPIRATORY TRACT INFECTION

The common aetiology of URTI is viral but occasionally it can be bacterial. Clinically it may not be possible to differentiate one from the other. Signs and symptoms of bacterial and viral URTIs are similar. URTIs are identified by a variety of names, for example, *common cold, acute infective rhinitis, acute coryza* etc. The different upper respiratory tract syndromes, their definitions and common viral and bacterial aetiologies are listed in Table 41.2.

Laboratory Diagnosis

Diagnostic methods are selected based on differential aetiology (Flowchart 41.2).

Sample: Usually throat swabs/nasopharyngeal swabs/throat washings/nasal swab/ear discharge are collected on sterile cotton/decron-tipped (for molecular assays) swab. Swabs are transported to the laboratory as soon as possible at 4°C, except for *H. influenzae* and *S. pneumoniae* culture for which swabs are transported at room temperature.

For **bacterial aetiology** confirmation, swabs are inoculated on blood and chocolate agar plates or as desired for

TABLE 41.2 Upper Respiratory Tract Syndromes and Their Aetiologies

Clinical Syndromes	Definition and Clinical Features	Viruses (Common Causes)	Bacteria (Common Causes)
Rhinitis (common cold)	Inflammation of the nasal mucous membrane; rhinorrhoea, nasal obstruction, sneezing, sore throat and cough are the prominent symptoms	Rhinovirus (accounts for more than 50% of cases in adults), influenza viruses, parainfluenza viruses, adenovirus, picornavirus and coronavirus	*Chlamydia pneumoniae*, *Mycoplasma pneumoniae* and Group A streptococci; can get super infected by bacterial infections of the paranasal sinuses and the middle ear
Sinusitis	Inflammation of paranasal sinuses; pain and stuffiness of nose are the main symptoms	—	*Streptococcus pneumoniae*, *Haemophilus influenzae*, *M. catarrhalis*
Otitis media	An inflammatory condition of middle ear characterised by pain, fever, abnormalities of hearing and vertigo	—	*S. pneumoniae*, *H. influenzae*, *M. catarrhalis*
Pharyngitis	Acute pharyngitis is an inflammatory syndrome of the pharynx; sore throat with fever and pharyngeal inflammation are commonly seen	Adenovirus, rhinovirus, coronavirus, HSV-1 and HSV-2, influenza virus, cytomegalovirus, Epstein–Barr virus, HIV, RSV, human metapneumovirus	Group A β haemolytic streptococcus (most common), mixed anaerobic infections (Vincent's angina), *H. influenzae*, *Corynebacterium diphtheriae*, *Staphylococcus aureus*, *M. pneumoniae*, *Yersinia enterocolitica*, *Arcanobacterium haemolyticum*, *Corynebacterium ulcerans*, *C. pneumonia*
Laryngitis	Is usually associated with common cold or flu, characterised by hoarse voice often accompanied with a dry cough	Influenza viruses, paramyxovirus, adenovirus, picornavirus, coronavirus, rhinoviruses	*M. pneumoniae*, *C. pneumoniae*, Group A beta haemolytic streptococci: viral infections can be super infected with bacteria
Croup (acute laryngotracheobronchitis)	Acute infection of upper respiratory tract occurring usually in children between 6 months and 3 years of age and characterised by a barking cough, hoarseness of voice and stridor during inspiration.	Paramyxovirus, RSV, influenza viruses, adenovirus, picornavirus and coronavirus, rhinoviruses	Bacteria which can cause super infection are: Staphylococcus, Streptococcus, *H. influenzae*, *C. diphtheriae*
Epiglottitis	Localised invasive infection of supraglottic area	Cytomegalovirus, Epstein–Barr virus, Herpes simplex virus	*H. influenzae*, *S. pneumoniae*, *M. catarrhalis*, *Neisseria meningitidis*

specific aetiology. Few rapid diagnostic kits for Group A beta haemolytic streptococci, *S. pneumoniae* and *H. influenzae* are commercially available.

For **viral aetiology** confirmation, usually molecular assays are performed. Few rapid diagnostic kits/direct immunofluorescence test kits are also commercially available. Their use is recommended with caution as sensitivity and specificity of these tests are limited. Viral culture can be done only in reference labs.

LOWER RESPIRATORY TRACT INFECTION

LRTIs cause disease in bronchi/alveolar sacs, and the resulting infections are named depending on clinical manifestations/site involved/causative organism/underlying disease. Common LRTIs are bronchitis, bronchiolitis, pneumonia (community acquired [CAP]/hospital acquired [HAP] and acute exacerbation of chronic obstructive pulmonary disease (COPD). The different lower respiratory tract syndromes, their definitions and common viral and bacterial aetiologies are listed in Table 41.3.

Laboratory Diagnosis

Diagnostic methods are selected based on differential aetiology.

Sample: Sputum, endotracheal aspiration and bronchoalveolar lavage are commonly used samples. Sputum is usually a non-sterile sample. Tests are employed to judge colonisation from real infection especially in patients on ventilator. Samples are transported at 4°C to the laboratory, as soon as possible. For *H. influenzae* and *S. pneumoniae* culture swabs are transported at room temperature.

Samples are processed as samples from upper respiratory tract.

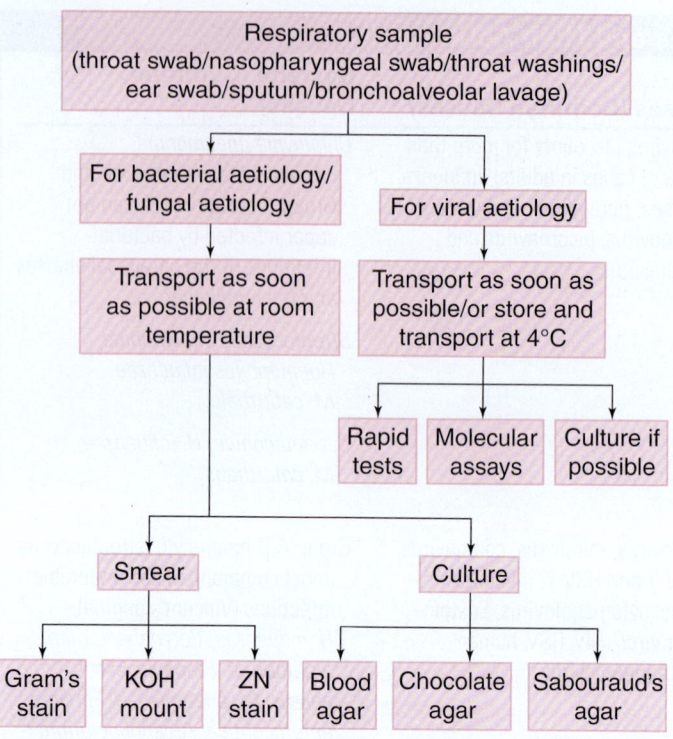

Flowchart 41.2 Laboratory Approach to a Case Presenting With Upper Respiratory Tract Syndrome.

URINARY TRACT INFECTION

DEFINITION

Urinary tract infection (UTI) is defined as an infection of the kidney, ureter, bladder or urethra. Common symptoms include an increased frequency of urination, increased urge to urinate and pain and burning in micturition; however, at times it may be asymptomatic.

TYPES

The types of UTI and their definitions are listed in Table 41.4.

PATHOGENESIS

Urine is normally a sterile body fluid. It can get contaminated with microbiota from the perineum, urethra, prostate or vagina. *Lactobacilli* spp. predominates as normal flora of the vagina. Skin and enteric flora, including *Staphylococcus* spp. and *E. coli*, are part of the microbial flora of the urethra and introitus. Most UTIs are the result of retrograde ascending infection by normal flora. Host defence factors in normal individuals do not allow all microbes to persist in urinary tract. Adaptive immunity of host, both cellular and humoral, have limited role in defence against UTI; however,

TABLE 41.3	Lower Respiratory Tract Syndromes and Their Aetiologies		
Clinical Syndrome	**Definition**	**Bacteria**	**Virus**
Acute exacerbation of COPD	Acute exacerbation of COPD is indicated by an acute change from patient's baseline dyspnoea, cough and/or sputum with increase in symptoms, and may warrant a change in regular medication	*H. influenzae, S. pneumoniae, M. catarrhalis* (super infection with Gram-negative rods may worsen the air function)	RSV, rhinovirus, influenza A, parainfluenza viruses
Bronchitis	Inflammation of bronchi; cough with or without sputum is cardinal sign	*M. pneumoniae, C. pneumoniae, Bordetella pertussis*	Rhinovirus, influenza virus, RSV, human metapneumovirus, coronavirus, adenovirus
Bronchiolitis	Small airway inflammation and obstruction	—	Mostly RSV
Pneumonia CAP	Pneumonia is an infection of the alveoli or the alveolar sacs. If the patient is not hospitalised and acquires infection outside the hospital, it is called CAP	*S. pneumoniae, M. pneumoniae, H. influenzae, Chlamydophila pneumoniae*	Influenza A and B viruses, adenoviruses, respiratory syncytial viruses, parainfluenza viruses
Pneumonia HAP	Pneumonia is an infection of the alveoli or the alveolar sacs. If the patient is hospitalised and is on ventilator for more than 24 h and acquires infection, it is called HAP	*S. pneumoniae, H. influenzae, S. aureus, Escherichia coli, Klebsiella pneumoniae, Proteus, Enterobacter, Serratia marcescens, Pseudomonas aeruginosa,* MRSA, *Acinetobacter, Legionella pneumophila, Burkholderia cepacia* and occasionally fungi-like *Aspergillus*	

TABLE 41.4 Types of UTI

	Commonly Affected Population
Uncomplicated UTI (infection is limited to bladder)	Dysuria, urgency and frequency in otherwise healthy non-pregnant women
Complicated UTI (infection extends beyond the bladder causing symptoms, such as flank pain or other symptoms of pyelonephritis, fever or sepsis)	Immunosuppressed individuals, urinary tract obstruction and urinary retention, renal failure, renal transplantation, individuals with risk factors that predispose to persistent or relapsing infection, e.g., calculi, indwelling catheters or other drainage devices, men, pregnant women, hospitalised individuals
Catheter-associated UTI	Presence of indwelling urinary catheters with signs and symptoms of UTI and no other source of infection, and Presence of $\geq 10^3$ CFU/mL in a single catheter urine specimen or in a midstream urine, despite removal of urinary catheter in the previous 48 h

CFU = colony-forming unit
(*Source*: Adapted from PSAP 2018 Book 1, *Infectious Diseases*, Helen S. Lee and Jennifer Le, Urinary Tract Infections, Table 1–3, p.11)

antibacterial properties of urine and flow and force of urine during micturition are highly effective in protecting from UTI. Functional and structural defects in urinary tract make it more susceptible for infection. Some microbes are capable of causing UTI and have genetically determined virulence factors (uropathogens). Virulence factors include bacterial capsules, adhesins, haemolysin, cytotoxic factors, necrotising factors, aerobactin etc.

AETIOLOGY

Aetiology may differ in case of outpatients (community acquired) and hospitalised patients (hospital acquired); however, in more than 80% cases in both the groups, it is *E. coli* (Table 41.5).

LABORATORY DIAGNOSIS

Microscopic Examination of Urine

It is done to demonstrate the presence of leucocytes in the sample. Normal urine may contain leucocytes in amount ranging from few to up to $10^6/24$ h. Generally, 10 leucocytes/mL (in uncentrifuged urine) are required to call it pyuria,

which corresponds to 1 leukocyte per 7 HPF. Wet mount also helps in reporting the presence of RBCs, epithelial cells, yeast cells, bacteria, cast and crystals (calcium oxalate, uric acid and triple phosphate).

Urine Culture

Sample Collection and Transportation

The most common method of collection of urine is 'midstream clean catch urine specimen'. Catheter specimen can be collected and processed in catheterised patients. In small children, suprapubic aspirate of urine can be collected. Sample should be collected in a sterile screw-capped universal container and should be transported to the laboratory immediately after collection latest within 2 h. If delay is expected, sample should be refrigerated at 2–4°C for up to 24 h and sent to laboratory maintaining the cold chain.

Culture

Culture media that can be used for plating of urine are either a combination of blood agar and MacConkey agar or a CLED agar. Normal urine sample is likely to get contaminated with microbial flora of perineum, urethra, prostate or vagina while collection. Hence, the mere growth of bacteria in urine culture does not indicate UTI. Presence of a single type of bacteria in amount $\geq 10^5$ colony-forming units (CFU)/mL in urine is essential to call it 'significant bacteriuria'. However, in catheterised patients, $\geq 10^3$ CFU is considered significant.

TABLE 41.5 Common Causes of UTI (in Order of Frequency)

Inpatients	Outpatients
E. coli	*E. coli*
Proteus spp.	Staphylococci (*S. saprophyticus*)
Pseudomonas spp.	*Klebsiella*
Coagulase negative staphylococci (*S. saprophyticus*)	*Enterobacter*
Klebsiella spp.	*Proteus*
Acinetobacter spp.	*Enterococci*
Enterobacter	

Dipstick Tests

Commercial dipstick tests are available to detect the presence of leukocyte esterase (an enzyme produced by neutrophils) and nitrite (bacterial nitrate reductase reduces nitrate) in urine.

TESTING ALGORITHM

An algorithm for approach to a patient with UTI is given in Flowchart 41.3.

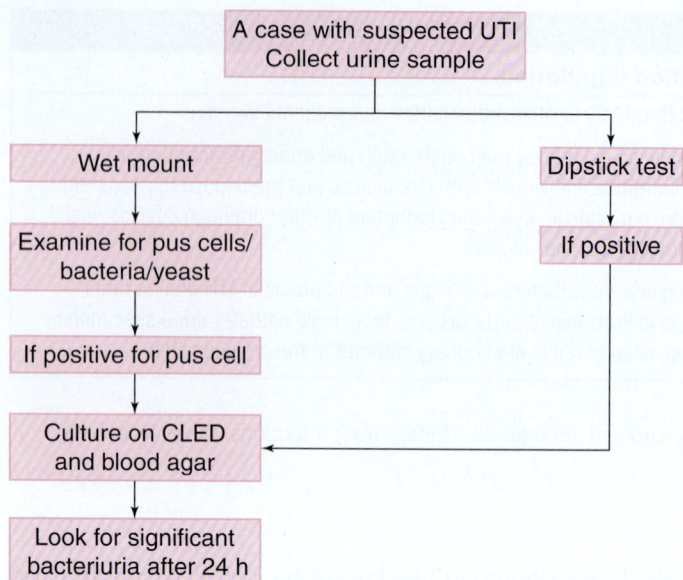

Flowchart 41.3 Approach to a Case of Suspected UTI.

LABORATORY DIAGNOSIS OF MENINGITIS AND ENCEPHALITIS

DEFINITION

Meningitis

It is the inflammation of the meninges. There are high number of white blood cells in CSF with onset of signs of meningeal irritation appearing in due course (Box 41.1). If the symptoms persist for more than 4 weeks, it is referred to as **chronic meningitis**.

Encephalitis

Encephalitis means inflammation of the brain. It is strictly a pathological diagnosis; but surrogate clinical/imaging markers may provide evidence of inflammation (Box 41.1).

Encephalopathy

Encephalopathy describes a clinical syndrome of altered mental status, manifesting as reduced consciousness or altered behaviour.

Meningoencephalitis

It is the inflammation of brain and meninges both.

Aseptic Meningitis

It is the inflammation of meninges with sterile CSF.

Acute Encephalitis Syndrome (World Health Organization's Definition)

Clinically, a case of AES is defined as a person of any age, at any time of year with the acute onset of fever and a change in mental status (including symptoms, such as confusion, disorientation, coma or inability to talk) and/or new onset of seizures (excluding simple febrile seizures).

Brain Abscess

It is a focal intracerebral infection, with collection of pus in a well-vascularised capsule.

AETIOLOGY OF MENINGITIS

The common infections causes of meningitis are bacterial (Table 41.6), viral (Table 41.7), fungal (Table 41.7), parasitic (Table 41.7) and tuberculosis.

TABLE 41.6	Common Bacterial Causes of Meningitis in Different Age Groups			
Neonates and Infants	**Children**	**Adults**	**Elderly**	**Many Other Causes Which Are Conventionally Not Listed**
E. coli	H. influenzae	N. meningitidis	S. pneumoniae	Tuberculosis (common in India)
Streptococcus agalactiae	N. meningitidis	S. pneumoniae	S. aureus	Rickettsia: scrub and murine typhus
S. aureus	S. pneumoniae		Gram-negative bacilli	Treponema pallidum
H. influenza				
S. pneumoniae				
K. spp.				
Listeria monocytogenes				

TABLE 41.7   Common Viral, Fungal and Parasitic Causes of Meningitis

Viral Causes	Fungal Causes	Parasitic Causes
Arboviruses (Japanese encephalitis virus)	*Cryptococcus neoformans*	*Entamoeba histolytica*
Paramyxoviruses (mumps, measles)	*Candida* spp.	*Naegleria fowleri*
Herpes viruses (Herpes simplex virus, varicella zoster virus)	*Aspergillus* spp.	*Acanthamoeba* spp.
Adenoviruses	*Histoplasma capsulatum*	*Toxoplasma gondii*
Enteroviruses (ECHO, coxsackie)	*Coccidioides immitis*	*Balamuthia* spp.

The non-infectious causes of aseptic meningitis are as follows (Box 41.2):
- Malignancy
- NSAID use
- Chemotherapy
- Excessive antibiotic use

BOX 41.2   Common Aetiological Types of Meningitis

Meningococcal meningitis: It is caused by bacteria called *N. meningitidis*.
Pneumococcal meningitis: It is caused by *S. pneumoniae*.
Haemophilus influenza type B (Hib) meningitis: It is caused by Hib.
Neonatal meningitis: This form affects newborn babies and is usually caused by Group B streptococcus bacteria and Gram-negative rods.
Staphylococcal meningitis: It is caused by staphylococci; usually develops as a complication of a diagnostic or surgical procedure and is rare and can be fatal.
Tubercular meningitis: Caused by Mycobacterium tuberculosis, it is Common in India.
Primary meningoencephalitis: It is caused by free living amoeba *N. fowleri* and *Acanthamoeba* spp. (Refer Chapter 32: Protozoa).

AETIOLOGY OF ACUTE ENCEPHALITIS SYNDROME

Most cases of acute encephalitis syndrome (AES) are caused by viruses (Table 41.8). Other causes include *Rickettsia* spp., *L. monocytogenes*, *M. pneumoniae*, *T. gondii* etc.

Most relevant agents of clinically important viral encephalitis in India (commonest listed first) are as follows:
- *Japanese encephalitis virus (JEV)*

TABLE 41.8   Viral Aetiology of AES

A. RNA viruses

1. Arboviruses

Flaviviridae	*Togaviridae*	*Bunyaviridae*
Japanese encephalitis virus (JEV)	Eastern equine encephalitis (EEE)	California virus (California Group), La Crosse Virus (LAC)
St Louis encephalitis virus (SLEV)	Western equine encephalitis (WEE)	Sandfly fever group—Toscana virus
West Nile virus (WNV)	Venezuelan equine encephalitis (VEE)	
Powassan virus (POW)		
Murray Valley encephalitis (MVE)		
Tick-borne encephalitis viruses (TBEV)		

2. *Picornaviridae* (enteroviruses, poliovirus)
3. *Retroviridae* (human immunodeficiency virus-1/2)
4. *Rabdoviridae* (rabies)
5. *Paramyxoviridae* (mumps, measles)

B. DNA viruses

Herpesviridae	*Polyomaviridae*	*Adenoviridae*
Herpes simplex virus (HSV)	JC virus	*Adenovirus*
Varicella zoster virus (VZV)	BK virus	
Epstein—Barr virus (EBV)		
Cytomegalovirus (CMV)		
Human herpesvirus-6 (HHV-6)		

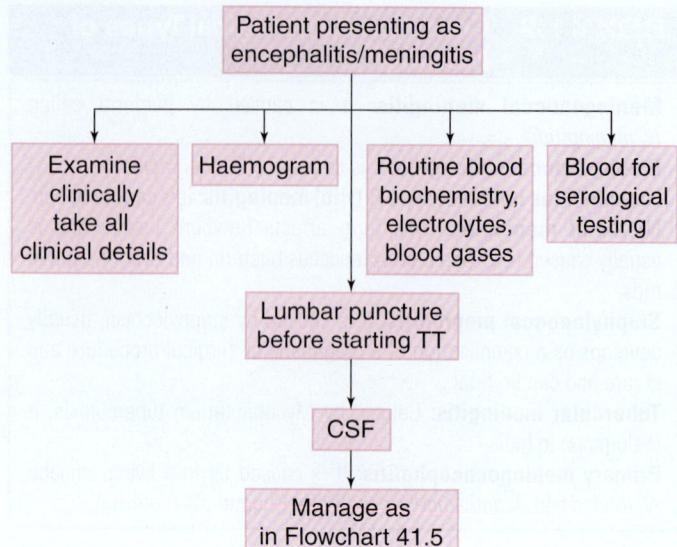

Flowchart 41.4 Diagnostic Algorithm for a Case of Acute Meningoencephalitis.

- *Dengue virus (DV)*
- *Enteroviruses*
- *Herpes simplex virus 1 (HSV-1)*
- *Rabies*
- *West Nile virus (WNV)*
- *Human immunodeficiency virus (HIV)*
- *Varicella zoster virus (VZV)*
- *Epstein–Barr virus (EBV)*
- *Human herpesvirus 6 (HHV-6)*
- *Measles virus* (also causes subacute sclerosing panencephalitis [SSPE])

LABORATORY APPROACH

The Laboratory approach to a case of meningitis/AES is given in Flowcharts 41.4 and 41.5.
- Collect history (will help in shortlisting causes)
- Clinically suspect an aetiology

Note: Specimen transport and processing for bacterial and viral causes is different. For bacterial culture, CSF is stored at 37°C or at room temperature; while for viral testing, it is stored at 4°C and is transported on ice.

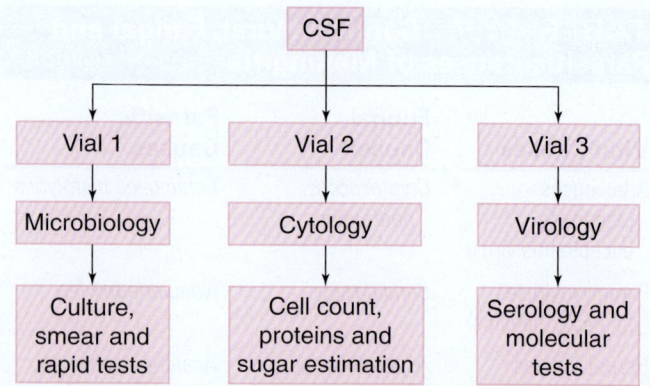

Flowchart 41.5 Management of CSF Collected From a Case of Acute Meningoencephalitis.

Samples required:
- CSF
- Blood and serum
- Others: For example, vesicular scraping in VZV/Herpes

CSF Handling (Flowchart 41.5)

CSF is collected in a screw-capped sterile container. It is difficult to collect and is limiting in volume. Care in handling should be taken. It may not be possible to get a repeat specimen. It should be sent to laboratory without any delay.
- Approximately 2–3 mL of CSF is collected. Part of it is used for:
  - Physical and cytological examination
  - Biochemical examination
  - Microscopic examination
- Remaining CSF is stored aseptically for:
  - serology
  - bacteriological or fungal examination
  - viral testing
- Part of it is used for molecular tests

The cell count (total and differential), glucose and protein levels of CSF often indicate, whether the aetiology of meningitis is viral, bacterial or fungal (Table 41.9).

Based on the findings of CSF cytology and biochemistry, tests for specific aetiology can be asked for as follows:

Tests for Bacterial Meningitis
- Smear examination: Gram's and ZN
- Culture on blood agar and chocolate agar

TABLE 41.9	CSF Findings in Patients With Infectious Causes of Meningitis			
Cause of Meningitis	White Blood Cell Count (Cells/mm³)	Primary Cell Type	Glucose (mg/dL)	Protein (mg/dL)
Viral	50–1000	Mononuclear	>45	<200
Bacterial	1000–5000	Neutrophilic	<40	100–500
Tuberculous	50–300	Mononuclear	<45	50–300
Cryptococcal	20–500	Mononuclear	<40	>45

(*Source:* Adapted from Approach to the patient with central nervous system infections; Mandell, Douglas, and Bennett's principles and practice of infectious diseases. 8th Edition, Page 1185, Chapter 83.)

- Culture and NAAT for mycobacteria
- Rapid test for demonstration of bacterial antigens by agglutination tests
- Nucleic acid amplification test

Test for Fungal Meningitis
- Smear examination: KOH and India ink
- Culture on Sabouraud's agar
- Rapid test for demonstration of cryptococcal antigens by agglutination tests
- Nucleic acid amplification test

Test for Parasitic Meningitis
- Smear examination
- Rapid test for demonstration of malaria antigens by immunochromatographic method
- Nucleic acid amplification test

Tests for Viral Encephalitis
- Over 100 viral agents can cause encephalitis
- Demonstration of specific antiviral antibodies in serum/CSF
- Nucleic acid amplification test
- Viral culture

AES due to JEV is common in certain parts of India. WHO has advocated the algorithm given in Flowchart 41.6 for JE testing.

FOOD-BORNE ILLNESSES

DEFINITION

Food-borne diseases are illnesses acquired through ingestion of contaminated food. Contamination of food with pathogenic microorganisms including viruses, bacteria, fungus, bacterial and fungal toxins and others may occur.

Common manifestations and their causes are shown in Table 41.10.

Table 41.11 shows the common pathogens associated with food-borne illnesses and associated features.

LABORATORY DIAGNOSIS

The laboratory diagnosis (Flowchart 41.7 and Table 41.11) of these infections is based on suspected aetiology. However, for bacterial (Table 41.11) and fungal causes, **stool microscopy** and **culture** is to be performed. In cases, where enteric fever is suspected, the **blood culture** may be necessary. Many of these cases may not require laboratory investigations as diagnosis is based on clinical ground.

Most of these cases do not require antibacterial drugs and are managed with electrolytes and fluid management. Those diagnosed with parasitic or some bacterial infections, such as *Salmonella* spp. need specific therapy.

OUTBREAK SITUATION

Outbreak of food-borne disease: As per CDC, it is defined as an incidence in which two or more persons experience a similar illness resulting from the ingestion of a common food.

Samples: Stool, vomitus, serum, plasma, blood of patient (as per the suspected aetiology) and food samples should be collected. If bacterial aetiology is suspected, the samples should be transported at 4°C in Cary Blair or any other suitable transport medium. If viral aetiology is suspected, the liquid stools should immediately be transported at 4°C.

Approach to aetiologic diagnosis in an outbreak situation is described in Flowchart 41.8.

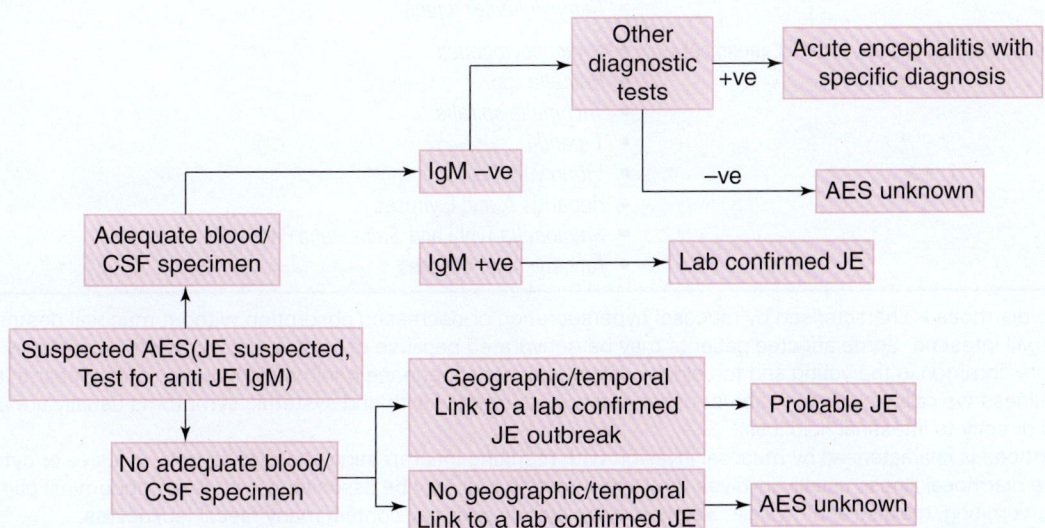

Flowchart 41.6 Algorithm for JE Testing. (Source: *WHO Manual for the Laboratory Diagnosis of Japanese Encephalitis Virus Infection: For Evaluation Purposes*, 30 March 2007 – FINAL DRAFT – For Evaluation Purposes, Figure 2, Page 11.)

TABLE 41.10 Aetiologic Agents Causing Various Manifestations of Food-Borne Illnesses

Clinical Presentation	Potential Food-Related Agents to Consider
Gastroenteritis (vomiting as primary symptom; fever and/or diarrhoea also may be present)	Viral gastroenteritis, most commonly rotavirus in an infant or norovirus and other caliciviruses in an older child or adult; or food poisoning due to preformed toxins (e.g. vomitoxin, *S. aureus* toxin, *Bacillus cereus* toxin) and heavy metals
Non-inflammatory diarrhoea (acute watery diarrhoea without fever/dysentery; some patients may present with fever)[a]	Can be caused by virtually all enteric pathogens (bacterial, viral, parasitic) but is a classic symptom of: • Enterotoxigenic *E. coli* • *Giardia* • *Vibrio cholerae* • Enteric viruses (astroviruses, noroviruses and other caliciviruses, enteric adenovirus, rotavirus) • *Cryptosporidium* • *Cyclospora cayetanensis*
Inflammatory diarrhoea (invasive gastroenteritis; grossly bloody stool and fever may be present)[b]	• *Shigella* spp. • *Campylobacter* spp. • *Salmonella* spp. • Enteroinvasive *E. coli* • Enterohaemorrhagic *E. coli* • *E. coli* O157:H7 • *Vibrio parahaemolyticus* • *Y. enterocolitica* • *Entamoeba histolytica*
Persistent diarrhoea (lasting ≥14 days)	Prolonged illness should prompt examination for parasites, particularly in travellers to mountainous or other areas where untreated water is consumed. Consider *C. cayetanensis*, *Cryptosporidium*, *Entamoeba histolytica*, and *Giardia lamblia*
Neurologic manifestations (e.g. paresthesias, respiratory depression, bronchospasm, cranial nerve palsies)	• Botulism (*Clostridium botulinum* toxin) • Organophosphate pesticides • Thallium poisoning • Scombroid fish poisoning (histamine, saurine) • Ciguatera fish poisoning (ciguatoxin) • Tetraodon fish poisoning (tetrodotoxin) • Neurotoxic shellfish poisoning (brevetoxin) • Paralytic shellfish poisoning (saxitoxin) • Amnesic shellfish poisoning (domoic acid) • Mushroom poisoning • Guillain–Barré syndrome (associated with infectious diarrhoea due to *Campylobacter jejuni*)
Systemic illness (e.g. fever, weakness, arthritis, jaundice)	• *L. monocytogenes* • *Brucella* spp. • *Trichinella spiralis* • *T. gondii* • *Vibrio vulnificus* • Hepatitis A and E viruses • *Salmonella* Typhi and *Salmonella* Paratyphi • Amoebic liver abscess

[a]Non-inflammatory diarrhoea is characterised by mucosal hypersecretion or decreased absorption without mucosal destruction and generally involves the small intestine. Some affected patients may be dehydrated because of severe watery diarrhoea and may appear seriously ill. This is more common in the young and the elderly. Most patients experience minimal dehydration and appear mildly ill with scant physical findings. Illness typically occurs with abrupt onset and brief duration. Fever and systemic symptoms usually are absent (except for symptoms related directly to intestinal fluid loss).

[b]Inflammatory diarrhoea is characterised by mucosal invasion with resulting inflammation and is caused by invasive or cytotoxigenic microbial pathogens. The diarrhoeal illness usually involves the large intestine and may be associated with fever, abdominal pain and tenderness, headache, nausea, vomiting, malaise and myalgia. Stools may be bloody and may contain many faecal leukocytes.

(*Source*: MMWR: Recommendations and Reports, Diagnosis and Management of Foodborne Illnesses, A Primer for Physicians and Other Health Care Professionals. Produced collaboratively by the American Medical Association, Table 1: Etiologic Agents to Consider for Various Manifestations of Foodborne Illness.)

TABLE 41.11 Common Pathogens Associated with Food-Borne Illnesses and Associated Features

Aetiology	Incubation Period	Signs and Symptoms	Associated Foods	Laboratory Testing
Bacillus anthracis	2 Days to weeks	Nausea, vomiting, malaise, bloody, diarrhoea, acute abdominal pain	Insufficiently cooked contaminated meat	Blood testing
Bacillus cereus (preformed enterotoxin)	1–6 h	Sudden onset of severe nausea and vomiting; diarrhoea may be present	Improperly refrigerated cooked or fried rice meats	Usually a clinical diagnosis
Bacillus cereus (diarrhoeal toxin)	10–16 h	Abdominal cramps, watery diarrhoea, nausea	Meats, slews, gravies, vanilla sauce	Testing not necessary, self-limiting
Brucella abortus, Brucella melitensis and *Brucella suis*	7–21 days	Fever, chills, sweating, weakness, headache, muscle and joint pain, diarrhoea, bloody stools during acute phase	Raw milk goat cheese made from unpasteurised milk, contaminated meats	Blood culture and positive serology
Campylobacter jejuni	2–5 days	Diarrhoea, cramps, fever and vomiting; Diarrhoea may be bloody	Raw and undercooked poultry, unpasteurised milk, contaminated water	Routine stool culture; *Campylobacter* required special media and incubation of 42°C to grow
Clostridium botulinum: children and adults (preformed toxin)	12–72 h	Vomiting diarrhoea, blood vision, diplopia, dysphagia and descending muscles weakness	Home-canned foods with a low-acid content, improperly canned commercial food, home-canned or fermented fish, herb-infused oils, baked potatoes in aluminium foil, cheese sauce, bottled garlic, food held warm for extended periods of time (e.g. in a warm oven)	Stool, serum and food can be tested for toxin
Clostridium botulinum: infants	3–30 days	In infants <12 months, lethargy, weakness, poor feeding, constipation, hypotonia, poor head control, poor gag and sucking reflex	Honey, home-canned vegetables and fruits, corn syrup	Stool serum and food can be tested for toxin
Clostridium perfringens toxin	8–16 days	Watery diarrhoea, nausea, abdominal cramps; fever is rare	Meats, poultry, gravy, dried or precooked foods, time- and/or temperature-abused food	Stools can be tested for enterotoxin and cultured for organism
Enterohaemorrhagic *E. coli (CHEC)* including *E. coli* O157:H7 *and* other shiga toxin-producing *E. coli (STEC)*	1–8 days	Severe diarrhoea that is often bloody, abnormal pain and vomiting, usually little or no fever is present; more common in children <4 years	Undercooked beef especially hamburger; unpasteurised milk and juice, raw fruits and vegetables (e.g. sprouts), salami (rarely) and contaminated	Stool culture: *E. coli* O157:H7 requires special media to grow

(*Source*: MMWR: Recommendations and Reports, Diagnosis and Management of Foodborne Illnesses, A Primer for Physicians and Other Health Care Professionals. Produced collaboratively by the American Medical Association, Table: Foodborne Illness.)

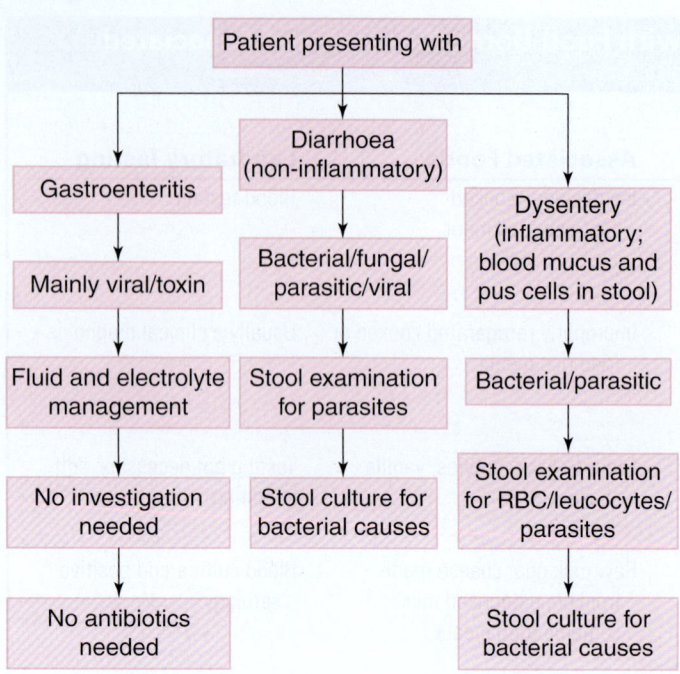

Flowchart 41.7 Approach to a Case of Food-Borne Illness.

SEXUALLY TRANSMITTED AND REPRODUCTIVE TRACT INFECTIONS

DEFINITIONS

Sexually transmitted infections (STI) are infections that are passed from one person to another through sexual contact. Reproductive tract infections (RTIs) are infections of the genital tract. They affect both women and men. RTIs are caused by organisms normally present in the reproductive tract (endogenous) or introduced from the outside during sexual contact (sexually transmitted) or medical procedures (iatrogenic; Table 41.12).

Most commonly, RTIs are seen in one or the other syndromic forms as shown in Table 41.13. Some people with an STI/RTI are asymptomatic or have minimal symptoms. Infections like HIV do not have a clear syndrome and hence is not listed here. For details on HIV, refer to Chapter 26 (Retroviruses). HIV transmission is more likely when STIs/RTIs are present for reasons listed below:

- HIV infection is facilitated by breech in skin and mucosa.
- HIV can easily infect lymphocytes present in genital discharge.
- Genital secretions have large copy numbers of HIV, which can easily spread.

The common syndromes and the most common organisms causing them are as follows:

- Urethral discharge in men: gonorrhoea and chlamydia
- Genital ulcers in men and women: syphilis and chancroid
- Lower abdominal pain (pelvic inflammatory disease [PID]) in women: gonorrhoea and chlamydia
- Vaginal discharge in women: bacterial vaginosis, trichomoniasis or yeast infection

LABORATORY APPROACH

Samples required:

Men: Urethral swab or voided urine, serum
Women: Vaginal swab, endocervical swab, serum

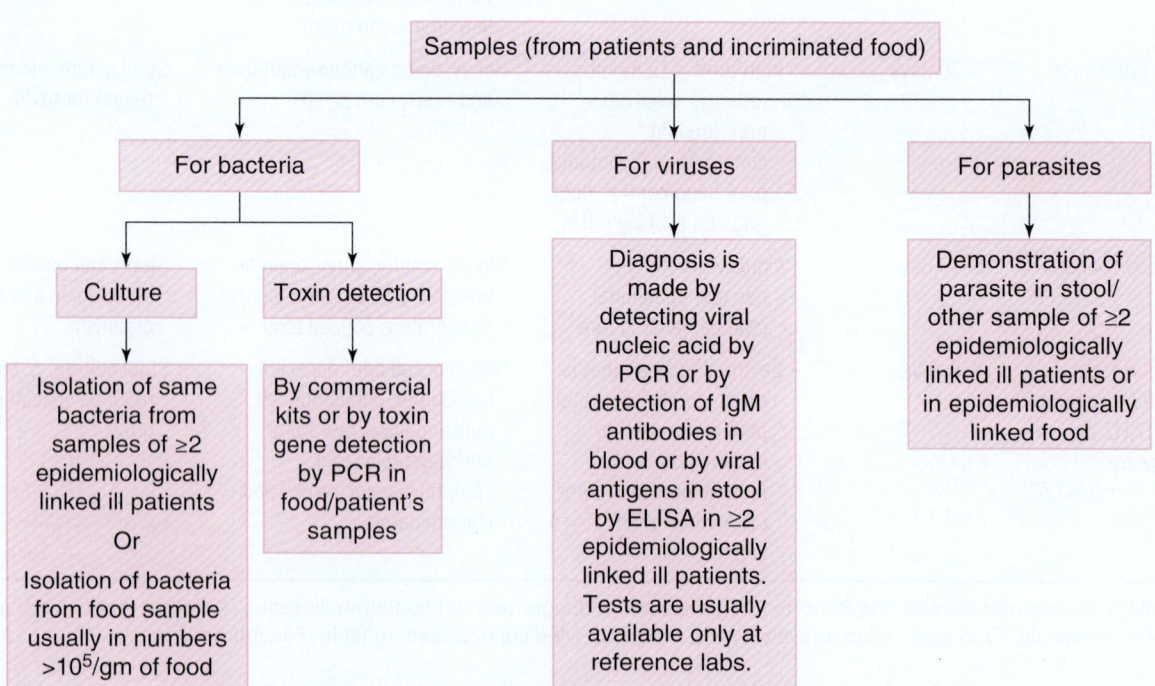

Flowchart 41.8 Approach for Aetiological Confirmation in an Outbreak of Food-Borne Illness.

TABLE 41.12 Common Modes of Transmission of RTI

	Where They Come From	How They Spread	Common Examples
Endogenous infections	Organisms normally found in vagina	Usually not spread from person to person, but overgrowth can lead to symptoms	Yeast infection, bacterial vaginosis
Sexually transmitted infections	Sexual partners with STI	Sexual contact with infected partner	Gonorrhoea, chlamydia, syphilis, chancroid, trichomoniasis, genital herpes, genital warts, HIV
Iatrogenic infections	Inside or outside the body: • Endogenous (vagina) • STI (cervix or vagina) • Contamination from outside	By medical procedures or following examination or intervention during pregnancy, childbirth, the postpartum period or in family planning (e.g. IUD insertion) and gynaecology settings. Infection may be pushed through the cervix into the upper genital tract and cause serious infection of the uterus, fallopian tubes and other pelvic organs; contaminated needles or other instruments (e.g. uterine sounds) may transmit infection if infection control is poor	Pelvic inflammatory disease (PID) following abortion or other transcervical procedure. Also, many infectious complications of pregnancy and postpartum period

(*Source*: Integrating STI/RTI Care for Reproductive Health Sexually transmitted and other reproductive tract infections: A guide to essential practice, Reproductive Health and Research, Copyright World Health Organization 2005, Table 1.1. Types of STI/RTI, P.14.)

TABLE 41.13 Syndrome and Aetiology of RTIs/STIs

Syndrome	STI/RTI	Organism	Type	Sexually Transmitted	Curable
Genital ulcer	• Syphilis	• *T. pallidum*	• Bacterial	• Yes	• Yes
	• Chancroid	• *Haemophilus ducreyi*	• Bacterial	• Yes	• Yes
	• Herpes	• Herpes simplex virus (HSV-2)	• Viral	• Yes	• Difficult
	• Granuloma inguinale (donovanosis)	• *Klebsiella granulomatis*	• Bacterial	• Yes	• Yes
	• Lymphogranuloma venereum	• *Chlamydia trachomatis*	• Bacterial	• Yes	• Yes
Discharge	• Bacterial vaginosis	• Multiple	• Bacterial	• No	• Yes
	• Yeast infection	• *Candida albicans*	• Fungal	• No	• Yes
	• Gonorrhoea	• *Neisseria gonorrhoeae*	• Bacterial	• Yes	• Yes
	• Chlamydia	• *C. trachomatis*	• Bacterial	• Yes	• Yes
	• Trichomoniasis	• *Trichomonas vaginalis*	• Protozoal	• Yes	• Yes
Other	• Genital warts	• Human papillomavirus (HPV)	• Virus	• Yes	• No

(*Source*: Integrating STI/RTI Care for Reproductive Health Sexually transmitted and other reproductive tract infections: A guide to essential practice, Reproductive Health and Research, Copyright World Health Organization 2005, Table 1.2. Types of STI/RTI, P.15.)

For aetiological diagnosis of a case with RTI/STI, the approach is shown in Flowchart 41.9. The lab methods for each disease will vary as per clinical diagnosis.

SKIN AND SOFT TISSUE INFECTIONS

INTRODUCTION

Skin infections are one of the most common diseases being referred to dermatologists. Presentation may vary from very benign folliculitis to life-threatening necrotising fasciitis. There are many diseases of skin involving different layers of skin as shown in Table 41.14. Various microbes which can cause skin infections are listed in Table 41.15 and Box 41.3.

AETIOLOGY OF VARIOUS SKIN INFECTIONS

For the ease of management, the Infectious Diseases Society of America has divided skin and soft tissue infections on the basis of the following:

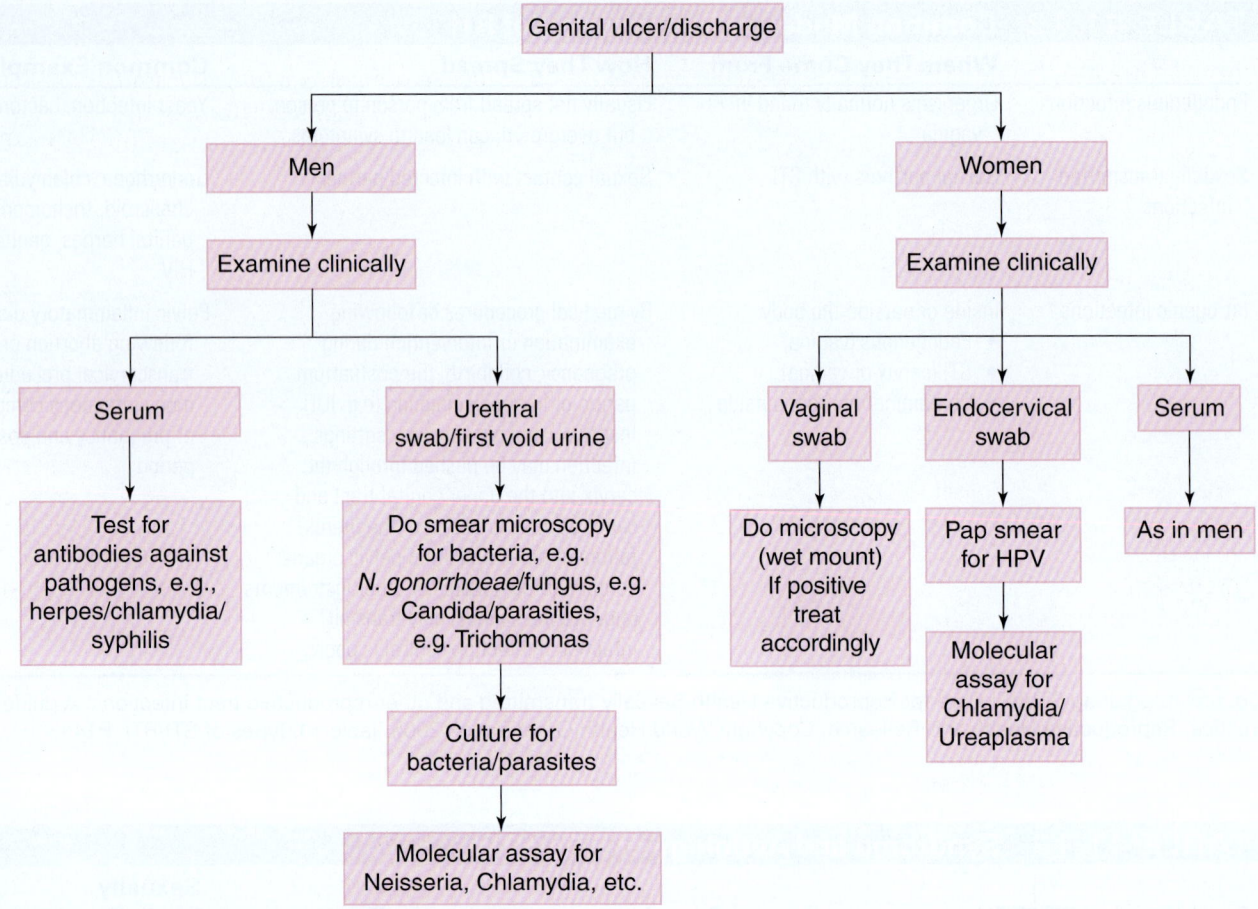

Flowchart 41.9 Approach to a Case of RTI/STI.

TABLE 41.14 Diseases Affecting Different Layers of Skin	
Layers and Appendages of Skin	**Diseases**
Epidermis	Impetigo, staphylococcal scalded skin syndrome
Dermis	Erysipelas, cellulitis
Subcutaneous fat	Cellulitis
Fascia and muscle	Necrotising fasciitis
Hair follicle	Folliculitis

TABLE 41.15 Bacterial Pathogens Causing Skin Diseases	
Organisms	**Diseases**
S. aureus	Impetigo, furuncule, boils, paronychia, impetigo, scalded skin syndrome (toxin mediated), toxic shock syndrome (toxin mediated), Toxic epidermal lysis (toxin mediated)
Streptococcus pyogenes	Cellulitis, erysipelas, impetigo, Scarlet fever (toxin mediated)
C. diphtheriae	Cutaneous diphtheria
Mycobacterium tuberculosis	Lupus vulgaris
Mycobacterium marinum	Chronic ulcers
Mycobacterium ulcerans	Buruli ulcer
P. aeruginosa	Colonisation and infection of burn wounds
Anaerobic bacteria	Deep abscess

1. *Skin extension:* Uncomplicated superficial skin infections (uSSTI) and complicated infections with deep involvement (SSTI)
2. *Rate of progression:* Acute and chronic wound infections
3. *Tissue necrosis:* Necrotising (pyomyositis, necrotising fasciitis, clostridial myonecrosis, Fournier's gangrene) and not-necrotising (impetigo, furunculous and carbuncles, animal and human bites, infected pressure ulcers)

- Fungal pathogens causing skin infections (dermatophytosis):
  - Epidermophyton
  - Microsporum
  - Trichophyton
  - Candida
  - *Malassezia furfur*
- Viral pathogens causing skin infections:
  - Herpes simplex
  - Papillomavirus
  - *Molluscum contagiosum*
  - Varicella zoster

infections. Acute bacterial skin and skin structure infection (ABSSSI) is the third category that has recently been added by FDA and is defined as a bacterial infection of the skin with a lesion involving ≥75 cm² of skin (erysipelas, cellulitis, surgical infections and cutaneous abscess).

LABORATORY DIAGNOSIS

Samples: As per suspicion of treating physician regarding aetiology of infection, laboratory decides for clinical sample to be collected and processed (Flowchart 41.10).

Pus: It can be collected (from wound/abscess/closed lesion) for Gram's stain and culture.

Swab: It can be collected from open wound for Gram's stain and culture. Swabs are not recommended for TB and fungal microscopy and culture.

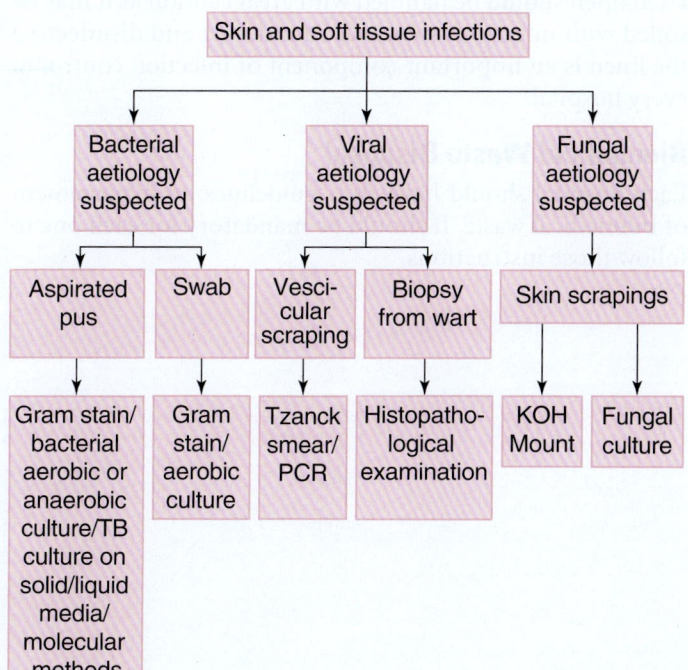

Flowchart 41.10 Plan for Collection and Processing of Samples in Different Skin Infections.

Blood: It can be collected and processed for culture, if bacteraemia is suspected.

Skin scrapings: For fungal examination. **Scrapings from base of ulcer:** For viral examination.

Biopsies from affected sites: For histology of skin warts.

Laboratory approach to a case with skin or soft tissue infections is shown in Flowchart 41.11.

PREVENTION OF HOSPITAL-ACQUIRED AND BLOOD-BORNE INFECTIONS

UNIVERSAL STANDARD SAFETY PRECAUTIONS

Universal standard safety precautions are meant to reduce the risk of transmission of blood-borne and other pathogens from both recognised and unrecognised sources. They help in lowering the incidence of hospital- and blood-acquired parenteral and other infections contracted through mucous membrane, non-intact skin and respiratory route. Vaccines like the one for protection against hepatitis B are useful and must be used as an adjunct.

The components of universal safety are discussed briefly as follows:[a]

Hand Hygiene

It means simply hand washing with soap and water, done for at least 40–60 sec. Hand hygiene is advocated on following occasions even if gloves are worn. Hand hygiene station, gloves, tissues and masks and PPE should be available at entry and exit of every ward.

1. Before and after direct patient contact
2. Between examining two patients
3. After removing gloves
4. Before doing any small procedure or handling patient device
5. After contact with any body fluid/blood/any clinical material/skin/mucous membrane/contaminated items of patient
6. After contact with any objects around patient

Personal Protective Equipment (PPE)

Must be used in clinical settings by all the health care workers to protect them and their patients against blood-borne and hospital-acquired infections. Parts of PPE are as follows:

1. **Gloves:** It is an important part of PPE. They must be worn during patient examination and when touching blood, body fluids, secretions, excretions, mucous membranes or non-intact skin. Gloves should be removed before leaving bedside of patient. Between examining two patients gloves must be changed.
2. **Mask and eye protection:** A surgical mask and an eye visor/goggles or a face shield protects health care worker

[a](*Source:* Adapted from AIDE-MEMOIRE: Standard Precautions in Health Care, Epidemic and Pandemic Alert and Response, © World Health Organization 2007.)

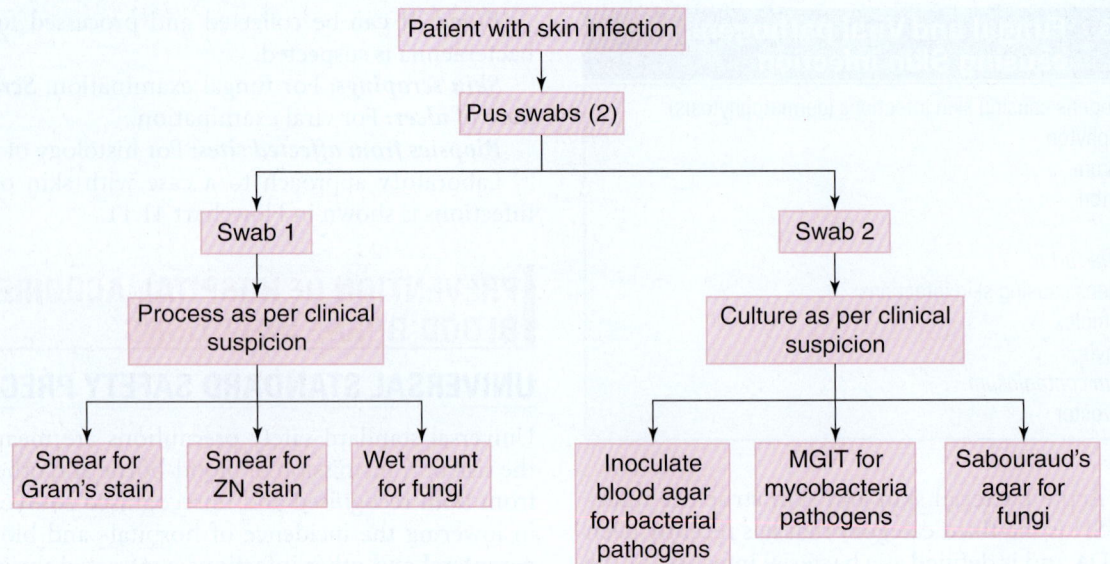

Flowchart 41.11 Laboratory Approach to a Case With Skin or Soft Tissue Infections.

from any splashes of blood or body fluid. It is again an important part of PPE.

3. **Gown:** It protects against soiling of clothing with splashes. Hence any procedure where there is a risk of splashing of blood/body fluids/secretions, should only be done after donning proper PPE, including gown. After removing the gown, hand hygiene should be performed.

4. **Shoe covers:** It protects against soiling of shoes and socks from any clinical material.

Prevention of Needle Stick and Injuries from Other Sharp Instruments

Used needles are a common source of blood-borne infections. Utmost care is required in handling cleaning and disposing used needles, sharp edges of equipment, such as razors and broken glass. Guidelines should be followed always.

Respiratory Hygiene and Cough Etiquette

Any health care worker/visitor/patient with respiratory symptoms should always cover their nose and mouth when coughing/sneezing with a cloth or tissue. If nothing is available, they should cover their mouth with their own elbow. Used towels/tissues should be disposed as per hospital guidelines and hand should be washed after that. Beds of patients with respiratory symptoms should be spaced at least 3–4 ft from each other.

Routine Cleaning of Floor, Walls and Surfaces

with disinfectants should be mandatory for each ward.

Linens

Used linen should be handled with great caution as it may be soiled with infectious material. Laundering and disinfecting the linen is an important component of infection control in every hospital.

Biomedical Waste Disposal

Each hospital should have clear guidelines on management of biomedical waste. It should be mandatory for each one to follow those instructions.

MULTIPLE CHOICE QUESTIONS

Section I: General Microbiology

1. Which of the following method uses a molecular method for testing of organisms?
   a. Biotyping
   b. Culture
   c. Plasmid fingerprinting
   d. Bacteriophage typing

2. In a medical microbiology laboratory, which of the following properties are used to identify an organism?
   a. Morphological appearance
   b. Biochemical reactions
   c. Molecular tests
   d. All of the above

3. The main function of a clinical microbiology laboratory is:
   a. Organism isolation
   b. Organism identification
   c. Antimicrobial testing
   d. All of the above

4. Who discovered polymerase chain reaction?
   a. Saiki
   b. Salk
   c. Watson and Crick
   d. Mekusick

5. The most commonly used method for counting bacteriophages is:
   a. Plaque assays
   b. Tissue culture
   c. Electron microscopy
   d. Immunological assays

6. Which of the following staining technique utilizes India ink?
   a. Simple staining
   b. Impregnation method
   c. Negative staining
   d. Differential staining

7. Chemical composition of which of the following bacterial structure determines reactions of the Gram and acid-fast stains?
   a. Plasma membrane
   b. Outer membrane
   c. Cell walls
   d. Cytoplasm

8. Gram-positive bacteria take the colour of which of the following stains?
   a. Malachite green
   b. Haematoxylin
   c. Gentian violet
   d. Safranin

9. The best setting of a microscope used to observe a wet film is:

   a. Light microscope with condenser in focus
   b. Dark field
   c. Fluorescence
   d. Light microscope with condenser out of focus

10. According to Koch's postulates, which one of the following is FALSE?
    a. The bacterium should be constantly associated with lesions of the disease
    b. It should be possible to isolate the bacterium in pure culture from the lesions
    c. Inoculation of such pure culture into suitable laboratory animals should reproduce lesions of the disease
    d. The bacterium may not produce the similar lesions in experimental animals

11. Which of the following statements is FALSE?
    a. Enrichment media is liquid media
    b. Nutrient agar is basal media
    c. Agar does not add nutrient to media
    d. Blood agar is a selective media

12. Which of the following statement is FALSE regarding culture media?
    a. Lowenstein–Jensen medium is used to isolate mycobacteria
    b. MacConkey agar prevents the growth of Gram-negative bacteria
    c. Sabouraud's dextrose agar is used for fungal culture
    d. Thayer–Martin's medium is used to isolate gonococcus

13. Size of a bacterium of medical importance is usually between:
    a. $2–5\,\mu m \times 0.2–1.5\,\mu m$
    b. $2–10\,\mu m \times 0.2–1.5\,\mu m$
    c. $2–5\,\mu m \times 2–15\,\mu m$
    d. $2–10\,\mu m \times 2–15\,\mu m$

14. Which of the following bacterial duo has defective cell wall?
    a. Mycoplasma and pneumococcus
    a. Mycoplasma and protoplasts
    b. Spheroplasts and *Vibrio cholerae*
    c. Spheroplasts and Chlamydia

15. The range of mutation rates of individual gene in bacteria is:
    a. 10^{-1} to 10^{-10} per bacterium per division
    b. 10^{-1} to 10^{-10} per bacterium per 10 division
    c. 10^{-4} to 10^{-10} per bacterium per division
    d. 10^{-4} to 10^{-10} per bacterium per 10 division

16. The cells having F plasmid in the chromosomes are called:
    a. Hfr
    b. F⁻
    c. Hbr
    d. C⁺

17. When a mutation results in substitution of a single amino acid, it is called:
    a. Frameshift mutation
    b. Point mutation
    c. Nonsense mutation
    d. Missense mutation

18. Detection of fimbriae can be done by which method?
    a. Hanging drop method
    b. Cragie's tube method
    c. Electron microscopy
    d. Differential staining

19. The initial phase of bacterial growth curve during which accumulation of nutrients occurs is:
    a. Decline phase
    b. Log phase
    c. Lag phase
    d. Stationary phase

20. Which of the following statements is FALSE regarding culture media?
    a. MacConkey agar is both differential and selective medium
    b. Wilson–Blair is an indicator medium
    c. Tetrathionate Broth is an enriched medium
    d. Stuart's Medium is a transport medium

21. The principle of Widal test is:
    a. Tube agglutination
    b. Latex agglutination
    c. Slide agglutination
    d. Precipitation

22. Which one of the following is NOT a selective medium?
    a. Thayer–Martin medium
    b. Deoxycholate citrate medium
    c. Sheep blood agar
    d. Wilson–Blair medium

23. Which of the following classes of antibiotics contains a β-lactum ring?
    a. Fluoroquinolones
    b. Penicillin
    c. Macrolides
    d. Aminoglycosides

24. Which of the following antibiotics is a macrolide?
    a. Erythromycin
    b. Imipenem
    c. Colistin
    d. Gentamicin

25. The predominant mechanism of mutation to ciprofloxacin is:
    a. Altered PBPs
    b. Efflux pump mechanism
    c. Mutation in DNA gyrase
    d. Non-penetration of antibiotic

26. Aggregation of bacteria in slime leading to antibiotic resistance occurs due to:
    a. Co-agglutination
    b. Biofilm formation

    c. Mutations leading to altered target site of antibiotics
    d. Mutation generating a target by pass mechanism

27. Which one of the following statements is TRUE in reference to transferable drug resistance:
    a. Usually results in a high degree of resistance
    b. May involve resistance to multiple drugs
    c. Use of antibiotic combinations cannot prevent it
    d. All the above

28. Which of the following is the CORRECT match regarding pathogenic bacteria and their route of entry into the body?
    a. *Corynebacterium diphtheriae*—food borne entry
    b. *Clostridium tetani*—inhalation
    c. *Rickettsia rickettsii*—contamination of wound with soil
    d. *Borrelia burgdorferi*—Arthropod vector-borne entry

29. DNA in a sample can be detected by which of the following techniques?
    a. Western blotting
    b. Northern blotting
    c. Southern blotting
    d. All of these

30. The transfer of DNA from one bacterium to another through the agency of phages is called:
    a. Transformation
    b. Transduction
    c. Mutation
    d. Conjugation

31. Hybridoma technology uses which of the following combination of cells?
    a. Myeloma cells and stem cells
    b. Plasma cells and myeloma cells
    c. Myeloma cells and bacterial cells
    d. Plasma cells and plasmids

32. A mutation that changes the whole sequence of amino acids occurring after it?
    a. Missense mutation
    b. Nonsense mutation
    c. Inversion
    d. Frameshift mutation

33. Most pathogenic organisms need moderate temperature to grow, typically between 20°C and 45°C. Such organisms are called:
    a. Mesophiles
    b. Psychrophiles
    c. Thermophiles
    d. Extremophiles

34. Which of the following action in NOT recommended in case of needle stick injury?
    a. Take off the gloves
    b. Squeeze puncture site gently to promote bleeding
    c. Wash the area with plenty of water and soap
    d. Clean the area with spirit

35. Biological oxygen demand (BOD) indicates:
    a. Extent to which water is polluted with inorganic compounds

b. Extent to which water is polluted with organic compounds
c. Amount of oxygen needed by green plants
d. Amount of oxygen released by green plants

36. In order to prevent contamination, the clinical specimens should be collected as per:
    a. Disinfectant policy
    b. Aseptic technique
    c. Infection control policy
    d. Sterilisation policy

37. Bleaching powder should be diluted to what concentration when cleaning a spill of blood or blood-containing fluids:
    a. 1/10 dilution
    b. 1/1 dilution
    c. 1/100 dilution
    d. 1/0.1 dilution

38. Which of the following activity is most appropriate in case of spills of blood/serum?
    a. Cover the spill with paper towels or cloth and pour on it a freshly prepared solution of 10,000 ppm of sodium hypochlorite
    b. Pour on the spillage site a freshly prepared solution of 10,000 ppm of sodium hypochlorite without covering the spill with paper towels or cloth
    c. Wipe the spillage site first and then pour a freshly prepared solution of 10,000 ppm of sodium hypochlorite
    d. Wash the site with warm water and soap without sodium hypochlorite

39. An endoscope can be best disinfected by:
    a. 2% glutaraldehyde for 10 min
    b. 2% glutaraldehyde for 20 min
    c. 1% carbolic acid for 10 min
    d. 1% carbolic acid for 20 min

40. The potency of a disinfectant depends on:
    a. Disinfectant concentration
    b. Time allowed for action
    c. pH of the medium
    d. All of the above

41. Which of the following is most appropriate sterilising cycle?
    a. 121°C for 15 min
    b. 121°C for 10 min
    c. 134°C for 10 min
    d. 134°C for 15 min

42. Which of the following method is sporicidal?
    a. Use of 1% Sodium hypochlorite
    b. Inspissation
    c. Autoclaving
    d. Boiling

43. Serum containing media are sterilised by:
    a. Filtration
    b. Pasteurisation
    c. Autoclaving
    d. Boiling

44. All are true in reference to autoclaving EXCEPT:
    a. Sterilisation is done by steam under pressure
    b. Method of choice for sterilising antisera
    c. *Bacillus stearothermophilus* is used as biological control
    d. Both bacterial spores and vegetative forms are killed

45. All the following are mechanisms of innate immunity EXCEPT:
    a. Skin and mucosal barrier
    b. Antibacterial substances in blood and tissue
    c. Acute phase proteins
    d. Immunological memory

46. Cell-mediated immunity can be detected by all the following methods EXCEPT:
    a. Schick test
    b. Skin test for delayed hypersensitivity
    c. Lymphocyte transformation test
    d. Migration inhibiting factor test

47. Antigen–antibody interaction is always:
    a. Specific
    b. Cross-reactive
    c. Irreversible
    d. Reversible

48. Differences in which regions of antibody molecule lead to different classes and subclasses of antibodies:
    a. Fc fragment
    b. Fab fragment
    c. Light chain
    d. Heavy chain

49. Which of the following statements is TRUE about innate immunity?
    a. It is the first line of defence against infections
    b. It is affected by prior immunisation
    c. It is strictly specific
    d. There is no difference in innate immunity exhibited by different individuals

50. Which one of following is NOT a function of macrophage?
    a. Antibody synthesis
    b. Antigen presentation
    c. Phagocytosis
    d. IL-1 synthesis

51. Variable part of an antibody molecule is:
    a. Fc fragment
    b. Fab fragment
    c. Light chain
    d. Heavy chain

52. Which of the following statements is TRUE for IgM?
    a. Monomer with 2 antigen binding sites
    b. Dimer with 4 antigen binding sites
    c. Tetramer with 8 antigen binding sites
    d. Pentamer with 10 antigen binding sites

53. All of the following reactions are antigen–antibody reactions EXCEPT:
    a. Agglutination

b. Polymerase chain reaction
c. ELISA
d. Complement fixation

54. Which statement best describes toll-like receptors?
    a. Cell-associated receptors that enhance phagocytosis of pathogen
    b. Acute phase proteins which activate complement system
    c. Receptors which activate a large number of T helper cells
    d. Receptors for polyclonal activation of B cells

55. Complement system:
    a. Is a group of highly specific polysaccharides
    b. Stimulates antibody production
    c. Has ability to lyse various cells
    d. Is found in invertebrates

56. EAC rosette is formed by:
    a. B lymphocytes
    b. T lymphocytes
    c. Monocytes
    d. Granulocytes

57. Which one among the following is an anaphylatoxin?
    a. C3a
    b. C3b
    c. C5b
    d. C1q

58. Deficiency of C3 and its regulatory protein C3b inactivator causes which syndrome?
    a. Hereditary angioneurotic oedema
    b. SLE and other collagen vascular diseases
    c. Recurrent pyogenic infections
    d. Bacteraemia due to toxoplasmosis

59. Which of the following diseases is due to hypersensitivity type III?
    a. Tuberculosis
    b. Glomerulonephritis
    c. Rh incompatibility
    d. Anaphylaxis

60. Major immunoglobulin present in the defence of external body surfaces is:
    a. IgA
    b. IgD
    c. IgE
    d. IgG

61. Elek's test is an example of:
    a. Agglutination test
    b. Neutralisation test
    c. Precipitation test
    d. Complement fixation test

62. The CD3 complex is an integral part of:
    a. B cell receptor
    b. T cell receptor
    c. MHC class I
    d. MHC class II

63. Which hypersensitivity reaction is responsible for rejection of an organ transplant?
    a. Immediate
    b. Delayed
    c. Allergy
    d. Antibody mediated

64. Which of the following hypersensitivity reaction is mediated by cell-mediated immunity?
    a. Type I reaction
    b. Type II reaction
    c. Type III reaction
    d. Type IV reaction

65. Penicillin hypersensitivity is due to:
    a. Type I hypersensitivity
    b. Type II hypersensitivity
    c. Type III hypersensitivity
    d. Type IV hypersensitivity

66. An epitope associates with which part of an antibody molecule?
    a. Constant region of both heavy and light chain
    b. Constant region of the heavy chain
    c. Variable regions of heavy and light chains combined
    d. Constant region of the light chain

67. What happens when B cells get activated by antigens? The B cells:
    a. Undergo clonal selection followed by clonal expansion
    b. May form immunological memory cells
    c. Divide continuously
    d. Both A and B

68. A person with hereditary angioneurotic oedema is most likely deficient in:
    a. T cells
    b. Phagocytes
    c. B cells
    d. C1 inhibitor

69. Ram Devi presented to emergency with generalised oedema, sweating, flushing tachycardia and oedema after a bee sting. This reaction is mediated by:
    a. IgM
    b. IgG
    c. IgE
    d. IgA

70. Decreased levels of serum C4 are indicative of:
    a. Type I hypersensitivity
    b. Immune complex mediated damage
    c. Immune deficiency
    d. Granuloma formation

71. High titres of antinuclear antibodies are indicative of:
    a. Parasitic infections
    b. Fungal infections
    c. Autoimmune diseases
    d. Bacterial diseases

72. In response to first exposure to an antigen, the first immunoglobulin class to be produced is:
    a. IgM
    b. IgG
    c. IgA
    d. IgE

73. The chromogen most commonly used to label antibodies in ELISA is:
    a. Malachite green
    b. Alkaline phosphatase
    c. Congo red
    d. Superdismutase

74. J chain is a molecule:
    a. which joins heavy chains with light chains of antibody molecule
    b. which joins heavy chains of antibody molecule
    c. which forms part of structure of pentameric IgM and dimeric IgA
    d. which joins hinge region of antibody molecule

75. The immune response to an antigen can be non-specifically enhanced by:
    a. Antibody
    b. MHC
    c. Antigen
    d. Adjuvant

76. Which part of the major histocompatibility complex contains the genes for HLA proteins?
    a. Chromosome 6 (p)
    b. Chromosome 10 (p)
    c. Chromosome 6 (q)
    d. Chromosome 10 (q)

77. The functions of immune cells such as differentiation, stimulation or inhibition are governed by the low molecular weight proteins known as:
    a. Complement
    b. Acute phase proteins
    c. Gamma globulins
    d. Cytokines

Section II: Aetiological Agents
Part A: Bacterial Pathogens

78. The presence of characteristic intracellular inclusion bodies indicates infection by:
    a. *Cryptococcus neoformans*
    b. *Chlamydia trachomatis*
    c. *Entamoeba coli*
    d. *Mycoplasma pnemoniae*

79. Choose the INCORRECT statement about E tests:
    a. The testing strip has a rising antibiotic concentration
    b. The test is qualitative
    c. The strip has a scale
    d. Principally similar to disc diffusion test

80. The MIC value is:
    a. The highest concentration that inhibits bacterial growth
    b. The highest concentration that allows bacterial growth
    c. The lowest concentration that inhibits bacterial growth
    d. The lowest concentration that allows bacterial growth

81. Which of the following mechanisms is responsible for development of drug resistance in tuberculosis?
    a. Phage mediated
    b. HFr mediated
    c. Conjugation
    d. Mutation

82. Atypical mycobacteria are classified in:
    a. Lancefield groups
    b. Runyon groups
    c. Kauffmann–White serogroups
    d. Heiberg's groups

83. XDR tuberculosis indicates resistance to which of the following combination of drugs?
    a. Rifampicin and Isoniazid
    b. MDR plus any fluoroquinolone plus any injectable second line anti-tubercular drug
    c. All second line anti-tubercular drugs
    d. Any fluoroquinolone plus all injectable second line anti-tubercular drugs

84. Cord growth seen on mycobacterial culture is characteristic of:
    a. Avirulent strains
    b. Virulent strains
    c. Saprophytic strains
    d. Atypical mycobacteria

85. An adult male presents to the OPD with hypopigmented patches on forearm, partial loss of cutaneous sensations in affected areas, the presence of thickened ulnar nerve and the presence of AFB in nasal smears. The most probable diagnosis is:
    a. Cutaneous tuberculosis
    b. Leprosy
    c. Atypical mycobacteriosis
    d. Actinomycosis

86. Which one of the following is a non-cultivable mycobacterium?
    a. *M. tuberculosis*
    b. *M. bovis*
    c. *M. leprae*
    d. *M. avium intracellulare*

87. Which of the following tests are banned by the World Health Organization for diagnosis of tuberculosis?
    a. Culture
    b. ELISA
    c. PCR
    d. LPA

88. Which of the following is a first line anti-tubercular drug?
    a. Amikacin
    b. Levofloxacin

c. Ethambutol
d. PAS

89. Mutations in katG gene of *Mycobacterium tuberculosis* are responsible for resistance to:
a. Isoniazid
b. Rifampicin
c. Pyrazinamide
d. Streptomycin

90. Following type of pulmonary TB is most likely to be associated with AFB sputum positivity:
a. Primary complex
b. Fibronodular
c. Cavitary
d. Pleural effusion

91. A 30-year-old male presents to the OPD with lumpy jaw, suppurative inflammation and formation of multiple abscesses and sinus tracts with discharge of pus containing sulphur granules. Most probable clinical diagnosis is:
a. Cutaneous tuberculosis
b. Leprosy
c. Atypical mycobacteriosis
d. Actinomycosis

92. In the above (Question no. 91) case following tests can be ordered to confirm the diagnosis EXCEPT:
a. Smear examination for Gram-positive bacteria
b. Microscopic examination of crush smear of granule
c. Smear examination for acid fast bacilli
d. KOH mount

93. In the above (Question no. 91) case, the best therapy would be:
a. Anti-tubercular drugs
b. Aminoglycosides
c. Penicillin
d. Co-trimoxazole

94. An outbreak of surgical site infection caused by *Staphylococcus aureus* has occurred in an ICU. To investigate it, which of the following sample would most likely yield the organism?
a. Rectal swab
b. Ear swab
c. Nasal swab
d. Throat swab

95. TRUE methicillin resistance among *Staphylococcus aureus* isolates is due to:
a. A mutation of penicillin-binding proteins (PBPs) that results in poor recognition of methicillin by transpeptidases
b. Loss of permeability of *S. aureus* to methicillin
c. An active efflux system that pumps methicillin out of the bacterium
d. A beta-lactamase enzyme that inactivates methicillin

96. The coagulase test is used to differentiate:
a. *Staphylococcus epidermidis* from *Neisseria meningitides*
b. *S. aureus* from *S. epidermidis*

c. *Streptococcus pyogenes* from *S. aureus*
d. *S. pyogenes* from *Enterococcus faecalis*

97. Which of the following antibiotic is NOT appropriate to treat MRSA?
a. Vancomycin
b. Imipenem
c. Teicoplanin
d. Linezolid

98. Susceptibility to scarlet fever is indicated by:
a. Dick test
b. Schick test
c. ASO test
d. β-Haemolysis

99. A γ-haemolytic Gram-positive cocci, resistant to vancomycin, growing in 6.5% NaCl, is non-bile sensitive. It is likely to be:
a. *Streptococcus agalactiae*
b. *S. pneumoniae*
c. *E. faecalis*
d. *S. bovis*

100. Toxic Streptococcal Shock syndrome occurs as the result of accumulation of dangerous levels of _____ in the blood.
a. Epidermolytic toxin
b. Streptococcal pyrogenic exotoxin (SPE)
c. Anti-M protein antibodies
d. Group 1 M protein molecules

101. Each of the following statements about the classification of streptococci is correct EXCEPT:
a. *Streptococcus pneumoniae* are alpha-haemolytic and can be serotyped on the basis of their polysaccharide capsule
b. Enterococci are group D streptococci and can be classified by their ability to grow in 6.5% sodium chloride
c. Viridans streptococci are identified by Lancefield grouping, which is based on the C carbohydrate in the cell wall
d. Although *Streptococcus pneumoniae* and the viridans streptococci are alpha-haemolytic, they can be differentiated by the bile solubility test and their susceptibility to optochin

102. A 5-year-old boy with a severe earache is examined and found to have a temperature of about 39°C (102°F) and a red bulging tympanic membrane. Exudate from the ear canal contains Gram-positive diplococci. Cultures of the exudate yield alpha-haemolytic colonies that are optochin-sensitive. The most likely cause of the ear infection is:
a. *Streptococcus agalactiae*
b. *S. pyogenes*
c. *S. aureus*
d. *S. pneumoniae*

103. Medusa head appearance of colonies is a characteristic feature of:
a. *Clostridium perfringens*
b. *Mycobacterium hominis*

c. *Ureaplasma urealyticum*

d. *Bacillus anthracis*

104. Characteristic malignant pustule is a feature of infection by:
    a. Actinomycetes
    b. Anthrax
    c. Carcinoma
    d. Diphtheria

105. Wool sorter's disease is:
    a. Cutaneous anthrax
    b. Pulmonary anthrax
    c. Gastrointestinal anthrax
    d. Bulbar poliomyelitis

106. Protection against which of the following three diseases is given in form of triple toxoid vaccine?
    a. Measles, rubella and mumps
    b. Tetanus, pertussis and rabies
    c. Pertussis, tetanus and diphtheria
    d. Pertussis, diphtheria and tuberculosis

107. All the following conditions result in high mortality rate of patients suffering from tetanus EXCEPT:
    a. Temperature below 36.7°C and above 38.9°C
    b. Site of infection near to CNS
    c. Long incubation time
    d. Absence of acquired immunity

108. Which type of *Clostridium botulinum* does not cause food poisoning?
    a. Type A and B
    b. Type E
    c. Type F
    d. Type D

109. Which of the following is one of the commonly used transport medium for *Vibrio cholerae*?
    a. Amie's charcoal medium
    b. Selenite F broth
    c. Venkatraman–Ramakrishnan medium
    d. Tetrathionate broth

110. Which of the following media is specifically used for *Bordetella pertussis*?
    a. Chocolate agar
    b. Tinsdale agar
    c. MacConkey agar
    d. Bordet-Gengou medium

111. Babes–Ernst granules are present in:
    a. *E. coli*
    b. *M. tuberculosis*
    c. *M. Leprae*
    d. *C. diphtheriae*

112. Fatty acid composition of anaerobic bacteria can be determined by which of the following tests?
    a. Phage typing
    b. Biotyping
    c. Gas-liquid chromatography
    d. Immunoassay

113. All the species of Clostridium cause gas gangrene in man EXCEPT:
    a. *C. perfringens*
    b. *C. novyi*
    c. *C. septicum*
    d. *C. botulinum*

114. All of the following are true of clostridial myositis EXCEPT:
    a. Clostridial myositis is also called gas gangrene
    b. Jaundice is one of the visible symptoms
    c. Infection may lead to anaemia
    d. Incubation period is 3 weeks

115. Which of the following bacterial species protect adult vagina from infections?
    a. *Peptococcus* spp.
    b. *Veillonella* spp.
    c. *Bacteroides* spp.
    d. *Lactobacillus* spp.

116. Which one of the following strains of *Escherichia coli* produces a toxin that is identical to the Shiga toxin produced by *Shigella dysenteriae*?
    a. Enterohaemorrhagic (EHEC)
    b. Enterotoxigenic (ETEC)
    c. Enteroaggregative (EAEC)
    d. Enteroinvasive (EIEC)

117. The best diagnostic method for shigellosis is:
    a. Stool culture
    b. ELISA
    c. Microscopic examination of stool
    d. Agglutination test

118. In patients with kidney infections caused by *Proteus mirabilis*, struvite kidney stones may develop as a result of the actions of which one of the following *Proteus* products?
    a. Capsule
    b. Cell wall
    c. Pilli
    d. Urease

119. A 25-year-old woman developed a urinary tract infection. She presented to the OPD with fever, painful urination and flank pain. Urine culture yielded lactose-fermenting, indole-positive, Gram-negative rods. Which of the following bacterial structure imparts infectivity to the organism?
    a. Endotoxins
    b. F-factor
    c. Biochemical properties
    d. P fimbriae

120. All of the following are halophilic vibrios EXCEPT:
    a. *V. alginolyticus*
    b. *V. cholerae*
    c. *V. vulnificus*
    d. *V. parahaemolyticus*

121. Which of the following is a Gram-negative non-motile bacillus?
    a. *Enterobacter cloacae*
    b. *Klebsiella pneumoniae*
    c. *Citrobacter freundii*
    d. *Escherichia coli*

122. Which of the following tests is used to discriminate pseudomonas species and *E. coli*?
    a. Motility
    b. Gram stain
    c. Oxidation fermentation test
    d. All of the above

123. Which of the following is a motile bacterium?
    a. *Salmonella typhi*
    b. *Klebsiella pneumoniae*
    c. *Bacillus anthracis*
    d. *Shigella sonnei*

124. Which one of the following factors present in chocolate agar is essential for the growth of *Haemophilus influenzae*?
    a. Lygase
    b. Nicotinamide adenine dinucleotide (NAD)
    c. Deoxynuclease
    d. Haematin

125. Of the organisms listed below, which one is the commonest cause of post-splenectomy infections:
    a. *H. influenzae*
    b. *S. pyogenes*
    c. *E. faecalis*
    d. *K. pneumoniae*

126. Best method to confirm the diagnosis of pertussis is by:
    a. Culturing organism from blood
    b. Culturing organism from exudate obtained from the posterior pharyngeal wall
    c. Detecting specific antibodies by ELISA
    d. Culturing organism from secretions that bubbles from the patient's nose and mouth during cough paroxysms

127. Phage typing is NOT useful as an epidemiological tool for typing of?
    a. *Salmonella typhi*
    b. *Staphylococcus aureus*
    c. *Vibrio cholerae*
    d. *Shigella dysenteriae*

128. Traveller's diarrhoea is caused by:
    a. EPEC
    b. EHEC
    c. ETEC
    d. EIEC

129. Species of *Shigella* which is the most tolerant to environmental stresses is:
    a. *S. dysenteriae*
    b. *S. flexneri*
    c. *S. boydii*
    d. *S. sonnei*

130. Which one of the following conditions of antibiotic treatment during acute infection will predispose patients to become carriers of the infective agents?
    a. Enteritis due to *Salmonella* serotype typhimurium
    b. Typhoid due to *Salmonella* serotype typhi
    c. Dysentery due to *Shigella dysenteriae*
    d. Enteritis due to *Campylobacter jejuni*

131. Kauffmann–White scheme divides Salmonella into different serovars on the basis of:
    a. O-antigens
    b. Vi-antigens
    c. H-antigens
    d. O, Vi and H antigens

132. Typhoid is diagnosed in the first week of illness by:
    a. WIDAL test
    b. Blood culture
    c. Stool culture
    d. Urine culture

133. Causative organism of Vincent's angina is:
    a. Chlamydia *pneumoniae*
    b. CMV
    c. HACEK
    d. *Borrelia vincentii*

134. Significant bacteriuria means following bacterial counts on a culture plate after 24 h of incubation:
    a. $\geq 10^4$/mL
    b. $\geq 10^5$/mL
    c. $\geq 10^6$/mL
    d. $\geq 10^7$/mL

135. Which of the following is the causative agent of donovanosis?
    a. *Klebsiella pneumoniae*
    b. *Leishmania donovani*
    c. *Calymmatobacterium granulomatis*
    d. *Streptococcus mutans*

136. When peptic ulcers occur in individuals who rarely or infrequently use non-steroidal anti-inflammatory drugs, the ulcers are usually due to a gastric infection with:
    a. *Campylobacter jejuni*
    b. *Helicobacter pylori*
    c. *Escherichia coli*
    d. *Hafnia alvei*

137. An elderly patient admitted in a centrally air conditioned hospital developed fever, chest pain and dry cough. An organism grew when sputum was cultured on charcoal yeast medium. The most likely organism is:
    a. *Haemophilus influenzae*
    b. *Pseudomonas aeruginosa*
    c. *Legionella pneumophila*
    d. *Streptococcus pneumoniae*

138. *Neisseria gonorrhoeae* is unlikely a cause of:
    a. Urethritis
    b. Conjunctivitis
    c. Arthritis
    d. Myocarditis

139. Blood cultures are most likely to yield positive results during the first weeks of acute disease in patients with which of the following infection?
    a. Dysentery due to *S. dysenteriae*
    b. Enteritis due to *Salmonella* serotype typhimurium
    c. Enteritis due to *Campylobacter jejuni*
    d. Enteric fever due to *Salmonella* serotype typhi

140. Following are specific tests for diagnosis of syphilis EXCEPT:
    a. Treponema pallidum haemagglutination test
    b. Fluorescent treponemal antibody absorption test
    c. Complement fixation test
    d. Rapid plasma reagin test

141. Secondary syphilis does not show:
    a. Vesicular rash
    b. Maculopapular rash
    c. Condyloma latum
    d. Condyloma acuminatum

142. Which of the following statements is FALSE about *Listeria*?
    a. Consumption of contaminated milk may transmit the disease
    b. Humans are accidental hosts
    c. May abort pregnancy
    d. Cause neonatal meningitis

143. Which of the following diseases is transmitted by an arthropod?
    a. Yaws
    b. Pinta
    c. Leptospirosis
    d. Relapsing fever

144. Which of the following is the causative agent of Rickettsial pox?
    a. *Rickettsia prowazekii*
    b. *R. typhi*
    c. *R. akari*
    d. *R. conori*

145. Vector for epidemic typhus in human beings is:
    a. Louse
    b. Tick
    c. Gamasid mite
    d. Trombiculid mite

146. The incubation period of Rocky mountain spotted fever is:
    a. 1–2 weeks
    b. 3–7 days
    c. 7–14 days
    d. 14–26 days

147. Leptospirosis is an occupational disease. Which one of the following workers are most likely to be exposed?
    a. Butchers
    b. Sewer workers
    c. Woolsorters
    d. Food handlers

148. Which of the following serotypes of *Chlamydia trachomatis* cause lymphogranuloma venereum?
    a. A, B, Ba, C
    b. D–K
    c. L1–L3
    d. Only A

149. In Loffler's syndrome:
    a. Acute pneumonia is accompanied with eosinophilia
    b. Gastrointestinal upset and passage of roundworms in faeces occurs
    c. Pruritis is accompanied with eosinophilia
    d. Asymptomatic worm infection occurs

150. All of the following are true statements about *Chlamydia trachomatis* EXCEPT:
    a. It infects several types of epithelial cells
    b. It has a number of serotypes that correlate with the syndrome produced on infection
    c. It can be detected by direct immunofluorescence on clinical material
    d. It has a reservoir in domestic fowl

151. The infection least likely to be caused by anaerobic bacteria is:
    a. Brain abscess
    b. Gas gangrene
    c. Urinary tract infection
    d. Breast abscess

Part B: Viral Pathogens

152. Viruses cannot be identified by which of the following techniques?
    a. Tissue culture
    b. Acid fast stain
    c. Molecular assays
    d. Microarray

153. Components of quality assurance are:
    a. Internal quality control + staff training + validation
    b. Internal quality control + staff training + external quality assessment
    c. Internal quality control + external quality assessment + regular validation of all the processes and instruments
    d. Internal quality control + external quality assessment + regular validation of all the processes and instruments and staff to ensure improvement

154. Which of the following serum sample should NOT be rejected for analysis?
    a. Haemolysed
    b. Lipaemic sample
    c. Improperly labelled sample
    d. Volume just enough for a single round of testing

155. Which is the best temperature for short duration of storage of serum samples (1 week)?
    a. 4°C
    b. –20°C
    c. –40°C
    d. –70°C

156. Which of the following is a NOT a killed vaccine?
    a. Plague
    b. Measles
    c. Typhoid
    d. Rabies

157. Glycoprotein spikes extending from the envelope of some viruses are also known as:
    a. Capsids
    b. Matrix proteins
    c. Peplomers
    d. Pentons

158. The only DNA viruses that are replicated within the host cell cytoplasm are:
    a. Poxviruses
    b. Adenoviruses
    c. Herpesviruses
    d. Papovaviruses

159. A non-immune child exposed to a patient who has herpes zoster is at risk of developing:
    a. Measles
    b. Chicken pox
    c. Rubella
    d. Roseola infantum

160. Which of the following disease is NOT vaccine preventable?
    a. Mumps
    b. Hepatitis B
    c. Hepatitis C
    d. Hepatitis A

161. The following viruses replicate in the nucleus:
    a. Herpesviruses
    b. Parvovirus
    c. Picornaviruses
    d. Paramyxoviruses

162. All of the following viruses cause zoonotic infections EXCEPT:
    a. Rabies virus
    b. Hanta virus
    c. Lassa virus
    d. Rubella virus

163. Which of the following viruses is NOT known to be transmitted by blood?
    a. Human immunodeficiency virus
    b. Hepatitis B virus
    c. Hepatitis C virus
    d. Measles virus

164. Genetic reassortment is seen in:
    a. Parvovirus
    b. Hepatitis B virus
    c. Influenza A virus
    d. Human immunodeficiency virus

165. A type of cell culture/line used for culturing viruses that can divide indefinitely is:
    a. Primary cell culture
    b. Continuous cell line
    c. Diploid fibroblast cell
    d. Tissue culture

166. What would be the phenotype of cells infected with HIV-1 during primary stage of the infection?
    a. CD4/CXCR4
    b. CD8/CXCR4
    c. CD4/CCR5
    d. CD8/CCR5

167. Which of the following methods is NOT suitable for diagnosing HIV infection in a baby, born to an infected mother?
    a. DNA PCR
    b. Viral culture
    c. HIV antibody detection by ELISA
    d. p24 assay

168. HIV belongs to which of the following families of viruses?
    a. Reovirus
    b. Lentivirus
    c. Togavirus
    d. Adenovirus

169. The two main targets most commonly used in anti-HIV therapy:
    a. Reverse transcriptase and protease
    b. Reverse transcriptase and CCR5
    c. Protease and CCR5
    d. The viral glycoproteins gp120 and gp41

170. Effectiveness of antiretroviral therapy is measured by:
    a. A decrease in the plasma viral load and a rise in the CD4 count
    b. An increase in the RBC count and haemoglobin level
    c. An increase in serum HIV antibodies level
    d. A decrease in opportunistic infections

171. The ability of the human immunodeficiency virus (HIV) protease to cleave the Gag and Pol polyproteins is inhibited by:
    a. Acyclovir
    b. Indinavir
    c. Foscarnet
    d. Nevirapine

172. Which of the following is FALSE regarding HIV transmission?
    a. Transmitted through semen
    b. Higher chances of transmission during Caesarean section than normal labour
    c. Less infectious than hepatitis B
    d. Female to male transmission is more efficient than male to female

173. Window period in HIV is defined as the time period between:
    a. Viral entry and appearance of symptoms
    b. Viral entry and appearance of detectable antibodies
    c. Appearance of symptoms and treatment
    d. Infection and death

174. In patients with acquired immunodeficiency syndrome (AIDS), the most prevalent opportunistic infection is interstitial pneumonitis caused by:
    a. *Coxiella burnetii*
    b. *Toxoplasma gondii*
    c. *Cryptosporidium parvum*
    d. *Pneumocystis carinii*

175. Which of the following statements regarding mycobacterium and HIV co-infection is TRUE?
    a. Montoux reading is hampered
    b. CSF findings are altered
    c. Management of TB is altered
    d. Atypical mycobacteria usually cause TB in India

176. Commonest cause of acute severe infantile diarrhoea is:
    a. Enterovirus
    b. Rotavirus
    c. Human deficiency virus
    d. Enteric adenoviruses

177. DNA sequencing is applied to study the exact sequence of which component of a DNA molecule:
    a. Deoxyribose sugar
    b. Nucleotide base
    c. Phosphate group
    d. Hydrogen bonds

178. Which of the following statement is FALSE about viruses?
    a. They are acellular
    b. They have no metabolic machinery of their own
    c. They cannot reproduce themselves
    d. They do not require a living cell to survive

179. A group of signalling glycoproteins released by living cells in response to viral attack through stimulation of toll-like receptors are called:
    a. CR1
    b. Kinins
    c. Interferons
    d. Calcitonin

180. Epstein–Barr virus (EBV) is associated with all of the following conditions EXCEPT:
    a. Nasopharyngeal carcinoma
    b. Infectious mononucleosis
    c. Burkitt's lymphoma
    d. Cervical carcinoma

181. Latency for several years (especially in neural cells) is known to occur with which of the following viruses?
    a. Picornaviruses
    b. Herpesviruses
    c. Rhabdoviruses
    d. Flaviviruses

182. A 5-year-old girl does not seem to be ill but has developed a bright red rash that is especially prominent on her face. The area around her mouth is pale, but her cheeks are bright red, giving the appearance that they have been slapped. The rash has begun to spread to the trunk, where it has a fine, lash-like appearance. The mother recalls that the child has a bout of fever,

malaise and itching 3 weeks earlier. The infected child has a 7-year-old brother who suffers from sickle cell anaemia. If he becomes infected with the same virus as his sibling, he is at significant risk of developing:
    a. Hemolyticanaemia
    b. Acute hepatitis
    c. Progressive multifunctional leukoencephalopathy
    d. Aplastic crisis

183. Which of the following is NOT a member of *Picornaviridae*?
    a. Enterovirus
    b. Coxsackievirus
    c. Rhinoviruses
    d. Mumps virus

184. Which of the following statement is TRUE in reference to hepatitis E virus infection?
    a. Infection is more severe in pregnant women
    b. Is a major cause of blood-borne hepatitis
    c. Highly associated with risk of hepatocellular carcinoma
    d. Usually capable of establishing chronic infections

185. A reverse transcriptase is used in the replication of retroviruses and:
    a. Hepadnaviruses
    b. Papovaviruses
    c. Adenoviruses
    d. Poxviruses

186. Which of the following statement is TRUE about hepatitis C virus?
    a. Is not associated with hepatocellular carcinoma
    b. Does not respond to interferon therapy
    c. Has several genotypes
    d. Does not cause chronic infection

187. The protective level of hepatitis B antibody after vaccination is considered to be:
    a. 0.1 IU/mL
    b. 1 IU/mL
    c. 0.05 IU/mL
    d. 10 IU/mL

188. Which of the following statement is FALSE regarding hepatitis D virus?
    a. Is a mutant of HBV
    b. Depends on HBV for virion formation
    c. Is an RNA virus
    d. Is not related to HCV

189. Which of the following statements is TRUE regarding dengue virus?
    a. Anopheles mosquitoes transmit the virus
    b. Secondary infection with a different serotype of the virus usually results in dengue haemorrhagic fever
    c. The reservoir is ardeid birds
    d. Dengue antibodies do not cross-react with Japanese encephalitis virus antibodies

190. A patient presented the OPD with jaundice. The results of his hepatitis panel were: HBsAg: Positive, anti-HBs: Negative, anti-HBc IgM: Positive, HBeAg: Positive, HBV DNA: Positive. How will you interpret the results?
    a. Chronic HBV infection with high infectivity
    b. Chronic HBV infection with low infectivity
    c. Acute HBV infection with high infectivity
    d. Acute HBV infection with low infectivity

191. A 53-year-old man presents with a series of vesicles extending around the left side of his face. He said that he experienced pain over the region for about 2 days before he developed 'some kind of rash' that quickly turned into 'little blisters'. The vesiculated area hurts intensely. Scrapings of the vesicles reveal that the lesions contain multinucleated giant cells. The standard procedure to confirm this diagnostic rapidly is to use:
    a. A direct florescent antibody test to show the presence of viruses in the lesions
    b. A cell culture to replicate viruses from the lesion
    c. A polymerase chain reaction assay to identify virus DNA in the lesion
    d. A radioimmunoassay to detect viral antigens in the blood

192. According to WHO, which of the following is a bad clinical practice in management of dengue cases?
    a. Assessment and follow-up of patients with non-severe dengue and carefully observing for warning signs
    b. Administration of aspirin
    c. Monitoring haemodynamic status before and after each fluid bolus
    d. Giving intravenous fluid volume just sufficient to maintain effective circulation

193. A male baby is born at 38 weeks gestation with a petechial rash, low birth weight, hepatosplenomegaly and bilateral cataracts. This is thought to be due to an infection acquired while the baby was still in utero. Which of the following condition is most likely to cause this clinical presentation?
    a. Cytomegalovirus
    b. Herpes simplex virus
    c. Rubella virus
    d. *Toxoplasma gondii*

194. Which one of the following families contain RNA as the nucleic acid?
    a. Herpesviridae
    b. Orthomyxoviridae
    c. Hepadnaviridae
    d. Parvoviridae

195. A non-immune individual who has been exposed to influenza virus but is not yet suffering from signs and symptoms of disease may be protected from developing influenza by immediately receiving:
    a. Penicillin G
    b. Oseltamivir
    c. Influenza vaccine
    d. Ribavirin

196. All of the following viruses are correctly paired with the type of RNA contained within their nucleocapsid EXCEPT:
    a. Poliovirus and multiple non-overlapping, double-standard RNA fragments
    b. Human immunodeficiency virus (HIV) and diploid single-stranded RNA
    c. Ebola virus and a single negative-sense, single stranded RNA fragment
    d. Norwalk virus and a single positive-sense, single stranded RNA fragment

197. Which one of the following viruses does NOT spread by direct contact with animals, animal tissues or arthropod vectors?
    a. Ebola virus
    b. Marburg virus
    c. Lymphocytic choriomeningitis virus
    d. Chickenpox virus

198. Which one of the following statement accurately describes prions?
    a. Proteinaceous infectious particles
    b. Have nucleic acid
    c. Heat labile
    d. Large particles

Part C: Parasitic Pathogens

199. Which one of the following test is BEST for diagnosing intestinal worm infestation?
    a. Modified ZN stain of stool smear
    b. ELISA for antigen detection
    c. ELISA for antibody detection
    d. Saline mount of stool sample

200. Presence of ingested RBCs is characteristic of:
    a. *Entamoeba coli*
    b. *Iodamoeba butcheli*
    c. *Entamoeba histolytica*
    d. *Dientamoeba fragilis*

201. The parasite protozoa that appears as tennis racket with a concave ventral surface and sucker is:
    a. *Chilomastix mesnili*
    b. *Giardia lamblia*
    c. *Trichomonas vaginalis*
    d. *Isospora hominis*

202. Which of the following statements concerning *Trichomonas vaginalis* is TRUE?
    a. Transmission occurs parenterally
    b. Can be diagnosed by visualising the trophozoite
    c. Can be treated with imipenem
    d. Causes curdy white vaginal discharge

203. Which one of the following is the causative agent of visceral leishmaniasis?
    a. *Leishmania tropica*
    b. *L. donovani*
    c. *L. brasiliensis*
    d. *L. hemidactyli*

204. Congenital toxoplasmosis manifest as:
    a. Microcephaly
    b. Micro-opthalmus
    c. Hepatosplenomegaly
    d. All of the above

205. *Entamoeba histolytica* can be cultured in:
    a. MacConkey agar
    b. CLED medium
    c. Locke-egg medium
    d. NNN medium

206. All are pathogenic free-living amoeba EXCEPT:
    a. *Entamoeba histolytica*
    b. *Acanthamoeba species*
    c. *Nagleria fowleri*
    d. *Balamuthia species*

207. The toxicity of *Plasmodium falciparum* is due to the presence of:
    a. Cytoadherence
    b. Its ability to affect young RBCs
    c. Antibody and immune complex mediated damage
    d. All of the above

208. Gametocytes of which plasmodium species are sickle-shaped?
    a. *Plasmodium vivax*
    b. *P. falciparum*
    c. *P. malariae*
    d. *P. ovale*

209. Which plasmodial species does not show all the stages (Trophozoites, Schizonts and gametocytes) in the peripheral blood smear?
    a. *Plasmodium vivax*
    b. *P. falciparum*
    c. *P. malariae*
    d. *P. ovale*

210. In malaria, the infectious stage of plasmodium for humans is:
    a. Trophozoite
    b. Sporozoite
    c. Schizont
    d. Gametocyte

211. Which is the right time to detect malarial parasite in blood sample from patients?
    a. Five hours after the temperature comes down to the normal
    b. At the time when the temperature becomes normal
    c. Just before the rise in temperature
    d. At the time when temperature rises with shivering

212. Casoni's test is used for the diagnosis of:
    a. Taeniasis
    b. Hydatidiasis
    c. Trichuriasis
    d. Onchocerciasis

213. Which of the following nematode is ovoviviparous?
    a. *Enterobius vermicularis*
    b. *Trichinella spiralis*
    c. *Strongyloides stercoralis*
    d. *Dracunculus medinensis*

214. *Schistosoma japonicum* may cause:
    a. Periportal fibrosis
    b. Splenomegaly
    c. Hematuria
    d. Lymphadenopathy

215. Larval stage of which of the following parasite may cause disease in humans?
    a. *Taenia saginata*
    b. *Trichuris trichura*
    c. *Echinococcus vermicularis*
    d. *Toxocara canis*

216. Humans are infected by eating improperly cooked meat containing:
    a. *Taenia solium*
    b. *Toxocara canis*
    c. *Clonorchis sinensis*
    d. *Paragonimus westermanii*

217. The filarial worm that can be seen in the conjunctiva is:
    a. *Mansonella perstans*
    b. *M. ozzardi*
    c. *Loa loa*
    d. *Onchocerca volvulus*

218. The larval form of *Taenia solium* is termed as:
    a. *Cysticercus cellulosae*
    b. *C. bovis*
    c. Hydatid cyst
    d. Larva migrans

219. Which of the following trematodes lives in the blood vessels of humans?
    a. *Fasciolopsis buski*
    b. *Paragonimus westermani*
    c. *Schistosoma haematobium*
    d. *Heterophyes heterophyes*

220. Eggs of which of the following parasite is NOT bile stained?
    a. *Ascaris lumbricoides*
    b. *Trichuris trichiura*
    c. *Hymenolepis nana*
    d. *Taenia solium*

221. Which of the following is FALSE about *Strongyloides intestinalis*?
    a. This is due to infection with human threadworm
    b. Low-grade abdominal pain, recurrent diarrhoea and eosinophilia may occur
    c. It may be diagnosed with duodenal biopsy or microscopic examination of faeces for larvae
    d. Ivermection and albendazole are found of no use in eliminating *S. intestinalis*

222. All of the following are true with *Trichuris trichiura* EXCEPT:
    a. It is a parasite of man and animals

b. Adult worms are 3–5 cm in length, and live in the caecum and colon

c. Eggs discharged in faeces reach water bodies where they hatch into infective eggs in 3–4 weeks

d. Infective eggs eaten by fishes reach man when he feeds on the fishes

223. Which is the smallest tapeworm infecting man?
   a. *Taenia saginata*
   b. *Taenia solium*
   c. *Hymenolepis nana*
   d. *Diphyllobothrium latum*

224. Which one of the following ova does not float in the saturated salt solution?
   a. Ova of *H. nana*
   b. Ova of *T. solium*
   c. Ova of *Ancylostoma duodenale*
   d. Fertilised eggs of *A. lumbricoides*

225. Pulmonary migration of larvae usually does NOT occur in life cycle of which of the following nematodes?
   a. *Trichuris trichiura*
   b. *Ascaris lumbricoides*
   c. *Ancylostoma duodenale*
   d. *Strongyloides stercoralis*

226. It is imperative to understand the life cycle of human parasites for:
   a. Treatment
   b. Vaccination design
   c. Plan interventions to control and eradicate parasites
   d. All of the above

227. Hydatid cysts in liver develop due to:
   a. Development of larvae of *E. granulosus*
   b. Multiplication of *E. granulosus*
   c. Migration of larvae from lungs to liver
   d. Development of intermediate form of *E. granulosus* in liver

228. The correct sequence in life cycle of *Clonorchis sinensis* is:
   a. Man→Snail→Fish→Man
   b. Man→Cyclops→Fish→Man
   c. Dog→Man→Fish→Dog
   d. Man→Fish→Crabs→Man

229. Which of the listed parasites is an ectoparasite?
   a. Lungworm
   b. Tapeworm
   c. Demodex mites
   d. Roundworm

Part D: Fungal Pathogens

230. Which of the following infective agents is identified by morphological studies examining the pseudohyphae or chlamydospores?
   a. Parasites
   b. Bacteria
   c. Viruses
   d. Fungi

231. Fungi can be stained by:
   a. Lactophenol cotton blue
   b. Periodic acid Schiff stain
   c. Gomori methenamine silver stain
   d. All of the above

232. Reserve food material in fungi is:
   a. Glucose
   b. Volutin granules
   c. Glycogen
   d. Amino acids

233. Find out the incorrect match:
   a. Paronychia—Fungal and candidal infection of digits
   b. Onchycomycosis—Fungal and candidal infection of nails
   c. Diaper candidiasis—Candidal infection of women
   d. Balanitis—Candidal infection of penis of men

234. *Cryptococcus neoformans* is most often found in:
   a. Soil containing pigeon faeces
   b. Water containing organic pollutants
   c. Faeces of poultry
   d. Moist atmosphere

235. Germ tube formation is characteristic feature of:
   a. *Candida glabrata*
   b. *C. albicans*
   c. *C. auris*
   d. *Cryptococcus gatti*

236. Which of the following statements is TRUE in reference to *Rhinosporidium seeberi*?
   a. Cultivated on corn meal agar
   b. Infects nail and hair
   c. Produces sporangia
   d. Causes eumycetoma

237. Which of the following is NOT TRUE about the rhinosporidiosis?
   a. It is caused by *Rhinosporidium seeberi*
   b. Stagnant water is the reservoir of the pathogen
   c. Mucous membranes of nose and mouth are the mostly infected sites
   d. It leads to rhinoscleroma

238. Which of the following is budding yeast-like organism?
   a. Aspergillus
   b. Candida
   c. Penicillium
   d. Cryptococci

239. A few hours after entering lung, the infectious particles of *Cryptococcus neoformans* start to produce:
   a. Capsular polysaccharides
   b. Enterogenic toxin
   c. Conidiospores on the condidiophore
   d. Fungal hyphae and mycelium

240. Dimorphic fungi are defined as:
   a. Fungi which produce two types of growth in two different media
   b. Grow as filamentous forms at 37°C and yeast form at 22–25°C

c. Grow as yeast form at 37°C and filamentous form at 22–25°C

d. Grow as yeast and filamentous form both at 22–25°C

241. Following is a thermally dimorphic fungus:
a. *Cryptococcus neoformans*
b. *Histoplasma capsulatum*
c. *Candida albicans*
d. *Malassezia furfur*

242. Aflatoxin is produced by:
a. *Clostridium botulinum*
b. *Aspergillus flavus*
c. *Alpha haemolytic streptococci*
d. *Histoplasma capsulatum*

243. Which site is most affected by aspergillosis?
a. CNS
b. Skin
c. Bones and joints
d. Lungs

244. Intrinsic resistance to Amphotericin B is seen with which of the following fungi?
a. *Aspergillus terreus*
b. *A. fumigatus*
c. *A. nidulans*
d. *A. casaliflavus*

245. The causative agents for zygomycosis are members of:
a. *Mucor* spp.
b. *Rhizopus* spp.
c. *Absidia* spp.
d. All of the above

246. Superficial fungal infection of hair shaft is called:
a. Pitryasis
b. Piedra
c. Ring worm
d. Thrush

247. Which is the ideal treatment for superficial mycosis?
a. Removal of infected hair
b. Cleaning with a surface cleaning agent
c. Personal hygiene
d. All of the above

248. The major cause for onychomycosis is:
a. Walking by bare foot
b. Wearing badly fitting shoes for a long time
c. Contact between people of rural areas
d. Walking for long periods

249. Tinea capitis is a skin disease of:
a. Scalp
b. Leg
c. Chin
d. Abdomen

250. Fungi-producing spores in a sac-like structure are called:
a. Phycomycetes
b. Ascomycetes
c. Basidiomycetes
d. Deuteromycetes

251. Pathogenic fungus that produces spindle shaped macroconidia is:
a. *Penicillium* spp.
b. *Microsporum* spp.
c. *Epidermophyton* spp.
d. *Trichophyton* spp.

252. A farmer pricked his finger while farming. He developed a local pustule that later developed into an ulcer. Later, he discovered several nodules along the local lymphatic drainage. Probable causative organism is:
a. *Mucor racemosus*
b. *Sporothrix schenckii*
c. *Apophysomyces elegans*
d. *Penicillium marneffi*

253. A 65-year-old farmer resident of Bahranpur, Madhya Pradesh, presents with difficulty in swallowing and noduloulcerative lesion on hard palate since last 6 months. He also gives history of weight loss, weakness, productive cough and shortness of breath. Biopsy of the oral lesion as well as bone marrow biopsy shows plenty of intracellular yeast, 2–4 μm in size, within macrophages. The patient was found to be negative for HIV. What can be the most probable aetiology?
a. *Histoplasma capsulatum*
b. *Coccidioidomyces immitis*
c. *Blastomyces dermatitidis*
d. *Paracoccidioides brasiliensis*

254. Which of the following is FALSE about the Phaeohyphomycosis?
a. Lesions may be cutaneous, subcutaneous and deeper tissues
b. Brown coloured hyphae are seen in the affected areas
c. Debilitated and immunodeficient persons are the most susceptible to this disease
d. Brain is most often damaged by the pathogen *Phialophora spinifera*

255. Which one of the following pathogenic fungi multiplies intracellulary in reticuloendothelial system?
a. *Paracoccidioides brasiliensis*
b. *Cryptococcus gatti*
c. *Histoplasma capsulatum*
d. *Monosporium apiospermum*

256. Grape-like clusters of subspherical microconidia on the terminal branches are the distinguishing feature of:
a. *Trichophyton mentagrophytes*
b. *T. tonsurans*
c. *T. rubrum*
d. *Microsporum nodosum*

Section III: Laboratory Approach to Common Clinical Syndromes

257. A man with multiple sexual partners complains of a urethral discharge and painful urination. On Gram staining of the discharge several neutrophils, some containing Gram-negative diplococci are seen. Most likely pathogen causing disease is:
a. *Trichomonas vaginalis*

b. *Neisseria gonorrhoeae*
c. *Mycoplasma hominis*
d. *Ureaplasma urealyticum*

258. In another case with similar presentation (Question no. 257) the Gram stain smear of the discharge shows several neutrophils but no morphologically identified bacteria. Most likely pathogen causing disease is:
a. *Trichomonas vaginalis*
b. *Neisseria gonorrhoeae*
c. *Mycoplasma hominis*
d. *Ureaplasma urealyticum*

259. Most common cause of urinary tract infection is:
a. *Pseudomonas aeruginosa*
b. *Acinetobacter baumannii*
c. *Escherichia coli*
d. *Mycobacterium tuberculosis*

260. A 15-year-old girl has non-bloody diarrhoea for the last 11 h. This illness is probably NOT caused by:
a. *Clostridium difficile*
b. *Streptococcus pneumoniae*
c. *Shigella dysenteriae*
d. *Vibrio cholerae*

261. Which of the following usually does not cause diarrhoea?
a. *Salmonella enteritidis*
b. *Vibrio cholerae*
c. *Enterococcus faecium*
d. *Escherichia coli*

262. A 2-year-old boy is taken to his paediatrician's office because he has a fever and diarrhoea, is irritable and appears to be suffering from stomach cramps. Blood and mucus are found in the boy's stool, and sigmoidoscopy reveals that the intestinal mucosa is hyperaemic and haemorrhagic. In several areas of the mucosa, there are ulcerations covered with fibrous pseudomembranes. Blood cultures are negative, but stool culture reveals the presence of non-motile Gram-negative bacilli that do not ferment lactose. The most likely cause of illness is:
a. *Campylobacter jejuni*
b. *Salmonella serotype typhi*
c. *Shigella dysenteriae*
d. *Enterotoxigenic Escherichia coli* (ETEC)

263. What empiric therapy is recommended for a case of acute purulent meningitis in an under 5-year-old child is?
a. Ampicillin plus cefotaxime or ceftriaxone
b. Penicillin G
c. Erythromycin
d. Cefotaxime or ceftriaxone

264. A 65-year-old male was admitted to an emergency room with respiratory distress. He was a chronic smoker and known patient of chronic obstructive airway disease. He was shifted on ventilator and empirical treatment started. Initial sputum culture was negative but after 7 days, he developed fresh consolidation patches in both lungs and had high-grade fever. A brochoalveolar lavage was positive for Gram-negative coccobacilli which were, oxidase negative and did not ferment sugars and resistant to most of the antibiotics except carbapenems and colistin. The most probable bacteria was:
a. *Campylobacter jejuni*
b. *Proteus mirabilis*
c. *Burkholderia cepacia*
d. *Acinetobacter baumannii*

265. A soldier posted in a remote forest area had sudden onset of fever with chills and headache. On examination an eschar was found in his groin region associated with generalised lymphadenopathy. Which is the most appropriate test result in Weil–Felix reaction for diagnosing Rickettsial disease?
a. High OX2
b. High OX19
c. High OX19 and OX2
d. High OXK

266. A 7-year-old boy, suffering from stiff neck, high fever and delirium, is admitted in an emergency. He has a petechial rash along his belt line. Cerebrospinal fluid (CSF) is collected. The sample contains many PMNs and Gram-negative diplococci. Probable causative organism of the boy's illness is:
a. *Neisseria gonorrhoeae*
b. *E. coli*
c. *N. meningitidis*
d. *Haemophilus influenzae* type b (Hib)

267. A 9-month-old girl in respiratory distress is rushed to the hospital. The parents report that the infant developed a fever and barking cough overnight, and now she is gasping for breath and turning blue. Examination reveals that the infant's epiglottis is beefy red and swollen and is blocking the airway. A tracheostomy is performed to establish a patent airway, and blood samples are drawn. Blood cultures subsequently yield Gram-negative coccobacilli on chocolate agar but no growth on blood agar. Probable cause of the infant's illness is:
a. *Bordetella pertussis*
b. *Streptococcus pyogenes*
c. *Haemophilus influenzae* type b (Hib)
d. *Moraxella catarrhalis*

268. Which one of the following diseases shows strawberry appearance of cervix and vagina in speculum/colposcopy examination?
a. Gonorrhoea
b. Syphilis
c. Trichomoniasis
d. Candidiasis

269. Splenomegaly/enlarged spleen occurs in which of the following conditions?
a. Hepatitis
b. Malaria
c. Infectious mononucleosis
d. All of these

270. Which is the most *common* specimen collected in cases of lower respiratory infections?
   a. Transtracheal aspirate
   b. Sputum
   c. Throat swab
   d. Brochoalveolar lavage

271. Which is the LEAST satisfactory specimen for anaerobic culture?
   a. Sample from base of lesion after debridement of debris
   b. Swabs
   c. Aspirates from closed absecesses
   d. Biopsies

ANSWERS

(1) c, (2) d, (3) d, (4) c, (5) a, (6) c, (7) c, (8) c, (9) d, (10) d, (11) d, (12) b, (13) a, (14) b, (15) c, (16) a, (17) b, (18) c, (19) c, (20) c, (21) a, (22) c, (23) b, (24) a, (25) c, (26) b, (27) d, (28) d, (29) a, (30) b, (31) b, (32) d, (33) a, (34) d, (35) b, (36) b, (37) a, (38) a, (39) b, (40) d, (41) a, (42) c, (43) a, (44) b, (45) d, (46) a, (47) c, (48) a, (49) a, (50) a, (51) b, (52) d, (53) b, (54) a, (55) c, (56) b, (57) a, (58) c, (59) b, (60) a, (61) c, (62) b, (63) b, (64) d, (65) a, (66) c, (67) d, (68) d, (69) c, (70) b, (71) c, (72) a, (73) b, (74) c, (75) d, (76) a, (77) d, (78) b, (79) b, (80) c, (81) d, (82) b, (83) b, (84) b, (85) b, (86) c, (87) b, (88) c, (89) a, (90) c, (91) d, (92) d, (93) c, (94) c, (95) a, (96) b, (97) b, (98) a, (99) c, (100) b, (101) c, (102) d, (103) d, (104) b, (105) b, (106) c, (107) c, (108) d, (109) c, (110) d, (111) d, (112) c, (113) d, (114) d, (115) d, (116) a, (117) a, (118) d, (119) d, (120) b, (121) b, (122) c, (123) a, (124) b, (125) a, (126) b, (127) a, (128) c, (129) d, (130) a, (131) d, (132) b, (133) d, (134) b, (135) c, (136) b, (137) c, (138) d, (139) d, (140) d, (141) d, (142) b, (143) d, (144) c, (145) a, (146) b, (147) b, (148) c, (149) a, (150) d, (151) c, (152) b, (153) d, (154) d, (155) a, (156) b, (157) c, (158) a, (159) b, (160) c, (161) c, (162) d, (163) d, (164) c, (165) b, (166) c, (167) c, (168) b, (169) a, (170) a, (171) b, (172) b, (173) b, (174) d, (175) a, (176) b, (177) b, (178) d, (179) c, (180) d, (181) b, (182) d, (183) d, (184) a, (185) a, (186) c, (187) d, (188) a, (189) b, (190) c, (191) a, (192) b, (193) c, (194) b, (195) b, (196) a, (197) d, (198) a, (199) d, (200) c, (201) b, (202) b, (203) b, (204) d, (205) c, (206) a, (207) d, (208) b, (209) b, (210) b, (211) d, (212) b, (213) c, (214) c, (215) a, (216) a, (217) c, (218) a, (219) c, (220) c, (221) d, (222) d, (223) c, (224) b, (225) a, (226) d, (227) a, (228) a, (229) c, (230) d, (231) d, (232) c, (233) c, (234) a, (235) b, (236) c, (237) d, (238) b, (239) a, (240) c, (241) b, (242) b, (243) d, (244) a, (245) d, (246) b, (247) d, (248) b, (249) a, (250) b, (251) b, (252) b, (253) a, (254) d, (255) c, (256) a, (257) b, (258) d, (259) c, (260) b, (261) c, (262) c, (263) a, (264) d, (265) d, (266) c, (267) c, (268) c, (269) d, (270) b, (271) b.

FURTHER READING

Section I: General Microbiology

1. Krieg NR. Bergey's manual of systematic bacteriology. Int J Syst Bact 1985;28:408.
2. Topley WWC, Graham S, Wilson S, Collier LH, Mahy BWJ, TerMeulen V. Topley and Wilson's microbiology and microbial infections. 10th ed. United States: ASM Press; 2005.
3. Brooks GH, Carroll KC, Butel JS, Morse SA, Mietzner TA, editors. Jawetz, Melnick, & Adelberg's medical microbiology. 26th ed. China: McGraw-Hill; 2013.
4. Delves PJ, Martin SJ, Burton DR, Roitt IM. Roitt's essential immunology. 13th ed. United States: Wiley Blackwell; 2017.
5. Abbas A, Lichtman AH, Pillai S. Cellular and molecular immunology. 8th ed. Canada: Elsevier; 2014.
6. Rutala WA, Weber DJ. Disinfection and sterilization: an overview. Am J Infect Control 2013;41:S2–5.
7. Rutala WA, Weber DJ, Healthcare Infection Control Practices Advisory Committee. Guideline for disinfection and sterilization in healthcare facilities. Available from: https://www.cdc.gov/infectioncontrol/pdf/guidelines/disinfection-guidelines.pdf; 2008.
8. Reybrouck G. The testing of disinfectants. Int Biodeterior Biodegrad 1998;41:269–72.
9. Gardner JF, Peel MM. Introduction to sterilization and disinfection control. 2nd ed. Melbourne: Churchill Livingstone; 1991.
10. Indian Dental Association. Sterilization and disinfection. Available from: https://www.ida.org.in/Membership/Details/SterilizationandDisinfection.
11. NIH Human Microbiome Project. Available from: https://hmpdacc.org.
12. Casadevall A, Pirofski LA. Host-pathogen interactions: basic concepts of microbial commensalism, colonization, infection, and disease. Infect Immun 2000;68(12):6511–8.
13. Willey J, Sherwood L, Woolverton CJ. Prescott's microbiology. 10th ed. New York: McGraw-Hill; 2016.

Section II: Aetiological Agents
Part A: Bacterial Pathogens

1. Tille P. Bailey and Scott's diagnostic microbiology. 14th ed. St. Louis, Missouri: Elsevier; 2017.
2. Collee JG, Fraser AG, Marmion BP, Simmons A. Mackie and Mccartney practical medical microbiology. New York: Churchill Livingstone; 1996.
3. Procop GW, Koneman EW. Koneman's color atlas and textbook of diagnostic microbiology. 7th ed. Philadelphia: Wolters Kluwer; 2016.
4. Benett JE, Dolin R, Blaser MJ, editors. Mandell, Douglas and Benett's principles and practice of infectious diseases. 8th ed. Canada: McGraw Hill; 2015.
5. Kasper DL, Braunwald E, Fauci AS, Hauser SL, Longo DL, Jameson JL. Harrison's principles of internal medicine. 20th ed. New York: McGraw Hill; 2018.
6. Infectious Diseases Society of America. Guidelines on the treatment of MRSA infections in adults and children. Am Fam Physician 2011;84(4):455–63. Available from: https://www.aafp.org/afp/2011/0815/p455.html.
7. Methicillin resistant Staphylococcus aureus. Available from: https://www.cdc.gov/mrsa/.
8. Bush LM, Perez MT. Streptococcal infections. MSD manual. Available from: https://www.msdmanuals.com/professional/infectious-diseases/gram-positive-cocci/streptococcal-infections.
9. Fraser SL, Donskey CJ, Salata RA. Enterococcal infections. In: Brusch JL, editor. Available from: https://emedicine.medscape.com/article/216993-overview; 2018.
10. VRE in healthcare settings. Available from: https://www.cdc.gov/hai/organisms/vre/vre.html.
11. Finegold SM, Song Y, Liu C. Taxonomy—general comments and update on taxonomy of Clostridia and Anaerobic cocci. Anaerobe 2002;8(5):283–5.
12. Diphtheria by World Health Organization. Available from: https://www.who.int/biologicals/vaccines/diphtheria/en.
13. Centers for Disease Control and Prevention. Diphtheria vaccination. Vaccines and preventable diseases. Available from: https://www.cdc.gov/vaccines/vpd/diphtheria/index.html; 2018.
14. Centers for Disease Control and Prevention. Anthrax. Available from: https://www.cdc.gov/anthrax/basics/index.html; 2018.
15. Hall GS. Anaerobic Gram-positive bacilli. In: Clinical microbiology procedures handbook. fourth ed. Washington DC: ASM Press.
16. RNTCP guidelines for diagnosis of TB. Available from: www.tbcindia.org.
17. National Tuberculosis Management Guidelines. TB online. Available from: http://www.tbonline.info/media/uploads/documents/ntcp_adult_tb-guidelines-27.5.2014.pdf; 2014.
18. Guidelines, Central Tuberculosis Division. Under Ministry of Health and Family Welfare. Available from: https://tbcindia.gov.in/index1.php?lang=1&level=1&sublinkid=4571&lid=3176.
19. Guidelines—National Leprosy Eradication Programme (NLEP). Available from: http://nlep.nic.in/guide.html.
20. Centers for Disease Control and Prevention. Meningococcal disease (Neisseria meningitidis). Available from: https://wwwnc.cdc.gov/travel/diseases/meningococcal-disease; 2018.
21. WHO guidelines for the treatment of Neisseria gonorrhoeae. Available from: https://www.who.int/reproductivehealth/publications/rtis/gonorrhoea-treatment-guidelines/en/.
22. Clinical and Laboratory Standards Institute. Performance standards for antimicrobial susceptibility testing. 28th ed. CLSI Supplement M100. Wayne, PA: Clinical and Laboratory Standards Institute; 2018. Available from: https://clsi.org/media/1930/m100ed28_sample.pdf.
23. European Committee on Antimicrobial Susceptibility Testing. Available from: http://www.eucast.org/ast_of_bacteria/guidance_documents/.
24. Salmonella, Foodsafety.gov. Available from: https://www.foodsafety.gov/poisoning/causes/bacteriaviruses/salmonella/index.html.
25. Centers for Disease Control and Prevention. Shigella-Shigellosis. Available from: https://www.cdc.gov/shigella/index.html; 2018.
26. Perilla MJ, Ajeelo G, Bopp C, Elliott J, Facklam R, Knapp JS, et al. Manual for the laboratory identification and antimicrobial

susceptibility testing of bacterial pathogens of public health importance in the developing world: *Haemophilus influenzae, Neisseria meningitidis, Streptococcus pneumoniae, Neisseria gonorrhoeae, Salmonella serotype Typhi, Shigella,* and *Vibrio cholerae.* WHO/CDS/CSR/RMD/2003.6; 2003. Available from: https://www.who.int/csr/resources/publications/drugresist/IAMRmanual.pdf.

27. Rahi M, Gupte MD, Bhargava A, Varghese GM, Arora R. DHR-ICMR guidelines for diagnosis & management of Rickettsial diseases in India. Indian J Med Res 2015;141(4):417–22.

28. Rathi N, Kulkarni A, Yewale V. IAP guidelines on Rickettsial diseases in children. 2017;54(3):223–29. Available from: https://www.indianpediatrics.net/mar2017/223.pdf.

Part B: Viral Pathogens

1. Fields BN. . In: Knipe DM, Peter M, editors. Fields virology. 6th ed. Philadelphia: Wolters Kluwer; 2013.

2. Flint J, Racaniello VR, Racaniello VR, Rall GF, Skalka MA. Principles of virology. 4th ed. Washington DC: ASM Press; 2015.

3. World Health Organization. Herpes simplex virus. Available from: https://www.who.int/news-room/fact-sheets/detail/herpes-simplex-virus; 2017.

4. European Association for the Study of the Liver. EASL 2017 clinical practice guidelines on the management of hepatitis B virus infection. J Hepatol 2017;67:370–98.

5. National Vector Borne Disease Control Programme (NVBDCP). Available from: http://www.nvbdcp.gov.in.

6. Menezes R. Rabies in India. CMAJ 2008;178(5):564–6.

7. National Rabies Control Programme. National Guidelines on Rabies Prophylaxis. National Center for Disease Control. Directorate General of Health Services. Available from: http://pbhealth.gov.in/guideline%20for%20rabies%20prophylasix.pdf.

8. Luo GG, Ou JH. Oncogenic viruses and cancer. Virol Sin 2015;30(2):83–4.

9. Qiu J, Söderlund-Venermo M, Young NS. Human parvoviruses. Clin Microbiol Rev 2016;30(1):43–113.

10. Influenza NICD Recommendations for the diagnosis, prevention, management and public health response. Available from: http://www.nicd.ac.za/wp-content/uploads/2017/03/Influenza-guidelines-final_25_05_2017.pdf.

Part C: Parasitic Pathogens

1. Chatterjee KD. Parasitology protozoology and helminthology. 13th ed. India: CBS Publishers; 2017.

2. Centers for Disease Control and Prevention. Parasites, amebiasis, entamoebahistolytica infection. Available from: https://www.cdc.gov/dpdx/amebiasis/index.html; 2018.

3. National Guideline for Kala-azar Case Management. 3rd ed. Available from http://kalacorebd.com/wp-content/uploads/2016/04/Kala-azar-Case-Managment-Guideline-2016_Latest-Draft.pdf; 2016.

4. Aronson N, Herwaldt BL, Libman M, Pearson R, Lopez-Velez R, Weina P, et al. Diagnosis and treatment of leishmaniasis: clinical practice guidelines by the Infectious Diseases Society of America (IDSA) and the American Society of Tropical Medicine and Hygiene (ASTMH). Clin Infect Dis 2016;63(12):e202–64.

5. World Health Organization. Human African trypanosomiasis. Available from: https://www.who.int/trypanosomiasis_african/diagnosis/en/.

6. Treatment of malaria. Available from: https://www.malariasite.com/treatment-of-malaria/.

7. Guidelines for the management of sexually transmitted infections. Essential Medicines and Health Products Information Portal A World Health Organization resource. Available from: http://apps.who.int/medicinedocs/en/d/Jh2942e/4.9.html; 2004.

8. World Health Organization. Guidelines for the treatment of malaria. 3rd ed. Available from: https://www.ncbi.nlm.nih.gov/books/NBK294441/.

9. Diagnosis and treatment of malaria in India. National Vector Borne Disease Control Programme. Available from: http://nvbdcp.gov.in/Doc/Diagnosis-Treatment-Malaria-2013.pdf.

Part D: Fungal Pathogens

1. Chander J. Textbook of medical microbiology. 4th ed. India: Jaypee Brothers Medical Publishers (P) Ltd; 2018.

2. Walsh TJ, Hayden RT, Larone DH. Larone's medically important fungi: a guide to identification. 6th ed. Washington DC, United States: American Society for Microbiology; 2018.

3. Sciortino CV Jr. Atlas of clinically important fungi. Chichester: Wiley Blackwell; 2017.

4. Avni T, Leibovici L, Paul M. PCR diagnosis of invasive candidiasis: systematic review and meta-analysis. J Clin Microbiol 2011;49(2):665–70.

5. Khawcharoenporn T, Apisarnthanarak A, Mundy LM. Non-neoformans cryptococcal infections: a systematic review. Infection 2007;35:51. Available from: https://doi.org/10.1007/s15010-007-6142-8.

6. Centers for Disease Control and Prevention. Fungal diseases. Available from https://www.cdc.gov/fungal/index.html.

Section III: Laboratory Approach to Common Clinical Syndromes

1. Hersch EC, Oh RC. Prolonged febrile illness and fever of unknown origin in adults. Am Fam Physician 2014;90(2):91–6.

2. Kon K, Rai M. 1st ed. The microbiology of respiratory system infections, vol. 1. London: Elsevier; 2016.

3. Najar MS, Saldanha CL, Banday KA. Approach to urinary tract infections. Indian J Nephrol 2009;19:129–39.

4. Revised Treatment Guidelines of AES including JE- NVBDCP. Available from: https://nvbdcp.gov.in/WriteReadData/l892s/Revised_guidelines_on_AES_JE.pdf.

5. Gastroenteritis in an institution control guideline. Available from: https://www.health.nsw.gov.au/Infectious/controlguideline/Pages/gastro.aspx.

6. Centers for Disease Control and Prevention. Sexually transmitted diseases treatment guidelines. Available from: https://www.cdc.gov/std/tg2015/default.htm; 2015.

7. Stevens DL, Bisno AL, Chambers HF, Dellinger EP, Goldstein EJC, Gorbach SL, et al. Practice guidelines for the diagnosis and management of skin and soft tissue infections: 2014 update by the Infectious Diseases Society of America. Clin Infect Dis 2014;59(2):e10–52. Available from: https://doi.org/10.1093/cid/ciu296.

8. Kwak YG, Choi SH, Kim T, et al. Clinical guidelines for the antibiotic treatment for community-acquired skin and soft tissue infection. Infect Chemother 2017;49(4):301–25.

9. Hospital Infection Control Guidelines. Indian Council of Medical Research. Available from: https://www.icmr.nic.in/sites/default/files/guidelines/Hospital_Infection_control_guidelines.pdf.

10. World Health Organization. Prevention of hospital-acquired infections: a practical guide. 2nd ed. Available from: http://apps.who.int/medicinedocs/documents/s16355e/s16355e.pdf.

11. World Health Organization. Guidelines on prevention and control of hospital associated infection. World Health Organization Regional Office for Southeast Asia, New Delhi. Available from: http://apps.searo.who.int/PDS_DOCS/B0007.pdf; 2002.

INDEX